AMERICAN METRIC CONSTRUCTION HANDBOOK

AMERICAN METRIC CONSTRUCTION HANDBOOK

By R. J. Lytle

Research Associate:
Dorothy T. Burrows

FIRST EDITION 1976

McGraw-Hill Book Company

New York St. Louis San Francisco Auckland
Bogotá Hamburg Johannesburg London Madrid Mexico
Montreal New Delhi Panama Paris São Paulo
Singapore Sydney Tokyo Toronto

Manufactured in the United States of America

Book designed by Richard Kinney.
Book edited by Shirley M. Horowitz.

Current Printing (last digit)
10 9 8 7 6 5 4 3 2

International Standard Book Number: 0-912336-14-5
Library of Congress Catalog Card Number: 82-42926

McGraw-Hill ISBN 0-07-039277-3

PREFACE

In publishing this book at this time, early in the development of metric sizes and standards for our building materials and methods, our purposes are twofold.

First, the involvement of the maximum number of product manufacturers, architects, engineers, builders, contractors, and every other segment of the construction industry, in the process of arriving at a consensus on the shape of metric construction. There is grave danger that the ideas of those few now preparing the metric groundwork may set up unchallenged rules which will dictate metric construction for years to come. This is not to fault those who have voluntarily for some years been working on organizing and planning the eventual changeover. Their work has been of immense benefit to our industry; we owe them much. But, with the imminence of metrication, it is important that the widest possible participation be encouraged.

Second, those manufacturers and planners who are now making investments in new tooling, in new products, in new applications for old products, and all types of advance planning, must be given the opportunity to make rational provision for the eventual metrication of their products or services. At the same time, educators must have an opportunity to relate curriculum plans not only to the bare bones of standard SI metrics, which are established, but to the realities of products and procedures which are yet in a state of flux.

Despite the preference of the American National Metric Council for the spelling liter and meter, there are those who make strong arguments for litre and metre. But construction people are simply not interested in this sort of minutiae. It makes little difference how the words are spelled; it is the facility of their application to the construction process with which we are concerned.

I beg the reader's indulgence if, for instance, he finds that by the time he reads this the 2 × 4 equivalent described herein as 38 × 89 mm has been changed to some other dimension. I have done the best that I could to insure that product dimensions are the result of a consensus of, or at least valid proposals from, industry members. To this end, I have made changes in the proofs to insure that the printed book reflects the latest thinking of those involved at the time of publication.

This first edition will be followed by subsequent editions and/or annual supplements which will bring into focus the final shape of metric construction.

R. J. Lytle
January, 1976

CONTENTS

1. Introduction

The prospects are that the United States will "go metric" by either a gradual, unplanned drift, or by a planned, coordinated changeover so that at the end of a ten to twelve year period the nation will be predominantly but not exclusively on the metric system of measurement. The reasons for this shift are twofold: the simplicity of the metric system, and conformation to its worldwide use. The U.S., as one of the few major exporters not utilizing the metric system, is at a competitive disadvantage. The only other countries not on the metric system are Brunei, Burma, Liberia, and Yemen—a less than distinguished group.

The cost of changing the construction industry to the SI (*Système Internationale d'Unites*) will be substantial. But as the conversion progresses we will begin to recover these costs in the form of reduced man-hours in education and calculation, and in reduction of errors due to the simplicity of the metric system. Even more important, by providing an opportunity to rationalize sizes and specifications of building products, we will not only recover the initial costs of the changeover, but reduce overall costs.

Canada has committed itself to SI units, but the conversion of certain sectors depends upon U.S. progress toward adopting the metric system. For example the Canadian lumber industry ships 75% of its product to the United States (25% of U.S. consumption), so that details of the metric changeover for this industry must necessarily be coordinated with the decisions and timing of the U.S. construction industry.

Compared with the rest of U.S. industry, the construction industry has few compelling reasons for the changeover; we export little. But U.S. industries are, in fact, going metric now. And it will be difficult for construction to remain on the traditional measurement system once the general population becomes educated to think in metric terms.

Metric legislation now signed by the President will provide for a changeover period to be coordinated by a U.S. Metric board. No time period is fixed. As in previous years, the controversy does not center around whether or not we should metricate. It concerns whether or not we should let the costs of the changeover "lie where they fall," or provide grants to affected skilled workers and small business for the purchase of metric tools and equipment. Many feel that during a gradual changeover, most of such tools and equipment would be replaced anyway, minimizing such costs. Subsidies paid to individuals and corporations might balloon the costs and impede the change because of bureaucratic inflation. In any event, increased metric education and publicity, and the steady drift of major industries into the metric system, will cause specific legislation to be passed, if not this year, then next year.

Opportunities abound for the construction industry in changing to the SI, a modern version of the metric system. It would be relatively simple to make a "soft" conversion by changing our modules, material and product sizes to metric numbers. But how much more productive would it be to make a "hard" conversion? A "hard" conversion changes physical dimensions and presents opportunities for standardization. The opportunity lies in discarding century-old practices and material sizes which continue to cost unknown billions in labor and materials every year. How much can be saved by modular and dimensional coordination? How many man-hours do we now spend on fractions of an inch, on irrational measurements like the acre and the quart? In elementary and secondary education alone, it is estimated that we can save 25% of the student and instructor time required to teach traditional measurement mathematics. For personnel destined for the construction industry, this time saved can be well spent in acquiring increasingly sophisticated technical knowledge.

The British experience in rationalizing sizes and dimensions under the SI indicates its potential. Window sizes were reduced from 1600 inch sizes to 200 metric sizes. Estimated copper savings in their relatively small electric insulated cable industry were 1800 metric tons annually. Reductions in sizes of 50% to 85% have been achieved for screw fasteners, nuts, and bolts. Since the U.S. has more standardized construction product sizes than has Britain, the opportunities are even greater. Construction suppliers are well aware of the potential cost savings if they can reduce inventory by reducing the number of standard product sizes. This will reduce the capital funds and warehouse space necessary to supply the industry.

Opportunities exist in wood construction also. A consensus seems to be developing for sheet sizes (plywood, insulation board, gypsum board, etc.) to be 1200 mm by 2400 mm (47-1/4″ × 94-1/2″). The 2400 mm sheet size will mandate a reduction in ceiling height with consequent savings on inside and outside finishing materials, and reduced energy requirements for heating and cooling.

Change always creates problems, and our inevitable changeover to the SI in the U.S. construction industry will be no exception. For this reason, a planned and coordinated change as the result of metric legislation is desirable, if not essential. Many of the problems arising will relate to the requirements of our remodeling industry in insuring products compatible with existing structures. For instance, the 1200 mm (47-1/4″) plywood size will cause demands for 48″ plywood to be stocked for remodeling, at least for a changeover period. This will increase the costs of maintaining inventories and cause many manufacturers to manufacture both metric and traditional sizes. This problem can be surmounted by using the various joint treatments described in a later chapter.

Further on we will look into metric masonry sizes and find that the metric brick is compatible with the traditional brick. But this may not be true of other building products, so careful planning and recognition of the needs of all segments of the industry is essential. A few manufacturers may find it desirable to produce nonmetric sizes for many years after the changeover has supposedly been completed.

Labor unions have been reluctant to support the change. They feel that instead of increasing the export of products made by U.S. labor, we will be exporting jobs because it may be less expensive to produce metric-sized products abroad. Construction unions fear that the changeover will be confusing and some construction will be postponed pending the availability of metric materials and building codes. They invoke the specter of increased jurisdictional disputes; work lost through industrialization of traditional on-site work; the need for training programs; cost of metric tool replacement; longstanding work rules which may have to be amended; union members pushing for premium pay on metric jobs in the early stages of a changeover, etc., etc. Of course, many of these objections are raised to improve their future bargaining opportunities.

But it is important for all sectors of the construction industry to become involved in the solution of the problems. As every part of the industry begins to participate in charting the course to metric construction, many advantages and disadvantages which are not now evident will be brought to light. Since it appears that the change is inevitable, we should have maximum participation by all industry groups at the earliest possible date to insure that economical and wise consumer-oriented choices are made. This participation should not be designed to accelerate the change, but to insure the development of a rational comprehensive plan for eventual use.

METRIC PROGRESS IN EUROPE

The following notes were made by the author while visiting building exhibitions in London (Interbuild) and Paris (Batimat) in November, 1975.

London

While government officials have pronounced that metric conversion in the construction industry was completed two years ago, conversation with exhibitors at Interbuild reveals a far different story. Following are some highlights.

Plywood exhibitor: No plywood is being produced in the 1200 × 2400 mm size frequently mentioned to be the standard metric size in this country. Virtually all plywood produced in the U.K., both for home consumption and for use on the continent, is being produced at 1220 × 2440 mm, which is a rounded conversion at 4 feet × 8 feet. Some plywood is being produced for export at 2550 mm. A review of a current Do-It-Yourself magazine shows all sheet sizes for hardboard, pegboard, plywood, chip board, asbestos board and formica to be 1200 × 2440 mm.

Paneling (board type) exhibitor: The interviewee claims to have been convinced by the government to issue "metric only" price lists and specifications, which resulted in substantial financial loss because customers were not ready to buy on these specifications. He then switched all literature back to a dual dimension basis where it remains today, and he expects to keep it this way for years to come.

Millwork producer: Drawings for existing products are now shown in dual dimensions. New products are expressed in millimeters only. For instance, garage doors are expressed at 7 feet × 7 feet (2134 × 2134 mm). Approximate weight 95 pounds (43.2 kg).

Steel roofing and cladding sheets producer (products similar to that used in farm and industrial construction in the U.S.): Their 3 inch corrugated is dimensioned to cover 762 mm with 76 mm center to center of corrugations. Their AL5 configuration is 750 mm coverage with 95 mm flats and 25 mm peaks, with 150 mm center to center of configuration. Both are 19 mm thick. Steel gauges are expressed as 20G, 22G and 24G.

Two load span tables are provided with loads expressed in pounds per square foot (lb/ft^2) and kilonewtons per square meter (kN/m^2). Stress value is expressed at 150 N/mm^2.

Aluminum association: Although they produce sheet materials on metric dimensions, all extrusions are on imperial dimensions.

Major door and window producer: Interior doors are usually specified by the size of the opening, with the actual size of the door being a function of trim and jamb thickness. Door width openings are usually specified as 600, 700, 800 mm, etc. Doors, including a fixed transom, are produced 2300, 2350, 2400, 2700 and 3000 mm high. Door thickness: 40 mm for walls 50, 57, 75 and 100 mm thick. Despite all the foregoing, 60 percent of their doors are being produced on imperial sizes.

This firm's window operation is still 75 to 80 percent imperial. Their window catalogue contains two pages of metric sizes and four pages of imperial sizes. Window widths are 600, 900, 1200, 1800 and 2400 mm. Heights are 600, 900, 1050, 1200 and 1500 mm. Opening lights are all casement type, opening either in or out.

Kitchen cabinets: Catalogues are in metric, with imperial dimensions in parentheses. Most orders are still received in imperial. Counter top height is 900 mm. Wall units 300 mm (11-3/4 inches) deep, 600 or 900 mm high and standard width of 500 or 1000 mm. Floor units are 500 mm (1′ 7-3/4 inches deep) with widths of 500, 1000 and 1500 mm.

Plumbing: Copper fittings are in metric sizes with adaptors to imperial sizes. Plastic DWV items have been soft-converted to metric sizes.

Measuring Instruments: This manufacturer was deluged with orders for metric-only rules, tapes, etc., at the outset of the conversion. However, at the present, almost all orders are for dual dimension equipment.

Carpenters Try Squares are produced 300 mm long with one edge graduated in mm, numbered every 10 mm, and the other edge is in inches subdivided in eights and sixteenths. Both faces of the rule are the same. Pocket rules are 2m, 3m and 5m with graduations as above on both faces.

Most sales of steel measuring tapes are in lengths of 10, 15, 20, 30 meters with readings in feet, inches and eighths on one edge and meters and millimeters on the other.

Folding wood rules are available with metric in millimeters on one side and feet and inches the other side; length, 2m.

Engineering: Only a few of the exhibitors were aware of newtons and pascals as units of force and pressure. One book publisher intends to use these SI units in the next revision of his technical books.

In General: The impression gained is that the metrication of the construction industry is going to take many more years before it is 100 percent. This is in a country where the construction codes are national, not local or regional as in the U.S. It is true that all government construction is metric, but conversion in the private sector is much slower than we have been led to believe.

Paris

France, of course, has used the metric system for over a century. But the SI, with newtons, pascals, joules, etc., is another matter. As indicated at

Batimat, the kilogram-force, bars, and centimeters still predominate. Knowledgeable persons seem to feel that it will be many years before the conversion is complete, but that eventually the SI will become the accepted standard.

Lessons for the U.S.

What lessons can we in the U.S. draw from the contacts outlined above? Keeping in mind that the legislation passed by the Senate has no "teeth" in the form of a time limit for conversion and remembering too that control of the construction industry is much more fragmented than it is in England and France, what will happen here?

Two possibilities: (1) The conversion will drag out over decades, the only compulsion being personnel entering the industry who have been educated in metrics, also publicly financed projects will exert some pressure, since Housing and Urban Development's Minimum Property Standards will be expressed in metric. (2) The industry itself may express a desire to "let's get it over with." In this case, the conversion may be surprisingly rapid.

Or somewhere in between!

2. Learning the SI/Modern Metric System

With our global economy, advanced technology and communications, we need a coherent, coordinated system of measurement. Our traditional or English system is about as irrational as can be imagined. The elements making up the system bear little relation to one another. There are 12 inches in a foot and 36 in a yard. We break the inch down into sixteenths, eighths and quarters. The mile is 5280 feet or 1760 yards. In area, we have the square foot, square yard, and an acre of 42,560 square feet! In volume, we measure in cubic feet and cubic yards. But we have another volume measurement, totally unrelated—the gallon—which is 231 cubic inches or 8.33 pounds of water. Our gallon is further subdivided into four quarts or eight pints. As builders we have board feet of lumber, which is supposed to measure the volume of wood—but, as we know, it really doesn't. In mass, or weight, we have 16 ounces to the pound and 2000 pounds to the short ton. And then we confuse everyone even more with the long ton of 2240 pounds.

All that and we have not even mentioned some of our real oddities like furlongs, fathoms, rods, chains, barrels, teaspoons, penny nails, cups, grains, drams, bushels, pecks and acre feet. In retrospect, how did we ever learn it all?

Metrication will change all that. Based on units of ten, everything relates to everything else. A meter (m), (39.37 inches), is divided into 1000 millimeters (mm), (.03937 inches). A thousand meters is a kilometer (km) or about six tenths of a mile. In area, we simply have square millimeters (mm^2), square meters (m^2), square kilometers (km^2), and the hectare (ha) or 10000 square meters. In volume, we use cubic meters (m^3) or the liter (l) which is one cubic decimeter (dm^3) or one thousandth of a cubic meter (m^3) or a cube 100 millimeters (mm) on a side. A liter (l) is a little over a quart and can be further divided into milliliters (ml).

From volume we make the transposition to mass (weight) by saying that one liter of water is one kilogram (kg). From here we can easily go to force by saying that a newton (N) force which is applied for one second to one kilogram (kg) of mass will produce a velocity of one meter per second, or an acceleration of one meter per second per second. From here, it is just a step to the pascal (Pa) for pressure, which is one newton per square meter (N/m^2); the joule* for energy, work or heat, which is newton meters ($N \cdot m$); or the watt** (W) for power, which is one joule per second (J/s).

Now, see how well that all hangs together? If we constructioneers had been using this system all our lives, how many man-hours could have been saved? How many errors prevented?

The easiest way to learn SI metrics would be to simply move from the traditional world of measurement to the metric world, forgetting everything we ever knew about inches, pounds, etc. But we can't do this. For a while, at least, our drawings and our thinking are going to have to be done in "dual dimensioning." Construction plans will show both traditional and metric dimensions. Gradually the traditional measurements will be phased out. We will then find it an irritation to think in the old terms, much as we now find it an irritation to think in terms of the SI.

*force acting thru a distance
**the rate of doing work

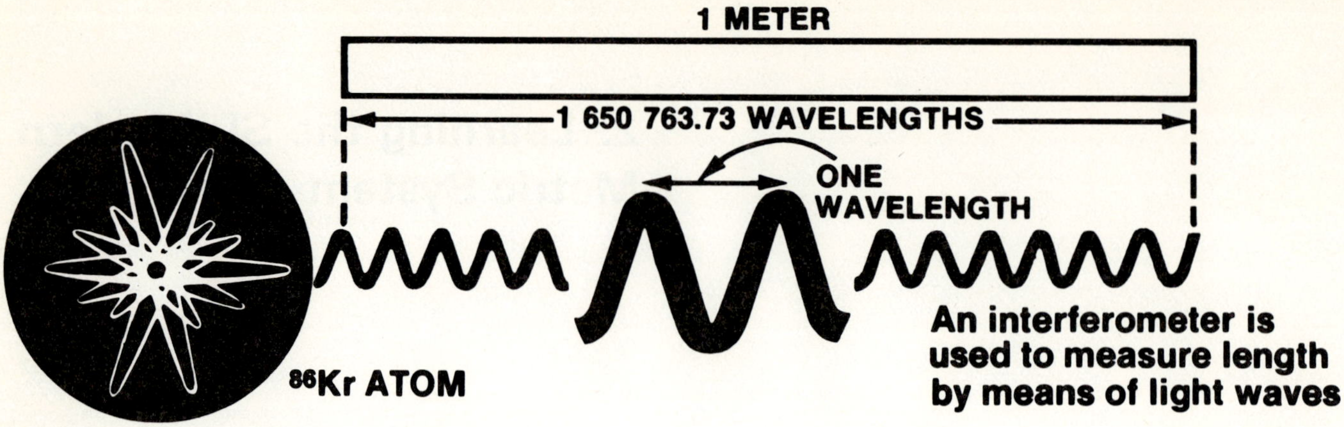

FIGURE 2-1. Metric Measurement. Photograph courtesy of the National Bureau of Standards.

FIGURE 2-2. Metric Signs. Photograph courtesy of the Michigan Department of State Highways.

RECOGNITION POINTS

We all have a mental data bank of familiar items with their measurements in terms of length, area, volume, weight, temperature and other characteristics. Here are some items selected to provide mental images in metric terms which can then be applied to relate or evaluate other quantities. In most of the conversions shown, approximations have been made, but the variation is slight.

Many constructioneers use the 3-4-5 relationship to construct or check right angles. A triangle with sides of 3 and 4 at 90° angle will have a hypotenuse of 5. This continues to be true in metrics. A similar but new relationship that we can all memorize is called the 10-11-12-13 relationship. Here's how it works:

10 meters (m)	=	11 yards
10 square meters (m^2)	=	12 square yards
10 cubic meters (m^3)	=	13 cubic yards

Length

1 millimeter (mm)	=	thickness of a thin U.S. dime
25 millimeters (mm)	=	1 inch
300 millimeters (mm)	=	1 foot
1.7 meters (m)	=	height of person (5 ft. 8 in.)
2 meters (m)	=	residential door height

Area

100 square meters (m^2)	=	floor area of small 3 bedroom home

Volume

1 liter (l)	=	1 cubic decimeter (dm^3) = 1 quart
1 liter (l)	=	contents of a cube 100 millimeters (mm) on a side
208 liters (l)	=	0.2 cubic meters (m^3) = 55 gallons (US) drum for oil, chemicals, etc.

Mass (Weight)

1 kilogram (kg)	=	mass of one liter (1) of water = 1 cubic decimeter (dm^3) of water
75 kilograms (kg)	=	mass (weight) of 165 pound man

Temperature, degrees Celsius

0° C	=	freezing point of water
20° C	=	mild, comfortable
30° C	=	hot
37° C	=	body temperature
100° C	=	boiling point of water

Other Points

15 kilowatts (kW)	=	20 horsepower
20 megapascals (MPa)	=	3000 psi compressive strength of concrete
1 kilopascal (kPa)	=	20 pounds per square foot (psf)
Residential floor load:		
2 kilopascals (kPa)	=	40 lbs. per square foot (psf)

UNITS

The SI system of measurement is composed of Base Units, Supplementary Units and Derived Units. In the following material, we shall not attempt to classify the units. Figure 9 in Chapter V does this well. Whether a unit is base or derived will make little difference in the construction industry. The seven base units are:

meter (m)	Length
kilogram (kg)	Mass
second (s)	Time
ampere (A)	Electric Current
kelvin (K)	Temperature (Thermodynamic)
candela (cd)	Luminous Intensity
mole (mol)	Amount of substance. This unit has to do with molecular physics and we will omit further reference because it will seldom, or never, be used in construction-related measurement.

The two supplementary units, the radian (rad) and the steradian (sr) also have little application in construction. Our preferred unit will be decimal degrees for efficiency, instead of minutes and seconds, but the radian will also be used in structural computations.

Derived units are derived from the base units such as square meter (m^2); cubic meter (m^3); density = kilogram per cubic meter (kg/m^3); or speed = meters per second (m/s). Others are force = newtons (N); energy, work and heat = joules (J); and watt (W) = power.

To help you learn the symbols, we will frequently use both the word and the symbol, although this would not be required in everyday use.

Length

The SI standard of length is the meter (m) which is 39.37 inches (approx.). The meter is further

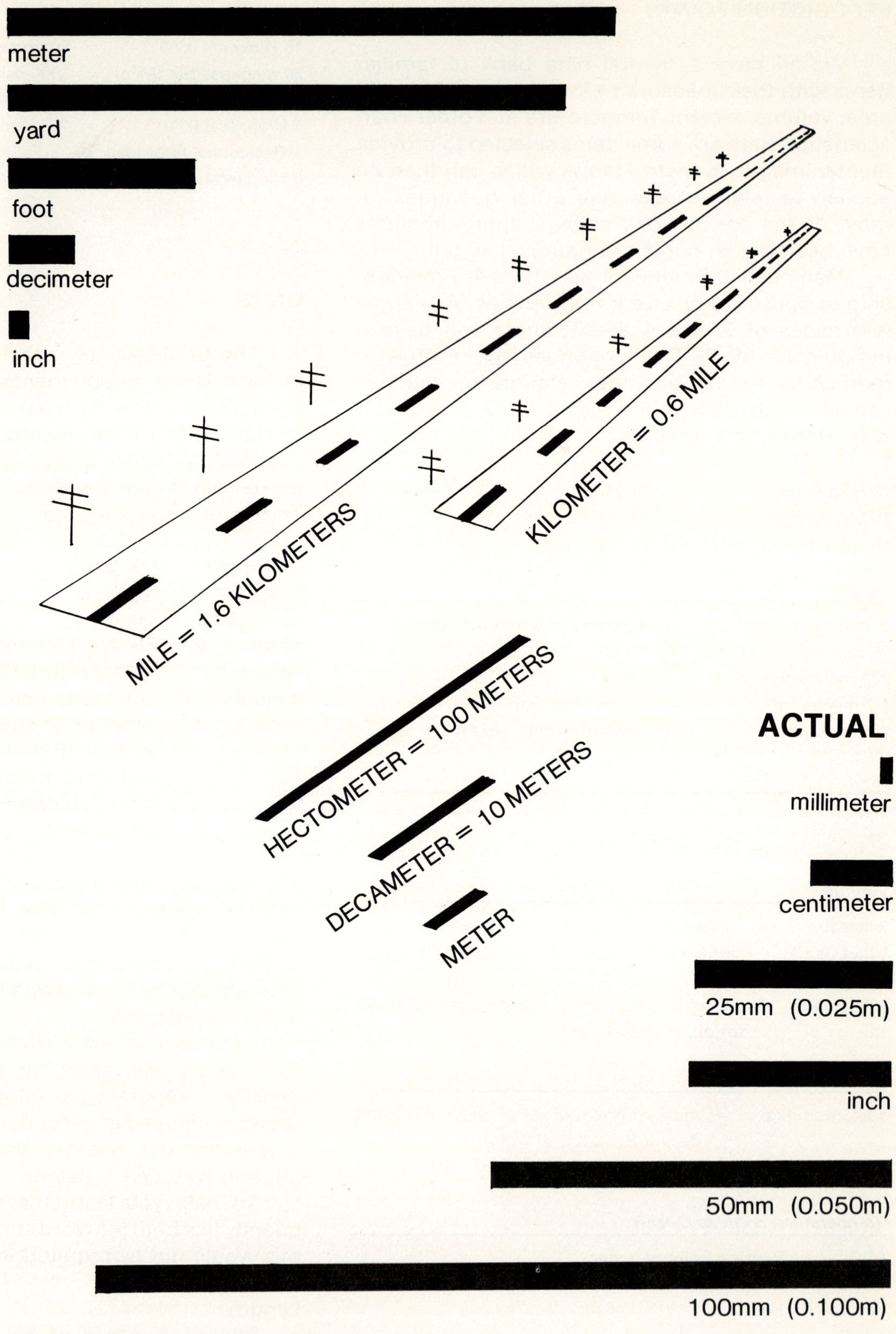

FIGURE 2-3. Visual Comparisons

subdivided into the millimeter, or one thousandth of a meter. Similarly, kilometers (km) are 1000 meters. In terms far too accurate for construction the meter is 1 650 763.73 wave lengths in vacuum of the orange-red line of the spectrum of krypton-86 (see Figure 2-1).

In clothing sizes and a few other areas, you will hear the term centimeter (cm), or one hundredth of a meter, used. But the construction industry has decided to confine its units to meters (m) and millimeters (mm). It is felt that centimeters may be confusing and inconsistent with other units which are in thousands (kilo) or thousandths (milli). Use of centimeters may also increase the likelihood of errors. Kilometers are primarily used for maps and road signs (see Figure 2-2).

As with the inch system, some use will be made of modules (M). The ISO, International Standards Organization, has adopted 100 mm as a module for architectural use. 100 mm is about four inches, which constitutes one of the modules used in the traditional system. Another module which may be used in construction is 300 mm, which is close to a foot and is sometimes called a "metric foot." Others are 400 mm, or about 16 inches; 600 mm, close to 24 inches; and 1200 mm, or about 47-1/4 inches, which will be the width of our panel products such as plywood, gypsum board, insulation board, etc. (See Figures 2-3 and 2-4 for visual comparisons.)

In building construction, dimensions of buildings shown on working drawings will be expressed in millimeters—such as 14 375 mm or, on a detailed drawing, 37 mm. A home 40 × 24 feet would be 12 190 × 7 315. Site plans will be expressed in meters with three decimal places, i.e., 28.050 m.

Area

The SI unit of area is the square meter (m^2). Other units encountered in the construction trade are the square kilometer (km^2) and the square millimeter (mm^2). A square hectometer (hm^2) and its acceptable alternate, the hectare (ha, equal to about 2-1/2 acres) will be used to describe land area (see Figure 2-4).

Volume

The basic unit for volume is the cubic meter (m^3). Thus we will have cubic meters (m^3) of concrete, earth, and lumber. A cubic decimeter (dm^3) is 1/1000 of a cubic meter (m^3), also known as a liter.

But the prefix chart shows a "deci" meaning 1/10. How can this be 1/1000? A side of a cubic decimeter is 1/10 of a meter, but when cubed it becomes 1/1000 (see Figure 2-5).

The volume represented by this cubic decimeter (dm^3) is a convenient measure for water, beer, oil, gasoline, etc., and has come to be called the liter (l). When typing this can be confusing because the one 1 and the "ell" (l) are often the same. So on the typewriter you may have to express liter by typing out the word. However, the preferred unit in engineering calculations will be the cubic decimeter (dm^3), which is the same as the liter. The liter (l) should be used only for fluids and particulate material such as sugar, salt, peas, etc. The liter equals a little more than a quart. A 1000 gallon fuel oil tank will contain 3800 liters. Similarly, 100 liters of wine will be 106 quarts. For smaller volumes, the milliliter (ml) is used.

Mass

If we fill with water the liter (l) or cubic decimeter (dm^3) that we developed above, it will have a mass of one kilogram (kg). We started with the meter (m), a measurement of length, and have now arrived step by step at mass, without any conversion factors.

But since water varies in purity and volume from temperature and atmospheric pressure, our National Bureau of Standards in Washington has a cylinder of platinum-iridium alloy that has a mass of *exactly* one kilogram (kg). This accuracy is far more than we need in the construction field; it is simpler to remember a kilogram as being the mass of one liter (l) of water.

This liter-kilogram (kg) relationship makes it easy to determine the mass of any known volume of water or any other liquid if its specific gravity is known. Also, we can determine the cubic content of any vessel by filling it with water and *either* measuring the quantity of water or "weighing" it.

Force

In the foregoing section on mass we have avoided the word "weight," except for the last sentence. Weight is defined as a measure of gravitational force acting on an object at a specified location. The force of gravity varies by about 0.5 percent over the earth's surface. In most construction-related applications this can be ignored.

A constant mass has an approximate constant weight on the earth's surface. The standard value

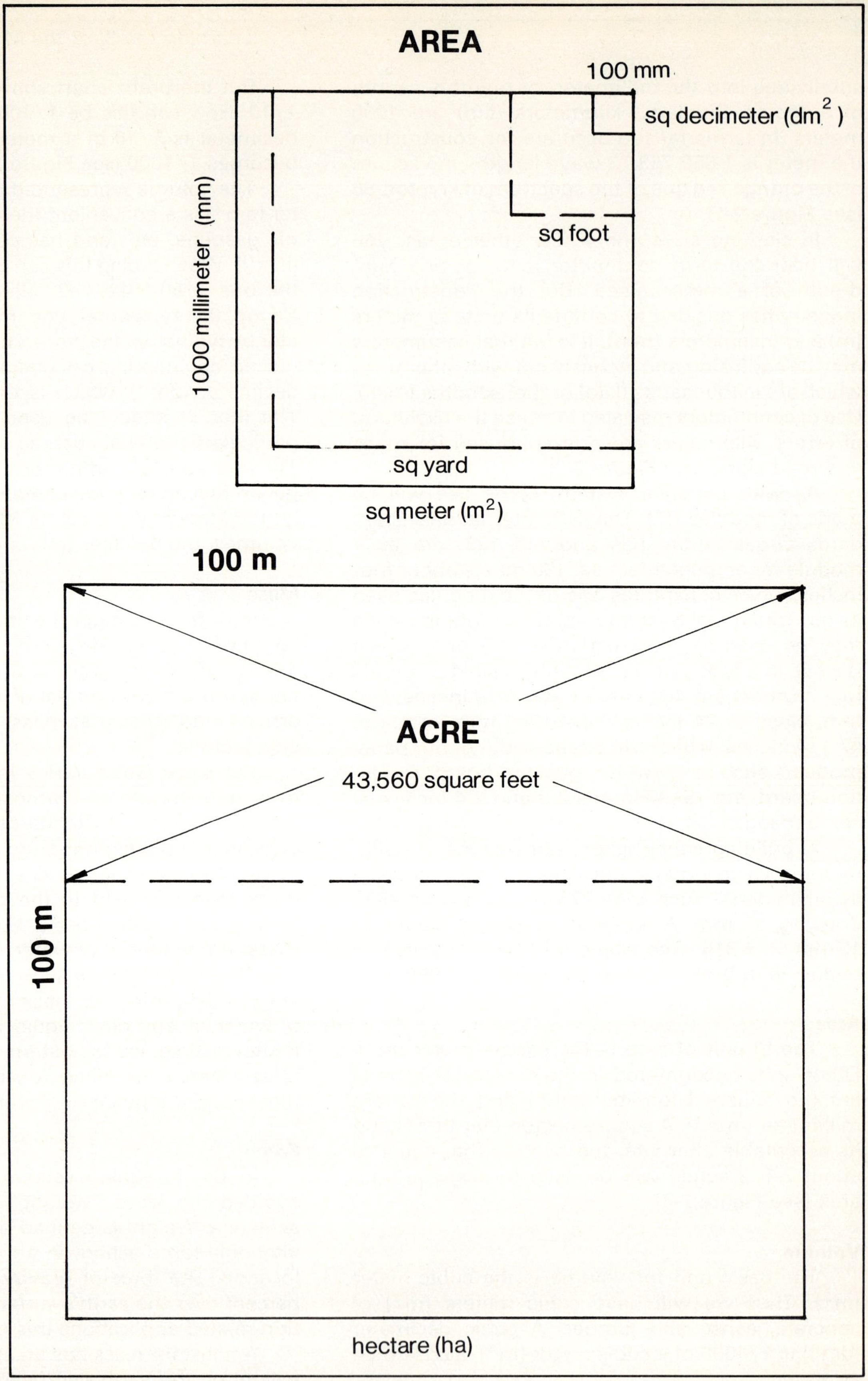

FIGURE 2-4. The Hectare and the Acre

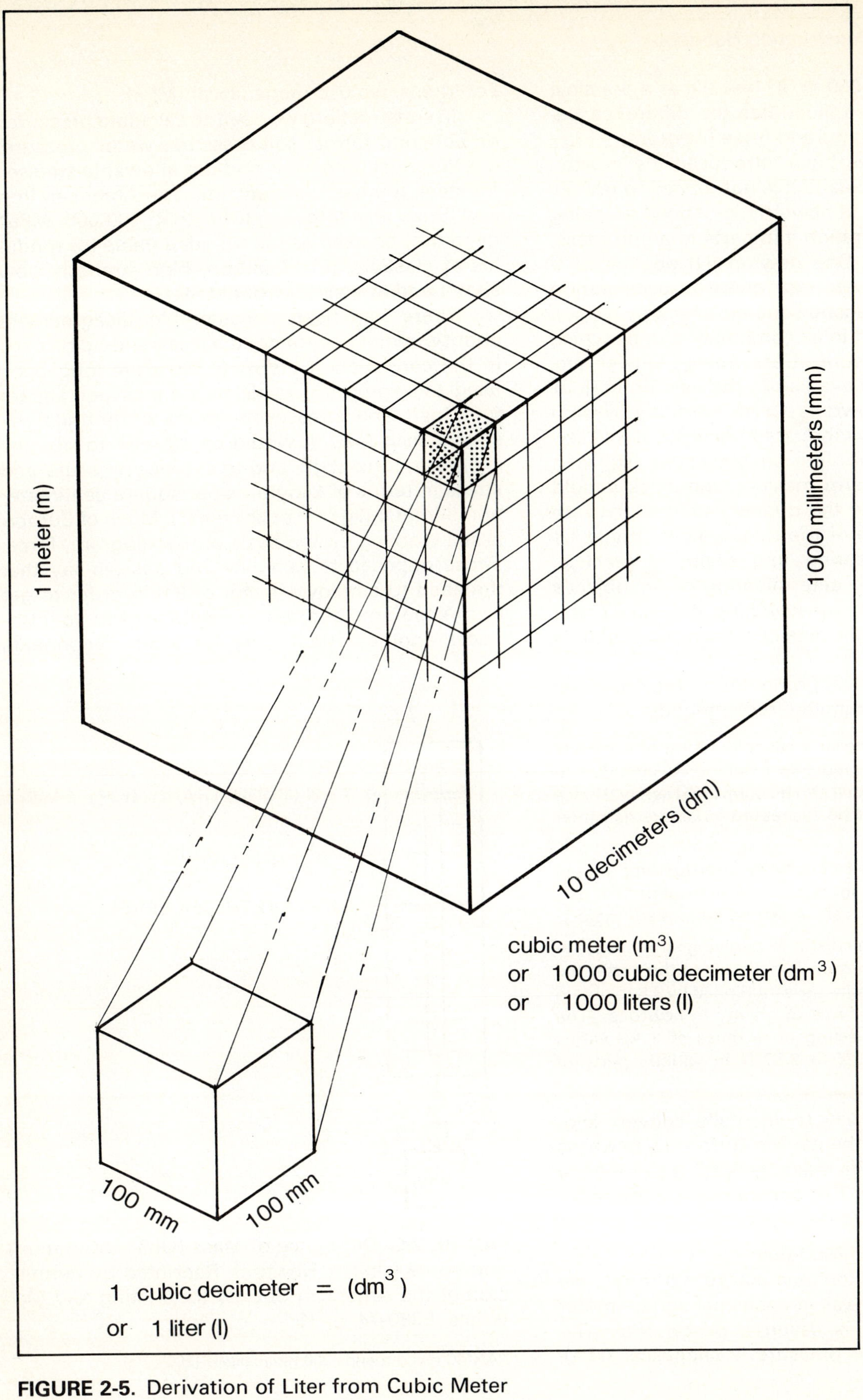

FIGURE 2-5. Derivation of Liter from Cubic Meter

of gravity is 9. 806 650 m/s^2 (use 9.8 as a practical average). Figure 2-6 illustrates the difference between mass (kilogram) and force (newton).

With the SI, we have introduced the newton (N) as a unit of force and it is not related to gravity or weight. A force in newtons is defined as being equal to the acceleration it imparts to a unit mass, the kilogram (kg). One newton (N) applied to a mass of one kilogram (kg) gives an acceleration of one meter per second squared ($N = kg \cdot m/s^2$). The newton is used in all combination units which include force: pressure, stress, energy, power, etc.

Purists would have us use the newton in floor and roof loads as newtons per square meter (N/m^2) as opposed to the more traditional "weight" expressed as kilograms per square meter (kg/m^2). In any event, the term mass or unit mass should be used to indicate the quantity of matter in an object, and the use of the term weight should be avoided in engineering and scientific practice. However, since the determination of an object's mass will be made by *weighing* the object, the use of the term weight instead of mass will probably continue.

The Metric Practice Committee of the American National Metric Committee recommends:

> A floor load rating or a capacity rating of a vehicle (or other load-supporting machine) is intended to define the mass that can be supported safely. Hence the rating should be expressed in kilograms rather than newtons.
>
> Similarly the area density of floor loading should be expressed in kilograms per square meter (kg/m^2) rather than in pascals (newtons per square meter).
>
> In engineering calculations involving structures, vehicles, or machines on the surface of the earth, the mass value in kilograms is multiplied by 9.8 to obtain the approximate force of gravity in newtons. (The force of gravity acting on a mass of 1 kg varies from about 9.77 N to 9.83 N in various parts of the world.)

A convenient way to mentally convert kilograms of mass on the earth's surface to newtons of force acting on the mass (gravity) is to multiply the kilograms by 10. This contains only a 2% error.

Pressure, Stress and Modulus

Having the newton as a measure of force, we can express pressure as newtons per square meter or N/m^2 and this is called a pascal (Pa). The convenient term for pressure is kilopascal (kPa). For stress, we use megapascal (MPa).

In construction, we need to calculate pressures on concrete forms, soil pressure, water pressure, etc. We must determine and use allowable stresses for steel, lumber, concrete, etc. The American Iron and Steel Institute suggests that 200 000 MPa* (N/mm^2) be used as the rounded value for modulus of elasticity (E) of carbon, high strength-, low alloy-, and low-alloy-wrought steels.

There has been considerable disagreement about whether newtons and pascals should be used in the construction industry. We have long been used to expressing loads as pounds per square foot (psf) and stresses in terms of pounds per square inch (psi). It would be simpler to use the kilogram/force (kgf), and to express pressure and stress in terms of kilograms per square centimeter or millimeter (kg/m^2 or kg/mm^2). Much of Europe is still working on the basis of the kilogram/force. Those opposed to newtons and pascals say that the need to employ a factor of 9.8 in order to get from kilograms of mass to newtons of force introduces another opportunity for error. Proponents

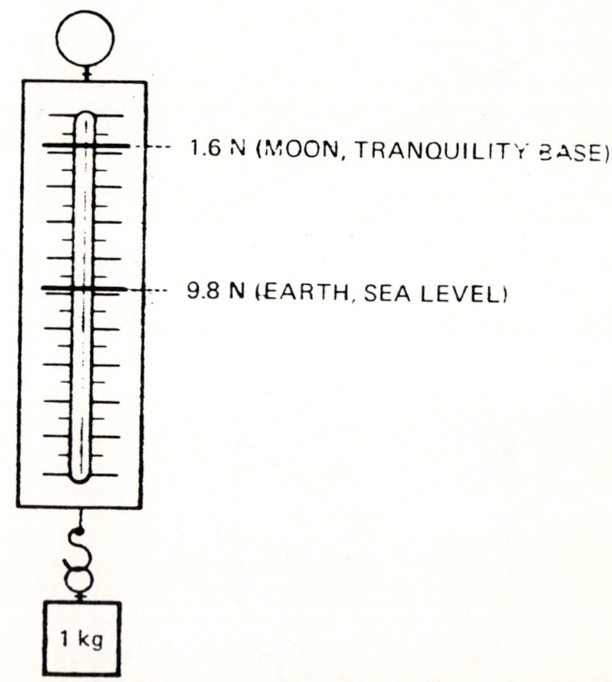

FIGURE 2-6. Difference of Mass (Unit = Kilogram) and Force (Unit = Newton). Reprinted by permission of the American Society for Testing and Materials, E380-74, p. 16.

*ANMC recommends 200 gigapascals (200 GPa)

of the pure SI point out that the English speaking countries that have recently "gone metric" have all gone to the pure SI with newtons and related units, and that European countries still using the older metric system have pledged to change to the SI.

Since it is vital that the entire construction industry be working in the same units, the resolution of this question is important although it may take a year or two to arrive at a consensus.**

Energy, Work, Heat and Power

The newton (N) is the basis for all of these units. A joule (J) is a newton meter or N · m. The joule then becomes the unit replacing a myriad of units that we are familiar with in construction. They include: British Thermal Unit, calorie, foot-pound force, foot-pound, and kilowatt-hour.

We express power as joules per second or J/s and we call this the watt (W). The watt replaces even more units, i.e., horsepower, calorie per second, foot-pound force per second, etc. We will use the term kilowatt (kW) for everyday descriptions, for example as the power of a tractor having 100 horsepower will be 74.57 kilowatts (kW). An electric saw rated at 10 horsepower will have power of 7.46 kilowatts (kW). But metric conversion for these ratings will probably result in rounding to 75 kW and 7.5 kW or other nominal ratings. In time, a more rational set of power ratings will probably develop such as 10, 20, 30, 100 kW, etc.

Temperature

For ordinary measurement of temperature, we use the degree Celsius (°C). This was formerly called Centigrade but the use of the term in some countries as a measurement of grade has made its use in temperature confusing. Also, the proposed use of the 400° circle with the 100° quadrant (centigrad) undoubtedly contributed to this confusion. On the Celsius scale, pure water at standard atmospheric pressure freezes at 0° and boils at 100°. Normal human body temperature is 37° and comfortable room temperature is about 22°.

But for engineering purposes, we should use the kelvin (K). (Note: kelvin (K) is not a degree, but a unit, hence we do *not* write degree kelvin or °K). The kelvin (K) is the same as one degree Celsius (°C) but the kelvin (K) scale starts at absolute zero

**But Andrew Lally of the AISE points out that the ASCE, ASME, industry groups, and a host of others agree on newtons and pascals.

TABLE 1: Most Used Metric Units and Symbols

Quantity	*Unit*	*Symbol*	*Accepted Alternate*
Length	kilometer	km	
	meter	m	
	centimeter	cm	
	millimeter	mm	
Area	square kilometer	km^2	
	square hectometer	hm^2	hectare
	square meter	m^2	
	square millimeter	mm^2	
Metric ton	ton	t	
Weight (Mass)	kilogram	kg	
	gram	g	
	milligram	mg	
Volume	cubic meter	m^3	
	cubic decimeter	dm^3	liter
Force	megapascal	MPa	
	kilonewton	kN	
	newton*	N	
Pressure	kilopascal	kPa	
	pascal	Pa	
Temperature	kelvin	K	
	degree Celsius	°C	
Time	second	s	
Electric current	ampere	A	
Amount of substance	mole	mol	
Luminous intensity	candela	cd	
Energy, work quantity of heat	joule	J	
Frequency	hertz	Hz	
Power	watt	W	
Electromotive force	volt	V	
Electric resistance	ohm	Ω	
Luminous flux	lumen	lm	
Illuminance	lux	lx	

*The newton meter (N·m) is also used to calculate moment or torque, but in this case the length (m) is perpendicular to the force.

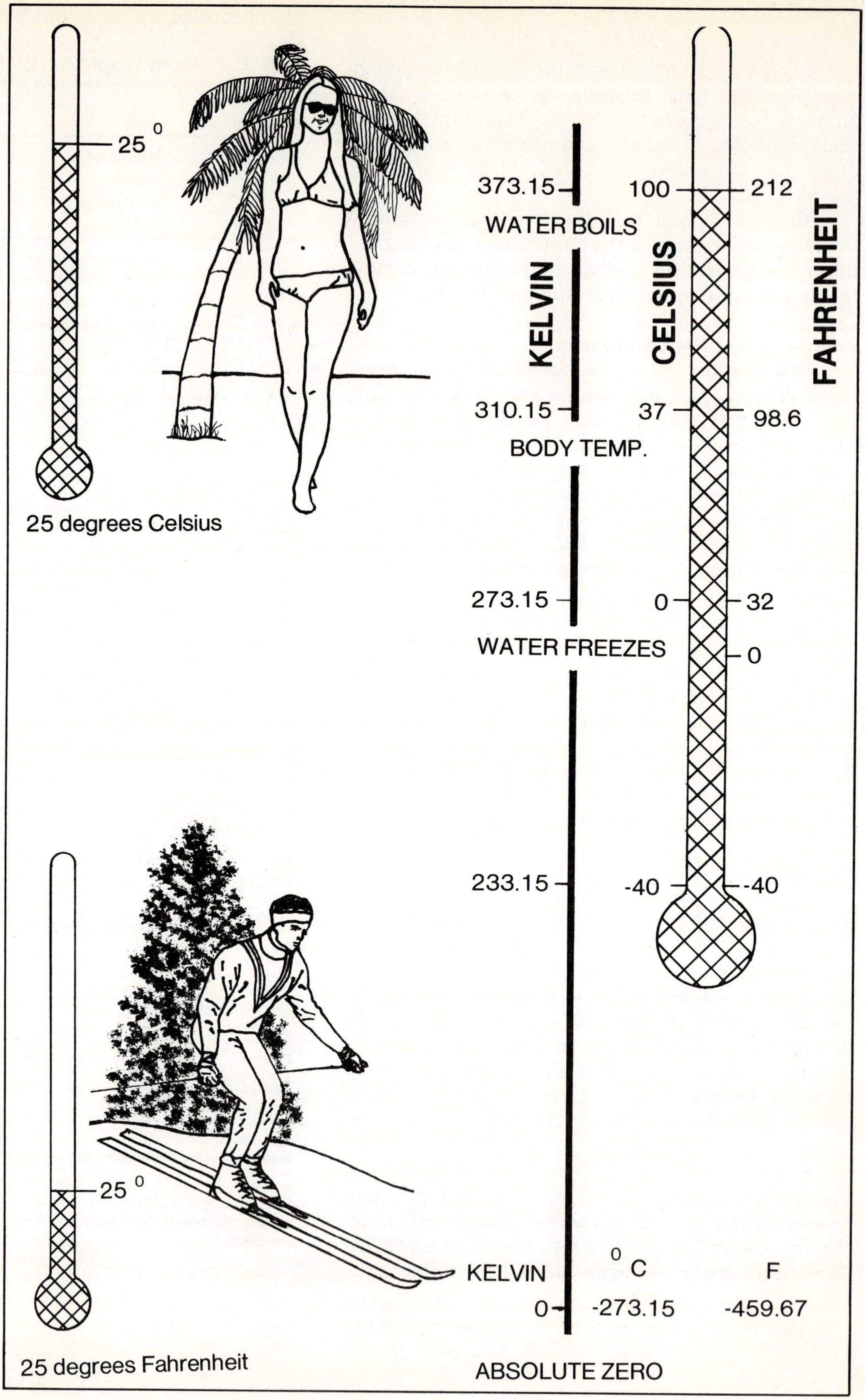

FIGURE 2-7. Temperature Equivalents

and water freezes at 273.15 K and boils at 373.15 K (see Figure 2-7).

Now that you have had a short course in SI metrics, the next thing to learn is how to speak and write the language of the meter. We should know which abbreviations to capitalize, for this is important, and when and where to use commas and decimal points. Table 1 gives the metric units and symbols you may encounter in your work or at home.

Rules for Writing Metric Quantities (as suggested by the National Bureau of Standards "Guidelines for the Use of the Metric System")

I. CAPITALS

Units: When written in full, the names of all units start with a lower case letter, except at the beginning of a sentence.

Example:	for meter,	write meter

Symbols: Symbols for units are written with lower case letters, except that the first letter is capitalized when the name of the unit is derived from the name of a person.

Example:	for meter	write m
	for pascal	write Pa
	for newton	write N
	for Celsius	write C

Prefixes: With three exceptions, all symbols for numerical prefixes are written with lower case letters. The exceptions are tera (T), giga (G), and mega (M). Likewise, when written out in full, all prefixes are written in lower case letters.

Example:	for kilogram	write kg
	for millimeter	write mm
	for megameter	write Mm

The importance of following the precise use of uppercase and lowercase letters is shown by the following examples:

G for giga; g for gram
K for kelvin; k for kilo
M for mega; m for milli
N for newton; n for nano
T for tera; t for metric ton

II. PLURALS

(a) When written in full, the names of units are made plural when appropriate. Fractions (both common and decimal) are always singular.

Example:	10 meters;	0.5 meter.

(b) Symbols for units are the same both in singular or plural. (No "s" is added to indicate a plural.)

Example:	Right	Wrong
	10 m	10 m's

III. PERIODS

A period is not used after a symbol, except at the end of a sentence.

Example:	Right	Wrong
	16-mm film	16 mm. film

IV. THE DECIMAL POINT

The dot is used as the decimal point and is placed on the line. When writing numbers less than one, a zero must be written before the decimal point.

Example:	Right	Wrong
	0.5 meter	.5 meter

V. GROUPING OF DIGITS

(a) Separate digits into groups of three, counting from the decimal sign. Instead of the present comma, a space is left between the groups of three digits (many countries use a comma for the decimal point).

Example:	Right	Wrong
	10 000 000 mm	10,000,000 mm

(b) In numbers of four digits, the space is not commonly used. Example: 1000, not 1 000. The exception is when four digit numbers are part of a column in which longer numbers appear.

VI. SPACING

(a) When writing symbols or names for units having prefixes, no space is left between letters. Example: kg, not k g.

(b) When writing a symbol after a number to which it refers, a space must be left between the number and the symbol. Example: 10 mg. (Excepted from this rule are degree, minute, second of angle, and ° C.)

VII. COMPOUND UNITS

When writing the symbol for a compound unit that is formed by the multiplication of two or more units, a centered dot is used. Example: N·m. However, when writing out the name for such a unit in full, the centered dot is never used. A hyphen is recommended, or a space is permissible. Example: newton-meter or newton meter.

TABLE 2: SI Prefixes

Factor	*Prefix*	*Symbol*	*Multiplication Factor*	*U.S.A. Pronunciation*
10^{12}	tera	T	1 000 000 000 000	as in *terra*pin
10^{9}	giga	G	1 000 000 000	jig' a (*a* as in *a*round)
10^{6}	mega	M	1 000 000	as in *mega*phone
10^{3}	kilo	k	1 000	as in *kilo*watt*
10^{2}	hecto	h	100	heck'toe
10^{1}	deka	da	10	deck-a
10^{-1}	deci	d	0.1	as in *deci*mal
10^{-2}	centi	c	0.01	as in *senti*nel
10^{-3}	milli	m	0.001	as in *mili*tant
10^{-6}	micro	μ	0.000 001	as in *micro*dot
10^{-9}	nano	n	0.000 000 001	nan-o (as in *Nan*cy)
10^{-12}	pico	p	0.000 000 000 001	peek oh
10^{-15}	femto	f	0.000 000 000 000 001	as in *femi*nine
10^{-18}	atto	a	0.000 000 000 000 000 001	as in *anato*my

*So that the prefix will retain its identity, the first syllable of each prefix should be accented. U.S. citizens already are drifting into an undesirable pronunciation for "kilometer," accenting the second syllable instead of the first.

VIII. THINGS TO NOTE

(a) In giving temperatures, the proper form is degree Celsius. Note that "degree," the unit, is lower case but the modifier "Celsius" is capitalized. (The "degree centigrade" is obsolete.)

(b) When units such as the meter are squared or cubed, the proper method is to write the symbol for the unit, followed by the superscript ² or ³.

Example: 10 m^2 15 $^{-3}$

(c) The term "per" when part of a unit (such as newtons per square meter) is represented by a slash mark (N/m^2). Such abbreviations as k.p.h. or kph (for kilometers per hour) are not considered desirable.

(d) Practical considerations require some adjustments. For, example, l is the accepted symbol for liter. But on most typewriters there is little or no difference between the letter "l" and the number "1." So generally the full word liter is written out, although ml (for milliliter) is common because it does not confuse the reader.

3. The Conversion and Presentation of Numerical Values*

GENERAL

Many of the customary U.S. design factors in current use have numerical values which are simple integers, e.g., 4″ × 2″, 8′-0″, 1 ton/ft², 3000 lbf/in² (psi), etc., to facilitate calculation and reference. It is recognized that some of these values are "nominal" only, and allow considerable percentage variations. Others may have a legal significance by representing a minimum statutory requirement, and such a limitation will introduce considerably greater restriction in effecting a sensible conversion to metric values.

Early in the transitional period many situations are likely to arise which call for a value judgment as to the appropriate metric number to replace an imperial numerical value. This judgment should be based on the degree of accuracy inherent in the quantity to be converted.

The following suggestions are made.

EXACT CONVERSIONS

Exact (direct conversion denotes a precise conversion from imperial to metric values, accurate to a number of decimal places to satisfy scientific, legal or statistical requirements (e.g., 1 U.S. gallon = 3.785 approx. liters).

High precision, necessitating exact conversions, is seldom required or achieved in the building and construction industry.

Exact conversions should be avoided wherever possible.

SOFT CONVERSIONS

Soft conversion implies that the conversion is being made in the software (on paper) only and that the actual product, material or structure being manufactured remains physically unaltered.

Minimal rounding to the nearest integer or sensible number, rather than a preferred number, is normally applied (e.g., 8″ becomes 203 mm, 20′ becomes 6.1 m, 1 ton/ft² becomes 96 kPa).

Soft conversions are likely to be used as an interim measure in the conversion of some product data, standards, regulations, codes of practice, and other numerical statements.

The practice of soft conversion with no attempt by product manufacturers or designers to adopt rational metric dimensions, will merely result in a "metric veneer"—the SI description of an unchanged imperial environment, with none of the advantages of true metric conversion.

HARD CONVERSION—USE OF RATIONALIZED VALUES

Hard conversion denotes a physical change in product characteristics from existing values to preferred or rationalized metric values (e.g., 2′-0″ is

*Adapted from *Metric Handbook: Metric Conversion in Building and Construction.* Published by Standards Association of Australia, 1972.

changed to 600 mm, 3000 lbf/in^2 is changed to 20 MPa).

As in countries with fully developed metric systems, it is anticipated that final metric values will become simple integer numbers in SI units. In some areas there has been considerable progress toward such an objective, and the substitution of 2400 mm for the current 8 foot ceiling height limitation allows for more rational design, manufacture, and construction than a value of 2440 mm, or even worse, the exact conversion of 2438.4 mm.

It is well to remember that current values were in many instances chosen for reasons of numerical simplicity. Minimum or limiting values have little scientific significance, although they acquire considerable economic and emotional significance if they are in existence in statutory requirements over a period of time.

The manufacturer is likely to gain more acceptance with rationalized metric values for his products although some of them may still be "nominal."

The designer should specify hard converted (rationalized) metric sizes wherever they are economically available in order to assist the construction industry in effecting a smooth and successful conversion.

SUGGESTED APPROACH TO RATIONALIZATION OF LINEAR DIMENSIONS

Where linear dimensions are to be replaced by metric measurement, the following preferential system for rationalization is suggested:

Very Small Dimensions
(up to about 50 mm):

First preference : Multiples of 10 mm
Second preference : Multiples of 5 mm
Third preference : Multiples of 2 mm
Fourth preference : Multiples of 1 mm

Note: In building and construction, fractions of the millimeter are rarely required, but where such is unavoidable the decimal fraction should be kept to one place (e.g., 3.2-mm thick vinyl flooring).

Small Dimensions
(up to about 500 mm):

First preference : Multiples of 100 mm
Second preference : Multiples of 50 mm
Third preference : Multiples of 25 mm
Fourth preference : Multiples of 10 mm

Medium Dimensions
(up to about 6000 mm):

First preference : Multiples of 600 mm
Second preference : Multiples of 300 mm
Third preference : Multiples of 100 mm
Fourth preference : Multiples of 50 mm
Fifth preference : Multiples of 25 mm
Sixth preference : Multiples of 10 mm

Note: These preferences apply essentially to building work in which a preferred dimensional framework is used.

Standard U.S. residential door sizes are given below with their metric equivalent and a possible rational metric size.

Present	*Soft Conversion*	*Rational Hard Conversion*
2'0"	609.6 mm	600
2'4"	711.2 mm	700
2'6"	762.0 mm	
2'8"	812.8 mm	800
3'0"	914.4 mm	900

These metric dimensions reflect other factors as well as the preferences given above such as use of doors for various purposes, building codes, etc.

Large Dimensions
(site layouts, roadworks, etc):

First preference : Multiples of 10 m
Second preference : Multiples of 5 m
Third preference : Multiples of 1 m
Fourth preference : Multiples of 0.5 m
Fifth preference : Multiples of 0.1 m

SUGGESTED APPROACH TO RATIONALIZATION OF AREA AND VOLUME

A similar approach can be taken toward the rationalization of areas and volumes. However, it is important to note that if the length dimensions of an area are changed by 1 percent, then the area variation is approximately 2 percent, and if the length dimensions of a volume are changed by 1 percent, then the volume variation is approximately 3 percent.

SUGGESTED APPROACH TO RATIONALIZATION OF OTHER NUMERICAL VALUES

It may be necessary to provide metric equivalents for quantities other than those based on length. Where a factor of 1000 exists between the basic unit and its multiples or submultiples, it is recommended that the following system of preferences be used:

Range 100 to 1000:
First preference : Multiples of 100
Second preference : Multiples of 50
Third preference : Multiples of 20
Fourth preference : Multiples of 10

Range 1 to 100:
First preference : Multiples of 10
Second preference : Multiples of 5
Third preference : Multiples of 2
Fourth preference : Multiples of 1

Example: If an exact conversion yields a value of 93.21, using the recommendations for establishing a rationalized value would give:

(1) A first preference of 90 (an error of −3.4 percent);
(2) A second preference of 95 (an error of +1.9 percent);
(3) A third preference of 94 (an error of +0.85 percent);
(4) A fourth preference of 93 (an error of −0.22 percent).

Where it is imperative that the rationalized value is not less than the exact conversion then, in the above example, either the second or third preference is appropriate.

The decision as to which preference to use must be based on considerations relating to the quantity being reviewed and the degree of accuracy necessary. It is misleading to convert an arbitrarily selected value to four significant figures, and to retain these figures in the name of accuracy. Likewise it would be a mistake to "round off" a number by 5 percent when the inherent accuracy of the value being converted is in the order of ± 1 percent.

This concept of inherent accuracy is very important, as it provides us with the criterion for the degree of rounding that can be tolerated.

DIMENSIONING OF PRODUCTS AND TRADE LITERATURE

The Building and Construction Advisory Committee has recommended that all plans and specifications be dimensioned in meters and millimeters only. The centimeter is not a preferred unit for the purpose of the building and construction industry, and should not be used, as outlined earlier. Statements in trade literature, although preferred in metric units only, may be given in dual dimensions during the interim period, but the metric value should be stated first. Where dimensions are given in meters they should be written to three places of decimals, e.g.:

2.100
1234.250
0.825
64.000

Where dimensions are given in millimeters they should be written thus:

2100
1 234 250
825
64 000

Symbols for units of linear measure may be discarded where there is no possibility of ambiguity, and the following rules should be applied:

(1) Whole numbers indicate millimeters.
(2) Decimalized expressions (to three places of decimals) indicate meters.

If the need should arise to express dimensions by using figures in a manner that does not conform to the above rules, then the figures must be followed by the appropriate symbol. For example, in fine engineering it may be necessary to express dimensions of less than 1 mm—e.g., 1.125 mm, 0.008 mm; or in surveys of very large areas it may suffice to express dimensions in whole meters—e.g., 7500 m, 125 725 m.

When symbols are not given on a drawing, it must be clearly labelled "All dimensions are in meters," or "All dimensions are in millimeters."

NUMERICAL CONVERSION METHODS

There are three avenues open for the individual who needs to convert from traditional measurements to metric measurements. He can (1) use the tables found in the appendix; (2) use numerical conversion factors—the "long" way; or (3) use one of the calculator/converter machines now on the market (see Figure 3-1). Results should be expressed to three decimal places for meters, or in millimeters.

PRODUCT SIZES

In the U.S., we customarily state sizes in width first and height second. For example: 2′6″ × 6′8″ door, 9′ × 7′ garage door, 24″ × 36″ window, 4′ × 8′ plywood, etc.* In Europe and the Commonwealth countries, the longest dimension is usually expressed first, such as 2400 × 1200 plywood or a 2100 × 900 door. Where there is doubt, the size

The package consists of computer programs developed by Caterpillar Tractor Company and General Motors Corporation, documentation explaining how to get the programs running on different computers and how to use them, and test problems to permit verification.

The package is priced at $500.00 to recover National Bureau of Standards costs. It may be ordered by writing Conversion Package, Room B-311, Chemistry Building, National Bureau of Standards, Washington, D.C. 20234, phone (301) 921-2045. Orders should specify ASCII, EBCDIC, or BCD code and 556, 800, or 1600 bpi density.

designation is followed by "wide," "long," "high," or "thick," as applicable, such as a concrete block 390 × 190 × 190 wide, or a floor tile at 300 × 300 × 2.5 thick.

The U.S. practice of expressing the width first is likely to be retained, unless substantial reasons for following the European practice are found, which seems unlikely at present.

*Note that sometimes the last dimension in plywood is used to designate the direction of the grain. Thus plywood listed as 8′ × 4′ would have the grain running in the 4′ direction.

FIGURE 3-1. The 203 Rockwell Standard Calculator/Convertor. Photograph courtesy of TWS Electronics, Inc.

COMPUTER PROGRAM PACKAGE FOR METRIC CONVERSION

In order to assist engineers and manufacturers in the transition from the traditional U.S. system of measurement to the metric system, the National Bureau of Standards is making available a computer program package to perform the conversion from metric to traditional with the carefully controlled accuracy necessary to maintain required tolerances at minimum cost.

The package responds to the need for two sets of dimensions on an engineering drawing by providing programs to produce a table of metric-traditional equivalents for all dimensions appearing on that drawing. The table can then be attached to the drawing before it is reproduced.

4. Planning and Managing Metrication in the Construction Industry and in Your Company

Like stopping smoking or plunging into a cold swimming pool, the change to the SI Metrics by individuals is sometimes best made abruptly. You come to work some morning and find your inch rule has been replaced by one reading in millimeters and meters only. Product literature on window and door sizes would be available only in metric. We can call this the "cold turkey" method.

This "sink or swim" approach is advantageous, as opposed to dual dimensioning or a gradual change, because people learn almost immediately and the crutch of familiar measurements is gone. This is illustrated by an experiment conducted by Professors Reese and Cathcart of the University of Alberta, as published in the Canadian *Metric Monitor.*

A group of teachers were asked to weigh themselves and to measure their height in metric units. One half of the group used a scale and a metric rule which also indicated traditional units. The other half had only the metric units available. One hour later, after a discussion on a completely different subject, they were asked if they honestly remembered their weights in kilograms and their heights in centimeters. Over 80 percent of those using the scale reading only in kilograms remembered their weight, compared with only about 40 percent in the other group. In the height measurement the percentages were even more striking—95 percent vs. 45 percent. The point is that when people are faced with both systems, the concentrate on the one they are familiar with and ignore the other. Human nature being what it is, we take the easy way out.

Although the "cold turkey" approach may be best for individuals, it simply will not work for a nation or an industry. Think of the resulting chaos if someone in government ruled that on the 15th of next month, everyone would begin to draw plans, produce building materials, and build, in metric units!

Moreover, the conversion in each industry must be coordinated with many other sectors. The construction industry uses steel, but so do most other industries. So the production of steel to metric standards must be timed and coordinated with all of the using industries. Even such a simple thing as drafting paper must be changed to the ISO (International Organization for Standardization) sizes on a timetable suitable to all users. The costly production of both metric and traditionally sized materials during a changeover period must be minimized by careful planning.

Once metric legislation has become law, some manufacturers may find it advantageous to produce all new sales and technical literature with dual dimensioning from that date forward. In this way the gradual change will be accomplished without any sudden, expensive printing costs. At a later date, as coordinated with the industry as a whole, the traditional measurements may be dropped on subsequent printings of the same or revised literature.

The following is one solution to such problems.

The author's Panel-Clip Company provides a truss design service for users of its truss connectors. Since the truss designs are maintained on file for the use of future customers, years later, the metric conversion would be an overwhelming job if attempted in a short period of time. Instead, the company plans to begin showing dual dimensions on truss designs very soon after the legislation is signed into law. Thus, when the day for the final conversion arrives, most of the work will have been done and from then on new plans will have dimensions only in millimeters and loads in kilograms per square meter.

We in the United States can benefit from the experience of countries currently or recently converted to the SI. These countries include Great Britain—which also had to convert its monetary system—Australia, New Zealand and South Africa. Another advantage is the American National Metric Council's work in various sector committees. This council, since 1973, has done much of the work of preparing for an eventual metric changeover. However, since many industries, companies, and institutions have not been involved in this voluntary work, much of it must be reaffirmed after metric legislation pases the Congress. The broadest possible involvement in the planning and scheduling of the changeover will insure wholehearted support for carrying out these plans.

OBJECTIVES OF CONVERSION

We must make every effort to insure that metric conversion is used as the catalyst for a new framework of sizes and dimensions. This requires review, rationalization, simplification, and improvement. Our concern must be directed primarily to the consumer of construction services; in other words, those who pay for and occupy buildings, or the taxpayers and users of roads and bridges. Only secondarily should we be concerned with the convenience of we construction people and our suppliers.

The objectives we should seek in metric conversion:

(1) To change to the most up-to-date version of the metric system with facets suitable to a consensus of the industry.
(2) To make the change in an efficient and economical manner.
(3) To simplify procedures and practices in the industry, and to eliminate unnecessary variety.
(4) To make possible the development and use of dimensionally coordinated products and systems by integration, simplification, and the improvement of interchangeability in all phases of the building process.
(5) To revise and improve building codes, standards and codes of practice.
(6) To rationalize the manufacture, stocking, handling, transportation and assembly or installation of building products.
(7) To prepare informative, concise and up-to-date technical data and trade literature.

Most of these objectives could be met without metric conversion but we would lack the stimulus of an enforced change as an opportunity for further review of our present methods.

INDIRECT CHANGES RESULTING FROM METRICATION

The accomplishment of the above objectives will be the major changes in the construction industry. But other changes will result from the changeover.

(1) Some manufacturing equipment will require modification or replacement in order to produce materials in metric sizes. Equipment which has a weighing or measurement function must be recalibrated in metric units.
(2) The production of metric sizes in existing plants with present equipment may result in either an increase or a decrease in productivity. Example: 1200 mm × 2400 mm plywood produced with equipment originally built for 48″ × 96″ will result in a loss of plant production capacity of about 3 percent.
(3) Data or records for cost-estimating will have to be rewritten in metric terms.
(4) Computer programs for design or other purposes will have to be rewritten.
(5) Educational institutions must prepare themselves to produce graduates fully acquainted with the new system. Responding to the schedule for metric conversion, persons graduating after the change to 100 percent SI need receive no training in the traditional system.

Building Codes

At this writing, the regional building codes have metrication plans as follows:

BUILDING OFFICIALS AND CODE ADMINISTRATORS INTERNATIONAL (BOCA), Chicago: Codes presently carry metric conversion tables. In their next edition in 1978 all tables will appear in both metric and U.S. Customary.

INTERNATIONAL CONFERENCE OF BUILDING OFFICIALS (ICBO), Los Angeles: The process of developing metric terminology for those areas that are unique to their organization has begun, with emphasis on limitation on building areas, heights, corridor widths, separation distances, etc. They plan a hard conversion in these areas.

SOUTHERN BUILDING CODES CONGRESS INTERNATIONAL, Birmingham, Ala.: They are presently participating with the National Conference of States on Building Codes and Standards and the American National Metric Council on metrication. They do not have any definite dates or plans to convert their Standards or Codes to the metric system.

TIMETABLE FOR CHANGE

In coordination with other industrial sectors, the construction industry must develop a time schedule for orderly metric conversion. This timetable can be divided into three phases.

During the *Planning* phase, predictions must be made as to the dates metric-planned projects will be on the drawing boards; the first production of metric materials; the final production of traditional materials; agreement for rationalization of sizes; and the revision of building codes.

The *Action* phase will involve everyone in the industry from architects and engineers to carpenters and masons. During this phase everyone will be working hard and trying new approaches.

The *Residual* phase will be characterized by cleaning up and finishing off uncompleted changeovers. During this phase, the phasing out of traditional sizes, regulations, and methods will be completed.

The recent experience of the British Commonwealth nations should help us recognize that, once started, the faster the transition from old to new, the less overall cost and dislocation there will be. Therefore we need a plan that has some definite target dates to prevent years of "discussion" and procrastination.

Such a timetable is the one shown in Figure 4-1, which was developed for Australia. But our problems are different and we must develop our own timetable. What's more, we must do it with dispatch.

CONVERSION WITHIN YOUR COMPANY

Individual companies in the construction industry subsectors such as architecture, engineering, lumber, concrete, masonry, steel, plumbing, electrical, etc., will be involved in arriving at a consensus and insuring coordination with other subsectors. At the same time, planning must begin within each company. At an appropriate point in the timetable, the company plans must be put into effect. There is no need to involve all personnel in the changeover until the schedule indicates.

The first step is for management to appoint a metric conversion project manager. In small organizations this individual may have other jobs. Larger firms may have a project manager with a staff of several people. If appointed early in the industry's metric conversion process, the project manager may be the company's representative within the industry sector, participating in decisions made at that level.

The project manager's activities can be divided into three phases timed to coincide with the construction industry timetable. These phases are *Investigation, Planning* and *Implementation.*

In the *Investigation* phase the project manager should:

(1) Investigate the implications of metric conversion for all parts of the organization, with particular emphasis on standardization and increased efficiency.

(2) Identify the needs of each part of the firm during the changeover, with emphasis on metric standards and units to be used, education and training, and equipment conversion.

(3) Identify the needs and modifications to internal information systems.

(4) Estimate the costs of alternative approaches to metric conversion.

(5) Prepare an initial schedule for the organization's metric conversion plan after considering plans of customers and suppliers.

(6) Investigate opportunities for increased productivity as a result of rationalization.

(7) Make recommendations to management prior to the date required.

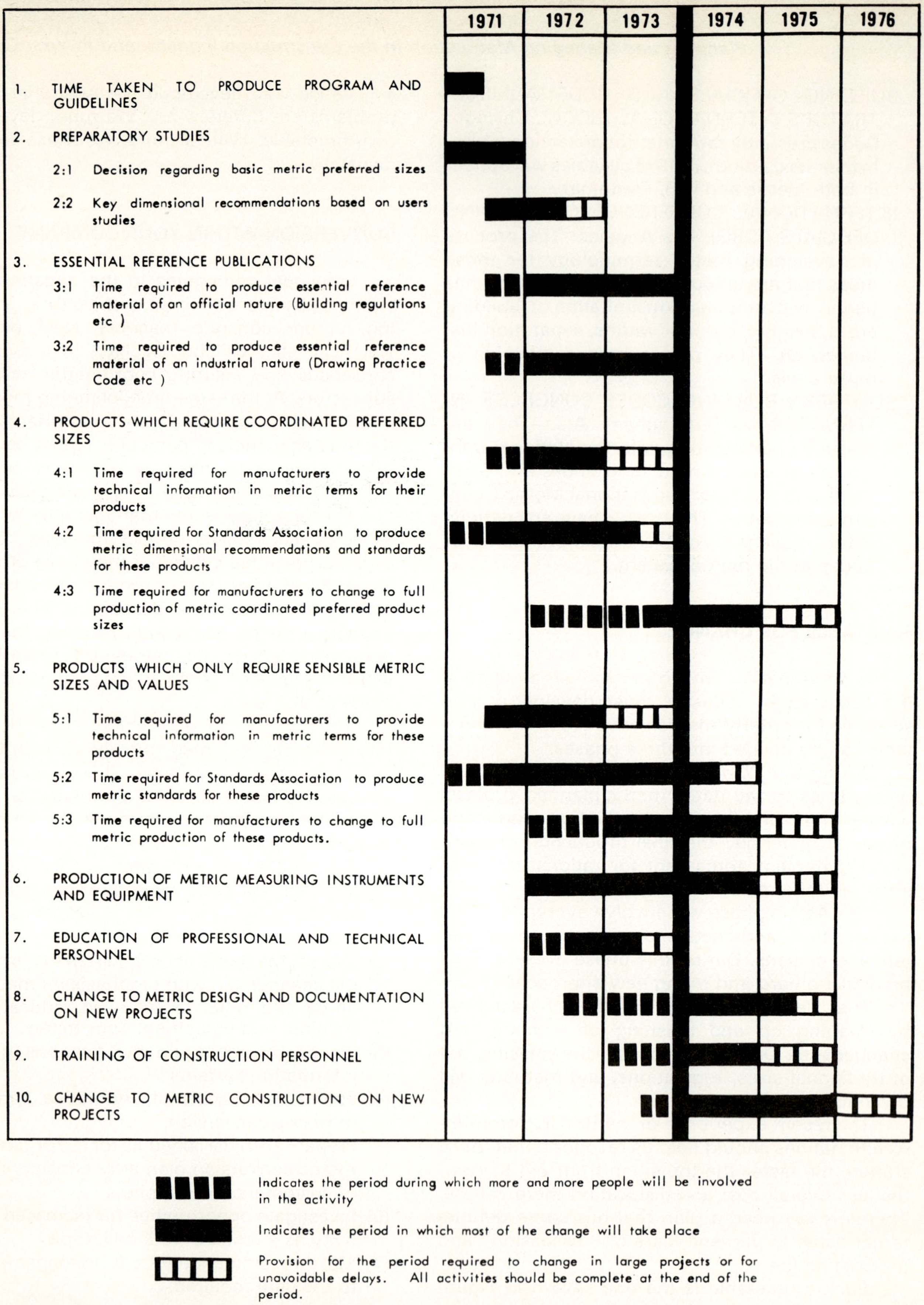

FIGURE 4-1. Metric Conversion Timetable for the Australian Construction Industry. Reprinted by permission of the Standards Association of Australia from *Metric Handbook* (MH1), 1972, p. 20.

In the *Planning* phase the project manager should:

(1) Work with the trade and professional organizations such as National Association of Home Builders, Associated General Contractors, American Institute of Architects, American Society of Civil Engineers, ASME, ASHREA, NBMDA, etc. etc., as his firm's business indicates. He should develop further plans through these sources.

(2) Work with local chapters of his firm's national affiliation to complete detailed local plans. At this time he may wish to consult with related trade organizations. For example, a contractor may discuss his plans with the local architects' chapter and concrete association.

(3) Complete plans for product changes and dates of manufacture or use in the firm's projects. Similarly, plan the phase-out of manufacture and/or use of traditionally sized materials.

(4) Assess training requirements for all employees. Schedule training sessions. Develop training aids.

(5) Check to see if any changes are required in local or state laws to facilitate the changeover. Work with local and national associations and politicians to implement these.

(6) In larger firms, write or rewrite a policy manual in metric terms.

(7) Investigate availability of metric materials, supplies and products. Schedule the purchase of these items.

(8) If a manufacturer, redesign your containers, packages, labels. Expedite the provision of new sales literature and technical data sheets.

(9) Determine and schedule changes in business administration and accounting procedures.

(10) Announce company policy and schedule to all personnel.

(11) Seek approval from management of a detailed plan incorporating who, what, when, and where.

The *Implementation* phase should put into effect all the plans and provide for modifications and rescheduling as indicated by experience.

5. Civil Engineering Calculations using SI[1] Units[3]

By R. Ernest Leffel,[2] F. ASCE

INTRODUCTION

The change from our customary and old metric units to the new SI metric units of measure will be more challenging than a simple translation or conversion to the new system. We have been using a gravitational system in the United States, particularly in civil engineering, while the SI is an absolute system. This will affect all our calculations having to do with mass, force, stress or work. It will simplify and rationalize the dynamic aspects especially. The following presentation is to clarify points needing better understanding in the use of the new absolute system.

Figure 5-1 presents five of the base SI units and the more useful units derived from them. Several derived units shown on the chart are still under debate and not yet completely accepted for U.S. practice. These include the hectare as a unit of area and metric ton as a unit of mass. Also, the spelling of meter and liter has fired considerable controversy. Hopefully, the Metric Practice Committee of the American National Metric Council will have completed their recommendations early this year and will provide all of us with a preferred basis for selection of units and their prefixes.

Table 1 also presents information supplementing that shown on Figure 9 concerning base and derived SI units, and some special units. (Tables 1–5 can be found in Appendix A.)

Table 2, with its accompanying notes, presents the prefixes and symbols used with SI units and how they are to be used.

Table 3 presents units and prefixes that a number of sanitary engineers including representatives of the ASCE, AAEE, AWWA and WPCF have suggested be recommended for preferred usage by environmental engineers.

Some of the more common equations and definitions used by civil engineers, adapted for use with SI units, are presented in Table 4.

A one page conversion table for civil engineers is given in Table 5.

There are several differences in the treatment of mass, weight, and force that must be remembered when using SI units rather than the customary American units.

Mass, Weight and Force

Mass can be defined as a quantity of matter measured by its resistance to change of motion or rest. The SI unit of mass is the kilogram. Force is defined as that external agency which changes or tends to change the condition of rest, or of steady linear motion, of a rigid body. The SI unit of force is the newton.

In the customary or English absolute system the "poundal" is the force needed to give one pound of mass an acceleration of one foot per second per second, or one poundal of force = 1 lb of

[1] *S*ysteme *I*nternational d'Unites (International System of Units).
[2] Vice President, Camp Dresser & McKee Inc., Boston, Massachusetts.
[3] Previously published by the American Society of Civil Engineers as Meeting Preprint 2377 of the ASCE Annual and National Environmental Engineering Convention, October 21–25, 1974.

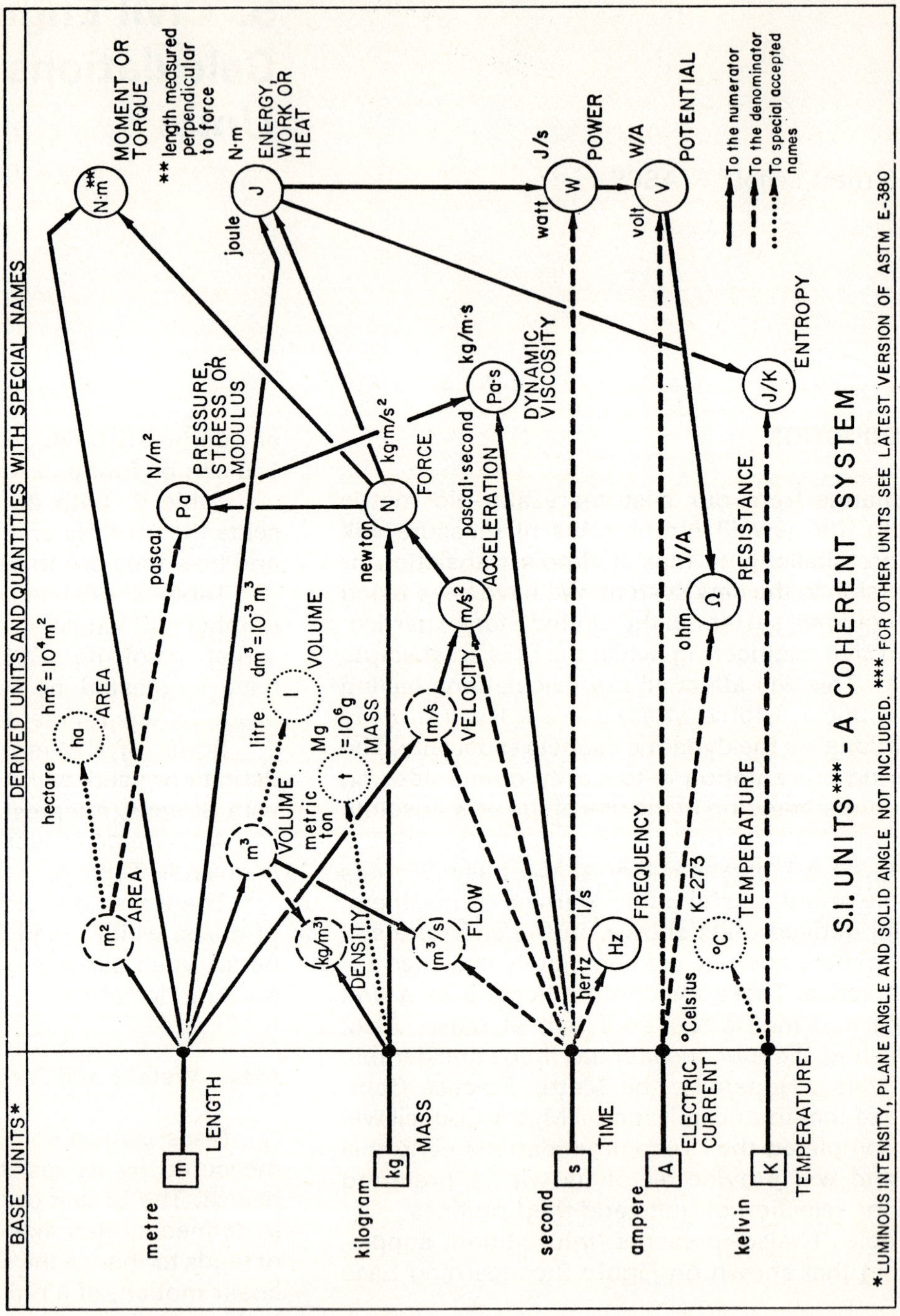

(Revised 5 August 1974)

FIGURE 5-1.

mass × 1 ft/sec². In SI units 1 newton of force = 1 kg of mass × 1 meter/sec².

On the earth the gravitational acceleration is 32.2 ft/sec² or 9.81 m/s². On the moon the gravitational acceleration is 5.25 ft/sec² or 1.6 m/s². Since weight is defined as the force exerted on a body by gravity, then a unit mass would weigh 32.2 poundals or 9.81 newtons on earth, and 5.25 poundals or 1.6 newtons on the moon.

Thus, a pound or kilogram of mass weighs less on the moon than it does on the earth. Therefore, since mass is constant, weight cannot be considered synonymous with mass in the absolute system. In the gravitational system, now customarily used by U.S. engineers, mass is normally equated to both weight and force.

To differentiate between mass and force units many engineers commonly use pound for mass and pound force for an equal number of force units. This is done by making 32.2 poundals divided by g or 32.2 equal to one pound force.

Whenever statics only is involved, no confusion has arisen in using pounds (force) in calculations. In dynamic calculations however the hidden factor of g = 32.2 has caused some problems: for instance, when the force of gravity occurs in the determination of potential energy in hydraulics.

Potential Energy = potential work possible
= force × distance
= Mgh foot poundals
= Mgh/32.2 foot pounds force
= Mh foot pounds force in the gravitational system

where M = mass in pounds
g = acceleration due to gravity in feet/second²
h = height above a datum in feet

which allows the factor g to be hidden. In SI, which is an absolute system, the equation becomes Potential Energy = Mgh in joules.

In the customary system the distance h traveled by an object having a mass of one pound after dropping in a vacuum for one second equals $v^2/2g$. The factor g is included to convert poundal units to pound force units. In SI the equation becomes $h = v^2/2$.

The practice of expressing, stating, or describing mass in terms of a particular force called weight is a procedure that will undoubtedly continue for a long time. However, in order to avoid confusion only the mass units in kilograms and force units in newtons, with the proper prefixes in SI, should be used in engineering calculations.

The old terms pounds per square inch and pounds per square foot are really pounds force per square inch, or square foot. Although in the past engineers have converted pounds force per square inch to kilograms force per square centimeter, the old metric unit kilograms force does *not* exist in SI.

For pressure the favored unit at this time is kilopascal (kPa) although its base unit equivalent kN/m² may be used in calculations for some time. For stress we find that the kilopascal or megapascal is generally recommended in the U.S. at this time, although the ISO Technical Committee #17 on Steel Products has suggested N/mm² as the preferred unit instead of megapascal (MPa). N/mm² actually is equal to MN/m² or MPa. The majority of engineers canvassed prefer to use the term MPa in discussions, converting to 10^6 N/m² for calculations.

FORCE AND PRESSURE

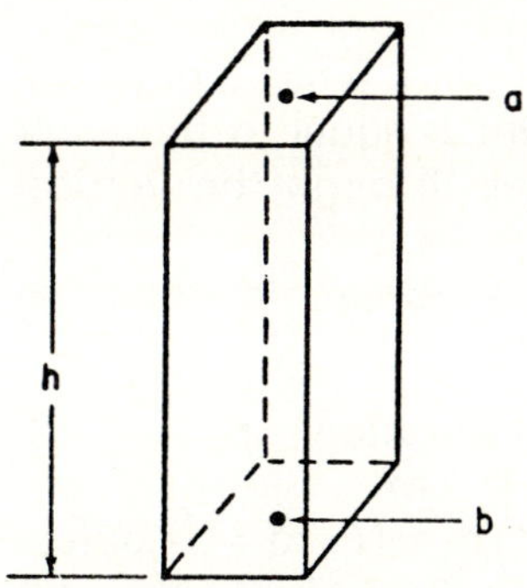

FIGURE 5-2. A Column of Water

$$\text{Force} = \text{mass} \cdot \text{acceleration due to gravity (g)}$$
$$\text{pressure}_b = \text{force/area}$$
$$P_b = \text{volume} \cdot \text{density} \cdot \text{g/area}$$
$$P_b = m^3 \cdot (kg/m^3) \cdot (m/s^2)/m^2$$
$$P_b = m \cdot (kg/m^3) \cdot (m/s^2) = h\rho g = 9.81\,\rho h \qquad (1)$$
$$\text{and also} = (kg \cdot m/s^2)/m^2 = N/m^2 = Pa$$

where g = acceleration due to gravity = 9.81 m/s²
ρ = mass density = kg/m³
h = height in m.

It here can be noted that if pound force had been used as in the gravitational system the equation would have been:

$$P_b = \text{lbf}/\text{ft}^2 \text{ or psf}$$

and also

$$P_b = wh = 62.4\ \text{lb}/\text{ft}^3 \cdot h\ \text{ft} \quad . \ . \ . \ . \ (2)$$

where w = weight density.

Note the lack of value of *g* in the gravitational equation.

PRESSURE AND ENERGY RELATIONSHIPS

The total energy equation and Bernoulli's equation are fundamental concepts basic to a civil engineer's professional education. The gravitational constant changes position when a move from customary to SI units is made. To illustrate the absolute system let us examine the hydraulics of a pipeline through which water is drawn by a pump from a reservoir for discharge through a nozzle. This arrangement can be seen on Figure 5-3 where the base datum is shown to be through the centerline of the pump.

A unit mass of water at point *a* has a potential energy with respect to point *b* equal to the work that would be done if it fell through the vertical distance z_1.

Potential energy at *a* = force · distance to datum

$$E_a = \text{mass} \cdot \text{acceleration} \cdot z_1$$
$$E_a = \text{kg} \cdot (\text{m}/\text{s}^2) \cdot \text{m}$$
$$E_a = (\text{kg} \cdot \text{m}/\text{s}^2)\,\text{m} = \text{N} \cdot \text{m} = \text{J (joules)}$$
$$\text{also } E_a = M \cdot g \cdot z_1 = 9.81\ Mz_1 \quad . \ . \ . \ . \ . \ (3)$$

where M = mass in kilograms.

The water at point *b* which is maintained at a constant depth has a pressure energy due to the force exerted by the column of water above point *b*

E_b = force · distance the water would move if released or from *b* to *a*.

$$E_b = z_1 \cdot \text{cross section area} \cdot \rho \cdot g \cdot z_1$$
$$E_b = \text{N} \cdot \text{m} = \text{J (joules)}$$
$$\text{also } E_b = 9.81\ Mz_1 \quad . \ . \ . \ . \ . \ . \ . \ . \ . \ . \ . \ . \ . \ . \ (4)$$

In the pipe between points c and d where the water is flowing to the pump there is a kinetic energy due to the velocity in the pipe. Some of the pressure energy in front of the pipe at b has been converted to kinetic energy. This can be illustrated at points x and y where the energy at x is due to both pressure and velocity while at point y the energy becomes solely velocity energy as the pressure becomes atmospheric within and outside the water as it leaves the nozzle.

Kinetic energy = force · distance
where force $F = ma$ = mass · gravity
and distance $s = V^2/2$

$$\text{therefore } E_v = M \cdot g \cdot v^2/2 \quad . \ . \ . \ . \ . \ . \ . \ (5)$$
$$= \text{kg} \cdot (\text{m}/\text{s}^2) \cdot (\text{m}^2/\text{s}^2)/2$$
$$= (\text{kg} \cdot \text{m}/\text{s}^2)(\text{m})(g/2)$$
$$= (g/2)(\text{N} \cdot \text{m}) = 4.9\ \text{J}$$

where E_v = kinetic energy, joules
h_v = equivalent velocity height.

Conservation of energy indicates that the total energy at one point in a continuous pipe system is equal to the total energy at any other point giving due consideration to any losses or gains in between.

Energy can be represented by head for each type of energy. Head can be obtained in each case by dividing by the number of mass units, since the gravity units is considered constant at 9.81.

The total head at any point is the sum of the potential head, pressure head and velocity head. For example in Figure 11 the total energy at f with respect to a datum through the pump would be $h_{po} + h_{pr} + h_v$. A plotting of the sum of the h_{po} and h_{pr} at each point in a pipeline represents the pressure or hydraulic gradient, elevations to which water would rise if it were not restricted.

A plotting of the sums of the potential, pressure and velocity heads results in the energy gradient.

In the gravitational system the total head is obtained in feet by:

$$h_t = Z + P/g\rho + v^2/2g,$$

while in the absolute system the total head is obtained by:

$$h_t = gZ + P/\rho + v^2/2 \quad . \ . \ . \ . \ . \ . \ . \ . \ (6)$$

where h_t = total head in J/kg
g = acceleration due to gravity in m/s^2
Z = height of point above datum in m
P = pressure in Pa
ρ = mass density in kg/m^3
v = velocity in m/s.

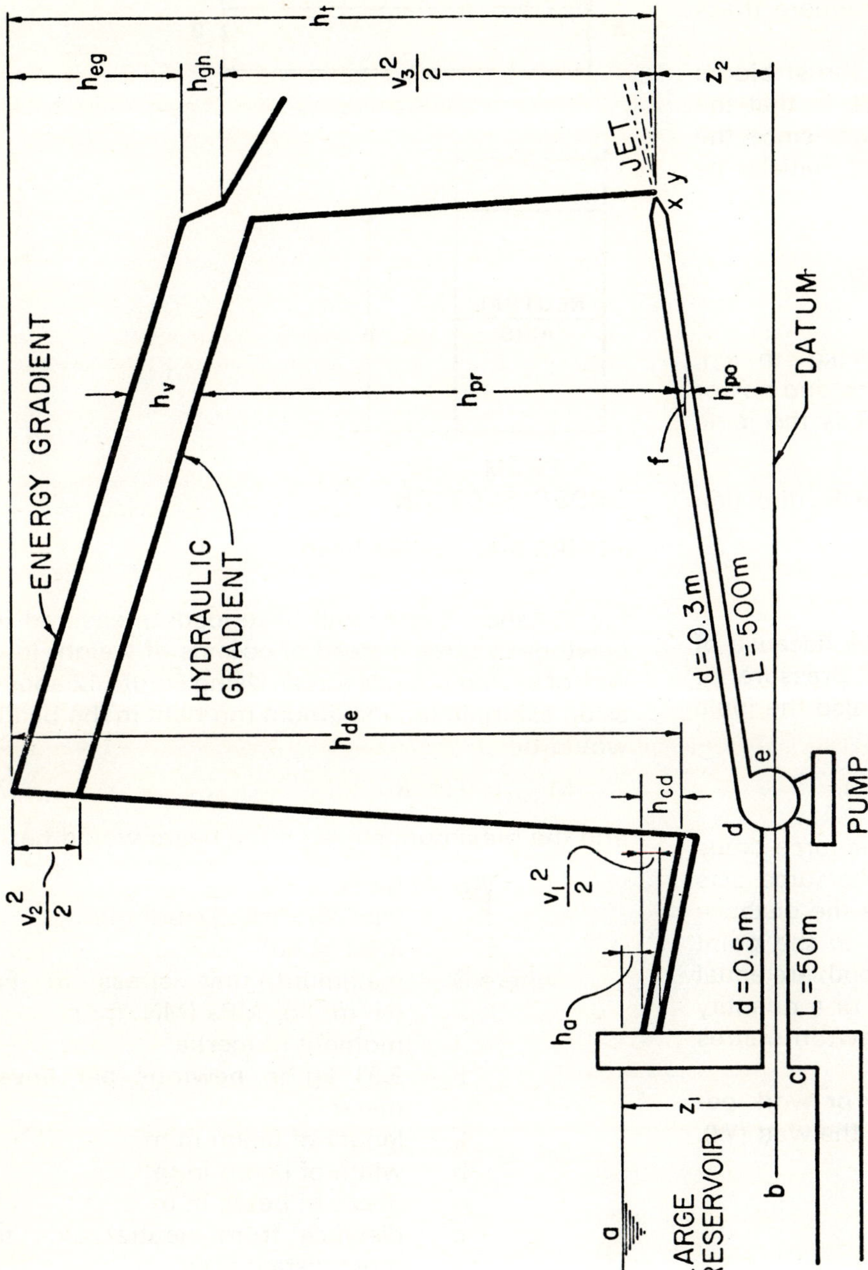

FIGURE 5-3. Hydraulic and Energy Gradients in SI

These factors must be remembered in specifying heads and pressures to be withstood by pipes and tanks or in ordering pumps. Hence, new tables must be prepared by pipe and equipment manufacturers converting not only to SI units but to the absolute system in those cases where mass and force units are not equal.

Designs for water hammer and thrust blocks on water lines will be different in SI, in that the value g will *not* enter the calculation, since the statement of the quantity of flow will initially be given in terms of mass units.

FORCE, ENERGY, WORK AND POWER RELATIONSHIPS

The unit of force, the newton (N) or (kg · m/s²), becomes the basis for all energy, work, and power determinations. The SI unit for work is the joule (J).

Work = force · distance moved in the direction of the force
= newton meter or N · m
= $(kg \cdot m/s^2)$ m.

Energy is the capacity to do work because of its height above a datum, its internal pressure, or its motion. The SI unit for energy is also the joule (J).

Energy = force · distance or N · m.

A torque is the product of a static force times the length of the radius at which the force acts and a moment is a static force times the distance from the point under consideration to the point where the force is applied. Although both are equal to N · m, they are not energy, work, or a quantity of heat, and so are referred to as newton metres (N · m) rather than joules.

Power is the amount of energy or work per unit of time. The SI unit for power is the watt (W).

Power = joules per second
= J/s.

FORCE, MOMENT AND STRESS RELATIONSHIPS

In structures, as in hydraulics, the changeover to SI will require care, for here also the change will not be only of units but of systems. The change will not be from force units in pounds force to kilograms force but from pounds force to newtons.

UNIFORM LOAD IN N/m

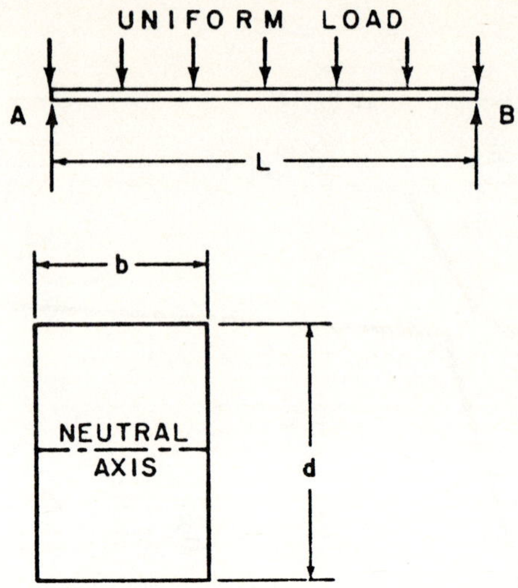

FIGURE 5-4. Loaded Beam

For instance, beams will have loads measured in newtons of force instead of pounds of weight (our way of saying pounds force). Using Figure 12 above as an example the maximum moment in the beam would be:

$$M_{max} = FL^2/8 \quad \text{. . . . .} \quad (7)$$

and the maximum stress in the beam would be:

$$S = Mc/I$$
$$= (NL^2/8) \cdot (d/2)/(bd^3/12)$$
$$= 3NL^2/4\ bd^2$$

where S = maximum unit stress in Pa (N/m^2) or MPa (MN/m^2)
I = moment of inertia
F = 9.81 kg in newtons per lineal meter
L = length of beam in m
b = width of beam in m
d = depth of beam in m
c = distance from neutral axis to most distant fiber.

There is some question as to whether the depth, width, and area components should be in mm, cm, or m. In mid-August 1974 a joint committee of AISI and AISC selected mm as the unit for all geometrical steel section properties shown in SI units. By listing A in mm², S in 10^3 mm³, I in

10^6 mm^4, and beam designations in kg/m, there appears to be less chance for error; and once we are using the 10^n power throughout, calculations are also simplified as is the determination of significant figures.

When specifying the allowable stress in steel, concrete, or wood it will not be simply a conversion from pounds force per square inch to kilograms force per square meter, but a changeover from pounds force per square inch to newtons per square meter. This is equally true of such moduli as the modulus of elasticity which changes from 29 $\times$ 10^6 psi to about 200 GPa (10^9 Pa) for steel and from 3 $\times$ 10^6 psi to 20 GPa (10^9 Pa) for concrete.

The prime rule in the changeover is to change all data over to SI base units before starting calculations, particularly where any dynamic rather than static forces come into the calculations.

CONCLUSION

No one set of rules will be entirely satisfactory for all groups of engineers. Therefore it is of the utmost importance to establish together the preferred usages employed by all engineers to obtain the coordinated effort which will be required to solve the interrelated and interdisciplinary problems of the coming "metric age."

REFERENCES

1. *Calculations in SI Units,* Ministry of Public Buildings and Works, England, 1970.
2. *Recommended Metric Units for Water Supply, Sewerage, Irrigation and Drainage,* Metric Conversion Board, Australia, 1974.
3. Fairweather, L. and Sliwa, J. A., *VNR Metric Handbook,* Van Nostrand Reinhold, 1969.
4. *Metric Practice Guide,* ASTM E380-72 American Society for Testing and Materials, 1972.
5. *SI Units and Recommendations for the Use of their Multiples and of Certain Other Units,* International Standards Organization, International Standard ISO 1000, 1973.
6. *Metric Conversion in Engineering and Manufacturing,* American National Metric Council, 1974.
7. *Metric Conversion in Building and Construction,* Metric Conversion Board, Australia, 1972.
8. *Metric Practice,* Metric Conversion Board, Australia, 1973.

6. The Architect and Metrication

The architectural profession must participate in the development and application of an industry-wide system of preferred dimensions and sizes for buildings and building components that will simplify the process and reduce the costs of construction. This should be coupled with a concerted effort to review and revise existing building codes, standards and practices, thus eliminating conflicting requirements.

In addition to this participation, the profession has these opportunities:

(a) to use the change to SI metrics to effect streamlining of design and specification, calculations, and supervision of construction;

(b) to review and rationalize current design practices and procedures and to improve written and drawn communication in the construction industry;

(c) to reduce costs by the use of simplified metric calculations as opposed to feet, inches, and fractions.

LINEAR DIMENSIONS ON DRAWINGS AND DEGREE OF ACCURACY

The following relationship between drawing types, units, and numbers of decimal places is recommended for all drawings:

(1) Locality plans and site plans (scales 1:500 to 1:2500) should be drawn in meters to three decimal places, with the place rounded to the nearest multiple of 5 or 10. Where the value is less than unity, the decimal marker is to be preceded by zero.
Example: 0.530 m

(2) All other drawings (scales between 1:1 and 1:200 inclusive) should be expressed in millimeters and no decimal places should be used, except for precision dimensions describing products which are manufactured to a decimal of a millimeter.

(3) Levels on drawings should be given in meters to three decimal places.

DRAWING SCALE RATIOS

It is important that any new series of drawing scale ratios include a range selected to satisfy the needs of architects, engineers, designers, contractors and manufacturers.

As part of the conversion to the metric system, the introduction of a new range of scales is needed. The change offers an excellent opportunity for reduction in variety, together with simplification and rationalization of graphical communication within the industry. Table 3 shows traditional and metric scales.

SPECIFICATION AND MEASUREMENT FOR BUILDING COMPONENTS

The face area of building components and cross-sectional dimensions should be shown in millimeters:

Gypsum Board, Plywood, etc.	1200 × 2400
Floor Tile	300 × 300
Studding	38 × 89

TABLE 3: Commonly Used Architectural Scales

U.S. Traditional Scales	*Closest ISO Scales*	*Corresponds to*
Site Plans		
1/16″ = 1′0″	1:200	5 mm = 1 m
1/8″ = 1′0″	1:100	10 mm = 1 m
1″ = 20′	1:200	5 mm = 1 m
1″ = 50′	1:500	2 mm = 1 m
—	1:100	1 mm = 1 m
Building Designs		
1/8″ = 1′	1:100	10 mm = 1 m
1/4″ = 1′	1:50	20 mm = 1 m
Details		
1/2″ = 1′0″	1:20	50 mm = 1 m
3/4″ = 1′0″	1:10	100 mm = 1 m
1″ = 1′0″		
1-1/2″ = 1′0″	1:10	100 mm = 1 m
3″ = 1′0″	1:5	200 mm = 1 m
1′ = 1′ (full size)	1:1 (full size)	

The abbreviation for millimeters (mm) behind the numerals may be deleted, provided that the notation used is consistent with that recommended for linear dimensions.

For both horizontal and vertical components, the width should be shown first, where no confusion about positioning can arise. (This differs from the policy adopted in Europe.)

Door	900 × 2100
Window	1400 × 300

Where misinterpretation is possible, such as in windows, the second dimension should be followed by an indication of whether height or width is intended.

Window	1400 × 300 high
Access Panel	600 × 900 high

Where length of products or members needs to be included for estimating or ordering purposes, it should be shown last with the unit of measurement clearly indicated. Lengths should be shown in millimeters except where an industry has decided to use the meter as a unit of length (as in lumber).

Studs	38 × 89 × 2.4 m long.
Precast Unit	190 × 190 × 4000 mm long.

DIMENSION LINES

Dimensioning as prescribed in Australia's Metric Handbook is shown in Figures 6-1 and 6-2.

STOCK PLANS

Many architects and material manufacturers will have conversion problems similar to those of Home Planners, Inc., a stock plan service.

If your firm operates a plan service or provides design service for roof trusses, you will want to consider getting ahead on the metric conversion by showing dual dimensions on the plans you are drawing now, assuming they have a useful life of 10 years or so. Consider, then, whether you should provide dual dimensions with the traditional measure as the base or with a rationalized metric dimension as the base. In Figures 6-3 and 6-4 we see that a width of a home which is 64′ 10″ converts exactly to 19 761.2 mm. This is not an easy figure to work with. In Figure 6-3 an exact conversion is shown, together with the original traditional dimensions. It is obvious that the numbers derived by an exact conversion are difficult to handle, or to remember.

In Figure 6-4, the decimals have been eliminated and the millimeters have been rounded wherever possible. The idea behind the figures on the dimension line running across the rear of the home through dining room and kitchen is of interest. First, consider the outside wall thickness. The exact conversion is 215.9 mm from 8-1/2″, obviously awkward to use. The new 210 mm figure for this masonry veneer wall is developed as follows, (without interior finish, as in the 8-1/2″ figure):

Studding, at this writing a rounded "soft" conversion of	89 mm
Sheathing—1/2″ equivalent	12 mm
Metric Brick—(see Chapter VI)	90 mm
"Finger" space for the masons	19 mm
	210 mm

Also, the interior stud wall thickness, without drywall or other finish material, is 89 mm.

In the bathroom, the traditional dimension of 5′ 0-1/2″ allows for a 60″ tub and 1/2″ to facilitate placement. It is unlikely that 5′ 0″ steel bath tubs will be changed in the metric conversion because

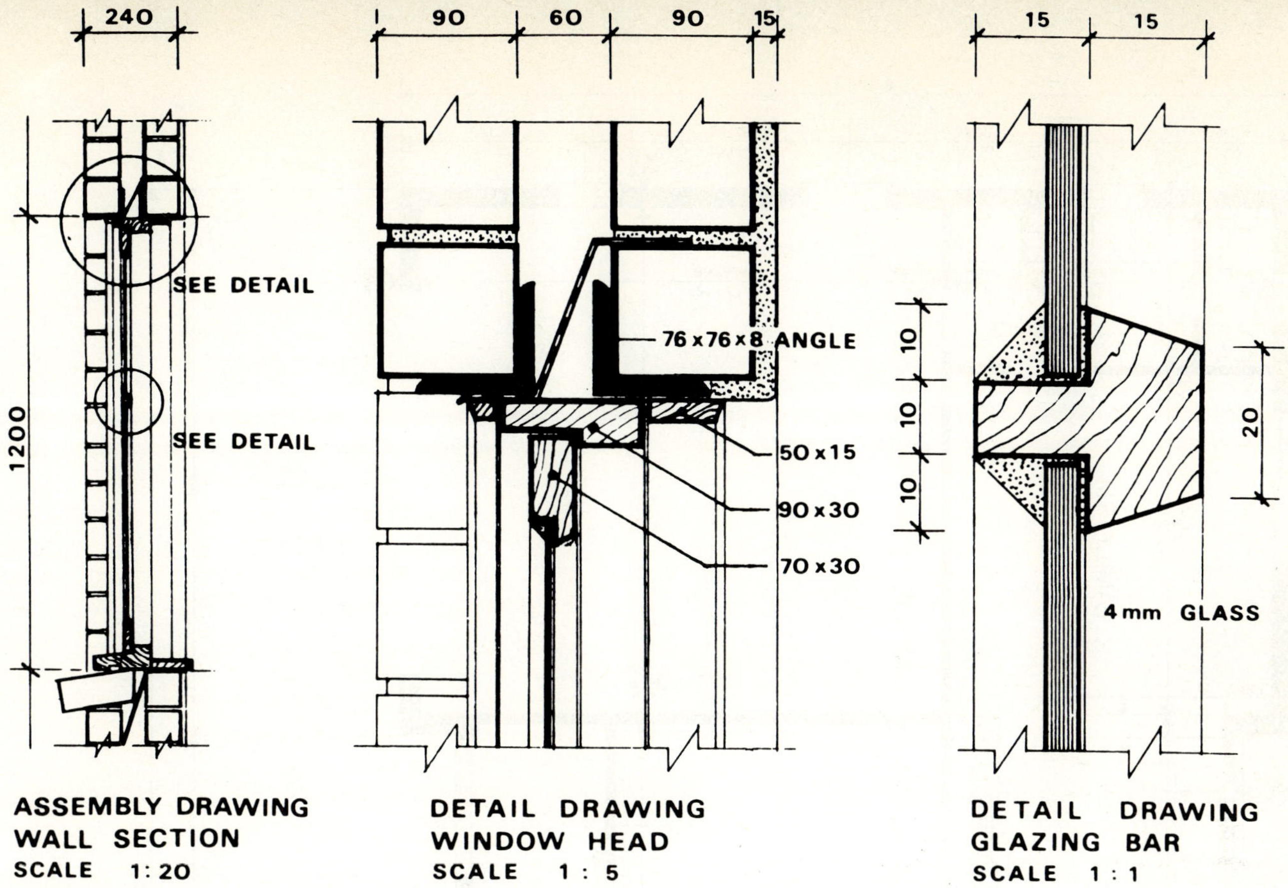

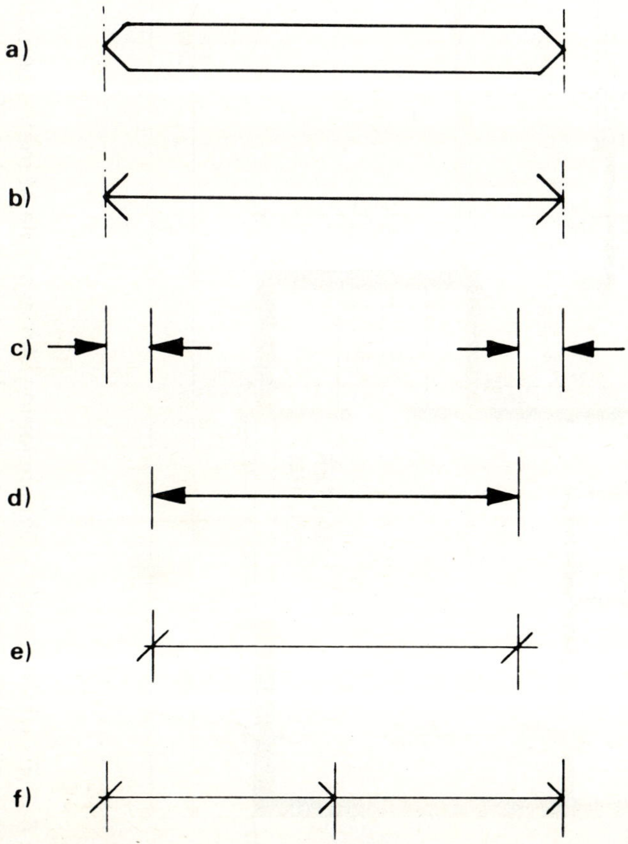

FIGURE 6-1. Working and Detailed Drawings. **FIGURE 6-2.** Terminations of Dimension Lines. Reprinted by permission of the Standards Association of Australia from *Metric Handbook* (MH1), 1972, p. 44.

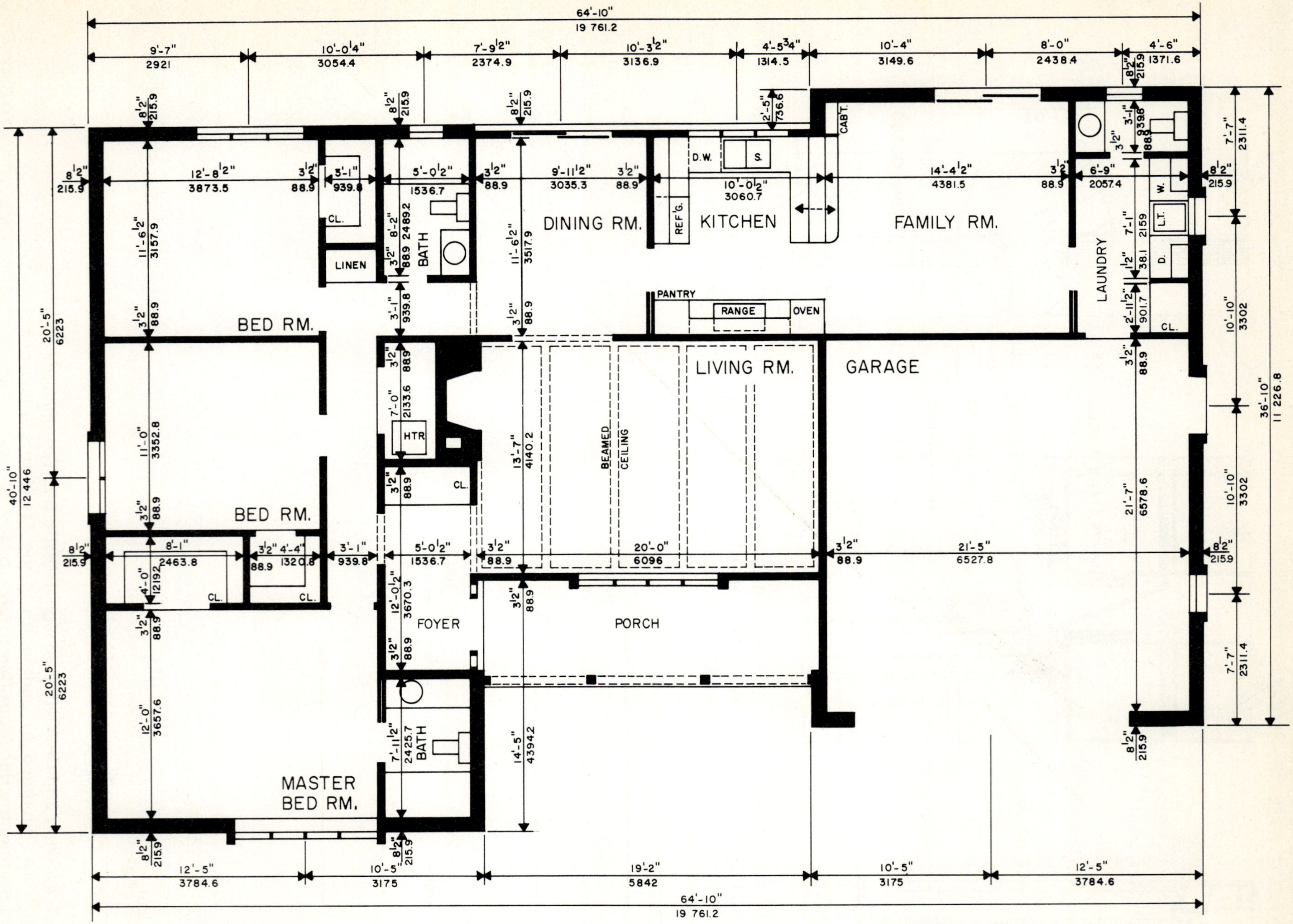

1769 SQUARE FEET
1/4" = 1'-0"

FIGURE 6-3. Exact Conversion Dimensioning. An example of metric conversion is this plan by a stock plan service, Home Planners, Inc. This floor

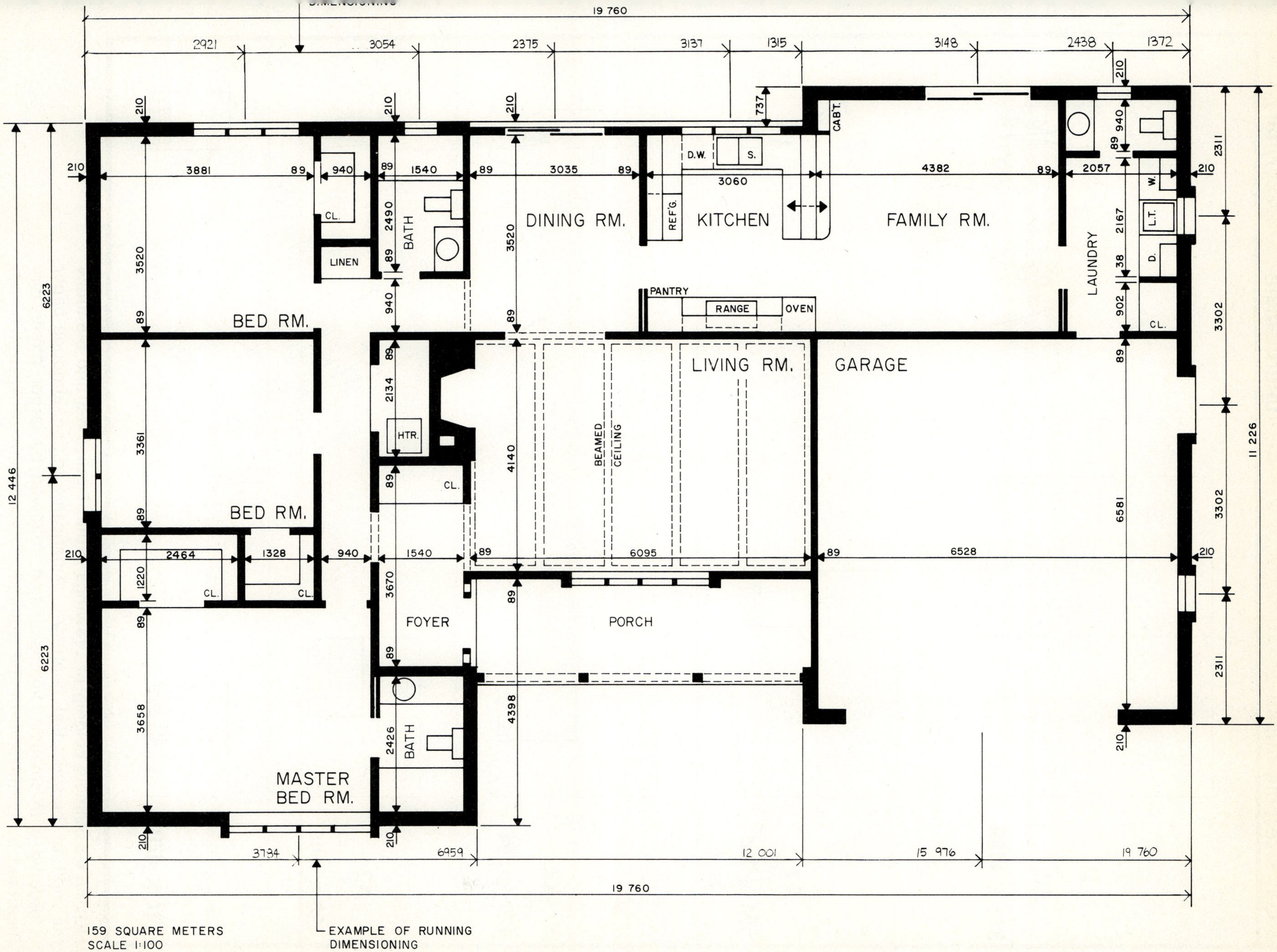

FIGURE 6-4. Rational and Rounded Dimensioning. Here is the same floor plan with rational metric conversion, eliminating decimals, and rounding wherever possible. See discussion in text.

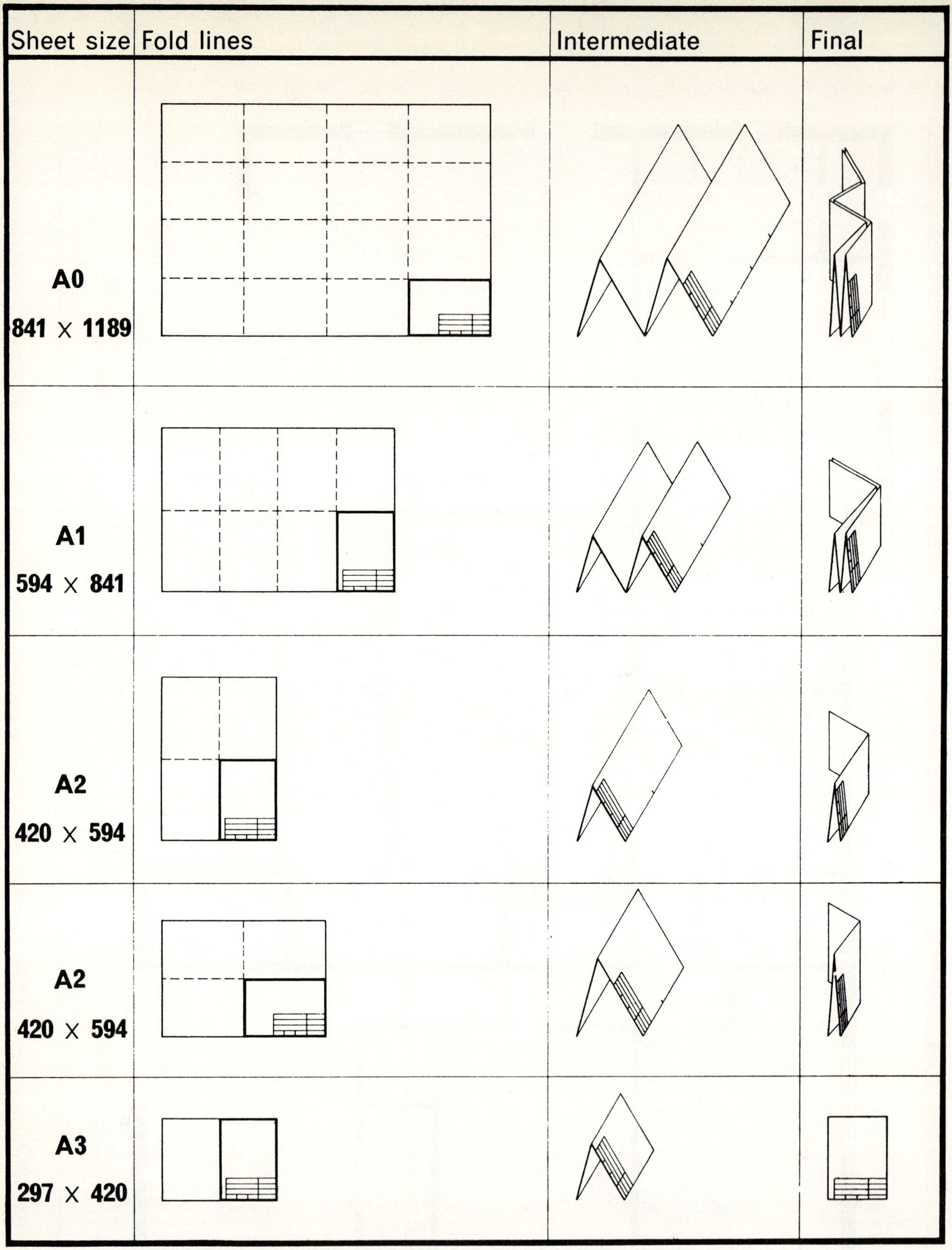

FIGURE 6-5. Simple Folding of Drawings. Note: Figures 6-5 and 6-6 show the width of the sheet last, contrary to American tradition, although paper is ordered with smallest dimension first, followed by larger dimension. Reprinted by permission of the Standards Association of Australia from *Metric Handbook* (MH2), 1975, pp. 18 and 19.

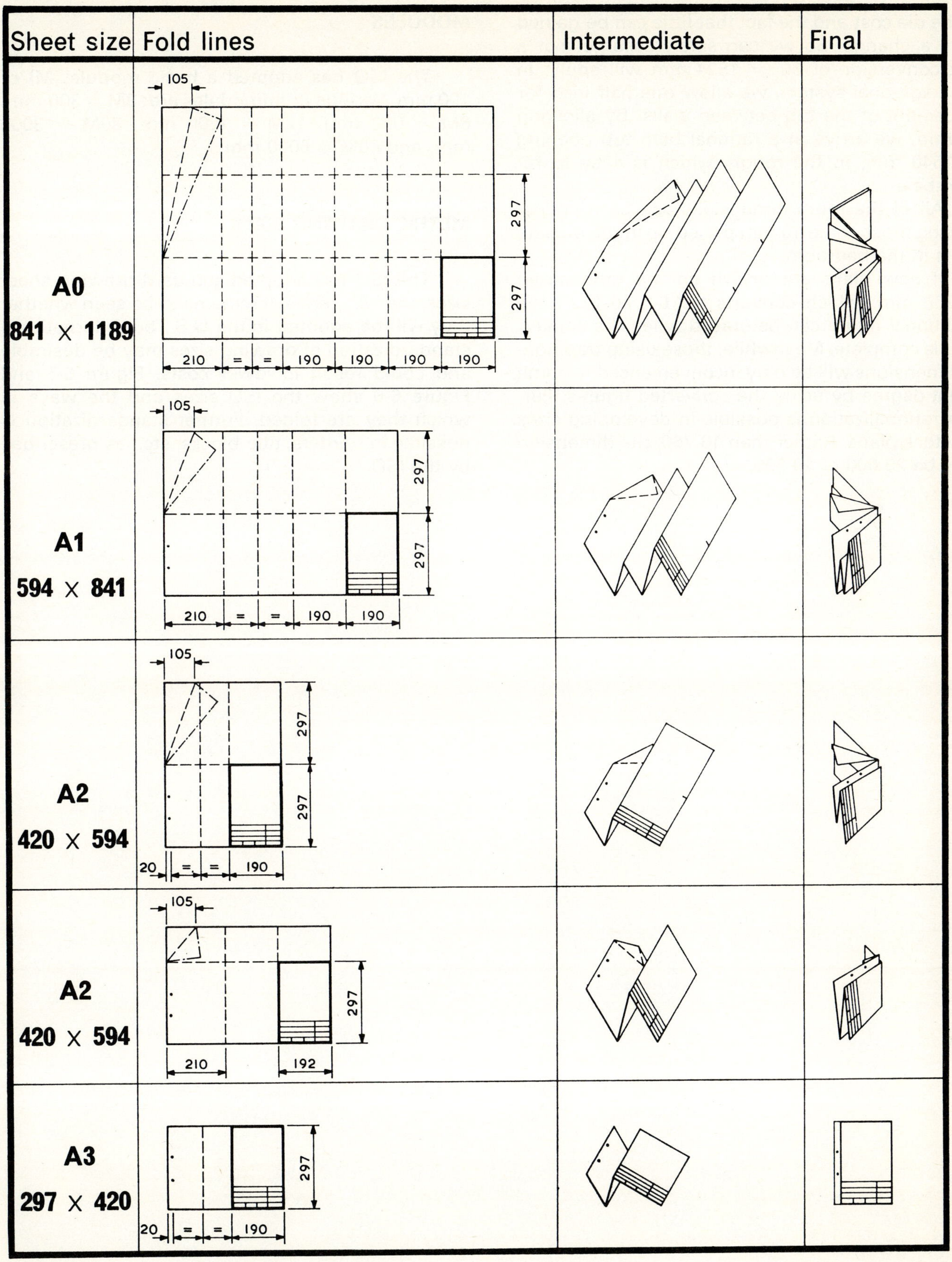

FIGURE 6-6. Folding of Drawings for Filing.

of the die cost and the fact that little can be gained from a change. So we can safely assume that a soft conversion of 60″ = 1524 mm will apply. In the traditional system, we allow one half inch for placement of the tub between walls. By allowing 16 mm, we arrive at a rational bath tub opening of 1540 mm, in the rough, which is easy to remember.

All of the room dimensions can be rounded, placing most of the 10 mm picked up in the outside walls in the bedroom.

If new plans are drawn on the rationalized 19 760 mm (which converts to: 64′ 10-1/2″) the traditional figure can be erased when the conversion is complete. Meanwhile, those using traditional dimensions will be only inconvenienced to some slight degree by using the converted figures. Further rationalization is possible in developing sizes for stock plans. Rather than 19 760, the dimension may be 20 000 or 19 500.

MODULES

The ISO has adopted a basic module (M) of 100 mm. Various multimodules are: 3M = 300 mm, 6M = 600 mm, 12M = 1200 mm, 30M = 3000 mm, and 60M = 6000 mm.

METRIC DRAWING SIZES

The ISO has adopted standard drawing sheet sizes, the "A" series. It remains to be seen whether they will be adopted in the U.S., but international standardization of drawing sizes may be desirable and could result in lower costs. Figure 6-5 and Figure 6-6 show the ISO sizes and the ways in which they are folded. Further standardization is possible in borders, title blocks, etc., as prescribed by the ISO.

7. Masonry and Concrete

METRIC BRICK

Our present modular standard brick is 4″ × 2-2/3″ × 8″ nominally; after deducting a 3/8″ mortar joint the actual size is 3-5/8″ × 2-1/4″ × 7-5/8″. The Brick Institute of America has decided that a conversion to a metric size of 100 mm × 67 mm × 200 mm nominally is within the present manufacturing tolerances specified by ASTM. With a joint size of 10 mm, the actual size will be 90 mm × 57 mm × 190 mm. This compares with the present actual size in millimeters of 92 × 57 × 193.7.

The change can therefore be considered a rounded soft conversion. Manufacturers will need little or no modification of existing manufacturing equipment. Since three courses high equals about 200 mm and the nominal length is 200 mm, the sizing works well with the architect's 100 mm module (see Figure 7-1). Other modular brick sizes can be converted just as easily, as indicated in the accompanying tables and drawings.

The British and Australians have adopted some different metric brick sizes, but since brick is seldom exported or imported, there is no need to coordinate the U.S. and British dimensions.

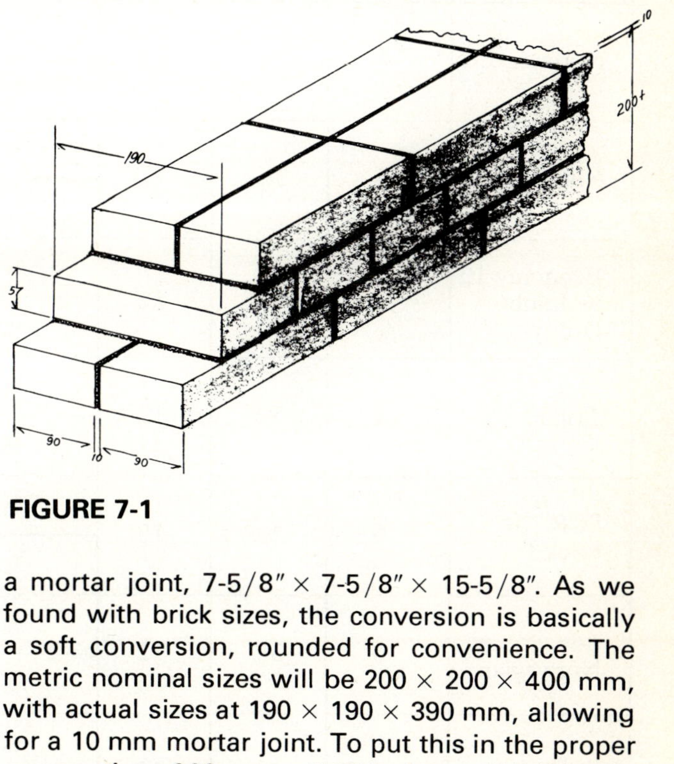

FIGURE 7-1

CONCRETE MASONRY UNITS

Our present 8″ × 8″ × 16″ nominal concrete block size is actually, after allowance of 3/8″ for a mortar joint, 7-5/8″ × 7-5/8″ × 15-5/8″. As we found with brick sizes, the conversion is basically a soft conversion, rounded for convenience. The metric nominal sizes will be 200 × 200 × 400 mm, with actual sizes at 190 × 190 × 390 mm, allowing for a 10 mm mortar joint. To put this in the proper perspective, 200 mm = 7.87 inches = 7-7/8″. We can see that this is not really a substantial change. The sizes are compatible with the brick sizes covered earlier. But the decrease in the height of the block from 193.75 mm (7-5/8″) to 190 mm will cause problems for steel door manufacturers as described in a later chapter.

TABLE 4: Traditional Sizes of Modular Brick

Unit Designation	*Nominal Dimensions, in.*			*Joint Thickness in.*	*Manufactured Dimensions in.*			*Modular Coursing in.*
	t	*h*	*l*		*t*	*h*	*l*	
Standard Modular	4	2⅔	8	⅜	3⅝	2¼	7⅝	3C = 8
				½	3½	2¼	7½	
Engineer	4	3⅕	8	⅜	3⅝	2¹³⁄₁₆	7⅝	5C = 16
				½	3½	2¹¹⁄₁₆	7½	
Economy 8 or Jumbo Closure	4	4	8	⅜	3⅝	3⅝	7⅝	1C = 4
				½	3½	3½	7½	
Double	4	5⅓	8	⅜	3⅝	4¹⁵⁄₁₆	7⅝	3C = 16
				½	3½	4¹³⁄₁₆	7½	
Roman	4	2	12	⅜	3⅝	1⅝	11⅝	2C = 4
				½	3½	1½	11½	
Norman	4	2⅔	12	⅜	3⅝	2¼	11⅝	3C = 8
				½	3½	2¼	11½	
Norwegian	4	3⅕	12	⅜	3⅝	2¹³⁄₁₆	11⅝	5C = 16
				½	3½	2¹¹⁄₁₆	11½	
Economy 12 or Jumbo Utility	4	4	12	⅜	3⅝	3⅝	11⅝	1C = 4
				½	3½	3½	11½	
Triple	4	5⅓	12	⅜	3⅝	4¹⁵⁄₁₆	11⅝	3C = 16
				½	3½	4¹³⁄₁₆	11½	
SCR brick [2]	6	2⅔	12	⅜	5⅝	2¼	11⅝	3C = 8
				½	5½	2¼	11½	
6-in. Norwegian	6	3⅕	12	⅜	5⅝	2¹³⁄₁₆	11⅝	5C = 16
				½	5½	2¹¹⁄₁₆	11½	
6-in. Jumbo	6	4	12	⅜	5⅝	3⅝	11⅝	1C = 4
				½	5½	3½	11½	
8-in. Jumbo	8	4	12	⅜	7⅝	3⅝	11⅝	1C = 4
				½	7½	3½	11½	

[1] Available as solid units conforming to ASTM C 216- or ASTM C 62-, or, in a number of cases, as hollow brick conforming to ASTM C 652-.

[2] Reg. U.S. Pat. Off., SCPI.

TABLE 5: Metric Sizes of Modular Brick

Unit Designation	*Nominal Dimensions* mm *t*	*h*	*l*	*Joint Thickness* mm	*Manufactured Dimensions* mm *t*	*h*	*l*	*Modular Coursing* mm
Standard Modular	100	67	200	10	90	57	190	3C=200
				12	88	65	188	
Engineer	100	80	200	10	90	70	190	5C=400
				12	88	68	188	
200 Economy or Jumbo Closure	100	100	200	10	90	90	190	1C=100
				12	88	88	188	
Double	100	135	200	10	90	128	190	3C=400
				12	88	123	188	
Roman	100	50	300	10	90	40	290	2C=100
				12	88	38	288	
Norman	100	67	300	10	90	57	290	3C=200
				12	88	55	288	
Norwegian	100	80	300	10	90	70	290	5C=400
				12	88	68	288	
Economy 12 or Jumbo Utility	100	100	300	10	90	90	290	1C=100
				12	88	88	288	
Triple	100	135	300	10	90	125	290	3C=400
				12	88	123	288	
SCR brick TM	150	67	300	10	140	57	290	3C=200
				12	138	55	288	
6-in. Norwegian	150	80	300	10	140	70	290	5C=400
				12	138	68	288	
6-in. Jumbo	150	100	300	10	140	90	290	1C=100
				12	138	88	288	
8-in. Jumbo	200	100	300	10	190	90	290	1C=100
				12	188	88	288	

Here are the suggested metric sizes for all concrete masonry units, as developed by the Canadian Standards Association:

TABLE 6

Length (mm)		Width (mm)		Height (mm)	
Nominal	Actual	Nominal	Actual	Nominal	Actual
200	190	100	90	100	90
400	390	150	140	200	190
		200	190		
		250	240		
		300	290		

The Canadian Standards Association plans to use megapascals (MPa) to designate compressive strength, replacing the present pounds per square inch (psi). Density will be expressed as kilograms per cubic meter (kg/m^3), replacing the present pounds per cubic foot. Thus:

1000 psi	becomes	7.5 MPa
1800 psi	becomes	12.5 MPa
125 lb. per cu. ft.	becomes	2000 kg/m^3
105 lb. per cu. ft.	becomes	1700 kg/m^3

FACING TILE

The Facing Tile Institute has not, at this writing, developed a policy for metric conversion. However, it is reasonable to assume that nominal metric sizes will be compatible with brick and block sizes. Facing tile has usually been laid with a 1/4″ joint; the actual size will be larger.

The present nominal 8″ × 12″ is actually manufactured to 7-3/4″ × 12-3/4″ and laid with a 1/4″ joint. In metric terms, we can assume that the nominal dimension would be 200 × 300 mm. Actual size could be 194 × 294 mm with a 6 mm joint.

CONCRETE

Most conversions in concrete practice will be soft conversions. Concrete will be measured and sold in terms of the cubic meter (m^3) instead of the cubic yard. As we learned from the recognition points in Chapter 2, 10 cubic meters (10 m^3) = about 13 cubic yards. One cubic yard equals 0.7646 cubic meters.

Rationalization will produce conversions like our present 94 lb. bag of cement which converts to 42.3 kilograms (42.3 kg). But it is reasonable to assume that rounding would produce a figure of 42 kg. It might be easier to work with at 40 or 50 kg to the bag. Australia has selected the 40 kg bag, which is 25 bags per metric ton. Canada has also decided on a 40 kg bag of cement, plus 30 kg for a bag of mortar.

In another example, our codes often relate the thickness of a slab on the ground to a 2 × 4 used for forms. The requirement reads: "Thickness 4″ with minimum of any one point 3-1/2″." In metric language, it may read: "Thickness 100 mm with minimum thickness at any one point 90 mm."

In Chapter 2 we discussed the different approaches to expressing force and stress. Pure SI would have us use newtons (N) and pascals (Pa) which are not gravity related. Another group urges the use of the kilogram force per square meter, or centimeter, or millimeter. In the coming months this question will be resolved and the 28-day compressive strength of concrete which is now expressed in pounds per square inch (psi) will be expressed as megapascals (MPa) or kg/cm^2. For example, present 3000 psi concrete would be 20 MPa or 211 kg/cm^2. Tables 7 and 8 show compressive strength conversions, and slumps for various kinds of construction.

Despite the soft conversion in concrete, opportunities exist in the use of concrete for greater efficiency. The 12″ concrete wall, for instance, was not developed as a wall thickness for its efficiency in a specific group of uses, but as a matter of convenience. It may be that the family of uses for the present 12″ wall can be better served by a wall at 250 mm rather than 300 mm, which a soft conversion would require.

CONCRETE REINFORCING STEEL

The Concrete Reinforcing Steel Institute is studying several alternative size schedules for reinforcing bars. Bar sizes would be designated by area in square millimeters, which will facilitate design, or, alternatively, by weight in terms of kilograms per meter. There would thus be no need for designers or estimators to check tables for the area of a section or for the weight per unit of length. There would also be a reduction in sizes from the present eleven to nine.

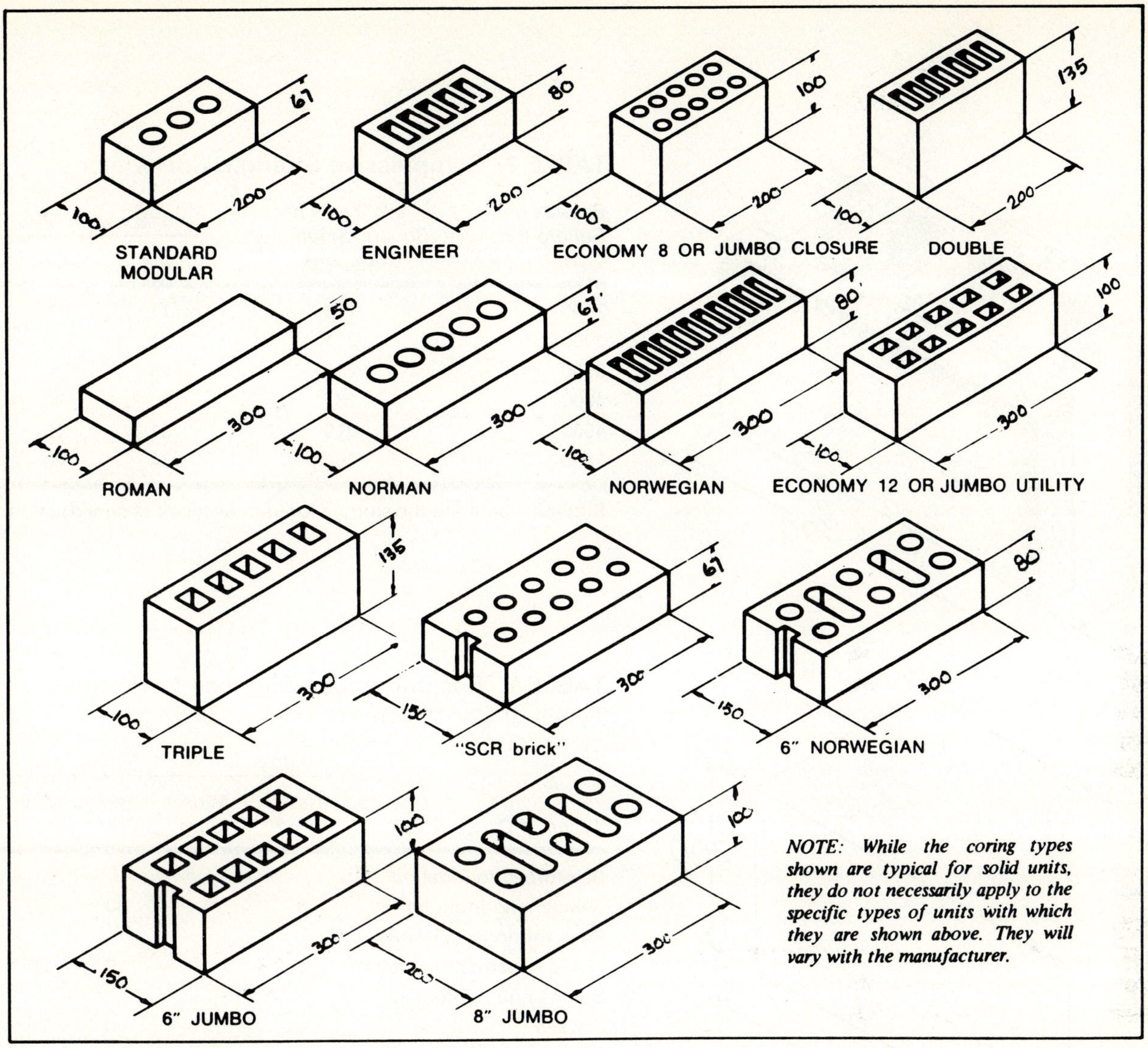

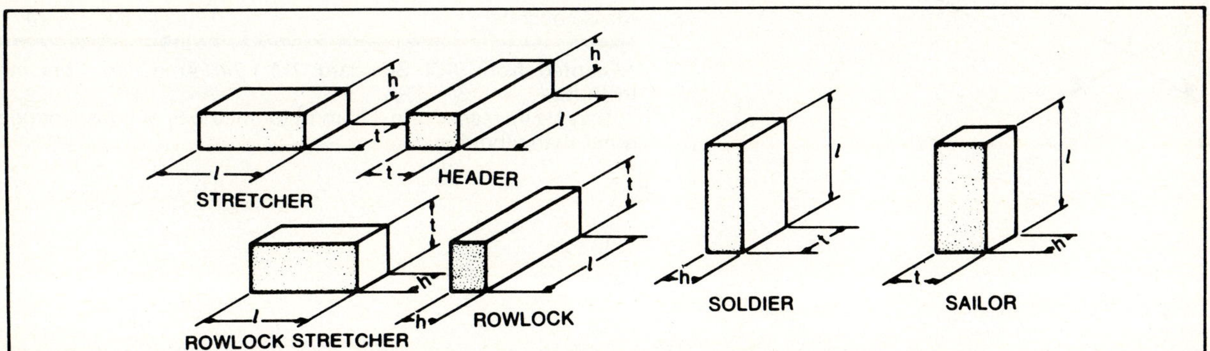

FIGURE 7-2. Modular Brick. **FIGURE 7-3.** Brick Positions in Wall. Both figures reprinted by permission of the Brick Institute of America from *Technical Notes on Brick and Tile Construction,* No. 10B, October 1971.

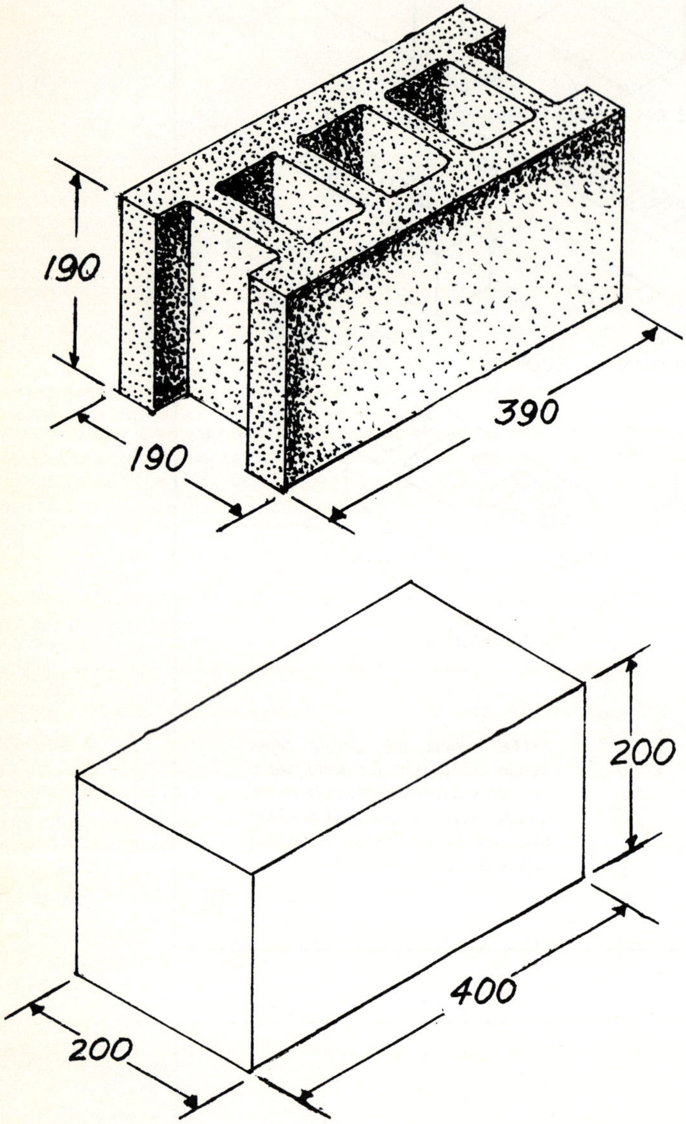

FIGURE 7-4

TABLE 7. Compressive Strength Conversions

Pounds per Square Inch psi	*kilograms per Square centimeter kg/cm²*	*megapascals MPa*	
		Soft	*Hard*
2500	176	17	15
3000	211	21	20
3500	246	24	25
4000	281	28	30
4500	316	31	35
5000	352	34	40

Similarly, here are the slumps for various types of construction:

TABLE 8. Recommended Slumps* for Various Types of Construction

	Slump, in.		*Slump, mm*	
Types of Construction	*Maximum***	*Minimum*	*Maximum***	*Minimum*
Reinforced foundation walls and footings	3	1	80	20
Plain footings, caissons, and substructure walls	3	1	80	20
Beams and reinforced walls	4	1	100	20
Building columns	4	1	100	20
Pavements and slabs	3	1	80	20
Mass concrete	2	1	80	20

*Adapted from ACI Standard 211.1-74, American Concrete Institute
**May be increased 1 in. (20 mm) for methods of consolidation other than vibration.

BASED ON CROSS-SECTIONAL AREA

TABLE A

Metric Sizes						*Current USA Sizes*	
Metric Ident. No.	*Area mm^2*	*Nom. Dia. mm*	*Weight kg/m*	*Area $in.^2$*		*Bar No.*	*Area $in.^2$*
C1	100	11	.784	.16	<	#3	.11
						#4	.20
C2	200	16	1.568	.31		#5	.31
C3	300	20	2.352	.47		#6	.44
C4	400	23	3.136	.62		#7	.60
C6	600	28	4.704	.93	<	#8	.79
						#9	1.00
C8	800	32	6.272	1.24		#10	1.27
C10	1000	36	7.840	1.55		#11	1.56
C15	1500	44	11.760	2.33		#14	2.25
C25	2500	56	19.600	3.88		#18	4.00

Table A and Table B differ only in C5, C6, C7 and C8 sizes.

TABLE B

Metric Ident. No.	*Area mm^2*	*Nom. Dia. mm*	*Weight kg/m*	*Area $in.^2$*		*Bar No.*	*Area $in.^2$*
C1	100	11	.784	.16	<	#3	.11
						#4	.20
C2	200	16	1.568	.31		#5	.31
C3	300	20	2.352	.47		#6	.44
C4	400	23	3.136	.62		#7	.60
C5	500	25	3.920	.78		#8	.79
C7	700	30	5.488	1.09	<	#9	1.00
						#10	1.27
C10	1000	36	7.840	1.55		#11	1.56
C15	1500	44	11.760	2.33		#14	2.25
C25	2500	56	19.600	3.88		#18	4.00

BASED ON MASS (WEIGHT)

TABLE C

Metric Size				*English Area $in.^2$*		*Current USA Sizes*	
Metric Ident. No.	*Weight kg/m*	*Nom. Dia. mm*	*Area mm^2*			*Bar No.*	*Area $in.^2$*
K1	1.00	13	127.5	.20		#3	.11
						#4	.20
						#5	.31
K2	2.00	18	255.0	.40		#6	.44
K3	3.00	22	382.5	.59		#7	.60
K4	4.00	25	510.0	.79		#8	.79
K5	5.00	28	637.6	.99		#9	1.00
K6	6.00	31	765.1	1.19		#10	1.27
K8	8.00	36	1020.1	1.58		#11	1.56
K11	11.00	42	1402.6	2.17		#14	2.25
K20	20.00	57	2550.2	3.95		#18	4.00

FIGURE 7-5. Proposed Metric Reinforcing Bar Sizes. Reprinted by permission of the Concrete Industry Board from the *Concrete Industry Bulletin,* October 1974 (article by Paul Rice).

The tables in Figure 7-5 show three alternatives. Table A and B would both combine the present #3 and #4 bars in one C1 size. Table A also combines #8 and #9 bars into C6. Table B combines #9 and #10 bars into C7 size. Table C is based on even multiples of weight, attractive for estimating purposes.

8. Wood Construction

Many of us can recall the protracted and sometimes bitter ten-year struggle through the 60's that finally resulted in the present American Lumber Standards being adopted in 1970. This may influence some in the wood manufacturing, distributing, and using industries to make a "soft" conversion of present lumber sizes. In other words, 1-1/2 inch lumber would be 38 mm (exact: 38.1 mm) rather than some other dimension. Yet since considerable opportunities exist for making a "hard" conversion, it is well that discussions leading to a consensus in this matter have already started under the auspices of the American National Metric Council, before metric conversion has begun being implemented.

As with any "hard" metric conversion, the most important of these opportunities is in the area of making sizes structurally more efficient in their normal uses. This would conserve our national lumber resources and make lumber a more competitive product in the building industry. For example, one of the proposals would change our present 2 × 4 (38 × 89 mm) to 35 × 90 mm. This would reduce the amount of wood fiber used in this size by 7%. However, some industry segments are opposed to this change, as we shall see later.

Second in importance, sizes would be specified in actual rather than nominal sizes. This would not only simplify the work of designers and users but would improve the image of the industry to its customer. Do-it-yourself consumers and even buyers of large construction projects are often suspicious and annoyed when they find that the 2 × 4 is actually only 1-1/2″ × 3-1/2″.

Finally, we have the opportunity to put to rest a most misunderstood term—the board foot. The metric designation will be in cubic meters and this will be finished product. One board foot is theoretically .0024 cubic meters, and one cubic meter of wood is equal to 423.8 board feet. But these relationships are true only if a board foot is a piece of wood with an actual content of one twelfth of a cubic foot, which is not the case.

Lumber will instead be inventoried and priced at so much a cubic meter of finished product. However, much retail business will be transacted with lumber priced at so much a lineal meter. For example, 3.0 meters of 38 × 89 mm @ 80¢ per meter = $2.40.

In mid-1975, the Softwood Lumber Subsector Committee Task Group of the American National Metric Council made the following recommendations:

(1) All dry dimension thicknesses and widths through 12 inch nominal should be converted to the nearest millimeter with related green sizes. Thus, a dry 2 × 4 (38.1 mm × 88.9 mm) would become 38 × 89 mm.

(2) Dimension widths for 14 inch and 16 inch nominal dry sizes should be two millimeters less than at present.

(3) Thicknesses and widths of timbers through 16 inch nominal should be the same as green dimension sizes, with 18-, 20-, 22-, and 24-inch timber sizes converted to the nearest multiple of five millimeters from present sizes.

(4) Boards and finish should be converted to the nearest millimeter from present sizes.

(5) Select and shop sizes should be established

by each rules-writing agency as presently done.

In terms of surfaced dry dimension lumber, recommendation (1) above would mean:

Nominal Inches	Present Millimeters	Metric Conversion
2	38.10	38
4	88.90	89
6	139.70	140
8	184.15	184
10	234.95	235
12	285.75	286

The recommendations of the Western Wood Products Association (WWPA) and the West Coast Lumber Inspection Bureau vary from the above by only one millimeter on several of the sizes.

For nomenclature, the "dry" size, as given above, would be used with a slightly larger "green" size to compensate for shrinkage in drying. We may figure "that is that" as far as sizing, but it may not be so. It may be that the determination of the above sizes by the Task Group has been dominated by the proposing manufacturers, rather than by users and consumers. Some other considerations:

- Homebuilders would like to have a 2 × 8 equivalent that can be used 16″ o.c. as floor joists for a 28′ home. This equivalent was lost in the 1970 changes, forcing us to go to the 2 × 10 in most species and grades. The result was increased costs in the most popular home plan dimension.
- The Truss Plate Institute is in favor of adding a 2″ × 5″ equivalent (38 × 114 mm) to the list of standard sizes. This would increase the economy of the light wood roof trusses because current design criteria often require slightly more than a 2 × 4. Now the designer must then increase the size of the member to 2 × 6. Are there other uses in framing, concrete form work, etc., for a 2 × 5 equivalent?
- Earlier, there was a proposal for lumber of 35 mm (about 1-3/8″) thickness. Objections were made by the Truss Plate Institute on the grounds that resistance to buckling would be decreased; edge area for nailing and metal connector plates in floor trusses would be reduced; the moment of inertia and section modulus would be reduced for lumber used in trusses; and other similar reasons.

Most lumber users and manufacturers can agree that the present lumber sizes are not considered perfect by any segment of the industry and metric conversion provides an opportunity to make changes which will accomplish the following objectives:

(1) more efficient end use performance capability;
(2) reduced sawmill waste;
(3) conservation of the natural resource;
(4) improved coordination with other building product lines; and
(5) better competitiveness with substitute materials.

LUMBER LENGTHS

The Western Wood Products Association recommends that lengths should be in 0.3 meter increments beginning at 1.8 meters. This would give lengths as follows with their approximate present equivalent sizes:

1.8 m	6 feet
2.1 m	7 feet
2.4 m	8 feet
2.7 m	
3.0 m	10 feet
3.3 m	
3.6 m	12 feet
3.9 m	
4.2 m	14 feet
4.5 m	
4.8 m	16 feet
5.1 m	
5.4 m	18 feet
5.7 m	
6.0 m	20 feet

Market demands will probably indicate that some of these sizes can be eliminated from both manufacturing and inventory. But before this is done, a thorough study should be made of potential uses of each length. (See the Appendix for strength properties of some commercially important woods in North America.)

Board thickness might approximate our present 25/32″ (19.84 mm), which when rounded would be 20 mm. The British have adopted a 19 mm board thickness which may be more efficient at the 400 and 600 mm spans. Another consideration may be the resawn 38 mm dimension, which would give us, after 5 mm for a kerf, 16 mm boards. In any event, there are many do-it-yourself applications for 16 mm board. The British have a 16 mm size.

In the following discussion of sheet sizes and their effect on wall heights, we develop a precut stud length of 2324 mm. The lumber distribution industry may well consider the elimination of the 2.4 m length in favor of the standard precut stud length.

PLYWOOD AND OTHER SHEET MATERIALS

Unlike lumber sizes, a consensus seems to have developed for a standard sheet size of 1200 × 2400 mm. This is about 47-1/4″ × 94-1/2″. Plywood, gypsum board, hardboard, particle board, asbestos board, insulation board, etc., would all be manufactured to this size. The 1200 mm width would be the *effective* coverage after considering joint treatments such as ship lap or tongue and groove joints.

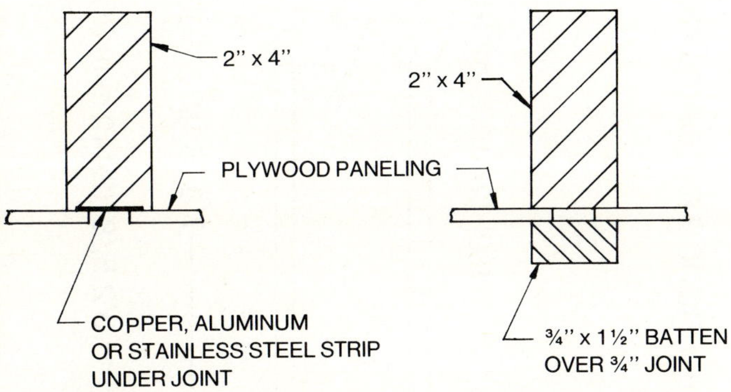

FIGURE 8-2. Joint Treatment for Metric Plywood used on Traditional Stud Spacing of 16″ o.c.

The 1200 mm dimension in sheet materials permits us to space studding at 300, 400, and 600 mm on centers, similar to our present 12″, 16″, and 24″ o.c. This 1200 mm dimension, being 3/4″ less than the present 48″ size, makes for easier handling because it is more easily grasped by the average man's arm reach.

The 2400 mm sheet dimension acts as a determinant of ceiling height, stud length and other dimensions related to the standard stud wall. Figure 8-1 shows this relationship and develops a stud length of 2324 mm.

When used as paneling on remodeling existing homes, a few problems may arise because studs will be spaced for 48″ plywood as opposed to the 1200 mm size. Two ways to overcome this are shown in Figure 8-2. During a changeover period, both 48″ and 1200 mm sizes may be available.

Thicknesses for plywood sheathing under consideration are 7, 12, and 17 mm, compared to our 3/8″, 1/2″, and 3/4″ present thicknesses. Interior sheet materials are being considered at 3, 4, 6, 9, 12, 16, and 19 mm.

CARPENTRY

Any discussion of the use of metric wood in construction would be incomplete without considering the effect on the man who actually puts the material in place—the carpenter—and his tools. Carpenters, as will all building tradesmen, must learn the elements of the metric system which apply to their work and materials handled. Their metric knowledge can be limited to meters, millimeters, kilograms, and conversions. Where materials were formerly expressed in "squares" (100 sq. ft.), as with roofing, the users must become familiar with square meters. Lumber and building products, door and window sizes, when manufactured in metric dimensions, will require thinking and layout in metric terms. The size of electric saws, once expressed in horsepower, will be indicated in watts.

Only those tools having a measurement function need change. The traditional six-foot folding rule would initially be continued at that length, with millimeters on one side and inches on the other. After the conversion had been completed, manufacturers would probably produce rules with millimeters on both sides. Tape measures would probably follow the same procedure.

The present carpenter's framing square, Figure 8-3, has a tongue of 1-1/2″ × 16″ and a blade of 2″ × 24″. With a 2 × 4 at 38 mm × 89 mm, 16″ o.c. being changed to 600 mm, then the square dimensions should be 38 mm × 400 mm on the tongue and 50 × 600 mm on the blade. This would facilitate the square's use in layout work just as it now is used. Graduations should be in millimeters.

Stanley Tools now has a carpenter square graduated every two millimeters, with all numerals in millimeters. They will soon have a combination square in English/metric, with the metric in millimeters. Later they will offer the combination square in millimeters only.

Tables on the square must be expressed in

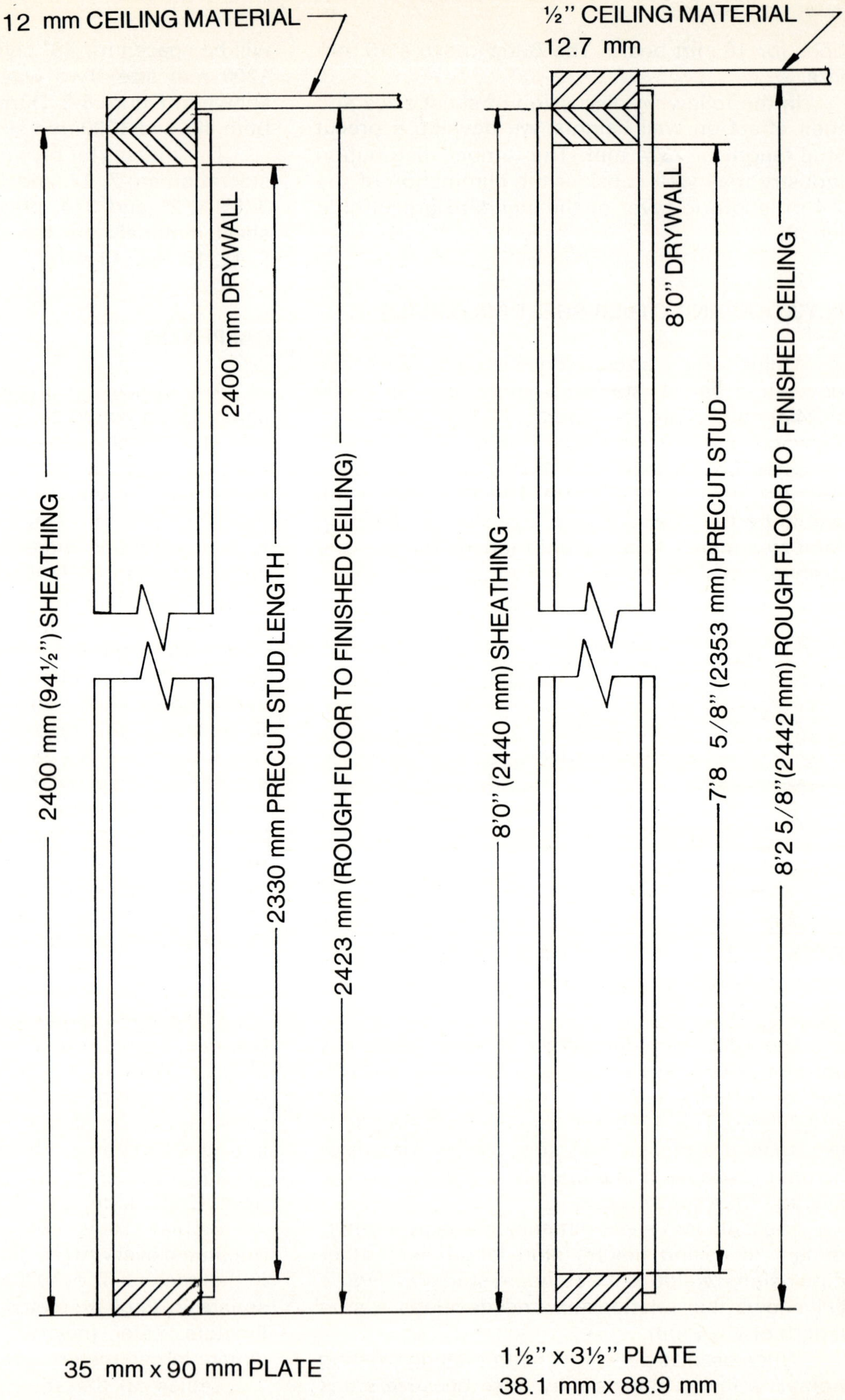

FIGURE 8-1. Wall Sections: Metric and Present Walls as Determined by Sheet Size

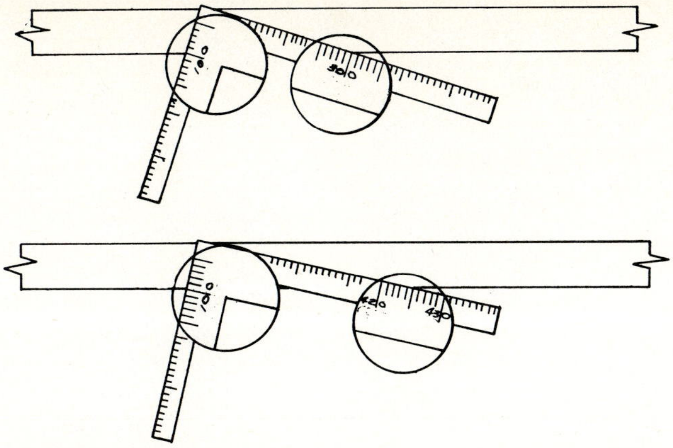

FIGURE 8-3. Carpenter's Square. (Top) Carpenter's square using 100/300 mm for marking common rafters. (Bottom) Square used to lay out hip rafter with 100 mm rise and 425 mm run (diagonal of 300 mm square).

metric and some of the present tables can be eliminated or replaced with more modern information. The rafter or framing table is used to determine the length of common, hip valley, and jack rafters, and their angles. This is a desirable feature and should be retained in metric terms, despite the widespread use of roof trusses. The Essex table used to determine board measure will be obsolete, since lumber will be measured in cubic meters of finished product. Although a new table could be designed in metric, this table is little used and might well be replaced with other information.

The octagon and brace scales are both seldom used by modern carpenters and can be eliminated. Other uses for the square can be continued, such as calculating proportions, making reductions or enlargements, using with circles, laying out an

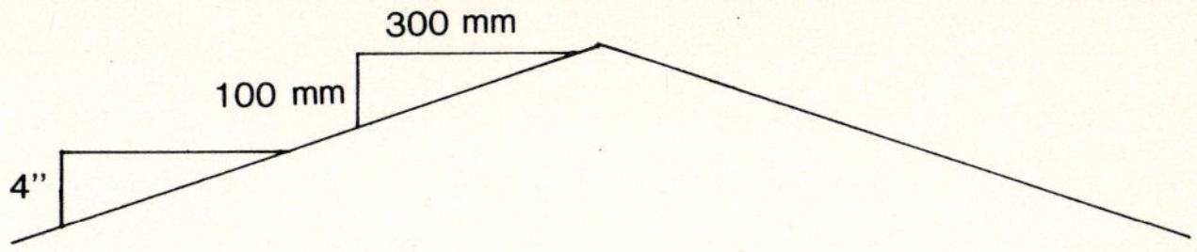

FIGURE 8-4. Roof Slopes. Roof slope expressed as traditional 4" rise to 12" and the same slope expressed in 100/300 mm terms.

ellipse for a pipe passing through a pitched roof, miter joints on polygons, hopper joints, and other work. Metric conversion tables might be added, but their use would be short lived.

Roof slopes are presently expressed in terms of inches rise to a foot of run—3/12, 4/12, 5/12, etc. With metrics this can be expressed as 100/300 mm, which is the same as 4/12. See Figure 8-4. Other metric pitches would be as follows:

Traditional	*Metric*
2/12	50/300
3/12	75/300
4/12	100/300
5/12	125/300
6/12	150/300
7/12	175/300
8/12	200/300
9/12	225/300
10/12	250/300
11/12	275/300
12/12	300/300

The diagonal of a 12" square is about 17", so carpenters use the 17" as the run when laying out hip and valley rafters. In other words, 4/17, 6/17, etc. In metric carpentry, the diagonal of a 300 mm square is approximately 425 mm, so the comparable figures would be 100/425 and 150/425. See Figure 8-3.

9. Building Products

The purpose of this chapter is not at this time to describe the metrication of all building products because many trade associations have not developed a consensus as to their metric conversion. But those areas where there have been some metric proposals developed may provide an example for other trade groups.

PAINTS AND LIQUID MATERIALS

Materials which are manufactured in liquid form will have to be packaged in liters, not gallons or quarts. Materials manufactured in powder form will be packaged in kilograms, not pounds. Coverage of materials will be indicated as square meters per liter or lineal meters per liter. Formulations will be changed from pounds per 100 gallons to kilograms per 1000 liters. Caulking compounds would be described as filling lineal mm per liter.

As far as official decisions made so far, the executive committee of the National Paint and Coatings Association recently approved five recommendations for the initial "transitional" entry of the paint industry into metric usage.

These recommendations were the result of a special NPCA metric conversion task force:

(1) Labels, and specification data on the labels and the sales aids, should contain soft-converted metric equivalents in addition to customary U.S. units, by January 1, 1978.

(2) The industry should consider a series of rounded metric container sizes "in the event the industry undertakes a hard conversion program"; 4 liters for 1 gallon; 1 liter for 1 quart; 500 ml for 1 pint; 250 ml for 1/2 pint; and 125 ml for 1/4 pint.

(3) Thought should be given as to the desirability of converting larger size containers such as pails and drums. Although the task force has not yet developed recommended standard sizes, it is expected that these would retain the same base diameters as in current, comparable U.S. containers.

BUILDING WIRE; ELECTRICAL

The American National Standards Institute has standards ANSI B-32 which are followed in the U.S.A. The ISO R388 preferred sizes are similar in some respects. Whether the ANSI or ISO sizes or a combination will prevail is yet to be worked out. The new sizes may effect a saving over the AWG sizes.

For instance: AWG 12 Gauge has a diameter of 0.0808″ (2.05 mm). The nearest ISO is 2.00 mm, a saving of 5.05% in copper. The industry believes that they can both do the same job. However, there are many related problems such as ratings, fuses, etc., which must be answered before wire sizes can be finally determined.

GLASS SIZES

At present, glass sizes are expressed in fractions of an inch or as Single Strength (SS) = 3/32″ and Double Strength (DS) = 1/8″.

With the change to metric measurement, a new range of thicknesses in millimeters will provide a single unit of measurement. Manufacturers expect no difficulties in the changeover because the thickness of flat glass can be adjusted during the actual manufacturing process. In addition, the nor-

TABLE 9: Guide to Architectural Glass Terms

Commercial Standard Glass thickness					*Products Where Glazing Materials are Commonly Used*		
Sheet Glass Term	*Weight (ounces)*	*Fraction (inches)*	*Decimal (inches)*	*Metric (millimeters)*	*Glass*	*Plastic Sheet*	*Organic coated*
	52	1/4	0.250	6 mm		Doors	
Heavy Sheet	40	3/16	0.190	5 mm	Patio Doors	Commercial Doors	
	32	5/32	0.160	4 mm	Patio Doors & Automobiles		
Double Strength	26	1/8	0.125	3 mm	Windows		Replacement Storm Doors and Replacement Windows
	20	3/32	0.100		Storm Doors	Replacement Storm Doors	
Single Strength	18		0.090		Windows		
	16		0.080	2 mm	Storm Windows		
		1/16	0.065			Storm Doors	

mal tolerances allow for considerable variation (plus or minus 10 percent).

Table 9 shows suggested metric thicknesses and the nearest equivalent in present thickness descriptions.

Sheet sizes have traditionally been cut to order. Therefore, the conversion to metric window sizes will make no difference. Glass sizes stocked by glass, hardware, and lumber dealers for repair and replacement can be the same as presently, expressed in millimeters, rounded to a convenient number.

SHEET METAL DUCT WORK

The Sheet Metal and Air Conditioning Contractors National Association, Inc. (SMACNA) has begun to investigate the effect of metrication on their specialty. Table 10 shows the present low velocity duct sizes with their millimeter equivalents. The drawings following are a good example of the way one sector's questions and answers must be coordinated with those of other sectors.

CABINETS

There has been little investigation into the possibilities in this field. But some questions need to be answered before standards can be developed for all types of cabinets, factory built and job built. Consider:

- Medicine and other cabinets now fit between studding 16″ o.c. with a space of 14-1/2″. With the metric 400 mm o.c. and 38 mm studs, the space available will be 362 mm. Some recessed medicine cabinets are now manufactured 14-1/8″ (357 mm). Will 5 mm be enough to permit easy installation?
- Kitchen cabinet countertops are now installed 36″ (914.4 mm) high. Has the height of the average housewife increased sufficiently to permit rounding to 920 mm? 930 mm? This would mean an increase of only a little more than a 1/4″ or 1/2″.
- In the above case, would whatever change is made affect appliance manufacturers who make equipment to fit under the countertop?
- In wall cabinets, the present 12″ depth is a derivation of 11-1/4″ (nominally 1″ × 12″) shelv-

TABLE 10: SMACNA Low Velocity Duct Size Ranges

Duct size	Millimeter equivalent
Thru 12 in	304.8
13 in – 18 in	330.2 – 457.2
19 in – 30 in	482.6 – 762.0
31 in – 42 in	787.4 – 1066.8
43 in – 54 in	1092.2 – 1371.6
55 – 60 in	1397.0 – 1524.0
61 – 84	1549.4 – 2133.6
85 – 96	2159.0 – 2438.4

To express these metric sizes in centimeters shift decimal one place to the left; in meters shift three places to the left. Example: 304.8 mm = 30.48 cm = 0.3048 m.

Reprinted from *Baking into Metrics? A Construction Industry Survey,* October 1974.

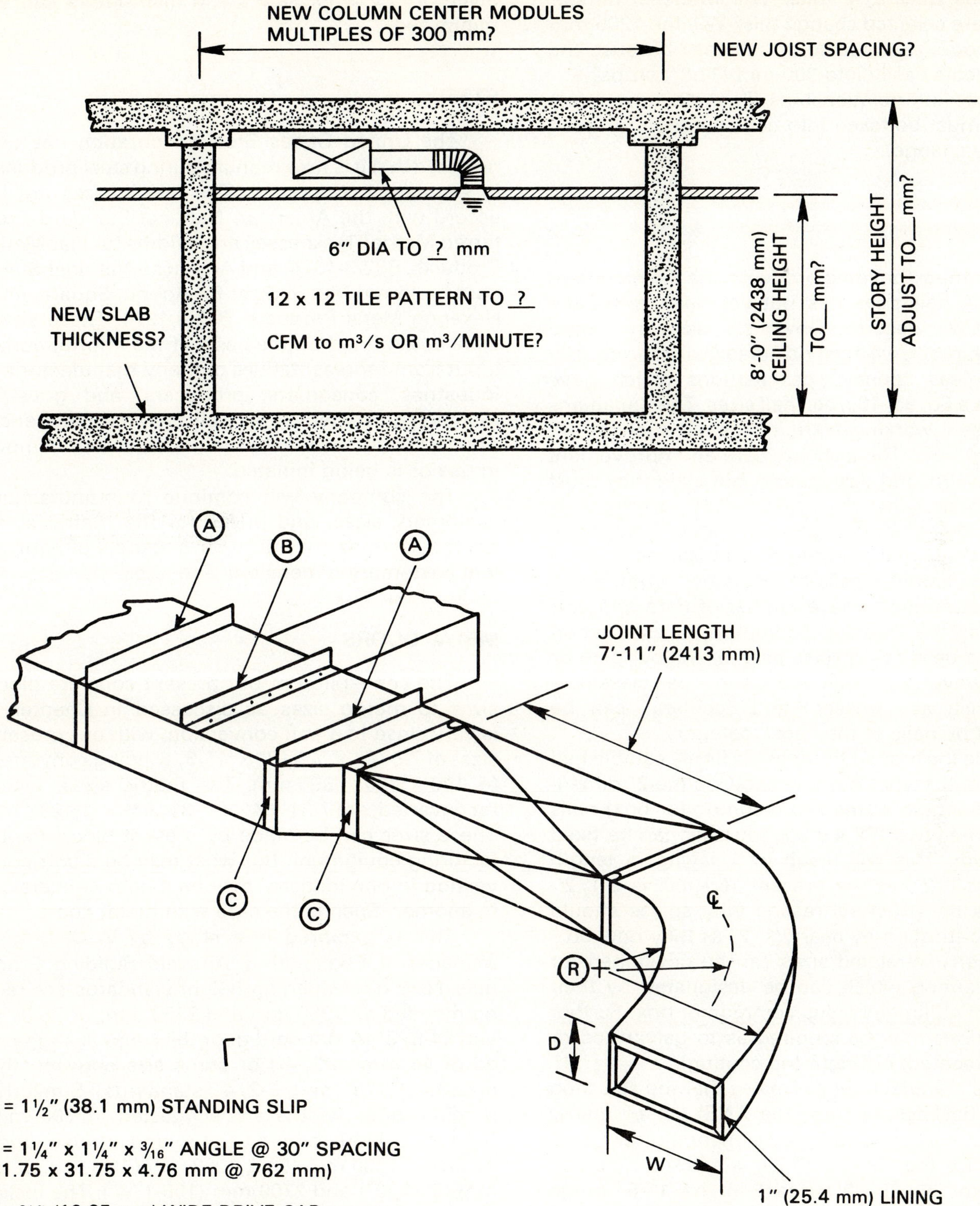

A = 1½" (38.1 mm) STANDING SLIP

B = 1¼" x 1¼" x 3/16" ANGLE @ 30" SPACING
(31.75 x 31.75 x 4.76 mm @ 762 mm)

C = ¾" (19.05 mm) WIDE DRIVE CAP

CFM = ? m³/s VELOCITY = ? m/s DUCT WEIGHT = ? kg

FIGURE 9-1. Illustration of Construction Changes. Reprinted by permission of the Sheet Metal and Air Conditioning Contractors' National Association, Inc., from *Backing Into Metrics? A Construction Industry Survey,* October 1971, p. 12.

ing plus stiles 3/4" thick. Will whatever lumber sizes are selected change this? Will the 1200 mm plywood size mandate an increase because the sheet cuts easily into 300 mm (11.8") strips?

These interrelationships illustrate how many factors must be taken into consideration for each possible change.

NAILS

Like many building products, the development of metric nail sizes has not yet been given the study it will deserve. Obviously, as with "board feet" in lumber, we must rid ourselves of the equally meaningless "penny" designations which have given us a 6d, 8d, 16d, etc. nail sizes. The Canadians have been working with inch sizes such as 3", 3-1/4", 4", etc. The inch sizes are an improvement over the 10d, 16d designation, but even they must go.

The designation "common" and "box" nails might be retained to differentiate between heavier or lighter diameter nails of the same length. Since we will continue to have lumber of hard and soft densities, the heavier "common" classification might be used for species such as yellow pine or oak. However, with little effort, some nail classifications such as "sinkers" and "coolers" can be handled by nails of the "box" category.

Once the lumber thickness is firmly established at 38 mm for what we now know as the 2" dimension, the length of the 16d common (or box) nails used to nail two 2 × 4 studs together can be fixed at 75 mm. This will result in a saving in length of about 1/2", as the present 16d nail is 3-1/2". There is no structural reason why spikes should protrude through by nearly 1/2" as they now do.

Other use-related sizes can be developed for wood framing which can be designated by their length in millimeters plus common or box. Further descriptions may be required as to galvanized or cement coated, or bright (no coating).

One manufacturer estimates that the selection of nail thicknesses from the ANSI list of metric preferred wire sizes and provision for multiple uses can reduce the number of items stocked by a nail distributor by 20–30 percent in the 1"–6" range alone. This will result in a substantial reduction in distribution costs because capital invested in inventory by wholesalers and distributors can be minimized.

STEEL

The United States Steel Corporation has announced that it is now manufacturing steel products in standard metric sizes. The metric sizes are in accord with the American National Standard Preferred Metric Thicknesses and Widths for Flat Metal Products, B32-3-1974, and American National Standard Preferred Metric Sizes for Round, Square, and Hexagon Metal Products, B32-4-1974. These sizes were determined on the basis of the informational input from representatives of many manufacturing industries, consumers, producers, and governmental representatives, as well as by the experience of other countries where the metric system is now in use or is being initiated.

The company will continue to manufacture customary sizes, and prices for the metric sizes are the same as the prices for materials of equivalent customary dimensions and sizes.

METAL DOORS

The conversion of our present concrete block sizes to metric sizes as discussed in Chapter 7 is very close to a soft conversion, with our present sizes of 7-5/8" × 7-5/8" × 15-5/8" being converted to 190 × 190 × 390 mm. The metric sizes, when reconverted, are 7-31/64" × 7-31/64" × 15-23/74". These sizes can be made by present block manufacturing equipment. But what may be a soft conversion in one industry may be a hard conversion in another. Such is the case with metal doors.

This is explored in a study by V. C. Braun, Manager of Engineering, Amweld Building Products. New door-opening height standards are recommended as 1950 mm and 2350 mm, or 76-3/4" and 84-5/8". A standard door thickness is suggested of 44 mm (1-3/4") or some size between the present 1-3/8" and 1-3/4" standards. Similarly, nominal door widths are suggested at 700 mm (27-1/2"), 1100 mm (43-1/4"), and double door widths of 1500 mm (59"), 1900 mm (74-3/4"), 2300 mm (90-1/2") and 2700 mm (106-1/4"). The inclusion of the 1950 mm double opening creates an additional single size, a nominal 975 mm.

10. Land Titles and Surveying

TITLES

Now that legislation has passed providing some type of official sanction and direction for metric conversion in the United States, one of the more difficult problems to be worked out will be the question of land titles, mapping, and surveying.

All elements in this sector of the economy indicate they will cooperate in such a conversion, and some individuals feel metrication is inevitable, but from their experience in a complicated field they point out there will be difficulties far transcending a simple switchover from yards or miles to meters.

Wesley J. Hampton of "Realty Facts" feels that land measure is one area that a metric system would greatly simplify. While no one argues with the logic of the metric measuring system, it appears few other men in the land measurement field are as confident a conversion in the United States will be simple.

Basic to the problem is that land represents a real economic value and its legal description is crucial in defining its limits and through that its value. Some published statements that conversion of English or traditional measurements to metric units can generally be approximations meets with little credence in this field. Real estate descriptions must be exact, not approximate. A variation of 1-1/2 feet is small, but anything but unimportant if it puts the wall of your house over your neighbor's lot line.

With this basic question of value understood, the United States has problems because of its diversity. Britain, a small country, made the transition over a period of years and Australia, a homogeneous country, is now conducting its real estate transactions in metrics after a dual-unit trial period of six months.

The United States, however, poses more complex problems. As a publication of the Washington State Potato Commission points out, some deeds in New Orleans are still written in the terms of the French foot of pre-Napoleonic times and there are land areas in the Far West described in varas, not acres, from the old Spanish land grant days. Many states of the United States are laid out under the rectangular Jefferson-type survey, first proposed by Thomas Jefferson (with many idiosyncratic variations), but the oldest settled areas are not. In any case, every state regulates land measurement and conveyancy in its own way.

First step in a metric conversion in land measurement will be agreement on terminology. It appears likely that mere familiarity will supercede acre with the European hectare (ha), rather than its more cumbersome equivalent square hectometer.

Several federal agencies, such as the space agency, have taken the lead in the use of metric measurements, and some of the proposed legislation aimed at an orderly conversion of the entire United States economy to metrics specifically requires federal agencies to take the initiative by converting their operations to metric measurement in all possible cases.

The Department of Agriculture has made no formal choice but in working with foreign statistics it uses the term "hectare" because of its common usage. The USDA Statistical Reporting Service has chosen hectare and now publishes supplementary tables in crop reports in hectares.

Statistical usage will be easy, and it will probably be equally easy for the public to hear large land areas such as farms described as so many hectares instead of so many acres. But conveying ownership of such farms and, even more so, of smaller urban lots, will involve much more complicated considerations.

There appears almost no chance that anyone will seriously suggest going back and describing each piece of "owned property" in the United States in metric terms. The man-hours and expense involved in changing records now existing in thousands of public offices, not to speak of title companies and real estate agencies, causes shudders.

Consensus seems to be that, if metrication comes, it will be on a "go-forward" basis, with each piece of property newly described only when it has occasion to change hands.

The American Land Title Association, which represents over 2,000 abstract and title companies all over the country, has been active in studying the problem since 1970. Its former president, Thomas J. Holstein, made a lengthy report after a survey of its members preparatory to reporting to the Secretary of Commerce.

Members replying to the association survey listed the following terms which will be replaced in a metric system: "inches, feet, yards, rods, miles, acres, chains, varas, arpents, links, poles, perches and furlongs."

In addition, the kinds of records they mentioned in their home communities which would be affected at some time, either in wholesale or "go-forward" conversion, included "county and city assessment maps; municipal engineering maps; highway maps and drawings; county plats; warranty deeds; quit claim deeds; deeds to secure debt; mortgage documents; county tax records; maps in the county recorder, treasurer, auditor, and assessor offices, in municipal planning and zoning offices, in state and federal land offices; records of county tax assessor, engineer, surveyor, recorder or register of deeds; drain commissioner; police, fire and water department records; city engineering records, township and public lighting commission records; easements; mechanics lien entries in prothonotary's office; circuit court deed books; other court records; land books."

Holstein pointed out that, from the standpoint of title insurance work, a piece of property being conveyed under a new metric description still would have to be traced back through all previous records in the original deed terms. He expressed concern that a too-hasty adoption of metric conversion could lead to confusion in public offices and title companies which would require extra time and expense, making land conveyance costs higher for land buyers and impeding business:

> The availability of public records dealing with real estate—and the means by which they are compiled, kept, and preserved—have throughout this nation's history led to the greatest diversity of private real estate ownership and mortgage investment ever known. Extreme care should be taken in contemplating any change that would inhibit the health of the real estate segment of the American economy.

Because of these considerations, the ALTA Board of Governors in April, 1970, expressed the view that it saw no advantage in converting land titles to the metric system, since all land transfer occurs solely within the United States. The Board, however, pledged cooperation should Congress decide to convert to the metric system and offered assistance in gathering information and making recommendations.

This remains essentially the position of the group today. Dale P. Riordan, the Association's Director of Research, said in January, 1975:

> "In brief, the Association will cooperate in any decision to convert to metric measurement. However, we feel that land title public records should be changed only on a go-forward basis. Changing back-dated records in both public offices and title company files would involve costs vastly exceeding potential benefits. In addition, we urge that land title records be converted only when it is advantageous to do so. Thus, a rule of reason should be followed when considering metric conversion.

Mapping and surveying are related fields. An interesting article entitled "Metrics and Maps—A Foul Deed?" in the March/April, 1975, issue of *Metric News* again points to the complexities involved.

The article is an interview with John Q. Cummings, president, and Robert Dummer, sales manager, of the Rockford Map Company of Rockford, Illinois. The Rockford company is the largest publisher of plat maps in the country, covering 500 counties in 14 midwestern states. These are the maps which divide townships into parcels as small as half an acre and list the landholder of record, revised every two years. Because of their detail, maps of this type and those published in other sections of the country are in constant use by all levels of government and by the business community.

Cummings is quoted as feeling that metrication is inevitable but that "only careful planning by groups which include representatives of city and county surveyors will prevent chaos. Faulty planning will result in a logjam of cases in courts, clouded real estate titles, and prohibitive surveying costs which must be absorbed by the landholder."

He points out that while his company operates only in states with the rectangular Jefferson-type survey, the theoretical congressional township six miles square is merely theoretical. There are many deviations and mile square sections are not all a mile square. The situation is even more confusing in states which measure land under a less geometric system.

He and Dummer are skeptical of metrication advocates who feel that metrication will be easy because it would be used only on new surveying and mapping projects as they become necessary. They point to recent computerized results of aerial surveys of the entire country. Exact as they are, they are essentially only resurveys and do not change the recording of the original survey which is kept in a county courthouse and is the basis for every deed in the county. Their mapmakers, armed with these new surveys, still must go deeply into old survey and tax records to get legal descriptions for each land parcel.

In the same way, they do not feel these new and more exact surveys can lead to a smooth metric conversion by using survey data and merely converting existing deeds to metric terms.

"A resurvey using either system of measurement might reveal that a farmer over a period of years has unintentionally encroached on land along a roadway that by legal description belongs to the county. That situation might arise over fencelines between separate properties where physical barriers had never been erected. No landholder is willing to surrender land merely on the basis of the single survey. He is being hit in the bank book. The survey will be challenged and may go to court."

In this area, the mapmakers echo Holstein of the American Land Title Association. That organization also has expressed qualms over an increase in litigation which might interfere with commerce if metric conversion disturbs traditional and accepted land descriptions.

These problems of resurvey and litigation, of course, are not basic to metric conversion. No matter what system of measurement is used, a resurvey would show variances from original descriptions, which could lead to dispute. But it is important to realize that men in the field are aware that metric conversion will focus attention on such problems. A land description transaction in totally new terms is certain to spotlight such variations and discrepancies.

Whatever reservations they have about the problems, none of these men consider them insuperable barriers to metric conversion in the United States. They do feel it is a most complex problem and plead for time and careful consultation if such a plan becomes a reality.

SURVEYS

In the field of surveying, more basic technical changes are being considered as the days of metrication approach.

There is strong advocacy for a change to decimal degrees of latitude and longitude, replacing the old system of degrees, minutes, seconds and tenths of seconds. This, of course, would be a fundamental change for both surveyors and mapmakers.

Decimal degree notation has been in use for many years for weather observation locations, both by ships and planes, and some oil and gas drilling maps are expressed in these terms. It has the advantage of greater simplicity and precision. G. Stanley Doore, Chairman of the National Weather Service Metrication Working Group, says locations expressed to two decimal places are about 67% more precise than the same location expressed to the nearest minute. If carried to four decimal places, precision improves 178%.

The decimal system has obvious advantages in today's computer-oriented technology and would permit the dropping of two symbols, minute (′) and second (″), replacing them with the simple decimal point.

The disadvantages are those shared with many other scientific and economic fields facing metrication—the need for new maps, differently calibrated instruments, retraining of personnel, etc.

11. Metric Legislation and the American National Metric Council

Metric legislation has been introduced in all recent Congresses and for several years advocates have confidently predicted passage "soon." However, it was not until Dec. 22, 1975, when President Gerald Ford signed the Metric Conversion Act of 1975, that a bill actually passed into law.

This was achieved when the House of Representatives and the Senate made certain changes in earlier proposals, eliminating some of the areas of controversy which stalled previous efforts. Sectors of the economy which had previously been in opposition withdrew their objections in return for one major change from previous proposals. Most of the earlier bills considered proposed a 10-year plan for conversion to the metric system; this 10-year target has now been removed and voluntary conversion without any set time schedule has been accepted. In addition, the goal will be a nation predominantly metric, but not exclusively so, as some of the more enthusiastic advocates had advocated. This bill leaves quite a few areas open to interpretation; it is expected that more specific legislation will follow.

There is some feeling these changes may actually improve the prospects for quick conversion to the metric system, because those industries where its advantages are obvious may be ready to convert well before 10 years have passed.

One thing is not in dispute—metric use in America is already widespread and is going to increase rapidly. The only real question is whether or not to continue with the present unplanned company-by-company approach, or to have a more planned and coordinated effort.

Essentially, the world is now metric. International standards for trade are being set up on that basis. Since American industry certainly is not going to resign from the world, it is already going metric, and the list of major firms already producing at least partly in metric terms increases almost daily. Also agreed by all segments is that the construction industry will probably be among the last industries to metricate, since buildings are not transported across national boundaries. But suppliers of raw materials to industry also are the essential suppliers to construction and as they adapt to the needs of their international trading customers they will tend to become increasingly impatient about maintaining dual-measurement business.

Surveying the status of present legislation, it should be noted that one far-ranging piece of metric legislation is already fact. Public Law 93-380, known as Education Amendments of 1974, puts Congress on record as recognizing "a metric system will become the dominant system of weights and measures in the United States" and provides substantial financial aid in developing educational programs to train American children in metric mathematics. A number of states are already preparing to aim their future educational training into the metric system, dropping traditional measurements entirely. Educational planning is outside the scope of this book, but its results are not. The next generation of American workers in construction will be trained basically in metric measurements.

Two major bills to coordinate the changeover

of the United States to metric measurement were introduced almost as soon as the 94th Congress opened in 1975. They were House Resolution 254, introduced by Rep. Olin Teague (D-Tex.), and Senate Bill 100, introduced by Sen. Claiborne Pell (D-R.I.) Although somewhat different in content, the basic approach was the same in both bills, which in turn was very similar to legislation which had been before other recent sessions of Congress.

Essentially, they provided for a conversion over a period of 10 years, with a broadly based National Metric Conversion Board planning the changeover. This board would have drawn up a general plan for conversion and submitted it to Congress and the President in a year's time. During the remaining nine years of the board's life, detailed conversion schedules would have been gradually developed for each segment of United States industry. Both organized labor and small business had serious doubts about some features of this plan. Their objections fell chiefly in the economic area—costs and job (or business) security.

American industries which have already begun the process of metric conversion to a greater or lesser degree generally have picked up the costs of any required tools and training as a standard business expense. This principle that "costs shall lie where they fall" has been observed also in other major industrial nations converting to metric.

Backers advocating this approach argued that this plan would encourage conversion only where it is advantageous and necessary. Cost estimates vary widely and some facets of the British experience indicate that they did not run as high as experts expected.

Nevertheless, many individual workers and many small business owners are involved and not all have the cash flexibility of a major industry. Some of the earlier bills recognized this difficulty by providing "limited assistance to businesses and individuals substantially affected by metric conversion in bearing the cost of such conversion." Others opposed this vigorously on the grounds such legislation would open the door to heavy governmental costs. As now compromised, the national planning board will be authorized to study such financial assistance and to make recommendations as to the need.

Organized labor earlier had vigorously opposed any legislation which would tend to push and encourage metric conversion, and building unions led this opposition. The United Brotherhood of Carpenters and Joiners of America, the International Brotherhood of Electrical Workers, and the International Association of Machinists and Aerospace Workers joined some time back in messages to Congress urging a careful study of all possible results of metric conversion.

Removal of the 10-year time objective and the promise of consideration of financial assistance to mitigate any tool and training costs or loss of jobs resulting from metric conversion has succeeded in dispelling many of labor's worries.

Political and philosophical viewpoints probably determine how labor's objections are viewed. Many metric advocates quickly dismiss these objections as just more evidence of fear that anything which increases similarities in production here and abroad will likewise increase the "export of jobs." Louis F. Sokol, president of the U.S. Metric Association, says in the *Metric News* (May–June, 1974), "It has been rumored that labor's real opposition to metrication is based on the fact that they fear multinational manufacturing companies that have converted to metric could 'weather' a major labor strike by importing necessary component parts." Nevertheless, millions of Americans are unionized construction workers and they have their opinions. Nicholas R. Loope, director of research for the United Brotherhood of Carpenters and Joiners of America, outlined their views in a speech at the Building and Construction Sector Conference of the American National Metric Council in Washington on March 18, 1975.

Loope said, "In spite of what many people think, the AFL-CIO is not opposed to metric conversion. We feel there are too many unanswered questions regarding the impact of metric conversion on the U.S. economy." Pointing to the economic effects of the recession and high unemployment, he argued against "an irreversible policy of metric conversion according to a legislated schedule," and pleaded for flexibility until "we have a better understanding of its impact."

"The building and construction industry," he predicted, "will be slow to adopt the metric system, because of its extreme complexity, plus the lack of both external pressures and economic incentive."

Space will not permit a complete listing of his objections. They centered, however, about the loss of construction jobs, not only because of tendencies to build American plants abroad, but also because of delays in building investment here if construction appeared to be entering a period of confusing change. He also stressed construction delays, costly to both contractor and worker, if metric materials are not readily available on-site, the costs of tools and retraining, and the effects metric conversion would have on long-accepted building codes and safety regulations, as well as traditional bidding practices and union rules.

Likewise, some small business organizations have expressed their qualms. Ninety-five percent of the full-time businesses operating in the United States are classified as small, and certainly most building contractors are in this category.

In the magazine *Metric Reporter* for March 7, 1975, The National Federation of Independent Business (NFIB) which claims to represent 410,000 firms, is reported almost evenly divided on the metric question. NFIB says the very smallness of its firms poses an obstacle to metric conversion. Mistakes in costs and in judging timing of necessary conversion, which a major industry could hide in its budget, might exhaust the resources of a small businessman.

This organization emphatically does not support the "costs shall lie where they fall" philosophy, feeling that long-term, low-interest Small Business Administration loans should be available to small firms threatened by economic injury during a time of conversion.

The status of small business is so important that Carl Beck, chairman of the legislative action and policy committee of the National Small Businessman Association, believes that "the success of the metric conversion process in the United States will depend on whether or not it is done to the economic advantage of small business."

Quoted in the December, 1974, issue of the *News-Standard* of the Southern Pine Inspection bureau, Beck points to the dangers both of undue haste or an overly long changeover period. He is inclined to feel that an arbitrary timetable for total conversion would be premature at this time, but does feel that some expression of the "sense of Congress," on this point, if not actual legislation, would help businessmen in their planning for conversion.

This somewhat long discussion of what might appear to be "metric doubts" is not presented to sway opinion, but to give a brief overview of ideas being advanced. Legislation has been passed and, traditionally, American laws are finally drawn on a basis of compromise of differing viewpoints. While other legislation is being studied, Congress should have an input of ideas from all sectors of the economy which will be affected.

PRESENT EFFORTS

While any official metric conversion will depend upon action by Congress, it would be erroneous to think that nothing meanwhile is being done to centralize information and gather opinion.

A number of groups have taken an active interest in metrication for years. The central focus of activities in the private sector of interest to the construction industry is the American National Metric Council, which was established in 1972, and has its headquarters at 1625 Massachusetts Ave. N.W. in Washington, D.C.

This council was given impetus by the American National Standards Institute which in 1972 recognized the need for national coordination of metric conversion activities and provided the first funding for the Council. It now depends primarily on private memberships of businesses, organizations, and individuals, and is working through a series of coordinating and sector committees. Membership includes not only industry representatives but labor union spokesmen and others intimately concerned with particular segments of industry.

The coordinating Committee for the Construction Industry has subgroups on Standards & Codes, Design, Products Manufacturers, Users, Labor and Real Estate. This group is highly active and is a focus for the ideas being presented in connection with metric conversion in the construction industry.

In addition, another subcommittee on Lumber and Wood Products is exchanging ideas vital to construction. This subcommittee has representatives of forest management, softwood and hardwood lumber, panel products, packaging and handling, furniture, millwork, poles, piling and ties, labor and home building. It has a Canadian liaison member also, because of the intimate tieup between the U.S. and Canadian lumber industries.

The American National Metric council envisions the ultimate creation of some type of National Metric Conversion Board. Although strictly private and voluntary, it hopes to have much information already gathered and coordinated when official action is finally taken.

The Council, since it brings together representatives of major industry to discuss common problems, has been especially sensitive to anti-trust implications and has been careful to maintain safeguards in this area during its activities.

12. History of Measurement Systems

Why all the attention focused on a metric system of measurement? Why are scientists, engineers, businessmen, school children the world over involved in a discussion of meters, kilograms, etc. Because without standards of measurement, we are lost.

The U.S. Department of Commerce, in an excellent pamphlet, "Brief History of Measurement Systems," emphasizes the point by stating in its opening remarks that "weights and measures were among the earliest tools invented by man." Tools—just like the stone ax and the flint point.

Primitive man had to build shelters, make clothing, trade goods; to do so there had to be some agreement on size and shape. Early Babylonian and Egyptian records and the Bible all indicate that this standardization system involved using natural surroundings for measurement—parts of his body for length, plant seeds and stones for weight, the sun, moon, and planets for time (see Figure 12-1).

And it is instructive that the first known "international" system of weights and measures was devised by the Romans, simply because they had to standardize commerce in what was virtually the known civilized world of their time. Today's pressure for an international metric system comes from the same source—the need to standardize commerce in a day when trade is universal throughout the world.

A survey of how our present somewhat haphazard measurement system developed is a "window" on the history of man. One of the most amusing accounts appeared in "Shop Tips," a publication of the Ford Parts Division of the Ford Marketing Corporation, reprinted with their permission:

> You have to give the Romans credit. They were the ones who had standards for weights and measures which they kept dry and secure in a temple in Rome. And, in keeping with Roman law, men in all parts of the Roman Empire had to use weights and measures exactly like the standards. Men used the same ounce, the same foot, the same pound and the same mile in their calculations . . . whether they were in Britain or Egypt, Spain or Syria. Yet, when the Roman Empire fell, about 1500 years ago, these standards were lost.
>
> As a result, the Middle Ages became a confusing time for weighing and measuring things. Almost every town and every guild in that town had its own system. To illustrate the problem, we have only to look at a few examples.
>
> On the proper method to do surveying, an early German treatise instructed the surveyor to establish the length of the 'Rood' by getting 16 men (after they had left church services) and having them put their left foot, one behind the other. The length thus obtained was then the lawful 'Rood' with which to measure and survey the land, and the 16th part of it was to be considered the lawful foot measurement.
>
> Note: The 'Rood' is felt to be today's 'Rod' which is 16-1/2 feet.
>
> Many of today's other units of measurement indicate the crude method from which they originated. A Mile is from the Latin 'milia passuum,' which means 1,000 paces, with each pace being about 5 feet . . . in other words, a double step.
>
> Inch is from the Latin *uncia,* meaning a 12th part. The English yard was established by Henry I as the distance from the point of his nose to the

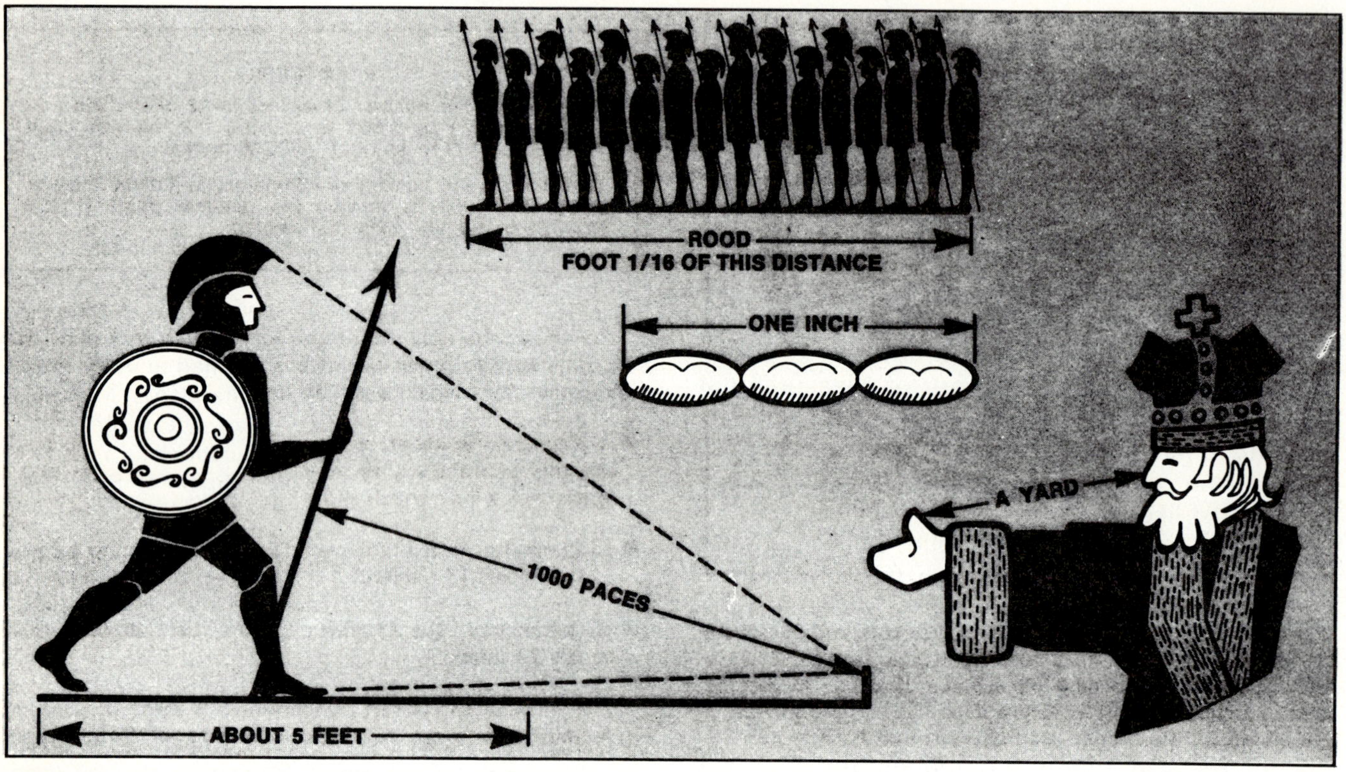

FIGURE 12-1. Early Measurements. Reprinted by permission of Ford Parts Division from "Shop Tips," courtesy of Ford Marketing Corporation.

end of his thumb. If he happened to have had a long nose or very short arms, our yard as we know it today could well have been only 33 inches!

An inch in the year 1324 was indicated by the length of three dry barleycorns (grains of barley) that were laid end to end.

Note: The 'barleycorn' is still used today as a unit of length in shoe sizes which are numbered in thirds of an inch, or 'barleycorns' in a system of 13's.

The pound, which is roughly the same as the Old Roman *libra* (we get the abbreviation 'Lb' from that), was long defined as the weight of 7,680 grains of wheat.

Therefore, the 'grain' is still the smallest unit of weight in today's common systems.

This little account shows how several cultures (and others not mentioned) contributed to our present measuring system. The whole story of early historical measurement is not known, but we do know that by the 18th century England had achieved a great degree of standardization, and it was the English system the colonists brought to America.

Interestingly, commerce was again the reason for standardization and for fairly widespread use of one measurement system. England colonized around the world and dominated commerce for three centuries, so its weights and measures became widely used.

The crucial importance of an agreed-upon system of weights and measures is recognized in the earliest documents of the United States. Both the Articles of Confederation and the later Constitution of the United States gave the power to fix standards of weights and measures to Congress. Today the National Bureau of Standards supplies standards to all the states and assures uniformity throughout the country.

The countries of continental Europe lagged behind England in achieving standardization, but it was here that the need for a really worldwide and coordinated system was recognized. Several intellectuals proposed such a plan over 300 years ago but it was a century before anything was actually done.

The development of the metric measure for temperature preceded the metric system itself by nearly a century. Anders Celsius, a Swedish astronomer (1701–1744), decided that the logical way to approach the measuring of temperature was to divide the temperature difference between the freezing point of water and its boiling point into 100 equal units, or degrees. The old name "centigrade" now becomes Celsius in his honor.

The metric system was first designed in the late eighteenth century. In the midst of the turmoil of the French Revolution, the French National Assembly asked the French Academy of Science to suggest standards for all measures and all weights. And it was the Commission appointed by this Academy which came up with the basic metric system.

The unit of length was scientifically based on a portion of the earth's circumference—one ten-millionth of the distance from the north pole to the equator along a meridian running through western Europe—and given the name "metre" from the Greek word "metron," meaning "a measure." Other base units for volume and weight were derived from the meter and similarly scientifically based, rather than capriciously on seeds and human feet.

And, most importantly, each unit was to be increased or divided in multiples of ten. This made possible easy calculations which could be performed by simply shifting a decimal point.

There was no great rush to the new system, but France made it compulsory in 1840. By that time technology was beginning to flower and scientists and engineers had found the standardized units and the decimal system of increase and decrease practical and convenient. It spread steadily throughout the world.

There have been several subsequent revisions and improvements of the international metric system, chiefly because of the increasing accuracy required of measurement science and devices. The most recent was in 1960, when the international General Conference of Weights and Measures adopted *Le Système International d'Unités* (International System of Units), which goes by the accepted abbreviation SI. This is the modern metric system.

The common idea that the metric system is something new and radical in the United States ignores history. Congress in 1866 made it lawful throughout the United States to use the metric system "in all contracts, dealings or court proceedings." And this country was one of the signers of the "Treaty of the Meter" in 1875, which set up the permanent machinery to supervise and refine the metric system. This permanent machinery still operates through an International Bureau of Weights and Measures in Sevres, France, which serves as the permanent secretariat of the signers of the "Treaty of the Meter"—known as the Metric Convention—in which the United States has been a participant since its beginning.

After the 1875 treaty metric standards were agreed upon by all signers, and since 1893 these international standards actually have served as the fundamental weights and measures standards of the United States.

Persistence of the more irrational "English system" of weights and measures in the United States and other countries influenced by British nineteenth century trade has probably led many Americans to think this is the "right" system, and metrication something of an oddity. Actually, England itself and all of the major areas once influenced by the Commonwealth already have abandoned or are in the process of discarding the traditional system. There is no prospect it will ever return—by the end of this decade, for instance, nothing but metric sizing and standards will be permitted in trade with the European Common Market.

Because United States industry currently is juggling two systems of measurement—traditional and metric—confusion is bound to increase. To end this confusion, the Secretary of Commerce after a three-year study recommended to Congress in 1971 that the United States change to the predominant use of the metric system through a coordinated national program.

APPENDICES

TABLE 1 - QUANTITIES AND SI UNITS

Unit Type	Quantity	Unit	Symbol	Expressions in Terms of Other Units
Base Units	Length	metre (a)	m	
	Mass	kilogram (a) (b)	kg	
	Time	second	s	
	Electric current	ampere	A	
	Thermodynamic temperature	kelvin (c)	K	
	Amount of substance	mole	mol	
	Luminous intensity	candela	cd	
Supplementary Units	Plane angle (g)	radian	rad	
	Solid angle	steradian	sr	
Derived Units Having Special Names	Frequency	hertz (d)	Hz	s^{-1}
	Force	newton	N	$kg \cdot m/s^2$
	Pressure, Stress	pascal	Pa	N/m^2
	Energy, Work, Quantity of heat	joule	J	N•m
	Power	watt	W	J/s
	Electric charge	coulomb	C	A•s
	Electric potential Potential difference Electromotive force	volt	V	W/A
	Electric resistance	ohm	Ω	V/A
	Electric conductance	siemens (e)	S	A/V
	Electric capacitance	farad	F	C/V
	Magnetic flux	weber	Wb	V•s
	Inductance	henry	H	Wb/A
	Magnetic flux density	tesla	T	Wb/m^2
	Luminous flux	lumen	lm	cd•sr
	Illuminance	lux	lx	lm/m^2
Some Other Derived Units	Area	square metre		m^2
	Volume	cubic metre		m^3
	Velocity---angular	radian per second		rad/s
	Velocity---linear	metre per second		m/s
	Acceleration---angular	radian per second squared		rad/s^2
	Acceleration---linear	metre per second squared		m/s^2
	Density(mass per unit volume)	kilogram per cubic metre		kg/m^3
	Moment of force & torque	newton metre		N•m
	Viscosity -- dynamic	pascal second		Pa•s
	Thermal conductivity	watt per metre kelvin		W/m•K
	Thermal flux density or irradiance	watt per square metre		W/m^2
	Thermal capacity or entropy	joule per kelvin		J/K
	Permeability	henry per metre		H/m
	Permittivity	farad per metre		F/m
	Magnetic field strength	ampere per metre		A/m
	Luminance	candela per square metre		cd/m^2
	Molar entropy	joule per mole kelvin		J/mol•K

Notes to Table 1

(a) The spelling shown in this Table is that used in ASTM 380-72, the ASCE accepted version of SI units.

(b) The kilogram is the only base unit whose name, for historical reasons, contains a prefix.

(c) The SI unit of temperature is the kelvin. The Celsius temperature scale (previously called Centigrade) is the commonly used scale for temperature measurements, except where the thermodynamic scale must be used as in pressure-volume relationships of compressible fluids. The Celsius scale is not part of the SI, but a difference of one degree Celsius equals one kelvin. The degree symbol ° is always associated with the Celsius temperature, but not with the kelvin. By definition: 0°C = 273.15K and 20°C = 293.15K.

(d) The SI unit of frequency, the hertz, was previously called "cycle per second". The reciprocal of the frequency is the period.

(e) The siemens was previously called the mho.

(f) The full names of all SI units are written with a small (lower case) initial letter except for Celsius. Only symbols of units named after persons are capitalized such as A,K,N,Pa,Hz,J and W.

(g) The ANMC recommends the use of decimalized degrees rather than minutes and seconds.

(h) Derived units are formed from the products or quotients of other unit quantities. The product of two or more units is indicated by means of a dot. This dot is preferably raised above the line as in N•m. The quotients are produced by placing one sloped line between the numerator and the denominator as in $kg/m^2 \cdot d$.

(i) Celcius is neither a basic nor a derived SI unit. The degree Celcius (°C) is an accepted alternative for a kelvin (K).

TABLE 2 --- PREFIXES

Multiplying factor	Prefix	Symbol	Pronunciation
1 000 000 000 000 = 10^{12}	tera	T	tĕr′ ȧ
1 000 000 000 = 10^{9}	giga	G	jĭg′ ȧ
1 000 000 = 10^{6}	mega	M	mĕg′ ȧ
1 000 = 10^{3}	kilo	k	kĭl′ ō
100 = 10^{2}	hecto	h	hĕk′ tō
10 = 10^{1}	deka	da	dĕk′ ȧ
0.1 = 10^{-1}	deci	d	dĕs′ ĭ
0.01 = 10^{-2}	centi	c	sĕn′ tĭ
0.001 = 10^{-3}	milli	m	mĭl′ ĭ
0.000 001 = 10^{-6}	micro	µ	mī′ kro
0.000 000 001 = 10^{-9}	nano	n	năn′ ō
0.000 000 000 001 = 10^{-12}	pico	p	pē′ cō
0.000 000 000 000 001 = 10^{-15}	femto	f	fĕm′ tō
0.000 000 000 000 000 001 = 10^{-18}	atto	a	ăt′ tō

Decimal multiples and sub-multiples of the SI units are formed by means of the prefixes given in Table 2.

SOME RULES GOVERNING THE USE OF UNIT AND PREFIX SYMBOLS

1. Symbols are printed in roman (upright) type without spacing between prefix symbol and the unit symbol.

2. Only one prefix is applied at one time to a given unit.
   Example:
   nanometre (nm),
   not millimicrometre (mµm).

3. In the particular case of the kilogram the root name to which the prefix is applied is gram because only one prefix should be used.
   Examples:
   milligram (mg), not microkilogram (µkg)
   megagram (Mg), not kilokilogram (kkg)

4. The prefix symbol is considered to be combined with the unit symbol which it immediately precedes, forming with it a new symbol. This new symbol can then be raised to a positive or negative power and can be combined with other symbols to form compound symbols.
   Examples:

   $1\ cm^3 = 1\ (cm)^3 = (10^{-2}m)^3 = 10^{-6}m^3$
   $1\ \mu s^{-1} = 1\ (\mu s)^{-1} = (10^{-6}s)^{-1} = 10^6 s^{-1}$
   $1\ mm^2/s = 1\ (mm)^2/s = (10^{-3}m)^2/s = 10^{-6}m^2{\cdot}s^{-1}$

5. It is recommended that only one prefix should be used in forming decimal multiples or sub-multiples of any SI unit. This prefix should be attached only to a unit in the numerator. An exception to this occurs when the base unit, kilogram, appears in the denominator.
   Examples:
   µN•m, not mN•mm
   mΩ/m, not Ω/km,
   J/kg showing the exception.

6. The choice of the appropriate multiple (decimal multiple or sub-multiple) of an SI unit is governed by convenience, the multiple chosen for a particular application being the one which will lead to numerical values within a practical range. The use of prefixes representing 10 raised to a power which is a multiple of 3 is recommended.

7. The multiple can usually be chosen so that the numerical values will be between 0.1 and 1000.

   Examples:

   12 kN for 12 000 N
   3.94 mm for 0.003 94 m
   14.94 kPa for 14 940 Pa

   However, in a table of values for the same quantity or in a discussion of such values within a given context, it will generally be better to use the same multiple for all items, even when some of the numerical values will be outside the range 0.1 to 1000. For certain quantities in particular applications, the same multiple is customarily used.

8. All prefixes are accented on the first syllable including kilometre.

9. Pronunciations are in accord with the 1973 edition of Webster's New Collegiate Dictionary.

10. Notes on Pronunciation of Prefixes

    ȧ is pronounced like "a" in abut.

    ă is pronounced like "a" in jack.

    ĕ is pronounced like "e" in eh or get.

    ĭ is pronounced like "i" in kill or mill.

11. Symbols are always in the singular: 25 kg not 25 kgs.

12. The period is not used after symbols of SI units, e.g. 25 kg, 76 ml, 2 Hz, as these are symbols and not abbreviations.

13. The product of two or more unit symbols is indicated by a multiplication dot between the unit symbols, e.g. N•m for newton metre.

14. Use a space not a comma to separate each group of three digits to the right or left of the decimal point (See Table 2).

15. Where a figure is less than unity, a zero should be inserted before the decimal point e.g. 0.4, 0.00231.

16. Where material to be typed contains exponential factors e.g. 10^6 or km^2, it is recommended that one and a half or double line spacing be used throughout.

TABLE 3. RECOMMENDED SI USAGE

QUANTITY & APPLICATION	UNITS TO BE MULTIPLIED — USUAL CUSTOMARY UNITS		BY MULTIPLIER	TO OBTAIN NO. OF RECOMMENDED SI UNITS		PERMISSIBLE ALTERNATE UNITS
	NAME	SYMBOL	FACTOR	NAME	SYMBOL	
LINEAR MEASUREMENT	mile	mi	1.609 34	kilometre	km	
	chain		20.116 8	metre	m	
	yard	yd	0.914 4	metre	m	
	foot	ft	0.304 8	metre	m	
	inch	in	25.40	millimetre	mm	
Pipe diameters				millimetre	mm	
				metre	m	
Pipe and channel lengths				metre	m	
				kilometre	km	
Radius of groundwater wells				millimetre	mm	
Land distances				metre	m	
				kilometre	km	
Engineering drawings				millimetre	mm	
				metre	m	
-				kilometre	km	
Contour elevation differences				metre	m	
Drawdown in wells				metre	m	
Height of capillary rise				millimetre	mm	
Depth of rainfall				millimetre	mm	
Depth of tanks, basins, reservoirs				metre	m	
Potentiometric head				metre	m	
Hydraulic head				metre	m	
Piezometric head				metre	m	
Level rods				metre & 0.01 m		
Staff gages				metre & 0.01 m		
ANGULAR MEASUREMENT						
Geometric measurement	degree	°	1	degree decimalized	°	
	minute	'	0.016 67	" "	°	
	second	"	0.000 2777	" "	°	
Rotational dynamics	radian		1	radian	rad	

TABLE 3. (Continued)

QUANTITY & APPLICATION	UNITS TO BE MULTIPLIED — USUAL CUSTOMARY UNITS		BY MULTIPLIER FACTOR	TO OBTAIN NO. OF RECOMMENDED SI UNITS		PERMISSIBLE ALTERNATE UNITS
	NAME	SYMBOL		NAME	SYMBOL	
AREA	square mile	sq mi	2.590 0	kilometre²	km²	
	acre	ac	0.404 7	hectare	ha	
		ac	4 046.9	metre²	m²	
	square yard	sq yd	0.836 1	metre²	m²	
	square foot	sq ft	0.092 9	metre²	m²	
	square inch	sq in	645.2	millimetre²	mm²	
Land area	square foot	sq ft	0.092 9	metre²	m²	
	acre	ac	0.404 7	hectometre²	hm²	hectare (ha)
				kilometre²	km²	
Surface area of lagoons, lakes				metre²	m²	
				hectometre²	hm²	hectare (ha)
				kilometre²	km²	
Surface area of reservoirs, tanks				metre²	m²	
x-section of small pipes				millimetre²	mm²	
x-section of large pipes				metre²	m²	
x-section of channels				metre²	m²	
x-section of steel shapes				millimetre²	mm²	
VOLUME	cubic yard	cu yd	0.764 6	metre³	m³	
	cubic foot	cu ft	0.028 3	metre³	m³	
			28.316 9	*litre	l	
	cubic inch	cu in	16 387	millimetre³	mm³	
			16.387	millilitre	ml	
	gallon, U.K.		4.546 1	litre	l	
	gallon, U.S.	gal	3.785 4	litre	l	
	quart, U.S.		0.946 4	litre	l	
	pint, U.S.		0.473 2	litre	l	
	fluid ounce, U.S.		29.573 5	millilitre	ml	
	bushel, U.S.		0.035 2	metre³	m³	
	barrel (petroleum)		0.159 0	metre³	m³	
Water distribution & billing				metre³	m³	
Domestic water supply				litre	l	
Wastewater				metre³	m³	
Irrigation diversions				metre³	m³	
Storage capacity				metre³	m³	
Pump displacement				millimetre³	mm³	
Engine displacement				litre	l	

*litre is generally used only to measure liquids; except that it is used to measure some granular materials, and engine displacements.

TABLE 3. (Continued)

QUANTITY & APPLICATION	UNITS TO BE MULTIPLIED USUAL CUSTOMARY UNITS NAME	SYMBOL	BY MULTIPLIER FACTOR	TO OBTAIN NO. OF RECOMMENDED SI UNITS NAME	SYMBOL	PERMISSIBLE ALTERNATE UNITS
TIME	second			second	s	
	minute					min
	hour					h
	day			day	d	
	year			year	a	
VELOCITY				metre/second	m/s	
Flowing water	foot/second		0.304 8	metre/second	m/s	
Vehicles	mile/hour	mph	1.609 3	kilometre/hour	km/h	
Angular	radian/second			radian/second	rad/s	
ACCELERATION	feet/sec/sec		0.304 8	metre/second^2	m/s^2	
Vehicles	mi/hr/sec	mph/sec	1.609 3	metre/second^2	m/s^2	
Gravity	32.2 ft/sec/sec		0.304 8	9.8067m/s^2		
MASS	ton, long(2240 lb)		1 016.047	kilogram	kg	
	ton,short(2000 lb)		907.185	kilogram	kg	
	ton, metric	t	1.000	megagram	Mg	metric ton(t)
	pound	lb	0.453 6	kilogram	kg	
	ounce	oz	0.028 35	kilogram	kg	
			28.3	gram	g	
Vehicles				megagrams	Mg	metric ton(t)
Loadings				kilograms	kg	
				megagrams	Mg	
MASS PER UNIT AREA	ton/sq mi		350.27	kilogram/kilometre^2	kg/km^2	
	pound/sq ft		4.882	kilogram/metre^2	kg/m^2	
Floor loadings	pound/sq ft		4.882	kilogram/metre^2	kg/m^2	
Soil loadings	ton/sq ft		9.765	megagram/metre^2	Mg/m^2	t/m^2
DENSITY (MASS PER UNIT VOLUME)	ton/cu yd		1 187	kilogram/metre^3	kg/m^3	
	pound/cu ft		16.02	kilogram/metre^3	kg/m^3	
	pound/cu in		27 680	kilogram/metre^3	kg/m^3	
Material: steel, sand, lime				kilo	kg/m^3	
Liquid: fuel, water					kg/l	

TABLE 3. (Continued)

QUANTITY & APPLICATION	UNITS TO BE MULTIPLIED USUAL CUSTOMARY UNITS NAME	SYMBOL	BY MULTIPLIER FACTOR	TO OBTAIN NO. OF RECOMMENDED SI UNITS NAME	SYMBOL	PERMISSIBLE ALTERNATE UNITS
RATE OF FLOW	cu ft/min	cfm	0.471 95	litre/second	1/s	
	cu ft/sec	cusec	0.028 32	metre^3/second	m^3/s	
	gal/day	gpd	0.003 79	metre^3/day	m^3/d	
	gal/hour		1.051 5	metre^3/second	m^3/s	
	gal/min	gpm	0.063 09	litre/second	1/s	
	gal/sec		3.785 4	litre/second	1/s	
Domestic water and wastewater	gal/capita/day		3.785 4	litre/capita day	1/cap•d	
	gal/day	gpd	0.003 79	kilolitre/day	kl/d	m^3/d
	million gal/day	mgd	0.043 81	metre^3/second	m^3/s	
Channel, stream or river flow				metre^3/second	m^3/s	
Sludge flow				metre^3/second	m^3/s	
Pipe flow				metre^3/second	m^3/s	
Pump discharge				metre^3/hour	m^3/h	
PUMP DISCHARGE CHARACTERISTICS						
Discharge vs. dynamic head				m^3/hr vs. m head		
Energy per unit volume				kilojoules /m^3	kJ/m^3	
Power vs. discharge				watt vs. m^3/h	W vs. m^3/h	
				%efficiency vs. m^3/h	% vs. m^3/h	
CONCENTRATION OF MATERIAL IN FLUID					mg/1	µg/1
					g/m^3	
VOLUME LOADING RATES	gal/acre/day		0.009 35	metre^3/hectare day	m^3/ha•d	
	gal/sq ft/day		0.040 75	metre^3/metre^2 day	m^3/m^2•d	
	gal/sq ft/min		0.679 1	litre/metre^2 sec	1/m^2•s	
	gal/cu ft/day		0.133 7	metre^3/metre^3 day	m^3/m^3•d	
	cu ft/1000 cu ft min		0.016 67	litre/metre^3 sec	1/m^3•s	
	cu ft/acre/sec		0.069 98	metre^3/hectare sec	m^3/ha•s	
Hydraulic load/unit area of filters				litre/m^2 day	1/m^2•d	
Hydraulic load/unit volume of filters				litre/m^3 day	1/m^3•d	

TABLE 3. (Continued)

QUANTITY & APPLICATION	UNITS TO BE MULTIPLIED USUAL CUSTOMARY UNITS NAME	SYMBOL	BY MULTIPLIER FACTOR	TO OBTAIN NO. OF RECOMMENDED SI UNITS NAME	SYMBOL	PERMISSIBLE ALTERNATE UNITS
MASS LOADING RATE	lb/cu ft/day		16.018		$kg/m^3 \cdot d$	
	lb/1000 cu ft/day		0.016 02		$kg/m^3 \cdot d$	
	lb/sq ft/day		4.882		$kg/m^2 \cdot d$	
	lb/ac/day		1.121		$kg/ha \cdot d$	
BOD loading in volumes					$mg/l \cdot d$	
BOD loading on areas					$g/m^2 \cdot d$	
					$mg/m^2 \cdot d$	
FORCE	pound force	lbf	4.448 22	newton	N	
	kip force		4.448 22	kilonewtons	kN	
	ton force (short ton)		9.964.44	kilonewtons	kN	
	dyne		0.01	kilonewtons	kN	
	poundal		0.138 26	newton	N	
	kilogram force		9.806 65	newton	N	
PRESSURE	lbf/sq.ft.	psf	0.047 88	kilopascal	kPa	kN/m^2
	lbf/sq.in.	psi	6.894 76	kilopascal	kPa	kN/m^2
	bar*		100	kilopascal	kPa	
Normal atmosphere*	atmosphere	atm	101.325	kilopascal	kPa	
	ton force/sq ft		95.76	kilopascal	kPa	
	ton force sq in		13790.	kilopascal	kPa	
head of water, 39.2°F	feet of head		2.988 98	kilopascal	kPa	
head of water, 4°C	metre of head		0.09 806	kilopascal	kPa	
head of mercury, 60°F	inch of head		3.376 85	kilopascal	kPa	
head of mercury, 0°C	millimetre of head		0.133 32	kilopascal	kPa	
Pipe system pressures				kilopascal	kPa	
Pumping heads (water)				metre of head	m	
STRESS	lbf/sq ft	psf	0.047 88	kilopascal	kPa	
	lbf/sq in	psi	6.894 78	kilopascal	kPa	
	kgf/sq cm		98.066 5	kilopascal	kPa	MPa, GPa

*Although the "bar" ($10^5 N/m^2$) is used in Europe and in meteorological and aerospace activities in the U.S., its further use is to be discouraged, since much confusion can result from the fact that the bar is 100 kPa and 0.1 MPa.

TABLE 3. (Continued)

QUANTITY & APPLICATION	UNITS TO BE MULTIPLIED: USUAL CUSTOMARY UNITS		BY	TO OBTAIN NO. OF RECOMMENDED SI UNITS		PERMISSIBLE ALTERNATE UNITS
	NAME	SYMBOL	MULTIPLIER FACTOR	NAME	SYMBOL	
HEAT, WORK OR ENERGY	British Thermal Unit	BTU	1.055 06	kilojoule	kJ	
	calorie	cal	4.186 8	joule	J	
	erg		10	micro joule	µJ	
	foot pound force		1.355 82	joule	J	
	kilowatt hour		3.6	megajoule	MJ	
	watt second		1.0	joule	J	
Electrical current measure	kilowatt hour		1.0	kilowatt hour	kW•h	
Quantity of heat	BTU		1.055 06	kilojoule	kJ	
POWER	BTU/hour		0.293 07	watt	W	
	ft lbf/hour		37.662	kilowatt	kW	
	ft lbf/sec		1.355 82	watt	W	
	horsepower	hp	0.745 7	kilowatt	kW	
TEMPERATURE	degree fahrenheit	°F	$(t_F+459.67)/1.8$	kelvin	K	
	degree Celsius	°C	$t_C+273.15$	kelvin	K	
	degree fahrenheit	°F	$(t_F - 32)/1.8$	degree Celsius	°C	
VISCOSITY						
Dynamic	poise		0.1	pascal second	Pa•s	
	lbf sec/ft		47.880 26	pascal second	Pa•s	
Kinematic	stokes		0.000 10	metre^2/second	m^2/s	
	sq ft/sec		0.092 90	metre^2/second	m^2/s	
BENDING MOMENT AND TORQUE	lbf foot		1.356	newton metre	N•m	
	lbf inch		0.113	newton metre	N•m	
	tonf foot		2.712	kilonewton metre	kN•m	
	tonf inch		0.2260	kilonewton metre	kN•m	
SECTION MODULUS	inch^3		16 390	millimetre^3	mm^3	
SECOND MOMENT OF AREA	inch^4		416 200	millimetre^4	mm^4	

TABLE 3. (Continued)

QUANTITY & APPLICATION	UNITS TO BE MULTIPLIED: USUAL CUSTOMARY UNITS		BY	TO OBTAIN NO. OF: RECOMMENDED SI UNITS		PERMISSIBLE
	NAME	SYMBOL	MULTIPLIER FACTOR	NAME	SYMBOL	ALTERNATE UNITS
GRADES AND GRADIENTS				slope ratio	m/m	
AIR FLOW RATES				kilogram/second	kg/s	
DIGESTION TANK CAPACITY	cu ft/capita		0.092 9	metre^3/person	m^3/per	
COEFFICIENT OF COMPRESSIBILITY	sq ft/tonf		0.010 44	metre^2/kilonewton	m^2/kN	
COEFFICIENT OF CONSOLIDATION	sq ft/year		1.075	millimetre^2/second	mm^2/s	
COEFFICIENT OF PERMEABILITY	ft/year		3.528	micrometre/second	μm/s	μ/s
MOMENTUM	pound ft/second		0,138 3	kilogram metre/s	kg•m/s	
MOMENT OF INERTIA	pound foot squared		0.042 14	kilogram metre^2	kg•m^2	
THERMAL CONDUCTIVITY	BTU/hr/ft/°F		0.532 4	watt/metre kelvin	W/m•K	
COEFFICIENT OF HEAT TRANSFER	BTU/hr/sq ft/°F		1.750 5	watt/metre^{2}kelvin	W/m^2•K	

FORCE

$\underline{F} = \underline{m} \cdot \underline{a}$ = (mass)(acceleration)
= (kilogram)(metre/second · second)
= $kg \cdot m/s^2$ = N = newtons

WEIGHT

$\underline{W}$ = gravitational force
= $\underline{m} \cdot \underline{g}$ = (mass)(gravitational attraction)
= $kg \cdot m/s^2$ = N = newtons

PRESSURE, STRESS OR MODULUS

$\underline{P},\underline{S},\underline{E} = \underline{m} \cdot \underline{a}/\underline{A}$ = (force)/(unit area)
= $(kg \cdot m/s^2)/m^2$
= N/m^2 = Pa = pascals

POTENTIAL ENERGY IN FLUIDS

$\underline{E}_{po} = \underline{m} \cdot \underline{g} \cdot \underline{z}$ = (force) (distance above datum)
= (mass at point)(gravity)(elev. of point)
= $kg \cdot m/s^2 \cdot m$
= N · m = J = joules

POTENTIAL HEAD IN FLUIDS

$\underline{H}_{po} = \underline{E}_{po}$/unit of mass = J/kg
= $\underline{m} \cdot \underline{g} \cdot \underline{z}/\underline{m}$
= kg · 9.81 · $\underline{z}$/kg
= 9.81 $\underline{z}$ = meters per unit of mass

PRESSURE ENERGY IN FLUIDS

$\underline{E}_{pr} = m \cdot \underline{g} \cdot z$ = (volume)(density)(gravity)(depth of fluid)
= $(m^3)(kg/m^3)(m/s^2)(m)$
= N·m = J = joules

PRESSURE HEAD IN FLUIDS

$\underline{H}_{pr} = E_{pr}$/unit of mass = J/kg
= $\underline{m} \cdot \underline{g} \cdot \underline{z}/\underline{m}$
= $\underline{g} \cdot \underline{z}$ = 9.81 $\underline{z}$ = metres per unit of mass

KINETIC (VELOCITY) ENERGY IN FLUIDS

$\underline{E}_v = \underline{m}\ \underline{g}\ \underline{z}$ (volume passing point in 1 second)·(density)
·(gravity)·(height z above point required
to reach v in 1 second)
= $(m^3 \cdot kg/m^3)(g)(V^2/2g)$ = $(kg)(v^2/2)$ = $\underline{m}\ \underline{v}^2/\underline{Z}$
= N·m = J

VELOCITY HEAD IN FLUIDS

$\underline{H}_v = \underline{E}_v$/unit of mass
= $(\underline{m} \cdot \underline{v}^2/2)/(\underline{m})$
= $\underline{v}^2/2$ = metres per unit of mass

WORK, ENERGY, HEAT

$\underline{W}$ = (force)(distance moved in direction of force)
= (newton)(metre)
= $(kg\ m/s^2)(m)$ = $(kg \cdot m^2/s^2)$
= J = joules

POWER

$\underline{P}$ = work done per unit of time
= $(kg \cdot m^2/s^2)/(s)$ = $kg \cdot m^2/s^3$
= J/s = W = watts

PUMP POWER REQUIREMENT

$\underline{P}$ = (Flow rate)(density)(gravity)(dynamic head)/(efficiency)
= $Q \cdot \rho \cdot \underline{g} \cdot H$/ratio
= $(m^3/s)(kg/m^3)(m/s^2)(m)$
= $kg \cdot m^2/s^3$ = W

TORQUE

$\underline{T}$ = (force)(radius at which torque acts)
= $\underline{F} \cdot \underline{r}$ = $(kg \cdot m/s^2)(m)$ = N · m*

MOMENT

$\underline{M}$ = (force)(length of moment arm)
= $\underline{F} \cdot \underline{x}$ = $(kg \cdot m/s^2)(m)$ = N · m*

BENDING MOMENT IN BEAM (uniformly distributed load)

$\underline{M}$ = (1/8)(force per unit of length)$(span^2)$
= $\underline{f}\ \underline{L}^2/8$ = $(N/m)(m^2)$ = N · m*

MANNING FORMULA FOR FLOW IN CHANNELS

$v = 0.01\ r^{0.67} s^{0.5}/n$

where v = mean velocity in m/s
n = coefficient of friction
r = hydraulic radius
= $\frac{\text{cross sectional area of flow}}{\text{wetted perimeter of flow}}$ in mm
s = slope of energy gradient

MANNING FORMULA FOR PIPES FLOWING FULL OR HALF FULL

$v = 0.00397\ D^{0.67} s^{-0.5}/n$

where v = mean velocity in m/s
n = coefficient of friction
D = diameter in mm
s = slope of energy gradient

HAZEN-WILLIAMS FORMULA FOR HEAD LOSS

$h = 2.97 \times 10^4\ (Q/C)^{1.85}/d^{4.87}$

where h = friction head in m/km
Q = flow in m^3/s
d = inside pipe diameter in mm
C = friction coefficient

HAZEN-WILLIAMS FORMULA FOR VELOCITY

$v = 10\ 930(C)(D/4)^{0.63}(s)^{0.54}$

where v = velocity in m/s
C = friction coefficient
D = inside diameter in m
s = slope of energy gradient

90° V-NOTCH WEIR

$Q = 1.38\ H^{2.5}$

where Q = flow in m^3/s
H = hydraulic head above bottom of notch in m

SUPRESSED RECTANGULAR SHARP-CRESTED WEIR

$Q = 1.84\ LH^{1.5}$

where Q = flow in m^3/s
L = length of weir in m
H = hydraulic head above weir in m

RECTANGULAR CONTRACTED WEIR

$Q = 1.84\ (L - 0.2H)\ H^{1.5}$

where Q = flow in m^3/s
L = length of weir in m
H = hydraulic head above weir in m

LENGTH OF SIDE WEIR

$L = 14.9\ wvH_1^{0.13}(H_2^{-0.5} - H_1^{-0.5})$

where L = length of weir in mm
w = width of channel in mm
V = mean velocity in m/s
H_1 = incoming head above weir
H_2 = outgoing head above weir in mm

ORIFICE

$Q = 4.43\ C_D\ AH^{0.5}$

where Q = flow in m^3/s
A = area of orifice in m^2
H = height of water above center of orifice in m

WALL THICKNESS OF PIPE

$t = Pd/2S$

where t = wall thickness in mm
P = internal pressure in kPa
d = internal diameter in mm
S = allowable tensile strength of pipe material in kPa

MOMENT OF INTERTIA OF RECTANGULAR SECTION

$I = \frac{bd^3}{12}$

where I = moment of intertia in mm^4
b = breadth in mm
d = depth in mm

SECTION MODULUS

$Z = I/y$

where Z = section modulus in mm^3
I = moment of inertia in mm^4
y = distance from the neutral axis to the extreme fiber in mm

RADIUS OF GYRATION OF SECTION

$k^2 = I/A$

where k = radius of gyration in mm
I = moment of inertia in mm^4
A = breadth x width in mm^2

EULER'S COLUMN FORMULA----- Critical Load

$P = \pi^2\ EI^1/L^2$

where P = column axial load in N
E = modulus of elasticity in Pa
I^1 = column area moment of inertia in m^4
L = unsupported length of column in m

*Moment and torque in this content are a product of a force and distance perpendicular to the line of action of the force, hence the result is not an energy measurement and may not be expressed in joules. If rotation occurs, as from the application of torque, then the rotation in radians is taken into account and the result (N·m/rad)(rad) is an energy expression and is measured in joules.

TABLE 5.
CONVERSIONS FROM CUSTOMARY UNITS TO SI METRIC

Units to be Multiplied	By	To Obtain Equivalent No. of
LENGTH (metre)		
mile, mi	1.609	kilometre
chain	20.116 8	metre, m
yard, yd	0.914 4	metre, m
foot, ft	0.304 8	metre, m
inch, in	25.40	millimetre, mm
AREA (metre^2 or hectare)		
square mile, sq mi	2.590	kilometre^2, km^2
acre, ac	0.404 7	hectare, ha
	4 046.9	metre^2, m^2
square yard, sq yd	0.836 1	metre^2, m^2
square foot, sq ft	0.092 9	metre^2, m^2
square inch, sq in	645.2	millimetre^2, mm^2
VOLUME (metre^3 or litre)		
cubic yard, cu yd	0.764 6	metre^3, m^3
cubic foot, cu ft	0.028 3	metre^3, m^3
	28.32	litre, l
cubic inch, cu in	16 387	millimetre^3, mm^3
	0.016 39	millilitre, ml
gallon, UK	4.546 1	litre, l
gallon, US	3.785 4	litre, l
quart, US	0.946 4	litre, l
pint, US	0.473 2	litre, l
fluid ounce, US	29.573 5	millilitre, ml
bushel, US	0.035 24	metre^3, m^3
barrel (petroleum)	0.159	metre^3, m^3
MASS (kilogram or tonne)		
ton (2,240 lb)	1 016.047	kilogram, kg
ton (2,000 lb)	907.185	kilogram, kg
ton (metric)	1.000	tonne, t; megagram, Mg
pound, lb	0.453 6	kilogram, kg
ounce, oz	0.028 35	kilogram, kg
MASS PER UNIT AREA		
ton/sq mi	350.27	kilogram/kilometre^2, kg/km^2
pound/sq ft	4.882	kilogram/metre^2, kg/m^2
DENSITY		
ton/cu yd	1186.6	kilogram/metre^3, kg/m^3
pound/cu ft	16.02	kilogram/metre^3, kg/m^3
pound/cu in	27 680	kilogram/metre^3, kg/m^3
VOLUME RATE OF FLOW		
cubic feet/min	0.471 95	litre/second, l/s
cu ft/sec (cusec)	0.028 32	metre^3/second (cumec), m^3/s
gallon/day (gpd)	0.003785	metre^3/day, m^3/d
gallon/hour	1.051 5	metre^3/sec, m^3/s
gallon/minute	0.063 09	litre/sec, l/s
gallon/second	3.785	litre/sec, l/s
VOLUME LOADING RATE		
million gallon/day (mgd)	0.043 81	metre^3/sec, m^3/s
gallon/acre	0.009 354	metre^3/hectare, m /ha
gallon/sq ft/day	0.040 746	metre^3/metre^2 day, m^3/m^2 d
gallon/sq ft/min	40.746	litre/metre^2 min, l/m^2 min
gallon/cu ft/day	0.133 7	metre^3/metre^3 day, m^3/m^3 d
cu ft/1,000 cu ft/min	0.016 67	litre/metre^3 sec, l/m^3·s
cu ft/acre/sec	0.069 97	metre^3/ha sec, m^3/ha·s
cu ft/sq ft	0.3048	metre^3/metre^2 m^3/m^2

Units to be Multiplied	By	To Obtain Equivalent No. of
MASS LOADING RATE		
lb/cu ft/day	16.018	kg/m^3 d
lb/sq ft/day	4.882	kg/m^2 d
lb/acre/day	1.1209	kg/ha d
lb/1,000 cu ft/day	0.016 02	kg/m^3 d
FORCE (newton)		
pound force, lbf	4.448	newton, N
kip force	4.448	kilonewton, kN
ton force (2,000 lbf)	8.896	kilonewton, kN
PRESSURE (pascal, newton/metre^2 or bar)		
lbf/sq ft (psf)	0.047 88	kilopascal, kPa (or kN/m^2)
lbf/sq in (psi)	6.895	kilopascal, kPa (or kN/m^2)
ton force/sq ft	95.76	kilopascal, kPa (or kN/m^2)
ton force/sq in	13 790	kilopascal, kPa (or kN/m^2)
bar	100	kilopascal, kPa (or kN/m^2)
ft head of water (39.2F)	2.988 98	kilopascal, kPa
mm head of Hg	0.133 3	kilopascal, kPa
STRESS (pascal or newton/metre^2)		
lbf/sq ft (psf)	0.047 88	kilopascal, kPa (or kN/m^2
lbf/sq in (psi)	6.895	kilopascal, kPa (or kN/m^2
HEAT, WORK OR ENERGY (joule)		
ft lbf	1.355 8	joule, J
kwhr	3.6	megajoule, MJ
calorie	4.187	joule, J
BTU	1.055	kilojoule, kJ
POWER (watt)		
ft lbf/sec	1.355 82	watt, W
BTU/hour	0.293 07	watt, W
horsepower, hp	0.745 7	kilowatt, kW
VISCOSITY (dynamic)		
poise	0.100	pascal second, Pa·s
lbf sec/ft^2	47.880	pascal second, Pa·s
VISCOSITY (kinematic)		
stokes	0.000 10	metre^2/second, m^2/s
BENDING MOMENT AND TORQUE		
pound force foot	1.356	newton metre, N·m
pound force inch	0.113 0	newton metre, N·m
ton force foot	2.712	kilonewton metre, kN·m
ton force inch	0.2260	kilonewton metre, kN·m
SECTION MODULUS		
inch^3	16 390	millimetre^3, mm^3
SECOND MOMENT OF AREA		
inch^4	416 200	millimetre^4, mm^4
COEFFICIENT OF COMPRESSIBILITY		
sq ft/ton	0.01044	m^2/kN
COEFFICIENT OF CONSOLIDATION		
sq ft/year	1.075	mm^2/s
COEFFICIENT OF PERMEABILITY		
ft/year	3.528	μm/s

RECOMMENDED METRIC UNITS IN CONSTRUCTION	Customary Unit	SI Unit	Working Units
I. Design			
A. Units			
1. Linear Measure	mile	m	km
	yd	m	m
	ft	m	m
	in	m	mm
2. Plane Angle	degree	rad	rad, decimal degree (degree-min-sec)
3. Area	$mile^2$	m^2	km^2
	acre	m^2	ha
	yd^2	m^2	m^2
	ft^2	m^2	m^2
	in^2	m^2	mm^2
4. Volume	yd^3, ft^3, in^3	m^3	m^3, liter (l)
5. Solid angle	sr	sr	sr
6. Mass	lb, [a]slug	kg	kg, Mg, t
7. Time	sec	s	s, h, d,
8. Force	lb,[b] kip, tonf	N	N, kN, Mn
9. Temperature	°F	K	K, °C
10. Energy	kWh	J	MJ
	Btu	J	kJ
	ft lb[b]	J	MJ
11. Luminous Intensity	cd	cd	cd
12. Electric Current	A	A	A, kA, MA
13. Power	ft lb[b]/sec, hp	W	W, kW, MW
B. General Application			
1. Loading			
a. Loads (mass)	lb[a]	kg	kg, Mg, t
Uniform Distrib.	lb[a]/ft; lb[a]/ft^2	kg/m^2	kg/m; kg/m^2
b. Computations (force)	lb[b]	N	kN, MN
Unit Force (pressure, stress)	lb[b]/in^2	Pa	(N/mm^2)*, Pa, kPa, MPa
2. Velocity	ft/sec, mph	m:s	m/s, (km/h)*, km/s
3. Density	lb[b]/ft^3	kg/m^3	kg/m^3
4. External Forces			
a. Reaction	kip	N	N, kN, MN
b. Shear	kip	N	N, kN, MN
c. Moment, Torque	kip-ft	N·m	N·m, kN·m, MN·m
5. Internal Resistance			
Tension, Compression, Shear	lb[b]/in^2	Pa	(N/mm^2),* MPa
C. Buildings and Bridges			
1. Cross-Sectional Properties			
a. Second Moment of Area (Moment of Inertia)	in^4	m^4	(mm^{4c}), m^4
b. Section Modulus	in^3	m^3	(mm^{3d})
c. Radius of Gyration	in	m	mm
d. Torsional Constant	in^4	m^4	mm^{4c}
e. Warping Constant	in^6	m^6	mm^{6**}
f. Modulus of Elasticity	lb[b]/in^2	Pa	(N/mm^2),* MPa,[e] GPa
2. Deflection	in	m	mm
3. End Rotation	rad	rad	rad
D. Foundations			
1. Pile Load	lb,[b] ton	N	N, kN, MN
2. Bond Stress	lb[b]/in^2	Pa	MPa
3. Settlement	in	m	mm

Reproduced by permission of the American Institute of Steel Construction; modified by the author.

RECOMMENDED METRIC UNITS IN CONSTRUCTION	Customary Unit	SI Unit	Working Units
E. *Hydraulics*			
1. Head	ft	m	m
2. Flow, quantity and discharge	gpm:ft^3/s	m^3/s	m^3/s, (l/s)*
3. Dynamic viscosity	lb sec/in^2	Pa·s	Pa·s, kPa·s, mPa·s
4. Kinematic viscosity	in^2/sec, ft^2/sec	m^2/s	(mm^2/s)* m^2/s
F. *Heat*			
1. Specific Heat	Btu/lb[a] °F	J/kg·K	J/kg·K, kj/kg·K
2. Coefficient of heat transfer and U-factor	Btu/hr ft^2 °F	W/(m^2·K)	W/m^2·K
3. Thermal conductivity, k	Btu in/hr ft^2 °F	W/(m·K)	W/m·K
4. Moisture content of air	lb/lb[b]	kg/kg	kg/kg
5. Gas constant	Btu/mol R°	J/(mol·K)	kJ/mol·K
G. *Soil Freezing and Thawing*			
1. Latent heat of fusion	Btu/lb[b]	J/kg	kJ/kg
2. Volumetric latent heat	Btu/ft^3	J/m^3	MJ/m^3
3. Volumetric specific heat capacity	Btu/ft^3 R°	J/m^3·K	MJ/(m^3·K)
H. *Electrical and Lighting*			
1. Potential Difference	V	V	V, kV, MV
2. Electric Resistance	Ω	Ω	μΩ, Ω, kΩ, MΩ, GΩ
3. Illumination	lum/ft^2	lx	lx
II. Materials			
A. *Basic*			
1. Asphalt Cutback liquids	gal, ft^3:ton	m^3	m^3, dm^3, (liter)*:(t)*
2. Bagged materials	ft^3;lb,[b] ton	kg	kg, Mg:(t)*
3. Concrete			
a. Admixtures	lb	kg	kg
b. Aggregates	ton:yd^3	kg:m^3	kg, Mg, (t)*:m^3
c. Cement	cwt, lb:bbl	kg:m^3	kg, Mg, (t)*:m^3
d. Volume	gal, ft^3, yd^3	m	m^3
e. Mass	lb[a]	kg	kg, Mg:(t)*
f. Water	gal, ft^3	m	dm^3, (liter),* m^3, km^3
g. Slump	in	m	mm
h. Concentration	lb[b]/ft^3	kg/m^3	kg/m^3, (mg/liter)*
4. Dampproofing	in; ft^2	m^2	mm, m; m^2
5. Glass	in, ft; ft^2 united inch (l + W)	m^2	mm, m; m^2
6. Insulation	lb:in; ft^2	m^2; m^3	kg:mm; m^2
7. Iron and Steel	lb, ton	kg	kg, Mg, (t)*
8. Lime	lb, ton	kg	kg, Mg, (t)*
9. Mortar	lb	kg	kg
10. Plaster	lb, ton:ft^2, ft^3	m^2:m^3	kg, Mg, (t)*:m^2; m^3
11. Rip-rap	ton:yd^2; yd^3	m^3	m^3:kg, Mg, (t)*
12. Stone	ton:yd^2; yd^3	m^2; m^3	kg, Mg, (t)*:mm, m;m^2; m^3
13. Timber	ft; board ft	m, m^3	mm, m, m^3
B. *Fabricated*			
1. Blocks	in	m	mm
2. Boards, panels, tiles, plywood	in; in^2; ft; ft^2	m^2	mm, m; mm^2, m^2
3. Bricks, blocks	in	m	mm
4. Coverings-floor, wall	in, ft; ft^2; yd^2	m^2	mm, m; m^2

RECOMMENDED METRIC UNITS IN CONSTRUCTION	*Customary Unit*	*SI Unit*	*Working Units*
5. Coverage			
1. Liquids	gal/ft², gal/yd²	m²; m³	m³/m², (liter/m²)*
2. Solids	lb/ft²	kg/m²	m²
6. Equipment	in, ft	m	mm, m
7. Fasteners, (nails, screws, bolts, etc.)	lb:in	kg:m	kg:mm
8. Lumber	ft; board foot	m², m³	mm, m; m²; m³
9. Reinforcing steel			
a. Bars	in, ft:lb	kg/m	mm:kg
b. Mesh	in:ft²:lb	m	mm; m²:kg
c. Wire	in, ft:lb	m	mm:kg
10. Roofing	ft; ft²:gal:lb; square = 100 ft²	m²	mm, m; m²; m³, (liter):kg
11. Sheets (membrane, flashing, metal, glass, etc.)	in, ft; ft²	m²	mm, m; mm², m²
12. Siding, shingles	in; ft²; square = 100 ft²	m²	mm; m²
13. Steel			
a. Rails	lb:lb/yd	kg/m	kg, Mg, (t)*:kg/m
b. Spikes	lb	kg	kg, mm
c. Structural	lb/ft:ton	kg/m	kg/m:kg, Mg (t)*
14. Wire			
a. Diameter	AWG, in	m	mm
b. Length	ft	m	mm, m

*Units in ( ) are acceptable but not preferred.
[a] lbm implied.
[b] lbf implied.
** 10^8mm⁶ steel, 10^4mm⁶ wood.
[c] 10^6mm⁴ steel, 10^3mm⁴ wood.
[d] 10^3mm³ steel, mm³ wood.
[e] 200 000 MPa steel, 69 000 MPa aluminum.
[f] 7850 kg/m³ steel.
Key to Punctuation:
Commas separate multiples of one type of measure.
Semicolons separate linear, square, and cubic units.
Colons separate linear, square, or cubic units from mass or force units.

Strength properties of some commercially important woods grown in the United State

Common names of species	Specific gravity	Static bending: Modulus of-- Rupture	Static bending: Modulus of-- Elasticity	Static bending: Work to maximum load	Impact bending--height of drop causing complete failure	Compression parallel to grain--maximum crushing strength	Compression perpendicular to grain--fiber stress at proportional limit	Shear parallel to grain--maximum shearing strength	Tension perpendicular to grain--maximum tensile strength	Side hardness--load perpendicular to grain
		Kilo-pascals	Mega-pascals	Kilo-joules per cu. m.	Mm.	Kilo-pascals	Kilo-pascals	Kilo-pascals	Kilo-pascals	Newtons
				HARDWOODS						
Maple:										
Bigleaf	.44	51,000	7,600	60	580	22,300	3,100	7,700	4,100	2,800
	.48	74,000	10,000	54	710	41,000	5,200	11,900	3,700	3,800
Silver	.44	40,000	6,500	76	740	17,200	2,600	7,200	3,900	2,600
	.47	61,000	7,900	57	640	36,000	5,100	10,200	3,400	3,100
Oak, red:										
Northern red	.56	57,000	9,300	91	1120	23,700	4,200	8,300	5,200	4,400
	.63	99,000	12,500	100	1090	46,600	7,000	12,300	5,500	5,700
Oak, white:										
Bur	.58	50,000	6,100	74	1120	22,700	4,700	9,300	5,500	4,900
	.64	71,000	7,100	68	740	41,800	8,300	12,500	4,700	6,100
				SOFTWOODS						
Cedar:										
Alaska	.42	44,000	7,900	63	690	21,000	2,400	5,800	2,300	2,000
	.44	77,000	9,800	72	740	43,500	4,300	7,800	2,500	2,600
Atlantic white	.31	32,000	5,200	41	460	16,500	1,700	4,800	1,200	1,300
	.32	47,[illegible]00	6,400	28	330	32,400	2,800	5,500	1,500	1,600
Northern white	.29	29,000	4,400	39	380	13,700	1,600	4,300	1,700	1,000
	.31	45,000	5,500	33	300	27,300	2,100	5,900	1,700	1,400
Western redcedar	.31	35,900	6,500	34	430	19,100	1,700	5,300	1,600	1,200
	.32	51,700	7,700	40	430	31,400	3,200	6,800	1,500	1,600
Douglas-fir:										
Coast	.45	53,000	10,800	52	660	26,100	2,600	6,200	2,100	2,200
	.48	85,000	13,400	68	790	49,900	5,500	7,800	2,300	3,200
Interior North	.45	51,000	9,700	56	560	23,900	2,500	6,600	2,300	1,900
	.48	90,000	12,300	72	660	47,600	5,300	9,700	2,700	2,700
Fir:										
Balsam	.34	34,000	6,600	32	410	16,500	1,200	4,200	1,200	1,300
	.36	52,000	8,500	35	510	31,200	2,100	4,900	1,200	1,800
Noble	.37	43,000	9,500	41	480	20,800	1,900	5,500	1,600	1,300
	.39	74,000	11,900	61	580	42,100	3,600	7,200	1,500	1,800

White	.37	41,000	8,000	39	560	20,000	1,900	5,200	2,100	1,500
	.39	68,000	10,300	50	510	40,100	3,700	7,600	2,100	2,100
Hemlock:										
Eastern	.38	44,000	7,400	46	530	21,200	2,500	5,900	1,600	1,800
	.40	61,000	8,300	47	530	37,300	4,500	7,300	-----	2,200
Western	.42	46,000	9,000	48	560	23,200	1,900	5,900	2,000	1,800
	.45	78,000	11,300	57	580	49,000	3,800	8,600	2,300	2,400
Larch, western	.48	53,000	10,100	71	740	25,900	2,800	6,000	2,300	2,300
	.52	90,000	12,900	87	890	52,700	6,400	9,400	3,000	3,700
Pine:										
Loblolly	.47	50,000	9,700	57	760	24,200	2,700	5,900	1,800	2,000
	.51	88,000	12,300	72	760	49,200	5,400	9,600	3,200	3,100
Lodgepole	.38	38,000	7,400	39	510	18,000	1,700	4,700	1,500	1,500
	.41	65,000	9,200	47	510	37,000	4,200	6,100	2,000	2,100
Longleaf	.54	59,000	11,000	61	890	29,800	3,300	7,200	2,300	2,600
	.59	100,000	13,700	81	860	58,400	6,600	10,400	3,200	3,900
Ponderosa	.38	35,000	6,900	36	530	16,900	1,900	4,800	2,100	1,400
	.40	65,000	8,900	49	480	36,700	4,000	7,800	2,900	2,000
Shortleaf	.47	51,000	9,600	57	760	24,300	2,400	6,300	2,200	2,000
	.51	90,000	12,100	76	840	50,100	5,700	9,600	3,200	3,100
Redwood:										
Old-growth	.38	52,000	8,100	51	530	29,000	2,900	5,500	1,800	1,800
	.40	69,000	9,200	48	480	42,400	4,800	6,500	1,700	2,100
Young-growth	.34	41,000	6,600	39	410	21,400	1,900	6,100	2,100	1,600
	.35	54,000	7,600	36	380	36,000	3,600	7,600	1,700	1,900
Spruce:										
Engelmann	.33	32,000	7,100	35	410	15,000	1,400	4,400	1,700	1,150
	.35	64,000	8,900	44	460	30,900	2,800	8,300	2,400	1,750
White	.37	39,000	7,400	41	560	17,700	1,700	4,800	1,500	1,400
	.40	68,000	9,200	53	510	37,700	3,200	7,400	2,500	2,100

Source: Wood Handbook: Wood as an Engineering Material; Agriculture Handbook No. 72, Revised August 1974.

GALVANIZED SHEET THICKNESS TOLERANCES BASED ON ASTM A-525

Gage	Thickness In Inches			Weight		Thickness In Millimeters		
	Min.	Max.	Nom.	$\frac{lb.}{Ft^2}$	$\frac{kg}{m^2}$	Nom.	Min.	Max.
30	.0127	.0187	.0157	.656	3.20	.3988	.3188	.4788
29	.0142	.0202	.0172			.4369	.3569	.5169
28	.0157	.0217	.0187	.781	3.81	.4750	.3950	.5550
27	.0172	.0232	.0202			.5131	.4331	.5931
26	.0187	.0247	.0217	.906	4.42	.5512	.4712	.6312
25	.0207	.0287	.0247			.6274	.5274	.7274
24	.0236	.0316	.0276	1.156	5.64	.7010	.6010	.8010
23	.0266	.0346	.0306			.7772	.6772	.8772
22	.0296	.0376	.0336	1.406	6.86	.8534	.7534	.9534
21	.0326	.0406	.0366			.9296	.8296	1.0296
20	.0356	.0436	.0396	1.656	8.08	1.006	.906	1.106
19	.0406	.0506	.0456			1.158	1.028	1.288
18	.0466	.0566	.0516	2.156	10.52	1.311	1.181	1.441
17	.0525	.0625	.0575			1.461	1.331	1.591
16	.0575	.0695	.0635	2.656	12.96	1.613	1.463	1.763
15	.0650	.0770	.0710			1.803	1.653	1.953
14	.0705	.0865	.0785	3.281	16.01	1.994	1.784	2.204
13	.0854	.1014	.0934			2.372	2.162	2.582
12	.0994	.1174	.1084	4.531	22.21	2.753	2.523	2.983
11	.1143	.1323	.1233			3.132	2.902	3.362
10	.1292	.1472	.1382	5.781	28.21	3.510	3.280	3.740
9	.1442	.1622	.1532			3.891	3.661	4.121
8	.1591	.1771	.1681			4.270	4.040	4.500

Note:

Galvanized sheet gage, nominal thickness is .0037 inch greater than Manufacturers Standard Gage thickness, which is an uncoated product. Galvanized sheet nominal weight is 2.5 oz/sf heavier than a corresponding M. S. gage (uncoated).

Source: *Backing Into Metrics? A Construction Industry Survey;* Sheet Metal and Air Conditioning Contractors' National Association, Inc., 1974, p. 8.

Designation: E 380 – 72ε

American National Standard Z210.1-1973
Approved March 15, 1973
By American National Standards Institute

AMERICAN SOCIETY FOR TESTING AND MATERIALS
1916 Race St., Philadelphia, Pa., 19103
Reprinted from the Annual Book of ASTM Standards, Copyright ASTM
If not listed in the current combined Index, will appear in the next edition.

Standard

METRIC PRACTICE GUIDE*

(A Guide to the Use of SI—the International System of Units)

This Standard is issued under the fixed designation E 380; the number immediately following the designation indicates the year of original adoption or, in the case of revision, the year of last revision. A number in parentheses indicates the year of last reapproval.

This standard has been approved for use by the Department of Defense and for listing in the DoD Index of Specifications and Standards. Proposed revisions should be coordinated with the Department of Defense through the Army Materials and Mechanics Research Center, AMXMR-MS, Watertown, Mass. 02172.

ε NOTE—Editorial corrections to several conversion factors were made in June 1973, marked with a dagger (†).

FOREWORD

The ASTM Board of Directors on 18 January 1971 adopted the following resolution: "It shall be the general policy of the American Society for Testing and Materials to work toward the simplification of units of measurement. To this end, the Society should assume a position of leadership, working with other organizations toward the general national and international use of the International System of Units, generally known as SI." Recognizing the world-wide trend to the use of this modernized metric system,[1] the Society is providing these guidelines for the use of SI in its standards as well as in the standards of commerce and industry. This policy involves no change in standard dimensions, tolerances, or performance specifications.

CONTENTS

* This guide is under the jurisdiction of the ASTM Committee on Metric Practice. A list of members may be found in the ASTM Yearbook.

Current edition approved May 1, 1972. Published June 1972. Originally published in 1964 without designation as *ASTM Metric Practice Guide*, revised 1966. Adopted as standard 1968. Last previous edition E 380 – 70. Detailed instructions particularly applicable to ASTM committees are given in Instructions to ASTM Technical Committees on Metric Conversion (Order No. 13-00009-00-0), issued February 1971, which includes certain material formerly contained in E 380, in revised form.

[1] An Act of Congress in 1866 declared that "it shall be lawful throughout the United States of America to employ the weights and measures of the metric system." In 1893 an Executive Order directed that "the office of weights and measures ... will in the future regard the international prototype meter and kilogram as fundamental standards, and the customary units, the yard and the pound, will be derived therefrom in accordance with the act of July 28, 1866."

ASTM **E 380**

1. Scope

1.1 This Metric Practice Guide deals with the conversion of quantities in various systems of measurement to the International System of Units,[2] which is officially abbreviated SI in all languages. It includes units most frequently used in the various fields of science and industry.

1.2 The recommendations contained herein are based upon the following premises, which are believed to represent the broadest base for general agreement among proponents of the major metrology systems:

1.2.1 That for most scientific and technical work SI is generally superior to other systems; and that this system is more widely accepted than any other as the common language in which scientific and technical data should be expressed. This is particularly true for the fields of electrical science and technology since the common electrical units (ampere, volt, ohm, etc.) are SI units.

1.2.2 That ASTM recognizes the system described and maintained by the General Conference on Weights and Measures (CGPM) as SI, and along with limited modifications carefully described in this guide will use this system.

1.2.3 That unit usage can and should be simplified; that one means toward such simplification is the elimination of obsolete and unneeded units; and that another is a better understanding of SI.

1.2.4 That certain customary units, both U.S. and "metric," will gradually be phased out of existence, and this and other countries will use the common international measurement language, SI.

2. SI Units and Symbols

2.1 SI consists of seven base units, two supplementary units, a series of derived units consistent with the base and supplementary units, and a series of approved prefixes for the formation of multiples and submultiples of the various units. A number of derived units are listed in 2.1.3 including all approved units with special names (see also Appendix A2.13). Additional derived units without special names are formed as needed from base units or other derived units, or both.

2.1.1 *Base Units:*[3]

Quantity	Unit	SI Symbol	Formula
length	metre	m	...
mass	kilogram	kg	...
time	second	s	...
electric current	ampere	A	...
thermodynamic temperature	kelvin	K	...
amount of substance	mole	mol	...
luminous intensity	candela	cd	...

2.1.2 *Supplementary Units:*[3]

plane angle	radian	rad	...
solid angle	steradian	sr	...

2.1.3 *Derived Units:*

acceleration	metre per second squared	...	m/s^2
activity (of a radioactive source)	disintegration per second	...	(disintegration)/s
angular acceleration	radian per second squared	...	rad/s^2
angular velocity	radian per second	...	rad/s

[2] The International System of Units (Système International d'Unités) (SI) is described in Appendix A2.
[3] The seven base and two supplementary units are defined in Appendix A2.

E 380

Quantity	Unit	SI Symbol	Formula
area	square metre	. . .	m^2
density	kilogram per cubic metre	. . .	kg/m^3
electric capacitance	farad	F	$A \cdot s/V$
electrical conductance	siemens	S	A/V
electric field strength	volt per metre	. . .	V/m
electric inductance	henry	H	$V \cdot s/A$
electric potential difference	volt	V	W/A
electric resistance	ohm	Ω	V/A
electromotive force	volt	V	W/A
energy	joule	J	$N \cdot m$
entropy	joule per kelvin	. . .	J/K
force	newton	N	$kg \cdot m/s^2$
frequency	hertz	Hz	(cycle)/s
illuminance	lux	lx	lm/m^2
luminance	candela per square metre	. . .	cd/m^2
luminous flux	lumen	lm	$cd \cdot sr$
magnetic field strength	ampere per metre	. . .	A/m
magnetic flux	weber	Wb	$V \cdot s$
magnetic flux density	tesla	T	Wb/m^2
magnetomotive force	ampere	A	. . .
power	watt	W	J/s
pressure	pascal	Pa	N/m^2
quantity of electricity	coulomb	C	$A \cdot s$
quantity of heat	joule	J	$N \cdot m$
radiant intensity	watt per steradian	. . .	W/sr
specific heat	joule per kilogram-kelvin	. . .	$J/kg \cdot K$
stress	pascal	Pa	N/m^2
thermal conductivity	watt per metre-kelvin	. . .	$W/m \cdot K$
velocity	metre per second	. . .	m/s
viscosity, dynamic	pascal-second	. . .	$Pa \cdot s$
viscosity, kinematic	square metre per second	. . .	m^2/s
voltage	volt	V	W/A
volume[4]	cubic metre	. . .	m^3
wavenumber	reciprocal metre	. . .	(wave)/m
work	joule	J	$N \cdot m$

2.1.4 *SI Prefixes* (*see 3.2*):

Multiplication Factors	Prefix	SI Symbol
1 000 000 000 000 = 10^{12}	tera	T
1 000 000 000 = 10^{9}	giga	G
1 000 000 = 10^{6}	mega	M
1 000 = 10^{3}	kilo	k
100 = 10^{2}	hecto*	h
10 = 10^{1}	deka*	da
0.1 = 10^{-1}	deci*	d
0.01 = 10^{-2}	centi*	c
0.001 = 10^{-3}	milli	m
0.000 001 = 10^{-6}	micro	μ
0.000 000 001 = 10^{-9}	nano	n
0.000 000 000 001 = 10^{-12}	pico	p
0.000 000 000 000 001 = 10^{-15}	femto	f
0.000 000 000 000 000 001 = 10^{-18}	atto	a

*To be avoided where possible. See 3.2.2 and 3.2.5.

3. Rules for SI Style and Usage

3.1 *General*—The established SI units—base, supplementary, derived, and combinations thereof (see Section 2, SI Units and Symbols) with appropriate multiple or submultiple prefixes—should be used. To main-

[4] In 1964 the General Conference on Weights and Measures adopted the name litre as a special name for the cubic decimetre but discouraged the use of the name litre for volume measurement of extreme precision.

tain the coherence of the system, multiples or submultiples of SI units should not be used in combinations.

3.1.1 *Mixed Units*—The use of mixed units, especially those compounded from different systems, should be avoided. For example,

use kilogram per cubic metre (kg/m^3), not kilogram per gallon (kg/gal),
use 0.1789 rad or 10.25 deg, not 10 deg 15 min, and
use 12.75 lb, not 12 lb 12 oz.

3.2 *Application of Prefixes:*

3.2.1 *General*—SI prefixes (2.1.4) should be used to indicate orders of magnitude, thus eliminating insignificant digits and decimals, and providing a convenient substitute for writing powers of 10 as generally preferred in computation. For example,

12 300 m or 12.3×10^3 m becomes 12.3 km, and
0.0123 μA or 12.3×10^{-9} A becomes 12.3 nA.

3.2.2 *Selection*—Multiple and submultiple prefixes representing steps of 1000 are recommended. For example, show force in mN, N, kN, and length in mm, m, km, etc. The use of centimetres should be avoided unless a strong reason exists. An exception is made in the case of area and volume used alone (see 3.2.5). When expressing a quantity by a numerical value and a unit, prefixes should preferably be chosen so that the numerical value lies between 0.1 and 1000, except where certain multiples or submultiples have been agreed to for particular use. The same unit, multiple, or submultiple is used for tabular values even though the series exceeds the preferred range 0.1 to 1000.

3.2.3 Double prefixes and hyphenated prefixes should not be used. For example,

use GW (gigawatt), not kMW,
use pF (picofarad), not μμF, and
use Gg (gigagram), not Mkg.

3.2.4 *Combination Units*—Prefixes should not be used in the denominator of compound units, except for the kilogram (kg). Since the kilogram is a base unit of SI, this particular multiple is not a violation and should be used in place of the gram; for example, use 200 J/kg, not 2 dJ/g. Prefixes may be applied to the numerator of a combination unit, thus, megagram per cubic metre (Mg/m^3) or for greater clarity, mega(gram per cubic metre), but *not* kilogram per cubic decimetre (kg/dm^3) *nor* gram per cubic centimetre (g/cm^3).

3.2.5 With SI units of higher order such as m^2 and m^3, the prefix is also raised to the same order; for example, mm^3 is 10^{-9} m^3 not 10^{-3} m^3. In such cases, the use of cm^2, cm^3, dm^2, dm^3, and similar nonpreferred prefixes is permissible.

3.2.6 Among the base units of SI, the unit for mass is the only one whose name, for historical reasons, contains a prefix. Names of decimal multiples and submultiples of the unit of mass are formed by attaching prefixes to the word "gram."

3.3 *Non-SI Units*—The use of units that are not SI should be limited to those for temperature, time, and angle, given in 3.4.3, 3.4.4, and 3.4.5, respectively.

3.4 *Usage for Selected Quantities:*

3.4.1 *Mass, Force, and Weight:*

3.4.1.1 The principal departure of SI from the gravimetric form of metric engineering units is the separate and distinct units for mass and force. The kilogram is restricted to the unit of mass. The newton is the unit of force and should be used in place of kilogram-force. Likewise, the newton instead of kilogram-force should be used in combination units which include force, for example, pressure or stress (N/m^2 = Pa), energy (N·m = J), and power (N·m/s = W).

3.4.1.2 Considerable confusion exists in the use of the terms *mass* and *weight*. Mass is a property of matter to which it owes its inertia. If a body or particle of matter at rest on the earth's surface is released from the forces holding it at rest, it will experience the acceleration of free fall (acceleration of gravity, g). The force required to restrain it against free fall is commonly called weight. This force is proportional to the mass of the body and is often expressed in mass units (kg), but as it is a force it should be expressed in force units (N). The acceleration of free fall (g) varies in time and space; weight (which is proportional to it) does too, although mass does not (see Fig. 1). Observed g differs by over 0.5 percent at various points on the earth's surface; therefore the difference of local g from the agreed standard value, 9.80665 m/s^2, must be taken into account for precise measurements where g is involved. Further confusion arises in the measuring of weight because of the buoyant effect of the medium in which the weighing is performed. In common parlance the term *weight*, as of a person, is used where the technically correct word is mass. Since this non-technical use of the term weight is very common, the term should be avoided in technical practice.

3.4.1.3 Gravity is involved in weighing on balances or scales in which a standard mass balances the measured mass, either directly or indirectly. In these devices, the effect of gravity on the two masses is cancelled. Therefore, these devices are calibrated in mass units.

3.4.1.4 The relationship between newtons, kilograms, metres, and seconds given by the first law of motion ($F = ma$) should be used to simplify units (that is, $N = kg \cdot m/s^2$). Thus, the combination unit of foot-pound-force per pound-mass used to measure efficiency may be converted to newton-metre per kilogram ($N \cdot m/kg$) or joule per kilogram (J/kg), but it must not be converted to metre since pound-force and pound-mass are not equivalent units.

3.4.1.5 The use of the same name for units of force and mass causes confusion. When non-SI units are used, a distinction should be made between (*1*) force and (*2*) mass, for example, gf, kgf, or lbf to denote force in gravimetric engineering units and g, kg, or lb for mass.

3.4.2 *Length:*

3.4.2.1 Nominal dimensions name the item; no SI equivalent is required (see Appendix A1 for definition of "nominal"). For example, there is nothing 1 in. about a nominal "1-in. pipe," the dimensions of which should be converted as follows:

Nominal Size, in.	Outside Diameter, in. (mm)	Wall Thickness, in. (mm) Sch 40	Sch 80	Sch 160
1	1.315 (33.40)	0.133 (3.38)	0.179 (4.55)	0.250 (6.35)

Likewise, a "2 by 4" is that in name only and refers to the approximate dimensions in inches of a rough-sawn, green piece of timber, the finished dimensions of which are considerably less. A ¼-20 UNC screw thread should continue to be identified in this manner. However, the controlling dimensions of the part, such as the pitch, major, and minor diameters of a screw thread, should be converted to SI values in accordance with 3.1 and 3.2. Threads per inch should be converted to millimetre pitch.

3.4.2.2 When a dimension such as wire diameter or thickness of sheet and strip is expressed by a gage number, the appropriate gage system (for example, American Wire Gage or Awg) and the corresponding SI value should be given.

3.4.2.3 Surface finish expressed in microinch should be converted to nanometre (nm).

3.4.3 *Temperature*—The SI temperature scale is the International Thermodynamic Temperature Scale, and the SI unit used is the kelvin; therefore kelvins should be used to express temperature and temperature intervals as the preferred unit. However, wide use is made of the degree Celsius, particularly in engineering and in non-scientific areas, and it is permissible to use the Celsius scale where considered necessary. The Celsius scale (formerly called the centigrade scale) is related directly to the kelvin scale as follows:

One degree Celsius equals one kelvin, exactly.
A Celsius temperature (t) is related to a kelvin temperature (T) as follows:

$$T = 273.15 + t, \text{ exactly}$$

In using temperature information of extreme precision, the International Practical Temperature Scale must be recognized, because much available data are based upon it. This consists of international agreement on certain fixed points to permit calibration of instruments. Some of these fixed points are known to contain errors, but the differences between the Thermodynamic and the Practical scales are of no consequence in almost all temperature considerations. The basis of the practical scale is given in Footnote 9 (see A2.12.5).

3.4.4 *Time*—The SI unit for time is the second. This unit is preferred and should be used when technical calculations are involved. In other cases use of the minute, hour, day, etc., is permissible.

3.4.5 *Angles*—The SI unit for plane angle is the radian. The use of the arc degree and its decimal submultiples is permissible when the radian is not a convenient unit. Solid angles should be expressed in steradians.

3.4.6 *Stress and Pressure*—The SI unit for

pressure and stress is the pascal (newton per square metre). The use of the metric engineering units—kilogram-force per square centimetre, kilogram-force per square millimetre, bar, torr, etc.—is discouraged and should be discontinued.

3.4.7 *Other Quantities*—The use of other non-SI units (for example, metric horsepower for power and calorie for energy) should be discontinued as rapidly as possible.

3.5 *Style:*

3.5.1 *Capitalization*—Symbols for SI units are not capitalized unless the unit is derived from a proper name; thus, Hz for H. R. Hertz, but m for metre. Unabbreviated units are not capitalized, for example, hertz, newton, and kelvin. Numerical prefixes and their symbols are not capitalized, except for the symbols T, G, and M (mega) (see 2.1.4).

3.5.2 *Plurals*—Unabbreviated SI units form their plurals in the usual manner. SI symbols are always written in singular form, for example:

50 newtons or 50 N, and
25 grams or 25 g.

3.5.3 *Punctuation*—Periods should not be used after SI unit symbols except at the end of a sentence.

3.5.4 *Number Grouping*—To facilitate the reading of numbers having four or more digits, the digits should be placed in groups of three separated by a space instead of commas counting both to the left and to the right of the decimal point. In the case of four digits the spacing is optional. This style also avoids confusion caused by the European use of commas to express decimal points. For example, use

1 532 or 1532 instead of 1,532
132 541 816 instead of 132,541,816
983 769.816 78 instead of 983,769.81678

3.5.5 *Equations*—When U.S. customary units appear in equations, the SI equivalents should be omitted. Instead of inserting the SI equivalents in parentheses as in the case of text or small tables, the equations should be restated using SI quantities or a sentence, paragraph, or note added stating the factor to be used to convert the calculated result in U.S. units to the preferred SI units.

3.5.6 *Derived Units*—In derived unit symbols, use the center dot to indicate multiplication and a slash to indicate division. Symbols to the left of the slash are in the numerator and those to the right in the denominator.

4. Rules for Conversion and Rounding

4.1 *General:*

4.1.1 Appendix A3 contains conversion factors that give exact or seven-figure accuracy for implementing these rules except where the nature of the dimension makes this impractical.

4.1.2 Conversion of quantities should be handled with careful regard to the implied correspondence between the accuracy of the data and the given number of digits. In all conversions, the number of significant digits retained should be such that accuracy is neither sacrificed nor exaggerated.

4.1.3 Proper conversion procedure is to multiply the specified quantity by the conversion factor exactly as given in Appendix A3 and then round to the appropriate number of significant digits. For example, to convert 11.4 ft to metres: $11.4 \times 0.3048 = 3.474\ 72$, which rounds to 3.47 m. Do not round either the conversion factor nor the quantity before performing the multiplication, as accuracy would be reduced.

4.2 *Accuracy and Rounding*—Accurate conversions are obtained by multiplying the specified quantity by the appropriate conversion factor given in Appendix A3. However, this product will usually imply an accuracy not intended by the original value. Proper conversion technique includes rounding this converted quantity to the proper number of significant digits commensurate with its intended precision. The practical aspects of measuring must be considered when using SI equivalents. If a scale divided into $\frac{1}{16}$ths of an inch is suitable for making the original measurements, a metric scale having divisions of 1 mm is obviously suitable for measuring in SI units, and the equivalents should not be reported closer than the nearest 1 mm. Similarly, a gage or caliper graduated in divisions of 0.02 mm is comparable to one graduated in divisions of 0.001 in., and equivalents should be similarly indicated. Analogous situations exist in mass, force, and other measurements. Many techniques are used to guide the determination of the proper number of significant digits in the converted values. Two different approaches to rounding of quantities are here described—one for general use and the other for conversion of dimensions involving mechanical interchangeability.

4.2.1 *General Conversion*—This approach depends on first establishing the intended precision or accuracy of the quantity as a necessary guide to the number of digits to retain. This precision should relate to the number of digits in the original, but in many cases this is not a reliable indicator. A figure 1.1875 may be a very accurate decimalization of a noncritical 1$\frac{3}{16}$ths that should have been expressed 1.19. On the other hand, the value 2 may mean "about 2" or it may mean a very accurate value of 2 which should have been written 2.0000. It is therefore necessary to determine the intended precision of a quantity before converting. This estimate of intended precision should never be smaller than the accuracy of measurement, and as a rule of thumb should usually be smaller than one tenth the tolerance if one exists. After estimating the precision of the dimension, the converted dimension should be rounded to a minimum number of significant digits (see 4.3) such that a unit of the last place is equal to or smaller than the converted precision.

Examples:

1. A stirring rod 6 in. long. Precision is estimated to be about ½ in. (± ¼ in.). Converted this is 12.7 mm. The converted dimension 152.4 mm should be rounded to the nearest 10 mm, or 150 mm.

2. 50 000 psi tensile strength. Precision is estimated to be about ±200 psi (±1.4 MPa) based on an accuracy of ±0.25 percent for the tensile tester and other factors. Therefore, the converted dimension, 344.7385 MPa, should be rounded to the nearest whole unit, 345 MPa.

3. Test pressure 200 ± 15 psi. Since one tenth of the tolerance is 1.5 psi (10.34 kPa), the converted dimension should be rounded to the nearest 10 kPa. Thus, 1378.9514 ± 103.421 35 kPa becomes 1380 ± 100 kPa.

4.2.2 *Special Cases:*

4.2.2.1 Converted values should be rounded to the minimum number of significant digits that will maintain the required accuracy, as discussed above. In certain cases deviation from this practice to make use of convenient or whole numbers may be feasible, in which case the word "approximate" must be used following the conversion. For example,

1⅞ in. = 47.625 mm exact
47.6 mm normal rounding
47.5 mm (approx) rounded to preferred number
48 mm (approx) rounded to whole number

4.2.2.2 A quantity stated as a limit, such as "not more than" or "maximum" must be handled so that the stated limit is not violated. For example, a specimen "at least 4 in. wide" requires a width of at least 101.6 mm, or at least 102 mm.

4.3 *Significant Digits*[5]:

4.3.1 When converting integral values of units, consideration must be given to the implied or required precision of the integral value to be converted. For example, the value "4 in." may be intended to represent 4, 4.0, 4.00, 4.000, 4.0000 in., or even greater accuracy. Obviously, the converted value must be carried to a sufficient number of digits to maintain the accuracy implied or required in the original quantity.

4.3.2 Any digit that is necessary to define the specific value or quantity is said to be significant. When measured to the nearest 1 m, a distance may be recorded as 157 m; this number has three significant digits. If the measurement had been made to the nearest 0.1 m, the distance may have been 157.4 m; this number has four significant digits. In each of these cases the value of the right-hand digit was determined by measuring the value of an additional digit and then rounding to the desired degree of accuracy. In other words, 157.4 was rounded to 157; in the second case, the measurement in hundredths, 157.36 was rounded to 157.4.

4.3.3 Zeros may be used either to indicate a specific value, like any other digit, or to indicate the magnitude of a number. The 1960 U.S. population figure rounded to thousands was 179 323 000. The six left-hand digits of this number are significant; each of them *measures* a value. The three right-hand digits

[5] See also ASTM Recommended Practice E 29, for Indicating Which Places of Figures Are to Be Considered Significant in Specified Limiting Values, *1974 Annual Book of ASTM Standards*, Part 41.

are zeros which merely indicate the *magnitude* of the number rounded to the nearest thousand. This point may be further illustrated by the following list of estimates and measurements, each of which is of different magnitude but each of which has only one significant digit:

1000
100
10
0.01
0.001
0.0001

It is also important to note that in the case of the first three numbers above, the identification of significant digits is only possible through knowledge of the circumstances. The number 1000 may be rounded from about 965, or it may be rounded from 999.7, in which case all three zeros are significant.

4.3.4 Occasionally data required for an investigation must be drawn from a variety of sources where they have been recorded with varying degrees of refinement. Specific rules must be observed when such data are to be *added, subtracted, multiplied, or divided.*

4.3.4.1 The rule for addition and subtraction is that the *answer* shall contain no significant digits farther to the right than occurs in the least accurate figure. Consider the addition of three numbers drawn from three sources, the first of which reported data in millions, the second in thousands, and the third in units:

163 000 000
217 885 000
96 432 768
———————
477 317 768

The total indicates a precision that is not valid. The numbers should first be rounded to *one significant digit* farther to the right than that of the least accurate number, and the sum taken as follows:

163 000 000
217 900 000
96 400 000
———————
477 300 000

The answer is then rounded to 477 000 000 as called for by the rule. Note that if the second of the figures to be added had been 217 985 000, the rounding before addition would have produced 218 000 000, in which case the 0 following 218 would have been a significant digit.

4.3.4.2 The rule for multiplication and division is that the product or quotient shall contain no more significant digits than are contained in the number with the fewest significant digits used in the multiplication or division. The difference between this rule and the rule for addition and subtraction should be noted; the latter rule merely requires rounding of digits that lie to the right of the last significant digit in the least accurate number. The following illustration highlights this difference:

Multiplication:	113.2 × 1.43 = 161.876, rounded to 162
Division:	113.2 ÷ 1.43 = 79.16 rounded to 79.2
Addition:	113.2 + 1.43 = 114.63, rounded to 114.6
Subtraction:	113.2 − 1.43 = 111.77, rounded to 111.8

The above product and quotient are limited to three significant digits since 1.43 contains only three significant digits. In contrast, the rounded answers in the addition and subtraction examples contain four significant digits.

4.3.4.3 Numbers used in the above illustrations have all been estimates or measurements. Numbers that are exact counts are treated as though they consist of an infinite number of significant digits. More simply stated, when a count is used in computation with a measurement the number of significant digits in the answer is the same as the number of significant digits in the measurement. If a count of 40 is multiplied by a measurement of 10.2, the product is 408. However, if 40 were an estimate accurate only to the nearest 10, and hence contained but one significant digit, the product would be 400.

4.4 *Rounding Values*[6]*:*

4.4.1 When a figure is to be rounded to fewer digits then the total number available, the procedure should be as follows:

4.4.1.1 When the first digit discarded is less than 5, the last digit retained should not be changed. For example, 3.463 25, if rounded to four digits, would be 3.463; if rounded to three digits, 3.46.

[6] Adapted from "American National Standard Practice for Inch-Millimeter Conversion for Industrial Use," ANSI B48.1-1933 (R1947), ISO R370—1964.

E 380

4.4.1.2 When the first digit discarded is greater than 5, or if it is a 5 followed by at least one digit other than 0, the last figure retained should be increased by one unit. For example 8.376 52, if rounded to four digits, would be 8.377; if rounded to three digits, 8.38.

4.4.1.3 When the first digit discarded is exactly 5, followed only by zeros, the last digit retained should be rounded upward if it is an odd number, but no adjustment made if it is an even number. For example, 4.365, when rounded to three digits, becomes 4.36. 4.355 would also round to the same value, 4.36, if rounded to three digits.

4.5 *Conversion of Linear Dimensions of Interchangeable Parts*—The use of the exact relation 1 in. ≏ 25.4 mm generally produces converted values containing more decimal places than are required for the desired accuracy. It is therefore, necessary to round these values suitably and at the same time maintain the degree of accuracy in the converted values compatible with that of the original values.

4.5.1 *General*—The number of decimal places given in Table 1 for rounding toleranced dimensions relates the degree of accuracy to the size of the tolerances specified. Two methods of using this table are given: Method A, which rounds to values nearest to each limit, and Method B, which rounds to values always inside the limits.

In Method A, rounding is effected to the nearest rounded value of the limit, so that, on the average, the converted tolerances remain statistically identical with the original tolerances. The limits converted by this method, where acceptable for interchangeability, serve as a basis for inspection.

In Method B, rounding is done systematically *toward the interior* of the tolerance zone so that the converted tolerances are never larger than the original tolerances. This method must be employed when the original limits have to be respected absolutely, in particular, when components made to converted limits are to be inspected by means of original gages.

Method A—The use of this method ensures that even in the most unfavorable cases neither of the two original limits will be changed by more than 2 percent of the value of the tolerance. Proceed as follows:

(*a*) Calculate the maximum and minimum limits in inches.

(*b*) Convert the corresponding two values exactly into millimetres by means of the conversion factor 1 in. = 25.4 mm (see Table 2).

(*c*) Round the results obtained to the nearest rounded value as indicated in Table 1, depending on the original tolerance in inches, that is, on the difference between the two limits in inches.

Method B—This method must be employed when the original limits may not be violated, for instance, certain critical mating parts. In extreme cases, this method may increase the lower limit a maximum of 4 percent of the tolerance and decrease the upper limit a maximum of 4 percent of the tolerance.

(*a*) Proceed as in Method A, steps (*a*) and (*b*).

(*b*) Round each limit toward the interior of the tolerance, that is, to the next lower value for the upper limit and to the next higher value for the lower limit,[7] for example:

A dimension is expressed in inches as	1.950 ± 0.016
The limits are	1.934 to 1.966
Conversion of the two limits into millimetres gives	49.1236 to 49.9364
Method A—The tolerance equals 0.032 in. and thus lies between 0.01 and 0.1 in. (see Table 1). Rounding these values to the nearest 0.01 mm, the values in millimetres to be employed for these two limits are	49.12 to 49.94
Method B—Rounding toward the interior of the tolerance, millimetre values for these two limits are	49.13 to 49.93

This reduces the tolerance to 0.80 instead of 0.82 mm given by Method A.

4.5.2 *Special Method for Plus and Minus Toleranced Dimensions*—In order to avoid accumulation of rounding errors, the two limits of size normally are converted separately: thus, they must first be calculated if the dimension consists of a basic size and two tolerances. However (except when Method B is specified) as an alternative, the basic size may be converted to the nearest rounded value and each of the tolerances converted toward the interior of the tolerance. This method, which sometimes makes conversion

[7] If the digits to be rounded are zeros, the retained digits remain unchanged.

easier, gives the same maximum guarantee of accuracy as Method A, but usually results in smaller converted tolerances.

4.5.3 *Special Methods for Limitation Imposed by Accuracy of Measurements*—If the increment of rounding given for the tolerances in Table 1 is too small for the available accuracy of measurement, limits that are acceptable for interchangeability must be determined separately for the dimensions. For example, where accuracy of measurement is limited to 0.001 mm, study shows that values converted from 1.0000 ± 0.0005 in. can be rounded to 25.413 and 25.387 mm instead of 25.4127 and 25.3873 mm with little disadvantage, since neither of the two original limits is exceeded by more than 1.2 percent of the tolerance.

4.5.4 *Positional Tolerance*—If the dimensioning consists solely of a positional tolerance around a point defined by a nontoleranced basic dimension, the basic dimension must be converted to the nearest rounded value and the positional variation (radius) separately converted by rounding downwards.

4.5.5 *Toleranced Dimension Applied to a Nontoleranced Position Dimension*—If the toleranced dimension is located in a plane, the position of which is given by nontoleranced basic or gage dimension, such as when dimensioning certain conical surfaces, proceed as follows:

(*a*) Round the converted reference gage arbitrarily, to the nearest convenient value.

(*b*) Calculate exactly, in the converted unit of measurement, new maximum and minimum limits of the specified tolerance zone, in the new plane defined by the new basic dimension.

(*c*) Round these limits in conformity with the rules given in 4.4. For example, a cone of taper 0.05 in./in. has a diameter of 1.000 ± 0.002 in. in a reference plane located by the nontoleranced dimension 0.9300 in. By virtue of the taper of the cone, the limits of the tolerance zone depend on the position of the reference plane. Consequently, if the dimension 0.9300 in. = 23.6220 mm is rounded to 23.600 mm, that is, a reduction of 0.022 mm, each of the two original limits, when converted exactly into millimetres, must be corrected by 0.022 × 0.05 = 0.0011 mm, in the appropriate sense, before being rounded.

4.5.6 *Consideration of Maximum and Minimum Material Condition*—The ability to assemble mating parts depends on a "go" condition at the maximum material limits of the parts. The minimum material limits, which are determined by the respective tolerances, are often not as critical from a functional standpoint. Accordingly, it may be desirable to employ a combination of Methods A and B in certain conversions by using Method B for the maximum material limits and Method A for the minimum material limits. Alternatively, it may be desirable to round automatically the converted minimum material limits outside the original limits to provide greater tolerances for manufacturing.

4.5.7 While the technique described here in 4.5 provides good accuracy of conversion, it will often result in dimensions that are impractical for actual production use. For conversions intended for production, it is usually necessary to round to fewer decimal places and apply design judgment to each dimension to assure interchangeability.

4.6 *Other Units:*

4.6.1 *Temperature*—General guidance for converting tolerances from degrees Fahrenheit to kelvins or degrees Celsius is given in Table 3. Normally, temperatures expressed in a whole number of degrees Fahrenheit should be converted to the nearest 0.5 kelvin (or degree Celsius). As with other quantities the number of significant digits to retain will depend upon implied accuracy of the original dimension.

100 ± 5°F implied accuracy estimated total 2°F. 37.7777 ± 2.7777°C rounds to 38 ± 3°C.

1000 ± 50°F implied accuracy estimated total 20°F. 537.7777 ± 27.7777°C rounds to 540 ± 30°C.

4.6.2 *Pressure or Stress*—As with other quantities, pressure or stress values may be converted by the principle given above. Values with an uncertainty of more than 2 percent may be converted without rounding by the approximate factors:

1 psi = 7 kN/m² = 7 kPa
1 ksi = 7 MN/m² = 7 MPa

For conversion factors and values see Appendix A3 and Tables 4 and 5.

E 380

4.7 *Use of Tables:*

4.7.1 As an alternative to calculating conversions, conversion tables and charts are frequently provided for the rapid determination of converted values of commonly used quantities by addition only. Table 2 is of this type, and provides values for converting inches to millimetres. In using this table, the inch value to be converted should be written to as many decimal places as accuracy requires. The value should then be split into groups of not more than two significant digits each. The inch equivalent of each group should be taken from Table 2 and tabulated as in the following example, proper regard being given to the position of the decimal point in each case. For example, to convert 2.4637 in. to millimetres,

2.0000 in.	=	50.800 00 mm exactly
0.4600 in.	=	11.684 00 mm exactly
0.0037 in.	=	0.093 98 mm exactly
2.4637 in.	=	62.577 98 mm exactly

or rounded to three decimal places:

2.4637 in. = 62.578 mm

In this example, to maintain accuracy during conversion without retaining an unnecessary number of digits, the rounded millimetre equivalent of each group is carried to one decimal place more than the inch value being converted. The sum of the group of equivalents is then rounded to one less decimal place than the original inch value.

4.7.2 Table 6 provides values for the conversion of decimal and fractional values of an inch to millimetres. Combinations of inch and fractional inch-millimetre equivalents may be tabulated to obtain the desired millimetre conversion. As in the example in 4.7.1 the millimetre value should be rounded to one less decimal place than the decimal equivalent of the inch fraction taken to the intended or implied accuracy.

E 380

TABLE 1 Rounding Tolerances
Inches to Millimetres

Original Tolerance, in.		Fineness of Rounding, mm
at least	less than	
0.000 01	0.000 1	0.000 01
0.000 1	0.001	0.000 1
0.001	0.01	0.001
0.01	0.1	0.01
0.1	1	0.1

TABLE 2 Inch-Millimetre Equivalents

NOTE—All values in this table are exact, based on the relation 1 in. = 25.4 mm. By manipulation of the decimal point any decimal value or multiple of an inch may be converted to its exact equivalent in millimetres.

in.	0	1	2	3	4	5	6	7	8	9
					mm					
0		25.4	50.8	76.2	101.6	127.0	152.4	177.8	203.2	228.6
10	254.0	279.4	304.8	330.2	355.6	381.0	406.4	431.8	457.2	482.6
20	508.0	533.4	558.8	584.2	609.6	635.0	660.4	685.8	711.2	736.6
30	762.0	787.4	812.8	838.2	863.6	889.0	914.4	939.8	965.2	990.6
40	1016.0	1041.4	1066.8	1092.2	1117.6	1143.0	1168.4	1193.8	1219.2	1244.6
50	1270.0	1295.4	1320.8	1346.2	1371.6	1397.0	1422.4	1447.8	1473.2	1498.6
60	1524.0	1549.4	1574.8	1600.2	1625.6	1651.0	1676.4	1701.8	1727.2	1752.6
70	1778.0	1803.4	1828.8	1854.2	1879.6	1905.0	1930.4	1955.8	1981.2	2006.6
80	2032.0	2057.4	2082.8	2108.2	2133.6	2159.0	2184.4	2209.8	2235.2	2260.6
90	2286.0	2311.4	2336.8	2362.2	2387.6	2413.0	2438.4	2463.8	2489.2	2514.6
100	2540.0	...	...	...	...	...	...	...	...	...

TABLE 3 Conversion of Temperature Tolerance Requirements

Tolerance, °F	±1	±2	±5	±10	±15	±20	±25
Tolerance, K or °C	±0.5	±1	±3	±5.5	±8.5	±11	±14

TABLE 4 Pressure and Stress Equivalents—Pounds-Force per Square Inch and Thousand Pounds-Force per Square Inch to Kilopascals (Kilonewtons per Square Metre) and Megapascals (Meganewtons per Square Metre)

NOTE 1—This table may be used to obtain SI equivalents of values expressed in psi or ksi. SI values are usually expressed in kPa (kN/m²) when original value is in psi and in MPa (MN/m²) when original value is in ksi.

NOTE 2—This table may be extended to values below 1 or above 100 psi (ksi) by manipulation of the decimal point and addition as illustrated in 4.7.1.

Conversion Relationships: 1 in. = 0.0254 m (exactly)
1 lbf = 4.448 221 615 260 5 N (exactly)

psi ksi	0	1	2	3	4	5	6	7	8	9
	kPa (kN/m²) MPa (MN/m²)									
0	0.0000	6.8948	13.7895	20.6843	27.5790	34.4738	41.3685	48.2633	55.1581	62.0528
10	68.9476	75.8423	82.7371	89.6318	96.5266	103.4214	110.3161	117.2109	124.1056	131.0004
20	137.8951	144.7899	151.6847	158.5794	165.4742	172.3689	179.2637	186.1584	193.0532	199.9480
30	206.8427	213.7375	220.6322	227.5270	234.4217	241.3165	248.2113	255.1060	262.0008	268.8955
40	275.7903	282.6850	289.5798	296.4746	303.3693	310.2641	317.1588	324.0536	330.9483	337.8431
50	344.7379	351.6326	358.5274	365.4221	372.3169	379.2116	386.1064	393.0012	399.8959	406.7907
60	413.6854	420.5802	427.4749	434.3697	441.2645	448.1592	455.0540	461.9487	468.8435	475.7382
70	482.6330	489.5278	496.4225	503.3173	510.2120	517.1068	524.0015	530.8963	537.7911	544.6858
80	551.5806	558.4753	565.3701	572.2648	579.1596	586.0544	592.9491	599.8439	606.7386	613.6334
90	620.5281	627.4229	634.3177	641.2124	648.1072	655.0019	661.8967	668.7914	675.6862	682.5810
100	689.4757									

TABLE 5 Pressure and Stress Equivalents—Metric Engineering to SI Units

NOTE 1—This table may be used for obtaining SI equivalents of quantities expressed in kgf/cm^2 by multiplying the given values by 10^{-2}, that is, by moving the decimal point two places to the left.

NOTE 2—This table may be extended to values below 1 or above 100 kgf/cm^2 by manipulation of the decimal point and addition as illustrated in 4.7.1.

Conversion Relationships: 1 mm = 0.001 m (exactly)
1 kgf = 9.80665 N (exactly)

	0	1	2	3	4	5	6	7	8	9
kgf/mm^2	MPa (MN/m^2)									
0		9.8066	19.6133	29.4200	39.2266	49.0332	58.8399	68.6466	78.4532	88.2598
10	98.0665	107.8731	117.6798	127.4864	137.2931	147.0998	156.9064	166.7130	176.5197	186.3264
20	196.1330	205.9396	215.7463	225.5530	235.3596	245.1662	254.9729	264.7796	274.5862	284.3928
30	294.1995	304.0062	313.8128	323.6194	333.4261	343.2328	353.0394	362.8460	372.6527	382.4594
40	392.2660	402.0726	411.8793	421.6860	431.4926	441.2992	451.1059	460.9126	470.7192	480.5258
50	490.3325	500.1392	509.9458	519.7524	529.5591	539.3658	549.1724	558.9790	568.7857	578.5924
60	588.3990	598.2056	608.0123	617.8190	627.6256	637.4322	647.2389	657.0456	666.8522	676.6588
70	686.4655	696.2722	706.0788	715.8854	725.6921	735.4988	745.3054	755.1120	764.9187	774.7254
80	784.5320	794.3386	804.1453	813.9520	823.7586	833.5652	843.3719	853.1786	862.9852	872.7918
90	882.5985	892.4052	902.2118	912.0184	921.8251	931.6318	941.4384	951.2450	961.0517	970.8584
100	980.6650									

E 380

TABLE 6 Inch-Millimetre Equivalents of Decimal and Common Fractions

From 1/64 to 1 in.

Inch	½'s	¼'s	8ths	16ths	32nds	64ths	Millimetres	Decimals of an Inch[a]
						1	0.397	0.015 625
					1	2	0.794	0.031 25
						3	1.191	0.046 875
				1	2	4	1.588	0.062 5
						5	1.984	0.078 125
					3	6	2.381	0.093 75
						7	2.778	0.109 375
			1	2	4	8	3.175[a]	0.125 0
						9	3.572	0.140 625
					5	10	3.969	0.156 25
						11	4.366	0.171 875
				3	6	12	4.762	0.187 5
						13	5.159	0.203 125
					7	14	5.556	0.218 75
						15	5.953	0.234 375
		1	2	4	8	16	6.350[a]	0.250 0
						17	6.747	0.265 625
					9	18	7.144	0.281 25
						19	7.541	0.296 875
				5	10	20	7.938	0.312 5
						21	8.334	0.328 125
					11	22	8.731	0.343 75
						23	9.128	0.359 375
			3	6	12	24	9.525[a]	0.375 0
						25	9.922	0.390 625
					13	26	10.319	0.406 25
						27	10.716	0.421 875
				7	14	28	11.112	0.437 5
						29	11.509	0.453 125
					15	30	11.906	0.468 75
						31	12.303	0.484 375
	1	2	4	8	16	32	12.700[a]	0.500 0
						33	13.097	0.515 625
					17	34	13.494	0.531 25
						35	13.891	0.546 875
				9	18	36	14.288	0.562 5
						37	14.684	0.578 125
					19	38	15.081	0.593 75
						39	15.478	0.609 375
			5	10	20	40	15.875[a]	0.625 0
						41	16.272	0.640 625
					21	42	16.669	0.656 25
						43	17.066	0.671 875
				11	22	44	17.462	0.687 5
						45	17.859	0.703 125
					23	46	18.256	0.718 75
						47	18.653	0.734 375
		3	6	12	24	48	19.050[a]	0.750 0
						49	19.447	0.765 625
					25	50	19.844	0.781 25
						51	20.241	0.796 875
				13	26	52	20.638	0.812 5

E 380

TABLE 6—*Continued*

Inch	½'s	¼'s	8ths	16ths	32nds	64ths	Millimetres	Decimals of an Inch[a]
						53	21.034	0.828 125
					27	54	21.431	0.843 75
						55	21.828	0.859 375
			7	14	28	56	22.225[a]	0.875 0
						57	22.622	0.890 625
					29	58	23.019	0.906 25
						59	23.416	0.921 875
				15	30	60	23.812	0.937 5
						61	24.209	0.953 125
					31	62	24.606	0.968 75
						63	25.003	0.984 375
1	2	4	8	16	32	64	25.400[a]	1.000 0

[a] Exact.

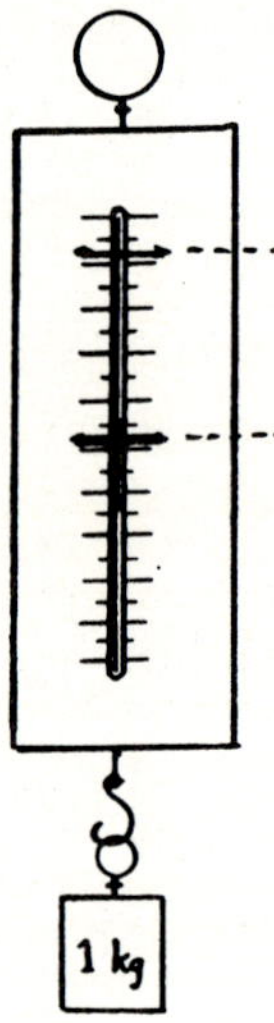

FIG. 1 **Illustration of Difference Between Mass (Unit = Kilogram) and Force (Unit = Newton).**

APPENDIXES

A1. TERMINOLOGY

A1.1 To help ensure consistently reliable conversion and rounding practices, a clear understanding of the related nontechnical terms is a prerequisite. Accordingly, certain terms used in this Guide are defined as follows:

accuracy (as distinguished from **precision**)—the degree of conformity of a measured or calculated value to some recognized standard or specified value. This concept involves the systematic error of an operation, which is seldom negligible.

approximate—describes a value that is nearly but not exactly correct or accurate.

deviation—variations from a specified dimension or design requirement, usually defining upper and lower limits. (See also **tolerance**.)

digit—one of the ten Arabic numerals (0 to 9) by which all numbers are expressed.

dimension—a geometric element in a design, such as length, angle, etc., or the magnitude of such a quantity.

feature—an individual characteristic of a part, such as screw-thread, taper, slot, etc.

figure (numerical)—an arithmetic value expressed by one or more digits.

nominal—describes a value assigned for the purpose of convenient designation; existing in name only.

precision (as distinguished from **accuracy**)—the degree of mutual agreement between individual measurements, namely repeatability and reproducibility.

significant (as applied to a digit)—any digit that is necessary to define a value or quantity (see Footnote 5).

tolerance—the total range of variation (usually bilateral) permitted for a size, position, or other required quantity; the upper and lower limits between which a dimension must be held.

U.S. customary units—units based upon the yard and the pound commonly used in the United States of America and defined by the National Bureau of Standards (Ref 14). Some of these units have the same name as similar units in the United Kingdom (British, English, or U.K. units) but are not necessarily equal to them.

Three terms that may be encountered in the literature dealing with the modernized metric system are "metricize," "metrication," and "rationalize." These have been defined as follows:

metricize—to convert any other unit to its metric (SI) equivalent. This may be an exact, a rounded, or a rationalized equivalent. For example, a 2-in. test specimen becomes 50.8 mm by exact conversion; 51 mm by rounding in accordance with Section 4, Rules for Conversion and Rounding, and 50 mm when rationalized.

metrication—any act tending to increase the use of the metric system (SI).

rationalize—to round completely a converted value to a popular standard figure compatible with noncritical mating components, interchangeable parts, or other nominal sizes in a series.

A2. DEVELOPMENT OF THE INTERNATIONAL SYSTEM OF UNITS

Le Système International d'Unités

A2.1 The decimal system of units was first conceived in the 16th century when there was a great confusion and jumble of units of weights and measures. It was not until 1790, however, that the French National Assembly requested the French Academy of Sciences to work out a system of units suitable for adoption by the entire world. This system, based on the metre as a unit of length and the gram as a unit of mass, was adopted as a practical measure to benefit industry and commerce. Physicists soon realized its advantages and it was adopted also in scientific and technical circles. The importance of the regulation of weights and measures was recognized in Article 1, Section 8, when the United States Constitution was written in 1787, but the metric system was not legalized in this country until 1866. In 1893, the international metre and kilogram became the fundamental standards of length and mass in the United States, both for metric and customary weights and measures.

A2.2 Meanwhile, international standardization began with an 1870 meeting of 15 nations in Paris that led to the May 20, 1875, International Metric Convention, and the establishment of a permanent International Bureau of Weights and Measures near Paris. A General Conference on Weights and Measures (CGPM) was also constituted to handle all international matters concerning the metric system. The CGPM nominally meets every sixth year in Paris and controls the International Bureau of Weights and Measures which preserves the metric standards, compares national standards with them, and conducts research to establish new standards. The National Bureau of Standards represents the United States in these activities.

A2.3 The original metric system provided a coherent set of units for the measurement of length, area, volume, capacity, and mass based on two fundamental units: the metre and the kilogram. Measurement of additional quantities required for science and commerce has necessitated development of additional fundamental and derived units. Numerous other systems based on these two metric units have been used. A unit of time was added to produce the centimetre-gram-second (cgs) system adopted in 1881 by the International Congress of Electricity. About 1900 practical measurements in metric units began to be based on the metre-kilogram-second (MKS) system. In 1935 Prof. Giovanni Giorgi recommended that the MKS system of mechanics be linked with the electromagnetic system of units by adoption of one of the units—ampere, coulomb, ohm, or volt—for the fourth base unit. This recommendation was accepted and in 1950 the ampere, the unit of electric current, was established as a base unit to form the MKSA system.

A2.4 The 10th CGPM in 1954 adopted a rationalized and coherent system of units based on the four MKSA units, plus the degree Kelvin as the unit of temperature and the candela as the unit of luminous intensity. The 11th CGPM in 1960 formally gave it the full title, International System of Units, for which the abbreviation is "SI" in all languages. Thirty-six countries, including the United States, participated in this 1960 conference. The 12th CGPM in 1964 made some refinements, and the 13th CGPM in 1967 redefined the second, renamed the unit of temperature as the kelvin (K), and revised the definition of the candela. The 14th CGPM in 1971 added a seventh base unit, the mole, and approved the pascal (Pa) as a special name for the SI unit of pressure or stress, the newton per square metre, and the siemens (S) as a special name for the unit of electrical conductance, the reciprocal ohm or the ampere per volt.

A2.5 SI is a rationalized selection of units from the metric system which individually are not new. It includes a unit of force (the newton) which was introduced in place of the kilogram-force to indicate by its name that it is a unit of force and not of mass. SI is a coherent system with seven base units for which names, symbols, and precise definitions have been established. Many derived units are defined in terms of the base units, symbols assigned to each, and, in some cases, given names, as for example, the newton (N).

A2.6 The great advantage of SI is that there is one and only one unit for each physical quantity—the metre for length (l), kilogram (instead of gram) for mass (m), second for time (t), etc. From these elemental units, units for all other mechanical quantities are derived. These derived units are defined by simple equations such as $v = dl/dt$ (velocity), $a = dv/dt$ (acceleration), $F = ma$ (force), $W = Fl$ (work or energy), $P = W/t$ (power). Some of these units have only generic names such as metre per second for velocity: others have special names such as newton (N) for force, joule (J) for work or energy, watt (W) for power. The SI units for force, energy, and power are the same regardless of whether the process is mechanical, electrical, chemical, or nuclear. A force of 1 newton applied for a distance of 1 metre can produce 1 joule of heat, which is identical with what 1 watt of electric power can produce in 1 second.

A2.7 Corresponding to the advantages of SI, which result from the use of a unique unit for each physical quantity, are the advantages which result from the use of a unique and well-defined set of symbols and abbreviations. Such symbols and abbreviations eliminate the confusion that can arise from current practices in different disciplines such as the use of *b* for both the *bar* (a unit of pressure) and *barn* (a unit of area).

A2.8 Another advantage of SI is its retention of the decimal relation between multiples and submultiples of the base units for each physical quantity—not that there is anything inherently superior in a number system to the base 10 but that SI conforms to the system of Arabic numericals. Prefixes are established for designating multiple and submultiple units from "tera" (10^{12}) down to "atto" (10^{-18}) for convenience in writing and talking.

A2.9 Another major advantage of SI is its coherence. A system of units is coherent if the product or quotient of any two unit quantities in the system is a unit of the resulting quantity. For example, in any coherent system, unit area results when unit length is multipled by unit length, unit velocity when unit length is divided by unit time, and unit force when unit mass is multiplied by unit acceleration. Thus, in a coherent system in which the foot is a unit of length, the square foot is the unit of area (but the acre is not). Similarly in a coherent system in which the foot, the pound, and the second are units of length, mass, and time, the unit of force is the poundal (and not the pound-weight or the pound-force).

A2.10 Whatever the system of units, whether it be coherent or noncoherent, magnitudes of some physical quantities must be arbitrarily selected and declared to have unit value. These magnitudes form a set of standards and are called base units. All other units are derived units related to the base units by definition. The seven base SI units are each very accurately defined in terms of physical measurements that can be made in a laboratory, except the kilogram, which is a particular mass preserved by the International Bureau of Weights and Measures.

A2.11 Various other units are associated with SI but are not a part thereof. They are related to units of the system by powers of 10 and are employed in specialized branches of physics. Examples of such units are the bar, a unit of pressure, approximately equivalent to one atmosphere and equal exactly to 100 kilonewtons per square metre. It is employed extensively by meteorologists. Another such unit is the gal (galileo) equal exactly to an acceleration of 0.01 metre per second squared. It is used in geodetic work. These, however, are not consistent units, that is to say, equations involving both these units and SI units cannot be written without a factor of proportionality even though the factor of proportionality is a simple multiple of 10.

A2.11.1 Originally (1795) the litre was intended to be identical with the cubic decimetre. The Third General Conference on Weights and Measures, meeting in 1901, decided to define the litre as the volume occupied by the mass of one kilogram of pure water at its maximum density under normal atmospheric pressure. Careful determinations subsequently established the litre so defined as being equivalent to 1.000028 dm^3. In 1964 the General Conference on Weights and Measures withdrew this definition of the litre, and declared that the word "litre" was a special name for the cubic decimetre. Thus its use is permitted in SI, but is discouraged, since it creates two units for the same quantity, and its use in precision measurements might conflict with measurements recorded under the old definition.

A2.12 Authorized translations of the original French definitions of the seven base and two supplementary units of the International System are given in the following paragraphs:

A2.12.1 *metre*—The metre is the length equal to 1 650 763.73 wavelengths in vacuum of the radiation corresponding to the transition between the levels $2p_{10}$ and $5d_5$ of the krypton-86 atom. (adopted at 11th CGPM 1960)

A2.12.2 *kilogram*—The kilogram is the unit of mass; it is equal to the mass of the international prototype of the kilogram. (adopted at 1st and 3rd CGPM 1889 and 1901)

A2.12.3 *second*—The second is the duration of 9 192 631 770 periods of the radiation corresponding to the transition between the two hyperfine levels of the ground state of the cesium-133 atom.[8] (adopted at 13th CGPM 1967)

A2.12.4 *ampere*—The ampere is that constant current which, if maintained in two straight parallel conductors of infinite length, of negligible cross section, and placed one metre apart in vacuum, would produce between these conductors a force equal to 2×10^{-7} newton per metre of length. (adopted at 9th CGPM 1948)

A2.12.5 *kelvin*—The kelvin, unit of thermodynamic temperature, is the fraction 1/273.16 of the thermodynamic temperature of the triple point of

[8] This definition supersedes the ephemeris second as the unit of time.

E 380

water.[9] (adopted at 13th CGPM 1967)

A2.12.6 *mole*—The mole is the amount of substance of a system which contains as many elementary entities as there are atoms in 0.012 kilogram of carbon-12.

NOTE—When the mole is used, the elementary entities must be specified and may be atoms, molecules, ions, electrons, other particles, or specified groups of such particles.

A2.12.7 *candela*—The candela is the luminous intensity, in the perpendicular direction, of a surface of 1/600 000 square metre of a blackbody at the temperature of freezing platinum under a pressure of 101 325 newtons per square metre. (adopted at 13th CGPM 1967)

A2.12.8 *radian*—The unit of measure of a plane angle with its vertex at the center of a circle and subtended by an arc equal in length to the radius. (adopted at 11th CGPM 1960)

A2.12.9 *steradian*—The unit of measure of a solid angle with its vertex at the center of a sphere and enclosing an area of the spherical surface equal to that of a square with sides equal in length to the radius. (adopted at 11th CGPM 1960)

---

[9] The International Practical Kelvin Temperature Scale of 1968 and the International Practical Celsius Temperature Scale of 1968 are defined by a set of interpolation equations based on the following reference temperatures[a]:

	K	°C
Hydrogen, solid-liquid-gas equilibrium	13.81	−259.34
Hydrogen, liquid-gas equilibrium at 33 330.6 Pa (25/76 standard atmosphere)	17.042	−256.108
Hydrogen, liquid-gas equilibrium	20.28	−252.87
Neon, liquid-gas equilibrium	27.102	−246.048
Oxygen, solid-liquid-gas equilibrium	54.361	−218.789
Oxygen, liquid-gas equilibrium	90.188	−182.962
Water, solid-liquid-gas equilibrium	273.16	0.01
Water, liquid-gas equilibrium	373.15	100.00
Zinc, solid-liquid equilibrium	692.73	419.58
Silver, solid-liquid equilibrium	1235.08	961.93
Gold, solid-liquid equilibrium	1337.58	1064.43

[a] Except for the triple points and one equilibrium hydrogen point (17.042 K) the assigned values of temperature are for equilibrium states at a pressure

$$p_0 = 1 \text{ standard atmosphere (101 325 Pa).}$$

In this scale the degree Celsius and the kelvin are identical, and are related by an exact difference of 273.15 degrees. It is seldom necessary to recognize the differences between the thermodynamic and practical scales, but the General Conference of Weights and Measures in 1960 established the following quantity symbols for use if desired:

Thermodynamic temperature:	kelvins	T
	degrees Celsius	t
Practical temperature:	kelvins	T_{int}
	degrees Celsius	t_{int}

A2.13 *Definitions of Derived Units of the International System Having Special Names:*

Physical Quantity	Unit and Definition
A2.13.1 Electric capacitance	The *farad* is the capacitance of a capacitor between the plates of which there appears a difference of potential of one volt when it is charged by a quantity of electricity equal to one coulomb.
A2.13.2 Electrical conductance	The *siemens* is the electrical conductance of a conductor in which a current of one ampere is produced by an electric potential difference of one volt.
A2.13.3 Electric inductance	The *henry* is the inductance of a closed circuit in which an electromotive force of one volt is produced when the electric current in the circuit varies uniformly at a rate of one ampere per second.
A2.13.4 Electric potential difference (electromotive force)	The *volt* (unit of electric potential difference and electromotive force) is the difference of electric potential between two points of a conductor carrying a constant current of one ampere, when the power dissipated between these points is equal to one watt.
A2.13.5 Electric resistance	The *ohm* is the electric resistance between two points of a conductor when a constant difference of potential of one volt, applied between these two points, produces in this conductor a current of one ampere, this conductor not being the source of any electromotive force.
A2.13.6 Energy	The *joule* is the work done when the point of application of a force of one newton is displaced a distance of one metre in the direction of the force.
A2.13.7 Force	The *newton* is that force which, when applied to a body having a mass of one kilogram gives it an acceleration of one metre per second per second.
A2.13.8 Frequency	The *hertz* is a frequency of one cycle per second.
A2.13.9 Illuminance	The *lux* is the illuminance produced by a luminous flux of one lumen uniformly distributed over a surface of one square metre.
A2.13.10 Luminous flux	The *lumen* is the luminous flux emitted in a solid angle of one steradian by a point source having a uniform intensity of one candela.
A2.13.11 Magnetic flux	The *weber* is the magnetic flux which, linking a circuit of one turn, produces in it an electromotive force of one volt as it is reduced to zero at a uniform rate in one second.

A2.13.12 Magnetic flux density	The *tesla* is the magnetic flux density given by a magnetic flux of one weber per square metre.
A2.13.13 Power	The *watt* is the power which gives rise to the production of energy at the rate of one joule per second.
A2.13.14 Pressure or stress	The *pascal* is the pressure or stress of one newton per square metre.
A2.13.15 Quantity of electricity	The *coulomb* is the quantity of electricity transported in one second by a current of one ampere.

A3. CONVERSION FACTORS[10]

A3.1 *General*—The following tables of conversion factors are intended to serve two purposes:

A3.1.1 To express the definitions of miscellaneous units of measure as exact numerical multiples of coherent "metric" units. Relationships that are exact in terms of the base unit are followed by an asterisk. Relationships that are not followed by an asterisk are either the results of physical measurements, or are only approximate.

A3.1.2 To provide multiplying factors for converting expressions of measurements given by numbers and miscellaneous units to corresponding new numbers and metric units.

A3.2 *Notation:*

A3.2.1 Conversion factors are presented for ready adaptation to computer readout and electronic data transmission. The factors are written as a number greater than one and less than ten with six or less decimal places. This number is followed by the letter E (for exponent), a plus or minus symbol, and two digits which indicate the power of 10 by which the number must be multiplied to obtain the correct value. For example.

3.523 907 E−02 is $3.523\ 907 \times 10^{-2}$
or 0.035 239 07.

Similarly,

3.386 389 E+03 is $3.386\ 389 \times 10^{3}$
or 3 386.389.

A3.2.2 An asterisk (*) after the sixth decimal place indicates that the conversion factor is exact and that all subsequent digits are zero. All other conversion factors have been rounded to the figures given in accordance with 4.4.

A3.3 *Organization:*

A3.3.1 The conversion factors are listed in two ways—alphabetically and classified by physical quantity. Both lists contain those units which have specific names and compounded units derived from these specific units. The classified list contains the more frequently used units for each physical quantity.

A3.3.2 The conversion factors for other compounded units can easily be generated from numbers given in the alphabetical list by the well-known rules for manipulating units. These rules, being adequately discussed in many science and engineering textbooks, are not repeated here.

[10] Based on E. A. Mechtly (Ref 20).

[11] Also available from Headquarters is a Conversion Slide Rule (Order No. 12-503800-00) providing some of the most frequently used conversions.

ALPHABETICAL LIST OF UNITS

(Symbols of SI units given in parentheses)

To convert from	to	Multiply by
abampere	ampere (A)	1.000 000*E+01
abcoulomb	coulomb (C)	1.000 000*E+01
abfarad	farad (F)	1.000 000*E+09
abhenry	henry (H)	1.000 000*E−09
abmho	siemens (S)	1.000 000*E+09
abohm	ohm (Ω)	1.000 000*E−09
abvolt	volt (V)	1.000 000*E−08
acre-foot	metre^3 (m^3)	1.233 482 E+03
acre	meter^2 (m^2)	4.046 856 E+03
ampere, international U.S. (A_{INT-US})[12]	ampere (A)	9.998 43 E−01
ampere, U.S. legal 1948 (A_{US-48})	ampere (A)	1.000 008 E+00

[12] In 1948 a new international agreement was reached on absolute electrical units, which changed the value of the volt used in this country by about 300 parts per million. Again in 1969 a new base of reference was internationally adopted making a further change of 8.4 parts per million. These changes (and also changes in ampere, joule, watt, coulomb) require careful terminology and conversion factors for exact use of old information. Terms used in this guide are:
volt as used prior to January 1948—volt, international U.S. (V_{INT-US})
volt as used between January 1948 and January 1969—volt, U.S. legal 1948 (V_{US-48})
volt as used since January 1969—volt (V)
Identical treatment is given the ampere, coulomb, watt, and joule. Since the henry, farad, and ohm were not changed in 1969, only conversions as listed first above are shown.

E 380

To convert from	to	Multiply by
ampere-hour	coulomb (C)	3.600 000*E+03
angstrom	metre (m)	1.000 000*E−10
astronomical unit	metre (m)	1.495 98 E+11
atmosphere (normal)	pascal (Pa)	1.013 25 E+05
atmosphere (technical = 1 kgf/cm²)	pascal (Pa)	9.806 650*E+04
bar	pascal (Pa)	1.000 000*E+05
barn	metre² (m²)	1.000 000*E−28
barrel (for petroleum, 42 gal)	metre³ (m³)	1.589 873 E−01
board foot	meter³ (m³)	2.359 737 E−03
British thermal unit (International Table)[13]	joule (J)	1.055 056 E+03
British thermal unit (mean)	joule (J)	1.055 87 E+03
British thermal unit (thermochemical)	joule (J)	1.054 350 E+03
British thermal unit (39°F)	joule (J)	1.059 67 E+03
British thermal unit (60°F)	joule (J)	1.054 68 E+03
Btu (thermochemical)/foot²-second	watt/metre² (W/m²)	1.134 893 E+04
Btu (thermochemical)/foot²-minute	watt/metre² (W/m²)	1.891 489 E+02
Btu (thermochemical)/foot²-hour	watt/metre² (W/m²)	3.152 481 E+00
Btu (thermochemical)/inch²-second	watt/metre² (W/m²)	1.634 246 E+06
Btu (thermochemical)·in./s·ft²·°F (*k*, thermal conductivity)	watt/metre-kelvin (W/m·K)	5.188 732 E+02
Btu (International Table)·in./s·ft²·°F (*k*, thermal conductivity)	watt/metre-kelvin (W/m·K)	5.192 204 E+02
Btu (thermochemical)·in./h·ft²·°F (*k*, thermal conductivity)	watt/metre-kelvin (W/m·K)	1.441 314 E−01
Btu (International Table)·in./h·ft²·°F (*k*, thermal conductivity)	watt/metre-kelvin (W/m·K)	1.442 279 E−01
Btu (International Table)/ft²	joule/metre² (J/m²)	1.135 653 E+04
Btu (thermochemical)/ft²	joule/metre² (J/m²)	1.134 893 E+04
Btu (International Table)/h·ft²·°F (*C*, thermal conductance)	watt/metre²-kelvin (W/m²·K)	5.678 263 E+00
Btu (thermochemical)/h·ft²·°F (*C*, thermal conductance)	watt/metre²-kelvin (W/m²·K)	5.674 466 E+00
Btu (International Table)/pound-mass	joule/kilogram (J/kg)	2.326 000*E+03
Btu (thermochemical)/pound-mass	joule/kilogram (J/kg)	2.324 444 E+03
Btu (International Table)/lbm·°F (*c*, heat capacity)	joule/kilogram-kelvin (J/kg·K)	4.186 800*E+03
Btu (thermochemical)/lbm·°F (*c*, heat capacity)	joule/kilogram-kelvin (J/kg·K)	4.184 000 E+03
Btu (International Table)/s·ft²·°F	watt/metre²-kelvin (W/m²·K)	2.044 175 E+04
Btu (thermochemical)/s·ft²·°F	watt/metre²-kelvin (W/m²·K)	2.042 808 E+04
Btu (International Table)/hour	watt (W)	2.930 711 E−01
Btu (thermochemical)/second	watt (W)	1.054 350 E+03
Btu (thermochemical)/minute	watt (W)	1.757 250 E+01
Btu (thermochemical)/hour	watt (W)	2.928 751 E−01
bushel (U.S.)	metre (m³)	3.523 907 E−02
caliber (inch)	metre (m)	2.540 000*E−02
calorie (International Table)	joule (J)	4.186 800*E+00
calorie (mean)	joule (J)	4.190 02 E+00
calorie (thermochemical)	joule (J)	4.184 000*E+00
calorie (15°C)	joule (J)	4.185 80 E+00
calorie (20°C)	joule (J)	4.181 90 E+00
calorie (kilogram, International Table)	joule (J)	4.186 800*E+03
calorie (kilogram, mean)	joule (J)	4.190 02 E+03
calorie (kilogram, thermochemical)	joule (J)	4.184 000*E+03
calorie (thermochemical)/centimetre²-minute	watt/metre² (W/m²)	6.973 333 E+02
cal (thermochemical)/cm²	joule/metre² (J/m²)	4.184 000*E+04
cal (thermochemical)/cm²·s	watt/metre² (W/m²)	4.184 000*E+04
cal (thermochemical)/cm·s·°C	watt/metre-kelvin (W/m·K)	4.184 000*E+02
cal (International Table)/g	joule/kilogram (J/kg)	4.186 800*E+03
cal (International Table)/g·°C	joule/kilogram-kelvin (J/kg·K)	4.186 800*E+03
cal (thermochemical)/g	joule/kilogram (J/kg)	4.184 000*E+03

[13] This value was adopted in 1956. Some of the older International Tables use the value 1.055 04 E+03. The exact conversion factor is 1.055 055 852 62*E+03.

E 380

To convert from	to	Multiply by
cal (thermochemical)/g·°C	joule/kilogram-kelvin (J/kg·K)	4.184 000*E+03
calorie (thermochemical)/second	watt (W)	4.184 000*E+00
calorie (thermochemical)/minute	watt (W)	6.973 333 E−02
carat (metric)	kilogram (kg)	2.000 000*E−04
centimetre of mercury (0°C)	pascal (Pa)	1.333 22 E+03
centimetre of water (4°C)	pascal (Pa)	9.806 38 E+01
centipoise	pascal-second (Pa·s)	1.000 000*E−03
centistokes	metre²/second (m²/s)	1.000 000*E−06
circular mil	metre² (m²)	5.067 075 E−10
clo	kelvin-metre²/watt (K·m²/W)	2.003 712 E−01
coulomb, international U.S. ($C_{INT\text{-}US}$)[12]	coulomb (C)	9.998 43 E−01
coulomb, U.S. legal 1948 ($C_{US\text{-}48}$)	coulomb (C)	1.000 008 E+00
cup	metre³ (m³)	2.365 882 E−04
curie	disintegration/second	3.700 000*E+10
day (mean solar)	second (s)	8.640 000 E+04
day (sidereal)	second (s)	8.616 409 E+04
decibar	pascal (Pa)	1.000 000*E+04
degree (angle)	radian (rad)	1.745 329 E−02
degree Celsius	kelvin (K)	$t_K = t_{°C} + 273.15$
degree centigrade	see 3.4.3	
degree Fahrenheit	degree Celsius	$t_{°C} = (t_{°F} - 32)/1.8$
degree Fahrenheit	kelvin (K)	$t_K = (t_{°F} + 459.67)/1.8$
deg F·h·ft²/Btu (thermochemical) (*R*, thermal resistance)	kelvin-metre²/watt (K·m²/W)	1.762 280 E−01
deg F·h·ft²/Btu (International Table) (*R*, thermal resistance)	kelvin-metre²/watt (K·m²/W)	1.761 102 E−01
degree Rankine	kelvin (K)	$t_K = t_{°R}/1.8$
dyne	newton (N)	1.000 000*E−05
dyne-centimetre	newton-metre (N·m)	1.000 000*E−07
dyne/centimetre²	pascal (Pa)	1.000 000*E−01
electron volt	joule (J)	1.602 19 E−19†
EMU of capacitance	farad (F)	1.000 000*E+09
EMU of current	ampere (A)	1.000 000*E+01
EMU of electric potential	volt (V)	1.000 000*E−08
EMU of inductance	henry (H)	1.000 000*E−09
EMU of resistance	ohm (Ω)	1.000 000*E−09
ESU of capacitance	farad (F)	1.112 650 E−12
ESU of current	ampere (A)	3.335 6 E−10
ESU of electric potential	volt (V)	2.997 9 E+02
ESU of inductance	henry (H)	8.987 554 E+11
ESU of resistance	ohm (Ω)	8.987 554 E+11
erg	joule (J)	1.000 000*E−07
erg/centimetre²-second	watt/metre² (W/m²)	1.000 000*E−03
erg/second	watt (W)	1.000 000*E−07
farad, international U.S. ($F_{INT\text{-}US}$)	farad (F)	9.995 05 E−01
faraday (based on carbon-12)	coulomb (C)	9.648 70 E+04
faraday (chemical)	coulomb (C)	9.649 57 E+04
faraday (physical)	coulomb (C)	9.652 19 E+04
fathom	metre (m)	1.828 800*E+00
fermi (femtometer)	metre (m)	1.000 000*E−15
fluid ounce (U.S.)	metre³ (m³)	2.957 353 E−05
foot	metre (m)	3.048 000*E−01
foot (U.S. survey)	metre (m)	3.048 006 E−01
foot³/minute	metre³/second (m³/s)	4.719 474 E−04
foot³/second	metre³/second (m³/s)	2.831 685 E−02
foot³	metre³ (m³)	2.831 685 E−02
foot²	metre² (m²)	9.290 304*E−02
foot/hour	metre/second (m/s)	8.466 667 E−05
foot/minute	metre/second (m/s)	5.080 000*E−03
foot/second	metre/second (m/s)	3.048 000*E−01
foot²/second	metre²/second (m²/s)	9.290 304*E−02
foot of water (39.2°F)	pascal (Pa)	2.988 98 E+03
footcandle	lumen/metre² (lm/m²)	1.076 391 E+01
footcandle	lux (lx)	1.076 391 E+01
footlambert	candela/metre² (cd/m²)	3.426 259 E+00
foot-pound-force	joule (J)	1.355 818 E+00

E 380

To convert from	to	Multiply by
foot-pound-force/hour	watt (W)	3.766 161 E−04
foot-pound-force/minute	watt (W)	2.259 697 E−02
foot-pound-force/second	watt (W)	1.355 818 E+00
foot-poundal	joule (J)	4.214 011 E−02
ft^2/h (thermal diffusivity)	metre^2/second (m^2/s)	2.580 640*E−05
foot/second^2	metre/second^2 (m/s^2)	3.048 000*E−01
free fall, standard	metre/second^2 (m/s^2)	9.806 650*E+00
gal (galileo)	metre/second^2 (m/s^2)	1.000 000*E−02
gallon (Canadian liquid)	metre^3 (m^3)	4.546 090 E−03
gallon (U.K. liquid)	metre^3 (m^3)	4.546 092 E−03
gallon (U.S. dry)	metre^3 (m^3)	4.404 884 E−03
gallon (U.S. liquid)	metre^3 (m^3)	3.785 412 E−03
gallon (U.S. liquid)/day	metre^3/second (m^3/s)	4.381 264 E−08
gallon (U.S. liquid)/minute	metre^3/second (m^3/s)	6.309 020 E−05
gamma	tesla (T)	1.000 000*E−09
gauss	tesla (T)	1.000 000*E−04
gilbert	ampere-turn	7.957 747 E−01
gill (U.K.)	metre^3 (m^3)	1.420 654 E−04
gill (U.S.)	metre^3 (m^3)	1.182 941 E−04
grad	degree (angular)	9.000 000*E−01
grad	radian (rad)	1.570 796 E−02
grain (1/7000 lbm avoirdupois)	kilogram (kg)	6.479 891*E−05
grain (lbm avoirdupois/7000)/gallon (U.S. liquid)	kilogram/metre^3 (kg/m^3)	1.711 806 E−02
gram	kilogram (kg)	1.000 000*E−03
gram/centimetre^3	kilogram/metre^3 (kg/m^3)	1.000 000*E+03
gram-force/centimetre^2	pascal (Pa)	9.806 650*E+01
henry, international U.S. (H_{INT-US})	henry (H)	1.000 495 E+00
horsepower (550 ft·lbf/s)	watt (W)	7.456 999 E+02
horsepower (boiler)	watt (W)	9.809 50 E+03
horsepower (electric)	watt (W)	7.460 000*E+02
horsepower (metric)	watt (W)	7.354 99 E+02
horsepower (water)	watt (W)	7.460 43 E+02
horsepower (U.K.)	watt (W)	7.457 0 E+02
hour (mean solar)	second (s)	3.600 000 E+03
hour (sidereal)	second (s)	3.590 170 E+03
hundredweight (long)	kilogram (kg)	5.080 235 E+01
hundredweight (short)	kilogram (kg)	4.535 924 E+01
inch	metre (m)	2.540 000*E−02
inch^2	metre^2 (m^2)	6.451 600*E−04
inch^3	metre^3 (m^3)	1.638 706 E−05
inch^3/minute	metre^3/second (m^3/s)	2.731 177 E−07
inch/second	metre/second (m/s)	2.540 000*E−02
inch of mercury (32°F)	pascal (Pa)	3.386 389 E+03
inch of mercury (60°F)	pascal (Pa)	3.376 85 E+03
inch of water (39.2°F)	pascal (Pa)	2.490 82 E+02
inch of water (60°F)	pascal (Pa)	2.488 4 E+02
inch/second^2	metre/second^2 (m/s^2)	2.540 000*E−02
joule, international U.S. (J_{INT-US})[12]	joule (J)	1.000 182 E+00
joule, U.S. legal 1948 (J_{US-48})	joule (J)	1.000 017 E+00
kayser	1/metre (1/m)	1.000 000*E+02
kelvin	degree Celsius	$t_C = t_K - 273.15$
kilocalorie (International Table)	joule (J)	4.186 800*E+03
kilocalorie (mean)	joule (J)	4.190 02 E+03
kilocalorie (thermochemical)	joule (J)	4.184 000*E+03
kilocalorie (thermochemical)/minute	watt (W)	6.973 333 E+01
kilocalorie (thermochemical)/second	watt (W)	4.184 000*E+03
kilogram-force (kgf)	newton (N)	9.806 650*E+00
kilogram-force-metre	newton-metre (N·m)	9.806 650*E+00
kilogram-force-second^2/metre (mass)	kilogram (kg)	9.806 650*E+00
kilogram-force/centimetre^2	pascal (Pa)	9.806 650*E+04
kilogram-force/metre^2	pascal (Pa)	9.806 650*E+00
kilogram-force/millimetre^2	pascal (Pa)	9.806 650*E+06
kilogram-mass	kilogram (kg)	1.000 000*E+00
kilometre/hour	metre/second (m/s)	2.777 778 E−01
kilopond-force	newton (N)	9.806 650*E+00
kilowatt-hour	joule (J)	3.600 000*E+06

E 380

To convert from	to	Multiply by
kilowatt-hour, international U.S. (kWh_{INT-US})[12]	joule (J)	3.600 655 E+06
kilowatt-hour, U.S. legal 1948 (kWh_{US-48})	joule (J)	3.600 061 E+06
kip (1000 lbf)	newton (N)	4.448 222 E+03
kip/inch^2 (ksi)	pascal (Pa)	6.894 757 E+06
knot (international)	metre/second (m/s)	5.144 444 E−01
lambert	candela/metre^2 (cd/m^2)	1/π* E+04
lambert	candela/metre^2 (cd/m^2)	3.183 099 E+03
langley	joule/metre^2 (J/m^2)	4.184 000*E+04
league (international nautical)	metre (m)	5.556 000*E+03
league (statute)	metre (m)	4.828 032*E+03
league (U.K. nautical)	metre (m)	5.559 552*E+03
light year	metre (m)	9.460 55 E+15
litre[14]	metre^3 (m^3)	1.000 000*E−03
lux	lumen/metre^2 (lm/m^2)	1.000 000*E+00
maxwell	weber (Wb)	1.000 000*E−08
microinch	metre (m)	2.540 000*E−08
micron	metre (m)	1.000 000*E−06
mil	metre (m)	2.540 000*E−05
mile (international nautical)	metre (m)	1.852 000*E+03
mile (U.K. nautical)	metre (m)	1.853 184*E+03
mile (U.S. nautical)	metre (m)	1.852 000*E+03
mile (U.S. statute)	metre (m)	1.609 344*E+03
mile^2 (U.S. statute)	metre^2 (m^2)	2.589 988 E+06
mile/hour (U.S. statute)	metre/second (m/s)	4.470 400*E−01
mile/minute (U.S. statute)	metre/second (m/s)	2.682 240*E+01
mile/second (U.S. statute)	metre/second (m/s)	1.609 344*E+03
mile/hour (U.S. statute)	**kilometre/hour**	1.609 344*E+00
millibar	pascal (Pa)	1.000 000*E+02
millimetre of mercury (0 °C)	pascal (Pa)	1.333 224 E+02
minute (angle)	radian (rad)	2.908 882 E−04
minute (mean solar)	second (s)	6.000 000 E+01
minute (sidereal)	second (s)	5.983 617 E+01
moment of inertia (lbm·ft^2)	kilogram-metre^2 (kg·m^2)	4.214 011 E−02†
moment of inertia (lbm·in^2)	kilogram-metre^2 (kg·m^2)	2.926 397 E−04†
moment of section[15] (second moment of area) (foot^4)	metre^4 (m^4)	8.630 975 E−03
moment of section[15] (second moment of area) (inch^4)	metre^4 (m^4)	4.162 314 E−07
month (mean calendar)	second (s)	2.268 000 E+06
oersted	ampere/metre (A/m)	7.957 747 E+01
ohm, international U.S. (Ω_{INT-US})	ohm (Ω)	1.000 495 E+00
ohm-centimetre	ohm-metre (Ω·m)	1.000 000*E−02
ounce-force (avoirdupois)	newton (N)	2.780 139 E−01
ounce-force-inch	newton-metre (N·m)	7.061 552 E−03
ounce-mass (avoirdupois)	kilogram (kg)	2.834 952 E−02
ounce-mass (troy or apothecary)	kilogram (kg)	3.110 348 E−02
ounce-mass/yard^2	kilogram/metre^2 (kg/m^2)	3.390 575 E−02
ounce (avoirdupois)/gallon (U.K. liquid)	kilogram/metre^3 (kg/m^3)	6.236 021 E+00†
ounce (avoirdupois)/gallon (U.S. liquid)	kilogram/metre^3 (kg/m^3)	7.489 152 E+00
ounce (avoirdupois) (mass)/inch^3	kilogram/metre^3 (kg/m^3)	1.729 994 E+03
ounce (U.K. fluid)	metre^3 (m^3)	2.841 307 E−05
ounce (U.S. fluid)	metre^3 (m^3)	2.957 353 E−05
parsec	metre (m)	3.083 74 E+16
peck (U.S.)	metre^3 (m^3)	8,809 768 E−03
pennyweight	kilogram (kg)	1.555 174 E−03
perm (0°C)	kilogram/pascal-second-metre^2 (kg/Pa·s·m^2)	5.721 35 E−11

[14] In 1964 the General Conference on Weights and Measures adopted the name litre as a special name for the cubic decimetre. Prior to this decision the litre differed slightly (previous value, 1.000028 dm^3) and in expression of precision volume measurement this fact must be kept in mind.

[15] This is sometimes called the moment of inertia of a plane section about a specified axis.

E 380

To convert from	to	Multiply by
perm (23°C)	kilogram/pascal-second-metre² (kg/Pa·s·m²)	5.745 25 E−11
perm-inch (0°C)	kilogram/pascal-second-metre (kg/Pa·s·m)	1.453 22 E−12
perm-inch (23°C)	kilogram/pascal-second-metre (kg/Pa·s·m)	1.459 29 E−12
phot	lumen/metre² (lm/m²)	1.000 000*E+04
pica (printer's)	metre (m)	4.217 518 E−03
pint (U.S. dry)	metre³ (m³)	5.506 105 E−04
pint (U.S. liquid)	metre³ (m³)	4.731 765 E−04
point (printer's)	metre	3.514 598*E−04
poise (absolute viscosity)	pascal-second (Pa·s)	1.000 000*E−01
poundal	newton (N)	1.382 550 E−01
poundal/foot²	pascal (Pa)	1.488 164 E+00
poundal-second/foot²	pascal-second (Pa·s)	1.488 164 E+00
pound-force (lbf avoirdupois)[16]	newton (N)	4.448 222 E+00
pound-force-inch	newton-metre (N·m)	1.129 848 E−01
pound-force-foot	newton-metre (N·m)	1.355 818 E+00
pound-force-foot/inch	newton-metre/metre (N·m/m)	5.337 866 E+01
pound-force-inch/inch	newton-metre/metre (N·m/m)	4.448 222 E+00
pound-force/inch	newton/metre (N/m)	1.751 268 E+02
pound-force/foot	newton/metre (N/m)	1.459 390 E+01
pound-force/foot²	pascal (Pa)	4.788 026 E+01
pound-force/inch² (psi)	pascal (Pa)	6.894 757 E+03
pound-force-second/foot²	pascal-second (Pa·s)	4.788 026 E+01
pound-mass (lbm avoirdupois)[17]	kilogram (kg)	4.535 924 E−01
pound-mass (troy or apothecary)	kilogram (kg)	3.732 417 E−01
pound-mass/foot²	kilogram/metre² (kg/m²)	4.882 428 E+00
pound-mass/second	kilogram/second (kg/s)	4.535 924 E−01
pound-mass/minute	kilogram/second (kg/s)	7.559 873 E−03
pound-mass/foot³	kilogram/metre³ (kg/m³)	1.601 846 E+01
pound-mass/inch³	kilogram/metre³ (kg/m³)	2.767 990 E+04
pound-mass/gallon (U.K. liquid)	kilogram/metre³ (kg/m³)	9.977 633 E+01†
pound-mass/gallon (U.S. liquid)	kilogram/metre³ (kg/m³)	1.198 264 E+02
pound-mass/foot-second	pascal-second (Pa·s)	1.488 164 E+00
quart (U.S. dry)	metre³ (m³)	1.101 221 E−03
quart (U.S. liquid)	metre³ (m³)	9.463 529 E−04
rad (radiation dose absorbed)	joule/kilogram (J/kg)	1.000 000*E−02
rhe	metre²/newton-second (m²/N·s)	1.000 000*E+01
rod	metre (m)	5.029 200*E+00
roentgen	coulomb/kilogram (C/kg)	2.579 760*E−04
second (angle)	radian (rad)	4.848 137 E−06
second (sidereal)	second (s)	9.972 696 E−01
section	metre² (m²)	2.589 988 E+06
section modulus (foot³)	metre³ (m³)	2.831 685 E−02
section modulus (inch³)	metre³ (m³)	1.638 706 E−05
shake	second (s)	1.000 000*E−08
slug	kilogram (kg)	1.459 390 E+01
slug/foot³	kilogram/metre³ (kg/m³)	5.153 788 E+02
slug/foot-second	pascal-second (Pa·s)	4.788 026 E+01
statampere	ampere (A)	3.335 640 E−10
statcoulomb	coulomb (C)	3.335 640 E−10
statfarad	farad (F)	1.112 650 E−12
stathenry	henry (H)	8.987 554 E+11
statmho	siemens (S)	1.112 650 E−12
statohm	ohm (Ω)	8.987 554 E+11
statute mile (U.S.)	metre (m)	1.609 344*E+03
statvolt	volt (V)	2.997 925 E+02
stere	metre³ (m³)	1.000 000*E+00
stilb	candela/metre² (cd/m²)	1.000 000*E+04
stokes (kinematic viscosity)	metre²/second (m²/s)	1.000 000*E−04
tablespoon	metre³ (m³)	1.478 676 E−05
teaspoon	metre³ (m³)	4.928 922 E−06
tex	kilogram/metre (kg/m)	1.000 000*E−06

[16] The exact conversion factor is 4.448 221 615 260 5*E+00.
[17] The exact conversion factor is 4.535 923 7*E−01.

E 380

To convert from	to	Multiply by
ton (assay)	kilogram (kg)	2.916 667 E−02
ton (long, 2240 lbm)	kilogram (kg)	1.016 047 E+03
ton (metric)	kilogram (kg)	1.000 000*E+03
ton (nuclear equivalent of TNT)	joule (J)	4.20 E+09
ton (register)	metre^3 (m^3)	2.831 685 E+00
ton (short, 2000 lbm)	kilogram (kg)	9.071 847 E+02
ton (short, mass)/hour	kilogram/second (kg/s)	2.519 958 E−01
ton (long, mass)/yard^3	kilogram/metre^3 (kg/m^3)	1.328 939 E+03
tonne	kilogram (kg)	1.000 000*E+03
torr (mm Hg, 0°C)	pascal (Pa)	1.333 22 E+02
township	metre^2 (m^2)	9.323 957 E+07
unit pole	weber (Wb)	1.256 637 E−07
volt, international U.S. ($V_{INT\text{-}US}$)[12]	volt (V)	1.000 338 E+00
volt, U.S. legal 1948 ($V_{US\text{-}48}$)	volt (V)	1.000 008 E+00
watt, international U.S. ($W_{INT\text{-}US}$)[12]	watt (W)	1.000 182 E+00
watt, U.S. legal 1948 ($W_{US\text{-}48}$)	watt (W)	1.000 017 E+00
watt/centimetre^2	watt/metre^2 (W/m^2)	1.000 000*E+04
watt-hour	joule (J)	3.600 000*E+03
watt-second	joule (J)	1.000 000*E+00
yard	metre (m)	9.144 000*E−01
yard^2	metre^2 (m^2)	8.361 274 E−01
yard^3	metre^3 (m^3)	7.645 549 E−01
yard^3/minute	metre^3/second (m^3/s)	1.274 258 E−02
year (calendar)	second (s)	3.153 600 E+07
year (sidereal)	second (s)	3.155 815 E+07
year (tropical)	second (s)	3.155 693 E+07

CLASSIFIED LIST OF UNITS

Acceleration

foot/second^2	metre/second^2 (m/s^2)	3.048 000*E−01
free fall, standard	metre/second^2 (m/s^2)	9.806 650*E+00
gal (galileo)	metre/second^2 (m/s^2)	1.000 000*E−02
inch/second^2	metre/second^2 (m/s^2)	2.540 000*E−02

Area

acre	metre^2 (m^2)	4.046 856 E+03
barn	metre^2 (m^2)	1.000 000*E−28
circular mil	metre^2 (m^2)	5.067 075 E−10
foot^2	metre^2 (m^2)	9.290 304*E−02
inch^2	metre^2 (m^2)	6.451 600*E−04
mile^2 (U.S. statute)	metre^2 (m^2)	2.589 988 E+06
section	metre^2 (m^2)	2.589 988 E+06
township	metre^2 (m^2)	9.323 957 E+07
yard^2	metre^2 (m^2)	8.361 274 E−01

Bending Moment or Torque

dyne-centimetre	newton-metre (N·m)	1.000 000*E−07
kilogram-force-metre	newton-metre (N·m)	9.806 650*E+00
ounce-force-inch	newton-metre (N·m)	7.061 552 E−03
pound-force-inch	newton-metre (N·m)	1.129 848 E−01
pound-force-foot	newton-metre (N·m)	1.355 818 E+00

(Bending Moment or Torque)/Length

pound-force-foot/inch	newton-metre/metre (N·m/m)	5.337 866 E+01
pound-force-inch/inch	newton-metre/metre (N·m/m)	4.448 222 E+00

E 380

Capacity (See Volume)

Density (See Mass/Volume)

Electricity and Magnetism[18]

To convert from	to	Multiply by
abampere	ampere (A)	1.000 000*E+01
abcoulomb	coulomb (C)	1.000 000*E+01
abfarad	farad (F)	1.000 000*E+09
abhenry	henry (H)	1.000 000*E−09
abmho	siemens (S)	1.000 000*E+09
abohm	ohm (Ω)	1.000 000*E−09
abvolt	volt (V)	1.000 000*E−08
ampere, international U.S. (A_{INT-US})[12]	ampere (A)	9.998 43 E−01
ampere, U.S. legal 1948 (A_{US-48})	ampere (A)	1.000 008 E+00
ampere-hour	coulomb (C)	3.600 000*E+03
coulomb, international U.S. (C_{INT-US})[12]	coulomb (C)	9.998 43 E−01
coulomb, U.S. legal 1948 (C_{US-48})	coulomb (C)	1.000 008 E+00
EMU of capacitance	farad (F)	1.000 000*E+09
EMU of current	ampere (A)	1.000 000*E+01
EMU of electric potential	volt (V)	1.000 000*E−08
EMU of inductance	henry (H)	1.000 000*E−09
EMU of resistance	ohm (Ω)	1.000 000*E−09
ESU of capacitance	farad (F)	1.112 650 E−12
ESU of current	ampere (A)	3.335 6 E−10
ESU of electric potential	volt (V)	2.997 9 E+02
ESU of inductance	henry (H)	8.987 554 E+11
ESU of resistance	ohm (Ω)	8.987 554 E+11
farad, international U.S. (F_{INT-US})	farad (F)	9.995 05 E−01
faraday (based on carbon-12)	coulomb (C)	9.648 70 E+04
faraday (chemical)	coulomb (C)	9.649 57 E+04
faraday (physical)	coulomb (C)	9.652 19 E+04
gamma	tesla (T)	1.000 000*E−09
gauss	tesla (T)	1.000 000*E−04
gilbert	ampere-turn	7.957 747 E−01
henry, international U.S. (H_{INT-US})	henry (H)	1.000 495 E+00
maxwell	weber (Wb)	1.000 000*E−08
oersted	ampere/metre (A/m)	7.957 747 E+01
ohm, international U.S. (Ω_{INT-US})	ohm (Ω)	1.000 495 E+00
ohm-centimeter	ohm-metre (Ω·m)	1.000 000*E−02
statampere	ampere (A)	3.335 640 E−10
statcoulomb	coulomb (C)	3.335 640 E−10
statfarad	farad (F)	1.112 650 E−12
stathenry	henry (H)	8.987 554 E+11
statmho	siemens (S)	1.112 650 E−12
statohm	ohm (Ω)	8.987 554 E+11
statvolt	volt (V)	2.997 925 E+02
unit pole	weber (Wb)	1.256 637 E−07
volt, international U.S. (V_{INT-US})[12]	volt (V)	1.000 338 E+00
volt, U.S. legal 1948 (V_{US-48})	volt (V)	1.000 008 E+00

Energy (Includes Work)

To convert from	to	Multiply by
British thermal unit (International Table)[13]	joule (J)	1.055 056 E+03
British thermal unit (mean)	joule (J)	1.055 87 E+03
British thermal unit (thermochemical)	joule (J)	1.054 350 E+03
British thermal unit (39°F)	joule (J)	1.059 67 E+03
British thermal unit (60°F)	joule (J)	1.054 68 E+03
calorie (International Table)	joule (J)	4.186 800*E+00
calorie (mean)	joule (J)	4.190 02 E+00
calorie (thermochemical)	joule (J)	4.184 000*E+00
calorie (15°C)	joule (J)	4.185 80 E+00

[18] ESU means electrostatic cgs unit. EMU means electromagnetic cgs unit.

E 380

To convert from	to	Multiply by
calorie (20°C)	joule (J)	4.181 90 E+00
calorie (kg, International Table)	joule (J)	4.186 800*E+03
calorie (kg, mean)	joule (J)	4.190 02 E+03
calorie (kg, thermochemical)	joule (J)	4.184 000*E+03
electron volt	joule (J)	1.602 19 E−19†
erg	joule (J)	1.000 000*E−07
foot-pound-force	joule (J)	1.355 818 E+00
foot-poundal	joule (J)	4.214 011 E−02
joule, international U.S. (J_{INT-US})[12]	joule (J)	1.000 182 E+00
joule, U.S. legal 1948 (J_{US-48})	joule (J)	1.000 017 E+00
kilocalorie (International Table)	joule (J)	4.186 800*E+03
kilocalorie (mean)	joule (J)	4.190 02 E+03
kilocalorie (thermochemical)	joule (J)	4.184 000*E+03
kilowatt-hour	joule (J)	3.600 000*E+06
kilowatt-hour, international U.S. (kWh_{INT-US})[12]	joule (J)	3.600 655 E+06
kilowatt-hour, U.S. legal 1948 (kWh_{US-48})	joule (J)	3.600 061 E+06
ton (nuclear equivalent of TNT)	joule (J)	4.20 E+09
watt-hour	joule (J)	3.600 000*E+03
watt-second	joule (J)	1.000 000*E+00

Energy/Area Time

Btu (thermochemical)/foot²-second	watt/metre² (W/m²)	1.134 893 E+04
Btu (thermochemical)/foot²-minute	watt/metre² (W/m²)	1.891 489 E+02
Btu (thermochemical)/foot²-hour	watt/metre² (W/m²)	3.152 481 E+00
Btu (thermochemical)/inch²-second	watt/metre² (W/m²)	1.634 246 E+06
calorie (thermochemical)/centimetre² minute	watt/metre² (W/m²)	6.973 333 E+02
erg/centimetre²-second	watt/metre² (W/m²)	1.000 000*E−03
watt/centimetre²	watt/metre² (W/m²)	1.000 000*E+04

Flow (See Mass/Time or Volume/Time)

Force

dyne	newton (N)	1.000 000*E−05†
kilogram-force	newton (N)	9.806 650*E+00
kilopond-force	newton (N)	9.806 650*E+00
kip	newton (N)	4.448 222 E+03
ounce-force (avoirdupois)	newton (N)	2.780 139 E−01
pound-force (lbf avoirdupois)[16]	newton (N)	4.448 222 E+00
poundal	newton (N)	1.382 550 E−01

Force/Area (See Pressure)

Force/Length

pound-force/inch	newton/metre (N/m)	1.751 268 E+02
pound-force/foot	newton/metre (N/m)	1.459 390 E+01

Heat

Btu (thermochemical)·in./s·ft²·°F (*k*, thermal conductivity)	watt/metre-kelvin (W/m·K)	5.188 732 E+02
Btu (International Table)·in./s·ft²·°F (*k*, thermal conductivity)	watt/metre-kelvin (W/m·K)	5.192 204 E+02
Btu (thermochemical)·in./h·ft²·°F (*k*, thermal conductivity)	watt/metre-kelvin (W/m·K)	1.441 314 E−01

E 380

To convert from	to	Multiply by
Btu (International Table)·in./h·ft²·°F (*k*, thermal conductivity)	watt/metre-kelvin (W/m·K)	1.442 279 E−01
Btu (International Table)/ft²	joule/metre² (J/m²)	1.135 653 E+04
Btu (thermochemical)/ft²	joule/metre² (J/m²)	1.134 893 E+04
Btu (International Table)/h·ft²·°F (*C*, thermal conductance)	watt/metre²-kelvin (W/m²·K)	5.678 263 E+00
Btu (thermochemical)/h·ft²·°F (*C*, thermal conductance)	watt/metre²-kelvin (W/m²·K)	5.674 466 E+00
Btu (International Table)/pound-mass	joule/kilogram (J/kg)	2.326 000*E+03
Btu (thermochemical)/pound-mass	joule/kilogram (J/kg)	2.324 444 E+03
Btu (International Table)/lbm·°F (*c*, heat capacity)	joule/kilogram-kelvin (J/kg·K)	4.186 800*E+03
Btu (thermochemical)/lbm·°F (*c*, heat capacity)	joule/kilogram-kelvin (J/kg·K)	4.184 000 E+03
Btu (International Table)/s·ft²·°F	watt/metre²-kelvin (W/m²·K)	2.044 175 E+04
Btu (thermochemical)/s·ft²·°F	watt/metre²-kelvin (W/m²·K)	2.042 808 E+04
cal (thermochemical)/cm²	joule/metre² (J/m²)	4.184 000*E+04
cal (thermochemical)/cm²·s	watt/metre² (W/m²)	4.184 000*E+04
cal (thermochemical)/cm·s·°C	watt/metre-kelvin (W/m·K)	4.184 000*E+02
cal (International Table)/g	joule/kilogram (J/kg)	4.186 800*E+03
cal (International Table)/g·°C	joule/kilogram-kelvin (J/kg·K)	4.186 800*E+03
cal (thermochemical)/g	joule/kilogram (J/kg)	4.184 000*E+03
cal (thermochemical)/g·°C	joule/kilogram-kelvin (J/kg·K)	4.184 000*E+03
clo	kelvin-metre²/watt (K·m²/W)	2.003 712 E−01
deg F·h·ft²/Btu (thermochemical) (*R*, thermal resistance)	kelvin-metre²/watt (K·m²/W)	1.762 280 E−01
deg F·h·ft²/Btu (International Table) (*R*, thermal resistance)	kelvin-metre²/watt (K·m²/W)	1.761 102 E−01
ft²/h (thermal diffusivity)	metre²/second (m²/s)	2.580 640*E−05

LENGTH

angstrom	metre (m)	1.000 000*E−10
astronomical unit	metre (m)	1.495 98 E+11
caliber (inch)	metre (m)	2.540 000*E−02
fathom	metre (m)	1.828 800*E+00
fermi (femtometer)	metre (m)	1.000 000*E−15
foot	metre (m)	3.048 000*E−01
foot (U.S. survey)[19]	metre (m)	3.048 006 E−01
inch	metre (m)	2.540 000*E−02
league (international nautical)	metre (m)	5.556 000*E+03
league (statute)	metre (m)	4.828 032*E+03
league (U.K. nautical)	metre (m)	5.559 552*E+03
light year	metre (m)	9.460 55 E+15
microinch	metre (m)	2.540 000*E−08
micron	metre (m)	1.000 000*E−06
mil	metre (m)	2.540 000*E−05
mile (international nautical)	metre (m)	1.852 000*E+03
mile (U.K. nautical)	metre (m)	1.853 184*E+03
mile (U.S. nautical)	metre (m)	1.852 000*E+03
mile (U.S. statute)	metre (m)	1.609 344*E+03
parsec	metre (m)	3.083 74 E+16
pica (printer's)	metre (m)	4.217 518 E−03
point (printer's)	metre (m)	3.514 598*E−04
rod	metre (m)	5.029 200*E+00
statute mile (U.S.)	metre (m)	1.609 344*E+03
yard	metre (m)	9.144 000*E−01

LIGHT

footcandle	lumen/metre² (lm/m²)	1.076 391 E+01
footcandle	lux (lx)	1.076 391 E+01

[19] The exact conversion factor is 1200/3937.

E 380

To convert from	to	Multiply by
footlambert	candela/metre² (cd/m²)	3.426 259 E+00
lux	lumen/metre² (lm/m²)	1.000 000*E+00

MASS

To convert from	to	Multiply by
carat (metric)	kilogram (kg)	2.000 000*E−04
grain	kilogram (kg)	6.479 891*E−05
gram	kilogram (kg)	1.000 000*E−03
hundredweight (long)	kilogram (kg)	5.080 235 E+01
hundredweight (short)	kilogram (kg)	4.535 924 E+01
kilogram-force-second²/metre (mass)	kilogram (kg)	9.806 650*E+00
kilogram-mass	kilogram (kg)	1.000 000*E+00
ounce-mass (avoirdupois)	kilogram (kg)	2.834 952 E−02
ounce-mass (troy or apothecary)	kilogram (kg)	3.110 348 E−02
pennyweight	kilogram (kg)	1.555 174 E−03
pound-mass (lbm avoirdupois)[17]	kilogram (kg)	4.535 924 E−01
pound-mass (troy or apothecary)	kilogram (kg)	3.732 417 E−01
slug	kilogram (kg)	1.459 390 E+01
ton (assay)	kilogram (kg)	2.916 667 E−02
ton (long, 2240 lbm)	kilogram (kg)	1.016 047 E+03
ton (metric)	kilogram (kg)	1.000 000*E+03
ton (short, 2000 lbm)	kilogram (kg)	9.071 847 E+02
tonne	kilogram (kg)	1.000 000*E+03

MASS/AREA

To convert from	to	Multiply by
ounce-mass/yard²	kilogram/metre² (kg/m²)	3.390 575 E−02
pound-mass/foot²	kilogram/metre² (kg/m²)	4.882 428 E+00

MASS/CAPACITY (SEE MASS/VOLUME)

MASS/TIME (INCLUDES FLOW)

To convert from	to	Multiply by
perm (0°C)	kilogram/pascal-second-metre² (kg/Pa·s·m²)	5.721 35 E−11
perm (23°C)	kilogram/pascal-second-metre² (kg/Pa·s·m²)	5.745 25 E−11
perm-inch (0°C)	kilogram/pascal-second-metre (kg/Pa·s·m)	1.453 22 E−12
perm-inch (23°C)	kilogram/pascal-second-metre (kg/Pa·s·m)	1.459 29 E−12
pound-mass/second	kilogram/second (kg/s)	4.535 924 E−01
pound-mass/minute	kilogram/second (kg/s)	7.559 873 E−03
ton (short, mass)/hour	kilogram/second (kg/s)	2.519 958 E−01

MASS/VOLUME (INCLUDES DENSITY AND MASS CAPACITY)

To convert from	to	Multiply by
grain (lbm avoirdupois/7000)/gallon (U.S. liquid)	kilogram/metre³ (kg/m³)	1.711 806 E−02
gram/centimetre³	kilogram/metre³ (kg/m³)	1.000 000*E+03
ounce (avoirdupois)/gallon (U.K. liquid)	kilogram/metre³ (kg/m³)	6.236 021 E+00†
ounce (avoirdupois)/gallon (U.S. liquid)	kilogram/metre³ (kg/m³)	7.489 152 E+00
ounce (avoirdupois) (mass)/inch³	kilogram/metre³ (kg/m³)	1.729 994 E+03
pound-mass/foot³	kilogram/metre³ (kg/m³)	1.601 846 E+01
pound-mass/inch³	kilogram/metre³ (kg/m³)	2.767 990 E+04
pound-mass/gallon (U.K. liquid)	kilogram/metre³ (kg/m³)	9.977 633 E+01†
pound-mass/gallon (U.S. liquid)	kilogram/metre³ (kg/m³)	1.198 264 E+02
slug/foot³	kilogram/metre³ (kg/m³)	5.153 788 E+02
ton (long, mass)/yard³	kilogram/metre³ (kg/m³)	1.328 939 E+03

E 380

Power

To convert from	to	Multiply by
Btu (International Table)/hour	watt (W)	2.930 711 E−01
Btu (thermochemical)/second	watt (W)	1.054 350 E+03
Btu (thermochemical)/minute	watt (W)	1.757 250 E+01
Btu (thermochemical)/hour	watt (W)	2.928 751 E−01
calorie (thermochemical)/second	watt (W)	4.184 000*E+00
calorie (thermochemical)/minute	watt (W)	6.973 333 E−02
erg/second	watt (W)	1.000 000*E−07
foot-pound-force/hour	watt (W)	3.766 161 E−04
foot-pound-force/minute	watt (W)	2.259 697 E−02
foot-pound-force/second	watt (W)	1.355 818 E+00
horsepower (550 ft·lbf/s)	watt (W)	7.456 999 E+02
horsepower (boiler)	watt (W)	9.809 50 E+03
horsepower (electric)	watt (W)	7.460 000*E+02
horsepower (metric)	watt (W)	7.354 99 E+02
horsepower (water)	watt (W)	7.460 43 E+02
horsepower (U.K.)	watt (W)	7.457 0 E+02
kilocalorie (thermochemical)/minute	watt (W)	6.973 333 E+01
kilocalorie (thermochemical)/second	watt (W)	4.184 000*E+03
watt, international U.S. (W_{INT-US})[12]	watt (W)	1.000 182 E+00
watt, U.S. legal 1948 (W_{US-48})	watt (W)	1.000 017 E+00

Pressure or Stress (Force/Area)

To convert from	to	Multiply by
atmosphere (normal = 760 torr)	pascal (Pa)	1.013 25 E+05
atmosphere (technical = 1 kgf/cm²)	pascal (Pa)	9.806 650*E+04
bar	pascal (Pa)	1.000 000*E+05
centimetre of mercury (0°C)	pascal (Pa)	1.333 22 E+03
centimetre of water (4°C)	pascal (Pa)	9.806 38 E+01
decibar	pascal (Pa)	1.000 000*E+04
dyne/centimetre²	pascal (Pa)	1.000 000*E−01
foot of water (39.2°F)	pascal (Pa)	2.988 98 E+03
gram-force/centimetre²	pascal (Pa)	9.806 650*E+01
inch of mercury (32°F)	pascal (Pa)	3.386 389 E+03
inch of mercury (60°F)	pascal (Pa)	3.376 85 E+03
inch of water (39.2°F)	pascal (Pa)	2.490 82 E+02
inch of water (60°F)	pascal (Pa)	2.488 4 E+02
kilogram-force/centimetre²	pascal (Pa)	9.806 650*E+04
kilogram-force/metre²	pascal (Pa)	9.806 650*E+00
kilogram-force/millimetre²	pascal (Pa)	9.806 650*E+06
kip/inch² (ksi)	pascal (Pa)	6.894 757 E+06
millibar	pascal (Pa)	1.000 000*E+02
millimetre of mercury (0°C)	pascal (Pa)	1.333 224 E+02
poundal/foot²	pascal (Pa)	1.488 164 E+00
pound-force/foot²	pascal (Pa)	4.788 026 E+01
pound-force/inch² (psi)	pascal (Pa)	6.894 757 E+03
psi	pascal (Pa)	6.894 757 E+03
torr (mm Hg, 0°C)	pascal (Pa)	1.333 22 E+02

Speed (See Velocity)

Stress (See Pressure)

Temperature

To convert from	to	Multiply by
degree Celsius	kelvin (K)	$t_K = t°_C + 273.15$
degree Fahrenheit	kelvin (K)	$t_K = (t°_F + 459.67)/1.8$
degree Rankine	kelvin (K)	$t_K = t°_R/1.8$
degree Fahrenheit	degree Celsius	$t°_C = (t°_F - 32)/1.8$
kelvin	degree Celsius	$t°_C = t_K - 273.15$

E 380

Time

To convert from	to	Multiply by
day (mean solar)	second (s)	8.640 000 E+04
day (sidereal)	second (s)	8.616 409 E+04
hour (mean solar)	second (s)	3.600 000 E+03
hour (sidereal)	second (s)	3.590 170 E+03
minute (mean solar)	second (s)	6.000 000 E+01
minute (sidereal)	second (s)	5.983 617 E+01
month (mean calendar)	second (s)	2.628 000 E+06
second (sidereal)	second (s)	9.972 696 E−01
year (calendar)	second (s)	3.153 600 E+07
year (sidereal)	second (s)	3.155 815 E+07
year (tropical)	second (s)	3.155 693 E+07

Torque (See Bending Moment)

Velocity (Includes Speed)

foot/hour	metre/second (m/s)	8.466 667 E−05
foot/minute	metre/second (m/s)	5.080 000*E−03
foot/second	metre/second (m/s)	3.048 000*E−01
inch/second	metre/second (m/s)	2.540 000*E−02
kilometre/hour	metre/second (m/s)	2.777 778 E−01
knot (international)	metre/second (m/s)	5.144 444 E−01
mile/hour (U.S. statute)	metre/second (m/s)	4.470 400*E−01
mile/minute (U.S. statute)	metre/second (m/s)	2.682 240*E+01
mile/second (U.S. statute)	metre/second (m/s)	1.609 344*E+03
mile/hour (U.S. statute)	kilometre/hour[20]	1.609 344*E+00

Viscosity

centipoise	pascal-second (Pa·s)	1.000 000*E−03
centistokes	metre^2/second (m^2/s)	1.000 000*E−06
foot^2/second	metre^2/second (m^2/s)	9.290 304*E−02
poise	pascal-second (Pa·s)	1.000 000*E−01
poundal-second/foot^2	pascal-second (Pa·s)	1.488 164 E+00
pound-mass/foot-second	pascal-second (Pa·s)	1.488 164 E+00
pound-force-second/foot^2	pascal-second (Pa·s)	4.788 026 E+01
rhe	metre^2/newton-second (m^2/N·s)	1.000 000*E+01
slug/foot-second	pascal-second (Pa·s)	4.788 026 E+01
stokes	metre^2/second (m^2/s)	1.000 000*E−04

Volume (Includes Capacity)

acre-foot	metre^3 (m^3)	1.233 482 E+03
barrel (oil, 42 gal)	metre^3 (m^3)	1.589 873 E−01
board foot	metre^3 (m^3)	2.359 737 E−03
bushel (U.S.)	metre^3 (m^3)	3.523 907 E−02
cup	metre^3 (m^3)	2.365 882 E−04
fluid ounce (U.S.)	metre^3 (m^3)	2.957 353 E−05
foot^3	metre^3 (m^3)	2.831 685 E−02
gallon (Canadian liquid)	metre^3 (m^3)	4.546 090 E−03
gallon (U.K. liquid)	metre^3 (m^3)	4.546 092 E−03
gallon (U.S. dry)	metre^3 (m^3)	4.404 884 E−03
gallon (U.S. liquid)	metre^3 (m^3)	3.785 412 E−03
gill (U.K.)	metre^3 (m^3)	1.420 654 E−04
gill (U.S.)	metre^3 (m^3)	1.182 941 E−04
inch^3, [21]	metre^3 (m^3)	1.638 706 E−05
litre[4]	metre^3 (m^3)	1.000 000*E−03
ounce (U.K. fluid)	metre^3 (m^3)	2.841 307 E−05

[20] Although speedometers may read km/h, the correct SI unit is m/s.
[21] The exact conversion factor is 1.638 706 4*E-05.

E 380

To convert from	to	Multiply by
ounce (U.S. fluid)	metre³ (m^3)	2.957 353 E−05
peck (U.S.)	metre³ (m^3)	8.809 768 E−03
pint (U.S. dry)	metre³ (m^3)	5.506 105 E−04
pint (U.S. liquid)	metre³ (m^3)	4.731 765 E−04
quart (U.S. dry)	metre³ (m^3)	1.101 221 E−03
quart (U.S. liquid)	metre³ (m^3)	9.463 529 E−04
stere	metre³ (m^3)	1.000 000*E+00
tablespoon	metre³ (m^3)	1.478 676 E−05
teaspoon	metre³ (m^3)	4.928 922 E−06
ton (register)	metre³ (m^3)	2.831 685 E+00
yard³	metre³ (m^3)	7.645 549 E−01

VOLUME/TIME (INCLUDES FLOW)

foot³/minute	metre³/second (m^3/s)	4.719 474 E−04
foot³/second	metre³/second (m^3/s)	2.831 685 E−02
inch³/minute	metre³/second (m^3/s)	2.731 177 E−07
yard³/minute	metre³/second (m^3/s)	1.274 258 E−02
gallon (U.S. liquid)/day	metre³/second (m^3/s)	4.381 264 E−08
gallon (U.S. liquid)/minute	metre³/second (m^3/s)	6.309 020 E−05

WORK (SEE ENERGY)

REFERENCES

American Society for Testing and Materials, 1916 Race St., Philadelphia, Pa. 19103:

(1) Definitions A 340, Terms, Symbols, and Conversion Factors Relating to Magnetic Testing
(2) Method D 2161, Conversion of Kinematic Viscosity to Saybolt Universal Viscosity or to Saybolt Furol Viscosity
(3) Petroleum Measurement Tables D 1250
(4) Rec. Practice D 1914, Conversion Units and Factors Relating to Atmospheric Analysis
(5) Rec. Practice E 29, for Indicating Which Places of Figures Are to Be Considered Significant in Specified Limiting Values

American National Standards Institute, Inc., 1430 Broadway, New York, N. Y. 10018:

(6) American National Standard Z10.1-1941, Abbreviations for Scientific and Engineering Terms
(7) American National Standard B48.1-1933 (R1947), Inch-Millimeter Conversion for Industrial Use, Practice for (ISO R370-1964)

International Organization for Standardization, Geneva, Switzerland (available from American National Standards Institute):

(8) ISO/R31/Part I—1956, Basic Quantities and Units of the SI
(9) ISO/R31/Part II—1958, Quantities and Units of Periodic and Related Phenomena
(10) ISO/R31/Part III—1960, Quantities and Units of Mechanics
(11) ISO/R31/Part IV—1960, Quantities and Units of Heat
(12) ISO/R31/Part V—1965, Quantities and Units of Electricity and Magnetism
(13) ISO/R1000—1969, Rules for the Use of Units of the International System of Units and a Selection of the Decimal Multiples and Sub-Multiples of the SI Units

National Bureau of Standards (documents available from Superintendent of Documents, U.S. Government Printing Office, Washington, D.C. 20402):

(14) "Unit of Weights and Measures (United States Customary and Metric): Definitions and Tables of Equivalents," *NBS Misc. Publ. MP233*, Dec. 20, 1960
(15) Judson, L. V., "Weights and Measures Standards of the the United States, a Brief History," *NBS Misc. Publ. 247*, October 1963
(16) "Units and Systems of Weights and Measures. Their Origin, Development, and Present Status," *NBS Letter Circ. LC 1035*, January 1960
(17) "Policy for NBS Usage of SI Units," *NBS Tech. News Bull.*, Vol 55, No. 1, January 1971

E 380

(**18**) Chart—"The Modernized Metric System," *NBS Spec. Publ. 304*, 1968
(**19**) Page, C. H. and Vigoureux, P., "The International System of Units (SI)," *NBS Special Publication 330*, 1972.

Miscellaneous:

(**20**) Mechtly, E. A., "The International System of Units—Physical Constants and Conversion Factors," National Aeronautics and Space Administration, Publication SP-7012. Available from Superintendent of Documents, U.S. Government Printing Office, Washington, D.C. 20402
(**21**) Palmer, J. F., "The International System of Units Handbook," Brown Engineering Co., Inc., Huntsville, Ala.
(**22**) McNish, A. G., "The International System of Units," *Materials Research and Standards*, Am. Soc. Testing Mats., Vol 5, No. 10, October 1965
(**23**) "Conversion Factors and Tables," British Standard 350, Part 1, 1959.
(**24**) "Conversion Factors and Tables," British Standard 350, Part 2, 1962.
(**25**) "Changing to the Metric System, Conversion Factors, Symbols, Definitions," Nat. Physical Lab., Her Majesty's Stationery Office, London, England, 1965
(**26**) *Metric Standards for Engineering*, British Standard Handbook No. 18 (1966)

INDEX OF TERMS

(by Section Number)

Units of Measurement—Conversion Factors*

Units of Length

To Convert from Centimeters

To	Multiply by
Inches	0.393 700 8
Feet	0.032 808 40
Yards	0.010 936 13
Meters	**0.01**

To Convert from Meters

To	Multiply by
Inches	39.370 08
Feet	3.280 840
Yards	1.093 613
Miles	0.000 621 37
Millimeters	**1 000**
Centimeters	**100**
Kilometers	**0.001**

To Convert from Inches

To	Multiply by
Feet	0.083 333 33
Yards	0.027 777 78
Centimeters	**2.54**
Meters	**0.025 4**

To Convert from Feet

To	Multiply by
Inches	**12**
Yards	0.333 333 3
Miles	0.000 189 39
Centimeters	**30.48**
Meters	**0.304 8**
Kilometers	**0.000 304 8**

* All boldface figures are exact; the others generally are given to seven significant figures.

In using conversion factors, it is possible to perform division as well as the multiplication process shown here. Division may be particularly advantageous where more than the significant figures published here are required. Division may be performed in lieu of multiplication by using the reciprocal of any indicated multiplier as divisor. For example, to convert from centimeters to inches by division, refer to the table headed "To Convert from *Inches*" and use the factor listed at "centimeters" (*2.54*) as divisor.

To Convert from Yards

To	Multiply by
Inches	**36**
Feet	**3**
Miles	0.000 568 18
Centimeters	**91.44**
Meters	**0.914 4**

To Convert from Miles

To	Multiply by
Inches	**63 360**
Feet	**5 280**
Yards	**1 760**
Centimeters	**160 934.4**
Meters	**1 609.344**
Kilometers	**1.609 344**

Units of Mass

To Convert from Grams

To	Multiply by
Grains	15.432 36
Avoirdupois Drams	0.564 383 4
Avoirdupois Ounces	0.035 273 96
Troy Ounces	0.032 150 75
Troy Pounds	0.002 679 23
Avoirdupois Pounds	0.002 204 62
Milligrams	**1 000**
Kilograms	**0.001**

To Convert from Kilograms

To	Multiply by
Grains	15 432.36
Avoirdupois Drams	564.383 4
Avoirdupois Ounces	35.273 96
Troy Ounces	32.150 75
Troy Pounds	2.679 229
Avoirdupois Pounds	2.204 623
Grams	**1 000**
Short Hundredweights	0.022 046 23
Short Tons	0.001 102 31
Long Tons	0.000 984 2
Metric Tons	**0.001**

To Convert from Metric Tons

To	Multiply by
Avoirdupois Pounds	2 204.623
Short Hundredweights	22.046 23
Short Tons	1.102 311 3
Long Tons	0.984 206 5
Kilograms	**1 000**

To Convert from Grains

To	Multiply by
Avoirdupois Drams	0.036 571 43
Avoirdupois Ounces	0.002 285 71
Troy Ounces	0.002 083 33
Troy Pounds	0.000 173 61
Avoirdupois Pounds	0.000 142 86
Milligrams	**64.798 91**
Grams	**0.064 798 91**
Kilograms	**0.000 064 798 91**

To Convert from Avoirdupois Ounces

To	Multiply by
Grains	**437.5**
Avoirdupois Drams	**16**
Troy Ounces	0.911 458 3
Troy Pounds	0.075 954 86
Avoirdupois Pounds	**0.062 5**
Grams	**28.349 523 125**
Kilograms	**0.028 349 523 125**

To Convert from Avoirdupois Pounds

To	Multiply by
Grains	**7 000**
Avoirdupois Drams	**256**
Avoirdupois Ounces	**16**
Troy Ounces	14.583 33
Troy Pounds	1.215 278
Grams	**453.592 37**
Kilograms	**0.453 592 37**
Short Hundredweights	**0.01**
Short Tons	**0.000 5**
Long Tons	0.000 446 428 6
Metric Tons	**0.000 453 592 37**

To Convert from Short Hundredweights

To	Multiply by
Avoirdupois Pounds	**100**
Short Tons	**0.05**
Long Tons	0.044 642 86
Kilograms	**45.359 237**
Metric Tons	**0.045 359 237**

To Convert from **Short Tons** To	Multiply by
Avoirdupois Pounds	**2 000**
Short Hundredweights	**20**
Long Tons	0.892 857 1
Kilograms	**907.184 74**
Metric Tons	**0.907 184 74**

To Convert from **Long Tons** To	Multiply by
Avoirdupois Ounces	**35 840**
Avoirdupois Pounds	**2 240**
Short Hundredweights	**22.4**
Short Tons	**1.12**
Kilograms	**1 016.046 908 8**
Metric Tons	**1.016 046 908 8**

To Convert from **Troy Ounces** To	Multiply by
Grains	**480**
Avoirdupois Drams	17.554 29
Avoirdupois Ounces	1.097 143
Troy Pounds	0.083 333 3
Avoirdupois Pounds	0.068 571 43
Grams	**31.103 476 8**

To Convert from **Troy Pounds** To	Multiply by
Grains	**5 760**
Avoirdupois Drams	210.651 4
Avoirdupois Ounces	13.165 71
Troy Ounces	**12**
Avoirdupois Pounds	0.822 857 1
Grams	**373.241 721 6**

Units of Capacity, or Volume, Liquid Measure

To Convert from **Milliliters** To	Multiply by
Minims	16.230 73
Liquid Ounces	0.033 814 02
Gills	0.008 453 5
Liquid Pints	0.002 113 4
Liquid Quarts	0.001 056 7
Gallons	0.000 264 17
Cubic Inches	0.061 023 74
Liters	**0.001**

To Convert from **Liters** To	Multiply by
Liquid Ounces	33.814 02
Gills	8.453 506
Liquid Pints	2.113 376
Liquid Quarts	1.056 688
Gallons	0.264 172 05
Cubic Inches	61.023 74
Cubic Feet	0.035 314 67
Milliliters	**1 000**
Cubic Meters	**0.001**
Cubic Yards	0.001 307 95

To Convert from **Cubic Meters** To	Multiply by
Gallons	264.172 05
Cubic Inches	61 023.74
Cubic Feet	35.314 67
Liters	**1 000**
Cubic Yards	1.307 950 6

To Convert from **Minims** To	Multiply by
Liquid Ounces	0.002 083 33
Gills	0.000 520 83
Cubic Inches	0.003 759 77
Milliliters	0.061 611 52

260-260 O - 67 - 2

To Convert from **Gills** To	Multiply by
Minims	**1 920**
Liquid Ounces	**4**
Liquid Pints	**0.25**
Liquid Quarts	**0.125**
Gallons	**0.031 25**
Cubic Inches	**7.218 75**
Cubic Feet	0.004 177 517
Milliliters	**118.294 118 25**
Liters	**0.118 294 118 25**

To Convert from **Liquid Ounces** To	Multiply by
Minims	**480**
Gills	**0.25**
Liquid Pints	**0.062 5**
Liquid Quarts	**0.031 25**
Gallons	**0.007 812 5**
Cubic Inches	1.804 687 5
Cubic Feet	0.001 044 38
Milliliters	29.573 53
Liters	0.029 573 53

To Convert from **Cubic Inches** To	Multiply by
Minims	265.974 0
Liquid Ounces	0.554 112 6
Gills	0.138 528 1
Liquid Pints	0.034 632 03
Liquid Quarts	0.017 316 02
Gallons	0.004 329 0
Cubic Feet	0.000 578 7
Milliliters	**16.387 064**
Liters	**0.016 387 064**
Cubic Meters	**0.000 016 387 064**
Cubic Yards	0.000 021 43

To Convert from **Liquid Pints** To	Multiply by
Minims	**7 680**
Liquid Ounces	**16**
Gills	**4**
Liquid Quarts	**0.5**
Gallons	**0.125**
Cubic Inches	**28.875**
Cubic Feet	0.016 710 07
Milliliters	**473.176 473**
Liters	**0.473 176 473**

To Convert from **Cubic Feet** To	Multiply by
Liquid Ounces	957.506 5
Gills	239.376 6
Liquid Pints	59.844 16
Liquid Quarts	29.922 08
Gallons	7.480 519
Cubic Inches	**1 728**
Liters	**28.316 846 592**
Cubic Meters	**0.028 316 846 592**
Cubic Yards	0.037 037 04

To Convert from **Cubic Yards** To	Multiply by
Gallons	201.974 0
Cubic Inches	**46 656**
Cubic Feet	**27**
Liters	**764.554 857 984**
Cubic Meters	**0.764 554 857 984**

To Convert from **Liquid Quarts**

To	Multiply by
Minims	**15 360**
Liquid Ounces	**32**
Gills	**8**
Liquid Pints	**2**
Gallons	**0.25**
Cubic Inches	**57.75**
Cubic Feet	0.033 420 14
Milliliters	**946.352 946**
Liters	**0.946 352 946**

To Convert from **Gallons**

To	Multiply by
Minims	**61 440**
Liquid Ounces	**128**
Gills	**32**
Liquid Pints	**8**
Liquid Quarts	**4**
Cubic Inches	**231**
Cubic Feet	0.133 680 6
Milliliters	**3 785.411 784**
Liters	**3.785 411 784**
Cubic Meters	**0.003 785 411 784**
Cubic Yards	0.004 951 13

Units of Capacity, or Volume, Dry Measure

To Convert from **Liters**

To	Multiply by
Dry Pints	1.816 166
Dry Quarts	0.908 082 98
Pecks	0.113 510 4
Bushels	0.028 377 59
Dekaliters	**0.1**

To Convert from **Dekaliters**

To	Multiply by
Dry Pints	18.161 66
Dry Quarts	9.080 829 8
Pecks	1.135 104
Bushels	0.283 775 9
Cubic Inches	610.237 4
Cubic Feet	0.353 146 7
Liters	**10**

To Convert from **Cubic Meters**

To	Multiply by
Pecks	113.510 4
Bushels	28.377 59

To Convert from **Dry Pints**

To	Multiply by
Dry Quarts	**0.5**
Pecks	**0.062 5**
Bushels	**0.015 625**
Cubic Inches	**33.600 312 5**
Cubic Feet	0.019 444 63
Liters	0.550 610 47
Dekaliters	0.055 061 05

To Convert from **Dry Quarts** To	Multiply by
Dry Pints	**2**
Pecks	**0.125**
Bushels	**0.031 25**
Cubic Inches	**67.200 625**
Cubic Feet	0.038 889 25
Liters	1.101 221
Dekaliters	0.110 122 1

To Convert from **Pecks** To	Multiply by
Dry Pints	**16**
Dry Quarts	**8**
Bushels	**0.25**
Cubic Inches	**537.605**
Cubic Feet	0.311 114
Liters	8.809 767 5
Dekaliters	0.880 976 75
Cubic Meters	0.008 809 77
Cubic Yards	0.011 522 74

To Convert from **Bushels** To	Multiply by
Dry Pints	**64**
Dry Quarts	**32**
Pecks	**4**
Cubic Inches	**2 150.42**
Cubic Feet	1.244 456
Liters	35.239 07
Dekaliters	3.523 907
Cubic Meters	0.035 239 07
Cubic Yards	0.046 090 96

To Convert from **Cubic Inches** To	Multiply by
Dry Pints	0.029 761 6
Dry Quarts	0.014 880 8
Pecks	0.001 860 10
Bushels	0.000 465 025

To Convert from **Cubic Yards** To	Multiply by
Pecks	86.784 91
Bushels	21.696 227

To Convert from **Cubic Feet** To	Multiply by
Dry Pints	51.428 09
Dry Quarts	25.714 05
Pecks	3.214 256
Bushels	0.803 563 95

Units of Area

To Convert from **Square Centimeters**

To	Multiply by
Square Inches	0.155 000 3
Square Feet	0.001 076 39
Square Yards	0.000 119 599
Square Meters	**0.000 1**

To Convert from **Square Meters**

To	Multiply by
Square Inches	1 550.003
Square Feet	10.763 91
Square Yards	1.195 990
Acres	0.000 247 105
Square Centimeters	**10 000**
Hectares	**0.000 1**

To Convert from **Hectares**

To	Multiply by
Square Feet	107 639.1
Square Yards	11 959.90
Acres	2.471 054
Square Miles	0.003 861 02
Square Meters	**10 000**

To Convert from **Square Inches**

To	Multiply by
Square Feet	0.006 944 44
Square Yards	0.000 771 605
Square Centimeters	**6.451 6**
Square Meters	**0.000 645 16**

To Convert from **Square Feet**

To	Multiply by
Square Inches	**144**
Square Yards	0.111 111 1
Acres	0.000 022 957
Square Centimeters	**929.030 4**
Square Meters	**0.092 903 04**

To Convert from **Square Yards**

To	Multiply by
Square Inches	**1 296**
Square Feet	**9**
Acres	0.000 206 611 6
Square Miles	0.000 000 322 830 6
Square Centimeters	**8 361.273 6**
Square Meters	**0.836 127 36**
Hectares	**0.000 083 612 736**

To Convert from **Acres**

To	Multiply by
Square Feet	**43 560**
Square Yards	**4 840**
Square Miles	**0.001 562 5**
Square Meters	**4 046.856 422 4**
Hectares	**0.404 685 642 24**

To Convert from **Square Miles**

To	Multiply by
Square Feet	**27 878 400**
Square Yards	**3 097 600**
Acres	**640**
Square Meters	**2 589 988.110 336**
Hectares	**258.998 811 033 6**

Special Tables

Length—Inches and Millimeters—Equivalents of Decimal and Binary Fractions of an Inch in Millimeters

From 1/64 to 1 Inch

½'s	¼'s	8ths	16ths	32ds	64ths	Millimeters	Decimals of an inch	Inch	½'s	¼'s	8ths	16ths	32ds	64ths	Millimeters	Decimals of an inch
					1	= 0.397	0.015625							33	=13.097	0.515625
				1	2	= .794	.03125						17	34	=13.494	.53125
					3	= 1.191	.046875							35	=13.891	.546875
			1	2	4	= 1.588	.0625					9	18	36	=14.288	.5625
					5	= 1.984	.078125							37	=14.684	.578125
				3	6	= 2.381	.09375						19	38	=15.081	.59375
					7	= 2.778	.109375							39	=15.478	.609375
		1	2	4	8	= 3.175	.1250				5	10	20	40	=15.875	.625
					9	= 3.572	.140625							41	=16.272	.640625
				5	10	= 3.969	.15625						21	42	=16.669	.65625
					11	= 4.366	.171875							43	=17.066	.671875
			3	6	12	= 4.762	.1875					11	22	44	=17.462	.6875
					13	= 5.159	.203125							45	=17.859	.703125
				7	14	= 5.556	.21875						23	46	=18.256	.71875
					15	= 5.953	.234375							47	=18.653	.734375
	1	2	4	8	16	= 6.350	.2500			3	6	12	24	48	=19.050	.75
					17	= 6.747	.265625							49	=19.447	.765625
				9	18	= 7.144	.28125						25	50	=19.844	.78125
					19	= 7.541	.296875							51	=20.241	.796875
			5	10	20	= 7.938	.3125					13	26	52	=20.638	.8125
					21	= 8.334	.328125							53	=21.034	.828125
				11	22	= 8.731	.34375						27	54	=21.431	.84375
					23	= 9.128	.359375							55	=21.828	.859375
		3	6	12	24	= 9.525	.3750				7	14	28	56	=22.225	.875
					25	= 9.922	.390625							57	=22.622	.890625
				13	26	=10.319	.40625						29	58	=23.019	.90625
					27	=10.716	.421875							59	=23.416	.921875
			7	14	28	=11.112	.4375					15	30	60	=23.812	.9375
					29	=11.509	.453125							61	=24.209	.953125
				15	30	=11.906	.46875						31	62	=24.606	.96875
					31	=12.303	.484375							63	=25.003	.984375
1	2	4	8	16	32	=12.700	.5	1	2	4	8	16	32	64	=25.400	1.000

LENGTH - INCHES TO MILLIMETERS 1 INCH = 25.4 MILLIMETERS

INCHES	MILLI-METERS	INCHES	MILLI-METERS	INCHES	MILLI-METERS	INCHES	MILLI-METERS	INCHES	MILLI-METERS
0.00	0.000	1.00	25.400	2.00	50.800	3.00	76.200	4.00	101.600
0.01	0.254	1.01	25.654	2.01	51.054	3.01	76.454	4.01	101.854
0.02	0.508	1.02	25.908	2.02	51.308	3.02	76.708	4.02	102.108
0.03	0.762	1.03	26.162	2.03	51.562	3.03	76.962	4.03	102.362
0.04	1.016	1.04	26.416	2.04	51.816	3.04	77.216	4.04	102.616
0.05	1.270	1.05	26.670	2.05	52.070	3.05	77.470	4.05	102.870
0.06	1.524	1.06	26.924	2.06	52.324	3.06	77.724	4.06	103.124
0.07	1.778	1.07	27.178	2.07	52.578	3.07	77.978	4.07	103.378
0.08	2.032	1.08	27.432	2.08	52.832	3.08	78.232	4.08	103.632
0.09	2.286	1.09	27.686	2.09	53.086	3.09	78.486	4.09	103.886
0.10	2.540	1.10	27.940	2.10	53.340	3.10	78.740	4.10	104.140
0.11	2.794	1.11	28.194	2.11	53.594	3.11	78.994	4.11	104.394
0.12	3.048	1.12	28.448	2.12	53.848	3.12	79.248	4.12	104.648
0.13	3.302	1.13	28.702	2.13	54.102	3.13	79.502	4.13	104.902
0.14	3.556	1.14	28.956	2.14	54.356	3.14	79.756	4.14	105.156
0.15	3.810	1.15	29.210	2.15	54.610	3.15	80.010	4.15	105.410
0.16	4.064	1.16	29.464	2.16	54.864	3.16	80.264	4.16	105.664
0.17	4.318	1.17	29.718	2.17	55.118	3.17	80.518	4.17	105.918
0.18	4.572	1.18	29.972	2.18	55.372	3.18	80.772	4.18	106.172
0.19	4.826	1.19	30.226	2.19	55.626	3.19	81.026	4.19	106.426
0.20	5.080	1.20	30.480	2.20	55.880	3.20	81.280	4.20	106.680
0.21	5.334	1.21	30.734	2.21	56.134	3.21	81.534	4.21	106.934
0.22	5.588	1.22	30.988	2.22	56.388	3.22	81.788	4.22	107.188
0.23	5.842	1.23	31.242	2.23	56.642	3.23	82.042	4.23	107.442
0.24	6.096	1.24	31.496	2.24	56.896	3.24	82.296	4.24	107.696
0.25	6.350	1.25	31.750	2.25	57.150	3.25	82.550	4.25	107.950
0.26	6.604	1.26	32.004	2.26	57.404	3.26	82.804	4.26	108.204
0.27	6.858	1.27	32.258	2.27	57.658	3.27	83.058	4.27	108.458
0.28	7.112	1.28	32.512	2.28	57.912	3.28	83.312	4.28	108.712
0.29	7.366	1.29	32.766	2.29	58.166	3.29	83.566	4.29	108.966
0.30	7.620	1.30	33.020	2.30	58.420	3.30	83.820	4.30	109.220
0.31	7.874	1.31	33.274	2.31	58.674	3.31	84.074	4.31	109.474
0.32	8.128	1.32	33.528	2.32	58.928	3.32	84.328	4.32	109.728
0.33	8.382	1.33	33.782	2.33	59.182	3.33	84.582	4.33	109.982
0.34	8.636	1.34	34.036	2.34	59.436	3.34	84.836	4.34	110.236
0.35	8.890	1.35	34.290	2.35	59.690	3.35	85.090	4.35	110.490
0.36	9.144	1.36	34.544	2.36	59.944	3.36	85.344	4.36	110.744
0.37	9.398	1.37	34.798	2.37	60.198	3.37	85.598	4.37	110.998
0.38	9.652	1.38	35.052	2.38	60.452	3.38	85.852	4.38	111.252
0.39	9.906	1.39	35.306	2.39	60.706	3.39	86.106	4.39	111.506
0.40	10.160	1.40	35.560	2.40	60.960	3.40	86.360	4.40	111.760
0.41	10.414	1.41	35.814	2.41	61.214	3.41	86.614	4.41	112.014
0.42	10.668	1.42	36.068	2.42	61.468	3.42	86.868	4.42	112.268
0.43	10.922	1.43	36.322	2.43	61.722	3.43	87.122	4.43	112.522
0.44	11.176	1.44	36.576	2.44	61.976	3.44	87.376	4.44	112.776
0.45	11.430	1.45	36.830	2.45	62.230	3.45	87.630	4.45	113.030
0.46	11.684	1.46	37.084	2.46	62.484	3.46	87.884	4.46	113.284
0.47	11.938	1.47	37.338	2.47	62.738	3.47	88.138	4.47	113.538
0.48	12.192	1.48	37.592	2.48	62.992	3.48	88.392	4.48	113.792
0.49	12.446	1.49	37.846	2.49	63.246	3.49	88.646	4.49	114.046

Continued - INCHES TO MILLIMETERS 1 INCH = 25.4 MILLIMETERS

INCHES	MILLI-METERS	INCHES	MILLI-METERS	INCHES	MILLI-METERS	INCHES	MILLI-METERS	INCHES	MILLI-METERS
0.50	12.700	1.50	38.100	2.50	63.500	3.50	88.900	4.50	114.300
0.51	12.954	1.51	38.354	2.51	63.754	3.51	89.154	4.51	114.554
0.52	13.208	1.52	38.608	2.52	64.008	3.52	89.408	4.52	114.808
0.53	13.462	1.53	38.862	2.53	64.262	3.53	89.662	4.53	115.062
0.54	13.716	1.54	39.116	2.54	64.516	3.54	89.916	4.54	115.316
0.55	13.970	1.55	39.370	2.55	64.770	3.55	90.170	4.55	115.570
0.56	14.224	1.56	39.624	2.56	65.024	3.56	90.424	4.56	115.824
0.57	14.478	1.57	39.878	2.57	65.278	3.57	90.678	4.57	116.078
0.58	14.732	1.58	40.132	2.58	65.532	3.58	90.932	4.58	116.332
0.59	14.986	1.59	40.386	2.59	65.786	3.59	91.186	4.59	116.586
0.60	15.240	1.60	40.640	2.60	66.040	3.60	91.440	4.60	116.840
0.61	15.494	1.61	40.894	2.61	66.294	3.61	91.694	4.61	117.094
0.62	15.748	1.62	41.148	2.62	66.548	3.62	91.948	4.62	117.348
0.63	16.002	1.63	41.402	2.63	66.802	3.63	92.202	4.63	117.602
0.64	16.256	1.64	41.656	2.64	67.056	3.64	92.456	4.64	117.856
0.65	16.510	1.65	41.910	2.65	67.310	3.65	92.710	4.65	118.110
0.66	16.764	1.66	42.164	2.66	67.564	3.66	92.964	4.66	118.364
0.67	17.018	1.67	42.418	2.67	67.818	3.67	93.218	4.67	118.618
0.68	17.272	1.68	42.672	2.68	68.072	3.68	93.472	4.68	118.872
0.69	17.526	1.69	42.926	2.69	68.326	3.69	93.726	4.69	119.126
0.70	17.780	1.70	43.180	2.70	68.580	3.70	93.980	4.70	119.380
0.71	18.034	1.71	43.434	2.71	68.834	3.71	94.234	4.71	119.634
0.72	18.288	1.72	43.688	2.72	69.088	3.72	94.488	4.72	119.888
0.73	18.542	1.73	43.942	2.73	69.342	3.73	94.742	4.73	120.142
0.74	18.796	1.74	44.196	2.74	69.596	3.74	94.996	4.74	120.396
0.75	19.050	1.75	44.450	2.75	69.850	3.75	95.250	4.75	120.650
0.76	19.304	1.76	44.704	2.76	70.104	3.76	95.504	4.76	120.904
0.77	19.558	1.77	44.958	2.77	70.358	3.77	95.758	4.77	121.158
0.78	19.812	1.78	45.212	2.78	70.612	3.78	96.012	4.78	121.412
0.79	20.066	1.79	45.466	2.79	70.866	3.79	96.266	4.79	121.666
0.80	20.320	1.80	45.720	2.80	71.120	3.80	96.520	4.80	121.920
0.81	20.574	1.81	45.974	2.81	71.374	3.81	96.774	4.81	122.174
0.82	20.828	1.82	46.228	2.82	71.628	3.82	97.028	4.82	122.428
0.83	21.082	1.83	46.482	2.83	71.882	3.83	97.282	4.83	122.682
0.84	21.336	1.84	46.736	2.84	72.136	3.84	97.536	4.84	122.936
0.85	21.590	1.85	46.990	2.85	72.390	3.85	97.790	4.85	123.190
0.86	21.844	1.86	47.244	2.86	72.644	3.86	98.044	4.86	123.444
0.87	22.098	1.87	47.498	2.87	72.898	3.87	98.298	4.87	123.698
0.88	22.352	1.88	47.752	2.88	73.152	3.88	98.552	4.88	123.952
0.89	22.606	1.89	48.006	2.89	73.406	3.89	98.806	4.89	124.206
0.90	22.860	1.90	48.260	2.90	73.660	3.90	99.060	4.90	124.460
0.91	23.114	1.91	48.514	2.91	73.914	3.91	99.314	4.91	124.714
0.92	23.368	1.92	48.768	2.92	74.168	3.92	99.568	4.92	124.968
0.93	23.622	1.93	49.022	2.93	74.422	3.93	99.822	4.93	125.222
0.94	23.876	1.94	49.276	2.94	74.676	3.94	100.076	4.94	125.476
0.95	24.130	1.95	49.530	2.95	74.930	3.95	100.330	4.95	125.730
0.96	24.384	1.96	49.784	2.96	75.184	3.96	100.584	4.96	125.984
0.97	24.638	1.97	50.038	2.97	75.438	3.97	100.838	4.97	126.238
0.98	24.892	1.98	50.292	2.98	75.692	3.98	101.092	4.98	126.492
0.99	25.146	1.99	50.546	2.99	75.946	3.99	101.346	4.99	126.746

Continued - INCHES TO MILLIMETERS 1 INCH = 25.4 MILLIMETERS

INCHES	MILLI-METERS	INCHES	MILLI-METERS	INCHES	MILLI-METERS	INCHES	MILLI-METERS	INCHES	MILLI-METERS
5.00	127.000	6.00	152.400	7.00	177.800	8.00	203.200	9.00	228.600
5.01	127.254	6.01	152.654	7.01	178.054	8.01	203.454	9.01	228.854
5.02	127.508	6.02	152.908	7.02	178.308	8.02	203.708	9.02	229.108
5.03	127.762	6.03	153.162	7.03	178.562	8.03	203.962	9.03	229.362
5.04	128.016	6.04	153.416	7.04	178.816	8.04	204.216	9.04	229.616
5.05	128.270	6.05	153.670	7.05	179.070	8.05	204.470	9.05	229.870
5.06	128.524	6.06	153.924	7.06	179.324	8.06	204.724	9.06	230.124
5.07	128.778	6.07	154.178	7.07	179.578	8.07	204.978	9.07	230.378
5.08	129.032	6.08	154.432	7.08	179.832	8.08	205.232	9.08	230.632
5.09	129.286	6.09	154.686	7.09	180.086	8.09	205.486	9.09	230.886
5.10	129.540	6.10	154.940	7.10	180.340	8.10	205.740	9.10	231.140
5.11	129.794	6.11	155.194	7.11	180.594	8.11	205.994	9.11	231.394
5.12	130.048	6.12	155.448	7.12	180.848	8.12	206.248	9.12	231.648
5.13	130.302	6.13	155.702	7.13	181.102	8.13	206.502	9.13	231.902
5.14	130.556	6.14	155.956	7.14	181.356	8.14	206.756	9.14	232.156
5.15	130.810	6.15	156.210	7.15	181.610	8.15	207.010	9.15	232.410
5.16	131.064	6.16	156.464	7.16	181.864	8.16	207.264	9.16	232.664
5.17	131.318	6.17	156.718	7.17	182.118	8.17	207.518	9.17	232.918
5.18	131.572	6.18	156.972	7.18	182.372	8.18	207.772	9.18	233.172
5.19	131.826	6.19	157.226	7.19	182.626	8.19	208.026	9.19	233.426
5.20	132.080	6.20	157.480	7.20	182.880	8.20	208.280	9.20	233.680
5.21	132.334	6.21	157.734	7.21	183.134	8.21	208.534	9.21	233.934
5.22	132.588	6.22	157.988	7.22	183.388	8.22	208.788	9.22	234.188
5.23	132.842	6.23	158.242	7.23	183.642	8.23	209.042	9.23	234.442
5.24	133.096	6.24	158.496	7.24	183.896	8.24	209.296	9.24	234.696
5.25	133.350	6.25	158.750	7.25	184.150	8.25	209.550	9.25	234.950
5.26	133.604	6.26	159.004	7.26	184.404	8.26	209.804	9.26	235.204
5.27	133.858	6.27	159.258	7.27	184.658	8.27	210.058	9.27	235.458
5.28	134.112	6.28	159.512	7.28	184.912	8.28	210.312	9.28	235.712
5.29	134.366	6.29	159.766	7.29	185.166	8.29	210.566	9.29	235.966
5.30	134.620	6.30	160.020	7.30	185.420	8.30	210.820	9.30	236.220
5.31	134.874	6.31	160.274	7.31	185.674	8.31	211.074	9.31	236.474
5.32	135.128	6.32	160.528	7.32	185.928	8.32	211.328	9.32	236.728
5.33	135.382	6.33	160.782	7.33	186.182	8.33	211.582	9.33	236.982
5.34	135.636	6.34	161.036	7.34	186.436	8.34	211.836	9.34	237.236
5.35	135.890	6.35	161.290	7.35	186.690	8.35	212.090	9.35	237.490
5.36	136.144	6.36	161.544	7.36	186.944	8.36	212.344	9.36	237.744
5.37	136.398	6.37	161.798	7.37	187.198	8.37	212.598	9.37	237.998
5.38	136.652	6.38	162.052	7.38	187.452	8.38	212.852	9.38	238.252
5.39	136.906	6.39	162.306	7.39	187.706	8.39	213.106	9.39	238.506
5.40	137.160	6.40	162.560	7.40	187.960	8.40	213.360	9.40	238.760
5.41	137.414	6.41	162.814	7.41	188.214	8.41	213.614	9.41	239.014
5.42	137.668	6.42	163.068	7.42	188.468	8.42	213.868	9.42	239.268
5.43	137.922	6.43	163.322	7.43	188.722	8.43	214.122	9.43	239.522
5.44	138.176	6.44	163.576	7.44	188.976	8.44	214.376	9.44	239.776
5.45	138.430	6.45	163.830	7.45	189.230	8.45	214.630	9.45	240.030
5.46	138.684	6.46	164.084	7.46	189.484	8.46	214.884	9.46	240.284
5.47	138.938	6.47	164.338	7.47	189.738	8.47	215.138	9.47	240.538
5.48	139.192	6.48	164.592	7.48	189.992	8.48	215.392	9.48	240.792
5.49	139.446	6.49	164.846	7.49	190.246	8.49	215.646	9.49	241.046

Continued - INCHES TO MILLIMETERS 1 INCH = 25.4 MILLIMETERS

INCHES	MILLI-METERS	INCHES	MILLI-METERS	INCHES	MILLI-METERS	INCHES	MILLI-METERS	INCHES	MILLI-METERS
5.50	139.700	6.50	165.100	7.50	190.500	8.50	215.900	9.50	241.300
5.51	139.954	6.51	165.354	7.51	190.754	8.51	216.154	9.51	241.554
5.52	140.208	6.52	165.608	7.52	191.008	8.52	216.408	9.52	241.808
5.53	140.462	6.53	165.862	7.53	191.262	8.53	216.662	9.53	242.062
5.54	140.716	6.54	166.116	7.54	191.516	8.54	216.916	9.54	242.316
5.55	140.970	6.55	166.370	7.55	191.770	8.55	217.170	9.55	242.570
5.56	141.224	6.56	166.624	7.56	192.024	8.56	217.424	9.56	242.824
5.57	141.478	6.57	166.878	7.57	192.278	8.57	217.678	9.57	243.078
5.58	141.732	6.58	167.132	7.58	192.532	8.58	217.932	9.58	243.332
5.59	141.986	6.59	167.386	7.59	192.786	8.59	218.186	9.59	243.586
5.60	142.240	6.60	167.640	7.60	193.040	8.60	218.440	9.60	243.840
5.61	142.494	6.61	167.894	7.61	193.294	8.61	218.694	9.61	244.094
5.62	142.748	6.62	168.148	7.62	193.548	8.62	218.948	9.62	244.348
5.63	143.002	6.63	168.402	7.63	193.802	8.63	219.202	9.63	244.602
5.64	143.256	6.64	168.656	7.64	194.056	8.64	219.456	9.64	244.856
5.65	143.510	6.65	168.910	7.65	194.310	8.65	219.710	9.65	245.110
5.66	143.764	6.66	169.164	7.66	194.564	8.66	219.964	9.66	245.364
5.67	144.018	6.67	169.418	7.67	194.818	8.67	220.218	9.67	245.618
5.68	144.272	6.68	169.672	7.68	195.072	8.68	220.472	9.68	245.872
5.69	144.526	6.69	169.926	7.69	195.326	8.69	220.726	9.69	246.126
5.70	144.780	6.70	170.180	7.70	195.580	8.70	220.980	9.70	246.380
5.71	145.034	6.71	170.434	7.71	195.834	8.71	221.234	9.71	246.634
5.72	145.288	6.72	170.688	7.72	196.088	8.72	221.488	9.72	246.888
5.73	145.542	6.73	170.942	7.73	196.342	8.73	221.742	9.73	247.142
5.74	145.796	6.74	171.196	7.74	196.596	8.74	221.996	9.74	247.396
5.75	146.050	6.75	171.450	7.75	196.850	8.75	222.250	9.75	247.650
5.76	146.304	6.76	171.704	7.76	197.104	8.76	222.504	9.76	247.904
5.77	146.558	6.77	171.958	7.77	197.358	8.77	222.758	9.77	248.158
5.78	146.812	6.78	172.212	7.78	197.612	8.78	223.012	9.78	248.412
5.79	147.066	6.79	172.466	7.79	197.866	8.79	223.266	9.79	248.666
5.80	147.320	6.80	172.720	7.80	198.120	8.80	223.520	9.80	248.920
5.81	147.574	6.81	172.974	7.81	198.374	8.81	223.774	9.81	249.174
5.82	147.828	6.82	173.228	7.82	198.628	8.82	224.028	9.82	249.428
5.83	148.082	6.83	173.482	7.83	198.882	8.83	224.282	9.83	249.682
5.84	148.336	6.84	173.736	7.84	199.136	8.84	224.536	9.84	249.936
5.85	148.590	6.85	173.990	7.85	199.390	8.85	224.790	9.85	250.190
5.86	148.844	6.86	174.244	7.86	199.644	8.86	225.044	9.86	250.444
5.87	149.098	6.87	174.498	7.87	199.898	8.87	225.298	9.87	250.698
5.88	149.352	6.88	174.752	7.88	200.152	8.88	225.552	9.88	250.952
5.89	149.606	6.89	175.006	7.89	200.406	8.89	225.806	9.89	251.206
5.90	149.860	6.90	175.260	7.90	200.660	8.90	226.060	9.90	251.460
5.91	150.114	6.91	175.514	7.91	200.914	8.91	226.314	9.91	251.714
5.92	150.368	6.92	175.768	7.92	201.168	8.92	226.568	9.92	251.968
5.93	150.622	6.93	176.022	7.93	201.422	8.93	226.822	9.93	252.222
5.94	150.876	6.94	176.276	7.94	201.676	8.94	227.076	9.94	252.476
5.95	151.130	6.95	176.530	7.95	201.930	8.95	227.330	9.95	252.730
5.96	151.384	6.96	176.784	7.96	202.184	8.96	227.584	9.96	252.984
5.97	151.638	6.97	177.038	7.97	202.438	8.97	227.838	9.97	253.238
5.98	151.892	6.98	177.292	7.98	202.692	8.98	228.092	9.98	253.492
5.99	152.146	6.99	177.546	7.99	202.946	8.99	228.346	9.99	253.746

Continued - INCHES TO MILLIMETERS 1 INCH = 25.4 MILLIMETERS

INCHES	MILLI-METERS	INCHES	MILLI-METERS	INCHES	MILLI-METERS	INCHES	MILLI-METERS	INCHES	MILLI-METERS
10.00	254.000	11.00	279.400	12.00	304.800	13.00	330.200	14.00	355.600
10.01	254.254	11.01	279.654	12.01	305.054	13.01	330.454	14.01	355.854
10.02	254.508	11.02	279.908	12.02	305.308	13.02	330.708	14.02	356.108
10.03	254.762	11.03	280.162	12.03	305.562	13.03	330.962	14.03	356.362
10.04	255.016	11.04	280.416	12.04	305.816	13.04	331.216	14.04	356.616
10.05	255.270	11.05	280.670	12.05	306.070	13.05	331.470	14.05	356.870
10.06	255.524	11.06	280.924	12.06	306.324	13.06	331.724	14.06	357.124
10.07	255.778	11.07	281.178	12.07	306.578	13.07	331.978	14.07	357.378
10.08	256.032	11.08	281.432	12.08	306.832	13.08	332.232	14.08	357.632
10.09	256.286	11.09	281.686	12.09	307.086	13.09	332.486	14.09	357.886
10.10	256.540	11.10	281.940	12.10	307.340	13.10	332.740	14.10	358.140
10.11	256.794	11.11	282.194	12.11	307.594	13.11	332.994	14.11	358.394
10.12	257.048	11.12	282.448	12.12	307.848	13.12	333.248	14.12	358.648
10.13	257.302	11.13	282.702	12.13	308.102	13.13	333.502	14.13	358.902
10.14	257.556	11.14	282.956	12.14	308.356	13.14	333.756	14.14	359.156
10.15	257.810	11.15	283.210	12.15	308.610	13.15	334.010	14.15	359.410
10.16	258.064	11.16	283.464	12.16	308.864	13.16	334.264	14.16	359.664
10.17	258.318	11.17	283.718	12.17	309.118	13.17	334.518	14.17	359.918
10.18	258.572	11.18	283.972	12.18	309.372	13.18	334.772	14.18	360.172
10.19	258.826	11.19	284.226	12.19	309.626	13.19	335.026	14.19	360.426
10.20	259.080	11.20	284.480	12.20	309.880	13.20	335.280	14.20	360.680
10.21	259.334	11.21	284.734	12.21	310.134	13.21	335.534	14.21	360.934
10.22	259.588	11.22	284.988	12.22	310.388	13.22	335.788	14.22	361.188
10.23	259.842	11.23	285.242	12.23	310.642	13.23	336.042	14.23	361.442
10.24	260.096	11.24	285.496	12.24	310.896	13.24	336.296	14.24	361.696
10.25	260.350	11.25	285.750	12.25	311.150	13.25	336.550	14.25	361.950
10.26	260.604	11.26	286.004	12.26	311.404	13.26	336.804	14.26	362.204
10.27	260.858	11.27	286.258	12.27	311.658	13.27	337.058	14.27	362.458
10.28	261.112	11.28	286.512	12.28	311.912	13.28	337.312	14.28	362.712
10.29	261.366	11.29	286.766	12.29	312.166	13.29	337.566	14.29	362.966
10.30	261.620	11.30	287.020	12.30	312.420	13.30	337.820	14.30	363.220
10.31	261.874	11.31	287.274	12.31	312.674	13.31	338.074	14.31	363.474
10.32	262.128	11.32	287.528	12.32	312.928	13.32	338.328	14.32	363.728
10.33	262.382	11.33	287.782	12.33	313.182	13.33	338.582	14.33	363.982
10.34	262.636	11.34	288.036	12.34	313.436	13.34	338.836	14.34	364.236
10.35	262.890	11.35	288.290	12.35	313.690	13.35	339.090	14.35	364.490
10.36	263.144	11.36	288.544	12.36	313.944	13.36	339.344	14.36	364.744
10.37	263.398	11.37	288.798	12.37	314.198	13.37	339.598	14.37	364.998
10.38	263.652	11.38	289.052	12.38	314.452	13.38	339.852	14.38	365.252
10.39	263.906	11.39	289.306	12.39	314.706	13.39	340.106	14.39	365.506
10.40	264.160	11.40	289.560	12.40	314.960	13.40	340.360	14.40	365.760
10.41	264.414	11.41	289.814	12.41	315.214	13.41	340.614	14.41	366.014
10.42	264.668	11.42	290.068	12.42	315.468	13.42	340.868	14.42	366.268
10.43	264.922	11.43	290.322	12.43	315.722	13.43	341.122	14.43	366.522
10.44	265.176	11.44	290.576	12.44	315.976	13.44	341.376	14.44	366.776
10.45	265.430	11.45	290.830	12.45	316.230	13.45	341.630	14.45	367.030
10.46	265.684	11.46	291.084	12.46	316.484	13.46	341.884	14.46	367.284
10.47	265.938	11.47	291.338	12.47	316.738	13.47	342.138	14.47	367.538
10.48	266.192	11.48	291.592	12.48	316.992	13.48	342.392	14.48	367.792
10.49	266.446	11.49	291.846	12.49	317.246	13.49	342.646	14.49	368.046

Continued - INCHES TO MILLIMETERS 1 INCH = 25.4 MILLIMETERS

INCHES	MILLI-METERS	INCHES	MILLI-METERS	INCHES	MILLI-METERS	INCHES	MILLI-METERS	INCHES	MILLI-METERS
10.50	266.700	11.50	292.100	12.50	317.500	13.50	342.900	14.50	368.300
10.51	266.954	11.51	292.354	12.51	317.754	13.51	343.154	14.51	368.554
10.52	267.208	11.52	292.608	12.52	318.008	13.52	343.408	14.52	368.808
10.53	267.462	11.53	292.862	12.53	318.262	13.53	343.662	14.53	369.062
10.54	267.716	11.54	293.116	12.54	318.516	13.54	343.916	14.54	369.316
10.55	267.970	11.55	293.370	12.55	318.770	13.55	344.170	14.55	369.570
10.56	268.224	11.56	293.624	12.56	319.024	13.56	344.424	14.56	369.824
10.57	268.478	11.57	293.878	12.57	319.278	13.57	344.678	14.57	370.078
10.58	268.732	11.58	294.132	12.58	319.532	13.58	344.932	14.58	370.332
10.59	268.986	11.59	294.386	12.59	319.786	13.59	345.186	14.59	370.586
10.60	269.240	11.60	294.640	12.60	320.040	13.60	345.440	14.60	370.840
10.61	269.494	11.61	294.894	12.61	320.294	13.61	345.694	14.61	371.094
10.62	269.748	11.62	295.148	12.62	320.548	13.62	345.948	14.62	371.348
10.63	270.002	11.63	295.402	12.63	320.802	13.63	346.202	14.63	371.602
10.64	270.256	11.64	295.656	12.64	321.056	13.64	346.456	14.64	371.856
10.65	270.510	11.65	295.910	12.65	321.310	13.65	346.710	14.65	372.110
10.66	270.764	11.66	296.164	12.66	321.564	13.66	346.964	14.66	372.364
10.67	271.018	11.67	296.418	12.67	321.818	13.67	347.218	14.67	372.618
10.68	271.272	11.68	296.672	12.68	322.072	13.68	347.472	14.68	372.872
10.69	271.526	11.69	296.926	12.69	322.326	13.69	347.726	14.69	373.126
10.70	271.780	11.70	297.180	12.70	322.580	13.70	347.980	14.70	373.380
10.71	272.034	11.71	297.434	12.71	322.834	13.71	348.234	14.71	373.634
10.72	272.288	11.72	297.688	12.72	323.088	13.72	348.488	14.72	373.888
10.73	272.542	11.73	297.942	12.73	323.342	13.73	348.742	14.73	374.142
10.74	272.796	11.74	298.196	12.74	323.596	13.74	348.996	14.74	374.396
10.75	273.050	11.75	298.450	12.75	323.850	13.75	349.250	14.75	374.650
10.76	273.304	11.76	298.704	12.76	324.104	13.76	349.504	14.76	374.904
10.77	273.558	11.77	298.958	12.77	324.358	13.77	349.758	14.77	375.158
10.78	273.812	11.78	299.212	12.78	324.612	13.78	350.012	14.78	375.412
10.79	274.066	11.79	299.466	12.79	324.866	13.79	350.266	14.79	375.666
10.80	274.320	11.80	299.720	12.80	325.120	13.80	350.520	14.80	375.920
10.81	274.574	11.81	299.974	12.81	325.374	13.81	350.774	14.81	376.174
10.82	274.828	11.82	300.228	12.82	325.628	13.82	351.028	14.82	376.428
10.83	275.082	11.83	300.482	12.83	325.882	13.83	351.282	14.83	376.682
10.84	275.336	11.84	300.736	12.84	326.136	13.84	351.536	14.84	376.936
10.85	275.590	11.85	300.990	12.85	326.390	13.85	351.790	14.85	377.190
10.86	275.844	11.86	301.244	12.86	326.644	13.86	352.044	14.86	377.444
10.87	276.098	11.87	301.498	12.87	326.898	13.87	352.298	14.87	377.698
10.88	276.352	11.88	301.752	12.88	327.152	13.88	352.552	14.88	377.952
10.89	276.606	11.89	302.006	12.89	327.406	13.89	352.806	14.89	378.206
10.90	276.860	11.90	302.260	12.90	327.660	13.90	353.060	14.90	378.460
10.91	277.114	11.91	302.514	12.91	327.914	13.91	353.314	14.91	378.714
10.92	277.368	11.92	302.768	12.92	328.168	13.92	353.568	14.92	378.968
10.93	277.622	11.93	303.022	12.93	328.422	13.93	353.822	14.93	379.222
10.94	277.876	11.94	303.276	12.94	328.676	13.94	354.076	14.94	379.476
10.95	278.130	11.95	303.530	12.95	328.930	13.95	354.330	14.95	379.730
10.96	278.384	11.96	303.784	12.96	329.184	13.96	354.584	14.96	379.984
10.97	278.638	11.97	304.038	12.97	329.438	13.97	354.838	14.97	380.238
10.98	278.892	11.98	304.292	12.98	329.692	13.98	355.092	14.98	380.492
10.99	279.146	11.99	304.546	12.99	329.946	13.99	355.346	14.99	380.746

Continued - INCHES TO MILLIMETERS 1 INCH = 25.4 MILLIMETERS

INCHES	MILLI-METERS	INCHES	MILLI-METERS	INCHES	MILLI-METERS	INCHES	MILLI-METERS	INCHES	MILLI-METERS
15.00	381.000	16.00	406.400	17.00	431.800	18.00	457.200	19.00	482.600
15.01	381.254	16.01	406.654	17.01	432.054	18.01	457.454	19.01	482.854
15.02	381.508	16.02	406.908	17.02	432.308	18.02	457.708	19.02	483.108
15.03	381.762	16.03	407.162	17.03	432.562	18.03	457.962	19.03	483.362
15.04	382.016	16.04	407.416	17.04	432.816	18.04	458.216	19.04	483.616
15.05	382.270	16.05	407.670	17.05	433.070	18.05	458.470	19.05	483.870
15.06	382.524	16.06	407.924	17.06	433.324	18.06	458.724	19.06	484.124
15.07	382.778	16.07	408.178	17.07	433.578	18.07	458.978	19.07	484.378
15.08	383.032	16.08	408.432	17.08	433.832	18.08	459.232	19.08	484.632
15.09	383.286	16.09	408.686	17.09	434.086	18.09	459.486	19.09	484.886
15.10	383.540	16.10	408.940	17.10	434.340	18.10	459.740	19.10	485.140
15.11	383.794	16.11	409.194	17.11	434.594	18.11	459.994	19.11	485.394
15.12	384.048	16.12	409.448	17.12	434.848	18.12	460.248	19.12	485.648
15.13	384.302	16.13	409.702	17.13	435.102	18.13	460.502	19.13	485.902
15.14	384.556	16.14	409.956	17.14	435.356	18.14	460.756	19.14	486.156
15.15	384.810	16.15	410.210	17.15	435.610	18.15	461.010	19.15	486.410
15.16	385.064	16.16	410.464	17.16	435.864	18.16	461.264	19.16	486.664
15.17	385.318	16.17	410.718	17.17	436.118	18.17	461.518	19.17	486.918
15.18	385.572	16.18	410.972	17.18	436.372	18.18	461.772	19.18	487.172
15.19	385.826	16.19	411.226	17.19	436.626	18.19	462.026	19.19	487.426
15.20	386.080	16.20	411.480	17.20	436.880	18.20	462.280	19.20	487.680
15.21	386.334	16.21	411.734	17.21	437.134	18.21	462.534	19.21	487.934
15.22	386.588	16.22	411.988	17.22	437.388	18.22	462.788	19.22	488.188
15.23	386.842	16.23	412.242	17.23	437.642	18.23	463.042	19.23	488.442
15.24	387.096	16.24	412.496	17.24	437.896	18.24	463.296	19.24	488.696
15.25	387.350	16.25	412.750	17.25	438.150	18.25	463.550	19.25	488.950
15.26	387.604	16.26	413.004	17.26	438.404	18.26	463.804	19.26	489.204
15.27	387.858	16.27	413.258	17.27	438.658	18.27	464.058	19.27	489.458
15.28	388.112	16.28	413.512	17.28	438.912	18.28	464.312	19.28	489.712
15.29	388.366	16.29	413.766	17.29	439.166	18.29	464.566	19.29	489.966
15.30	388.620	16.30	414.020	17.30	439.420	18.30	464.820	19.30	490.220
15.31	388.874	16.31	414.274	17.31	439.674	18.31	465.074	19.31	490.474
15.32	389.128	16.32	414.528	17.32	439.928	18.32	465.328	19.32	490.728
15.33	389.382	16.33	414.782	17.33	440.182	18.33	465.582	19.33	490.982
15.34	389.636	16.34	415.036	17.34	440.436	18.34	465.836	19.34	491.236
15.35	389.890	16.35	415.290	17.35	440.690	18.35	466.090	19.35	491.490
15.36	390.144	16.36	415.544	17.36	440.944	18.36	466.344	19.36	491.744
15.37	390.398	16.37	415.798	17.37	441.198	18.37	466.598	19.37	491.998
15.38	390.652	16.38	416.052	17.38	441.452	18.38	466.852	19.38	492.252
15.39	390.906	16.39	416.306	17.39	441.706	18.39	467.106	19.39	492.506
15.40	391.160	16.40	416.560	17.40	441.960	18.40	467.360	19.40	492.760
15.41	391.414	16.41	416.814	17.41	442.214	18.41	467.614	19.41	493.014
15.42	391.668	16.42	417.068	17.42	442.468	18.42	467.868	19.42	493.268
15.43	391.922	16.43	417.322	17.43	442.722	18.43	468.122	19.43	493.522
15.44	392.176	16.44	417.576	17.44	442.976	18.44	468.376	19.44	493.776
15.45	392.430	16.45	417.830	17.45	443.230	18.45	468.630	19.45	494.030
15.46	392.684	16.46	418.084	17.46	443.484	18.46	468.884	19.46	494.284
15.47	392.938	16.47	418.338	17.47	443.738	18.47	469.138	19.47	494.538
15.48	393.192	16.48	418.592	17.48	443.992	18.48	469.392	19.48	494.792
15.49	393.446	16.49	418.846	17.49	444.246	18.49	469.646	19.49	495.046

Continued - INCHES TO MILLIMETERS 1 INCH = 25.4 MILLIMETERS

INCHES	MILLI-METERS	INCHES	MILLI-METERS	INCHES	MILLI-METERS	INCHES	MILLI-METERS	INCHES	MILLI-METERS
15.50	393.700	16.50	419.100	17.50	444.500	18.50	469.900	19.50	495.300
15.51	393.954	16.51	419.354	17.51	444.754	18.51	470.154	19.51	495.554
15.52	394.208	16.52	419.608	17.52	445.008	18.52	470.408	19.52	495.808
15.53	394.462	16.53	419.862	17.53	445.262	18.53	470.662	19.53	496.062
15.54	394.716	16.54	420.116	17.54	445.516	18.54	470.916	19.54	496.316
15.55	394.970	16.55	420.370	17.55	445.770	18.55	471.170	19.55	496.570
15.56	395.224	16.56	420.624	17.56	446.024	18.56	471.424	19.56	496.824
15.57	395.478	16.57	420.878	17.57	446.278	18.57	471.678	19.57	497.078
15.58	395.732	16.58	421.132	17.58	446.532	18.58	471.932	19.58	497.332
15.59	395.986	16.59	421.386	17.59	446.786	18.59	472.186	19.59	497.586
15.60	396.240	16.60	421.640	17.60	447.040	18.60	472.440	19.60	497.840
15.61	396.494	16.61	421.894	17.61	447.294	18.61	472.694	19.61	498.094
15.62	396.748	16.62	422.148	17.62	447.548	18.62	472.948	19.62	498.348
15.63	397.002	16.63	422.402	17.63	447.802	18.63	473.202	19.63	498.602
15.64	397.256	16.64	422.656	17.64	448.056	18.64	473.456	19.64	498.856
15.65	397.510	16.65	422.910	17.65	448.310	18.65	473.710	19.65	499.110
15.66	397.764	16.66	423.164	17.66	448.564	18.66	473.964	19.66	499.364
15.67	398.018	16.67	423.418	17.67	448.818	18.67	474.218	19.67	499.618
15.68	398.272	16.68	423.672	17.68	449.072	18.68	474.472	19.68	499.872
15.69	398.526	16.69	423.926	17.69	449.326	18.69	474.726	19.69	500.126
15.70	398.780	16.70	424.180	17.70	449.580	18.70	474.980	19.70	500.380
15.71	399.034	16.71	424.434	17.71	449.834	18.71	475.234	19.71	500.634
15.72	399.288	16.72	424.688	17.72	450.088	18.72	475.488	19.72	500.888
15.73	399.542	16.73	424.942	17.73	450.342	18.73	475.742	19.73	501.142
15.74	399.796	16.74	425.196	17.74	450.596	18.74	475.996	19.74	501.396
15.75	400.050	16.75	425.450	17.75	450.850	18.75	476.250	19.75	501.650
15.76	400.304	16.76	425.704	17.76	451.104	18.76	476.504	19.76	501.904
15.77	400.558	16.77	425.958	17.77	451.358	18.77	476.758	19.77	502.158
15.78	400.812	16.78	426.212	17.78	451.612	18.78	477.012	19.78	502.412
15.79	401.066	16.79	426.466	17.79	451.866	18.79	477.266	19.79	502.666
15.80	401.320	16.80	426.720	17.80	452.120	18.80	477.520	19.80	502.920
15.81	401.574	16.81	426.974	17.81	452.374	18.81	477.774	19.81	503.174
15.82	401.828	16.82	427.228	17.82	452.628	18.82	478.028	19.82	503.428
15.83	402.082	16.83	427.482	17.83	452.882	18.83	478.282	19.83	503.682
15.84	402.336	16.84	427.736	17.84	453.136	18.84	478.536	19.84	503.936
15.85	402.590	16.85	427.990	17.85	453.390	18.85	478.790	19.85	504.190
15.86	402.844	16.86	428.244	17.86	453.644	18.86	479.044	19.86	504.444
15.87	403.098	16.87	428.498	17.87	453.898	18.87	479.298	19.87	504.698
15.88	403.352	16.88	428.752	17.88	454.152	18.88	479.552	19.88	504.952
15.89	403.606	16.89	429.006	17.89	454.406	18.89	479.806	19.89	505.206
15.90	403.860	16.90	429.260	17.90	454.660	18.90	480.060	19.90	505.460
15.91	404.114	16.91	429.514	17.91	454.914	18.91	480.314	19.91	505.714
15.92	404.368	16.92	429.768	17.92	455.168	18.92	480.568	19.92	505.968
15.93	404.622	16.93	430.022	17.93	455.422	18.93	480.822	19.93	506.222
15.94	404.876	16.94	430.276	17.94	455.676	18.94	481.076	19.94	506.476
15.95	405.130	16.95	430.530	17.95	455.930	18.95	481.330	19.95	506.730
15.96	405.384	16.96	430.784	17.96	456.184	18.96	481.584	19.96	506.984
15.97	405.638	16.97	431.038	17.97	456.438	18.97	481.838	19.97	507.238
15.98	405.892	16.98	431.292	17.98	456.692	18.98	482.092	19.98	507.492
15.99	406.146	16.99	431.546	17.99	456.946	18.99	482.346	19.99	507.746

Continued - INCHES TO MILLIMETERS 1 INCH = 25.4 MILLIMETERS

INCHES	MILLI-METERS	INCHES	MILLI-METERS	INCHES	MILLI-METERS	INCHES	MILLI-METERS	INCHES	MILLI-METERS
20.00	508.000	21.00	533.400	22.00	558.800	23.00	584.200	24.00	609.600
20.01	508.254	21.01	533.654	22.01	559.054	23.01	584.454	24.01	609.854
20.02	508.508	21.02	533.908	22.02	559.308	23.02	584.708	24.02	610.108
20.03	508.762	21.03	534.162	22.03	559.562	23.03	584.962	24.03	610.362
20.04	509.016	21.04	534.416	22.04	559.816	23.04	585.216	24.04	610.616
20.05	509.270	21.05	534.670	22.05	560.070	23.05	585.470	24.05	610.870
20.06	509.524	21.06	534.924	22.06	560.324	23.06	585.724	24.06	611.124
20.07	509.778	21.07	535.178	22.07	560.578	23.07	585.978	24.07	611.378
20.08	510.032	21.08	535.432	22.08	560.832	23.08	586.232	24.08	611.632
20.09	510.286	21.09	535.686	22.09	561.086	23.09	586.486	24.09	611.886
20.10	510.540	21.10	535.940	22.10	561.340	23.10	586.740	24.10	612.140
20.11	510.794	21.11	536.194	22.11	561.594	23.11	586.994	24.11	612.394
20.12	511.048	21.12	536.448	22.12	561.848	23.12	587.248	24.12	612.648
20.13	511.302	21.13	536.702	22.13	562.102	23.13	587.502	24.13	612.902
20.14	511.556	21.14	536.956	22.14	562.356	23.14	587.756	24.14	613.156
20.15	511.810	21.15	537.210	22.15	562.610	23.15	588.010	24.15	613.410
20.16	512.064	21.16	537.464	22.16	562.864	23.16	588.264	24.16	613.664
20.17	512.318	21.17	537.718	22.17	563.118	23.17	588.518	24.17	613.918
20.18	512.572	21.18	537.972	22.18	563.372	23.18	588.772	24.18	614.172
20.19	512.826	21.19	538.226	22.19	563.626	23.19	589.026	24.19	614.426
20.20	513.080	21.20	538.480	22.20	563.880	23.20	589.280	24.20	614.680
20.21	513.334	21.21	538.734	22.21	564.134	23.21	589.534	24.21	614.934
20.22	513.588	21.22	538.988	22.22	564.388	23.22	589.788	24.22	615.188
20.23	513.842	21.23	539.242	22.23	564.642	23.23	590.042	24.23	615.442
20.24	514.096	21.24	539.496	22.24	564.896	23.24	590.296	24.24	615.696
20.25	514.350	21.25	539.750	22.25	565.150	23.25	590.550	24.25	615.950
20.26	514.604	21.26	540.004	22.26	565.404	23.26	590.804	24.26	616.204
20.27	514.858	21.27	540.258	22.27	565.658	23.27	591.058	24.27	616.458
20.28	515.112	21.28	540.512	22.28	565.912	23.28	591.312	24.28	616.712
20.29	515.366	21.29	540.766	22.29	566.166	23.29	591.566	24.29	616.966
20.30	515.620	21.30	541.020	22.30	566.420	23.30	591.820	24.30	617.220
20.31	515.874	21.31	541.274	22.31	566.674	23.31	592.074	24.31	617.474
20.32	516.128	21.32	541.528	22.32	566.928	23.32	592.328	24.32	617.728
20.33	516.382	21.33	541.782	22.33	567.182	23.33	592.582	24.33	617.982
20.34	516.636	21.34	542.036	22.34	567.436	23.34	592.836	24.34	618.236
20.35	516.890	21.35	542.290	22.35	567.690	23.35	593.090	24.35	618.490
20.36	517.144	21.36	542.544	22.36	567.944	23.36	593.344	24.36	618.744
20.37	517.398	21.37	542.798	22.37	568.198	23.37	593.598	24.37	618.998
20.38	517.652	21.38	543.052	22.38	568.452	23.38	593.852	24.38	619.252
20.39	517.906	21.39	543.306	22.39	568.706	23.39	594.106	24.39	619.506
20.40	518.160	21.40	543.560	22.40	568.960	23.40	594.360	24.40	619.760
20.41	518.414	21.41	543.814	22.41	569.214	23.41	594.614	24.41	620.014
20.42	518.668	21.42	544.068	22.42	569.468	23.42	594.868	24.42	620.268
20.43	518.922	21.43	544.322	22.43	569.722	23.43	595.122	24.43	620.522
20.44	519.176	21.44	544.576	22.44	569.976	23.44	595.376	24.44	620.776
20.45	519.430	21.45	544.830	22.45	570.230	23.45	595.630	24.45	621.030
20.46	519.684	21.46	545.084	22.46	570.484	23.46	595.884	24.46	621.284
20.47	519.938	21.47	545.338	22.47	570.738	23.47	596.138	24.47	621.538
20.48	520.192	21.48	545.592	22.48	570.992	23.48	596.392	24.48	621.792
20.49	520.446	21.49	545.846	22.49	571.246	23.49	596.646	24.49	622.046

Continued - INCHES TO MILLIMETERS 1 INCH = 25.4 MILLIMETERS

INCHES	MILLI-METERS	INCHES	MILLI-METERS	INCHES	MILLI-METERS	INCHES	MILLI-METERS	INCHES	MILLI-METERS
20.50	520.700	21.50	546.100	22.50	571.500	23.50	596.900	24.50	622.300
20.51	520.954	21.51	546.354	22.51	571.754	23.51	597.154	24.51	622.554
20.52	521.208	21.52	546.608	22.52	572.008	23.52	597.408	24.52	622.808
20.53	521.462	21.53	546.862	22.53	572.262	23.53	597.662	24.53	623.062
20.54	521.716	21.54	547.116	22.54	572.516	23.54	597.916	24.54	623.316
20.55	521.970	21.55	547.370	22.55	572.770	23.55	598.170	24.55	623.570
20.56	522.224	21.56	547.624	22.56	573.024	23.56	598.424	24.56	623.824
20.57	522.478	21.57	547.878	22.57	573.278	23.57	598.678	24.57	624.078
20.58	522.732	21.58	548.132	22.58	573.532	23.58	598.932	24.58	624.332
20.59	522.986	21.59	548.386	22.59	573.786	23.59	599.186	24.59	624.586
20.60	523.240	21.60	548.640	22.60	574.040	23.60	599.440	24.60	624.840
20.61	523.494	21.61	548.894	22.61	574.294	23.61	599.694	24.61	625.094
20.62	523.748	21.62	549.148	22.62	574.548	23.62	599.948	24.62	625.348
20.63	524.002	21.63	549.402	22.63	574.802	23.63	600.202	24.63	625.602
20.64	524.256	21.64	549.656	22.64	575.056	23.64	600.456	24.64	625.856
20.65	524.510	21.65	549.910	22.65	575.310	23.65	600.710	24.65	626.110
20.66	524.764	21.66	550.164	22.66	575.564	23.66	600.964	24.66	626.364
20.67	525.018	21.67	550.418	22.67	575.818	23.67	601.218	24.67	626.618
20.68	525.272	21.68	550.672	22.68	576.072	23.68	601.472	24.68	626.872
20.69	525.526	21.69	550.926	22.69	576.326	23.69	601.726	24.69	627.126
20.70	525.780	21.70	551.180	22.70	576.580	23.70	601.980	24.70	627.380
20.71	526.034	21.71	551.434	22.71	576.834	23.71	602.234	24.71	627.634
20.72	526.288	21.72	551.688	22.72	577.088	23.72	602.488	24.72	627.888
20.73	526.542	21.73	551.942	22.73	577.342	23.73	602.742	24.73	628.142
20.74	526.796	21.74	552.196	22.74	577.596	23.74	602.996	24.74	628.396
20.75	527.050	21.75	552.450	22.75	577.850	23.75	603.250	24.75	628.650
20.76	527.304	21.76	552.704	22.76	578.104	23.76	603.504	24.76	628.904
20.77	527.558	21.77	552.958	22.77	578.358	23.77	603.758	24.77	629.158
20.78	527.812	21.78	553.212	22.78	578.612	23.78	604.012	24.78	629.412
20.79	528.066	21.79	553.466	22.79	578.866	23.79	604.266	24.79	629.666
20.80	528.320	21.80	553.720	22.80	579.120	23.80	604.520	24.80	629.920
20.81	528.574	21.81	553.974	22.81	579.374	23.81	604.774	24.81	630.174
20.82	528.828	21.82	554.228	22.82	579.628	23.82	605.028	24.82	630.428
20.83	529.082	21.83	554.482	22.83	579.882	23.83	605.282	24.83	630.682
20.84	529.336	21.84	554.736	22.84	580.136	23.84	605.536	24.84	630.936
20.85	529.590	21.85	554.990	22.85	580.390	23.85	605.790	24.85	631.190
20.86	529.844	21.86	555.244	22.86	580.644	23.86	606.044	24.86	631.444
20.87	530.098	21.87	555.498	22.87	580.898	23.87	606.298	24.87	631.698
20.88	530.352	21.88	555.752	22.88	581.152	23.88	606.552	24.88	631.952
20.89	530.606	21.89	556.006	22.89	581.406	23.89	606.806	24.89	632.206
20.90	530.860	21.90	556.260	22.90	581.660	23.90	607.060	24.90	632.460
20.91	531.114	21.91	556.514	22.91	581.914	23.91	607.314	24.91	632.714
20.92	531.368	21.92	556.768	22.92	582.168	23.92	607.568	24.92	632.968
20.93	531.622	21.93	557.022	22.93	582.422	23.93	607.822	24.93	633.222
20.94	531.876	21.94	557.276	22.94	582.676	23.94	608.076	24.94	633.476
20.95	532.130	21.95	557.530	22.95	582.930	23.95	608.330	24.95	633.730
20.96	532.384	21.96	557.784	22.96	583.184	23.96	608.584	24.96	633.984
20.97	532.638	21.97	558.038	22.97	583.438	23.97	608.838	24.97	634.238
20.98	532.892	21.98	558.292	22.98	583.692	23.98	609.092	24.98	634.492
20.99	533.146	21.99	558.546	22.99	583.946	23.99	609.346	24.99	634.746

Continued - INCHES TO MILLIMETERS 1 INCH = 25.4 MILLIMETERS

INCHES	MILLI-METERS	INCHES	MILLI-METERS	INCHES	MILLI-METERS	INCHES	MILLI-METERS	INCHES	MILLI-METERS
25.00	635.000	26.00	660.400	27.00	685.800	28.00	711.200	29.00	736.600
25.01	635.254	26.01	660.654	27.01	686.054	28.01	711.454	29.01	736.854
25.02	635.508	26.02	660.908	27.02	686.308	28.02	711.708	29.02	737.108
25.03	635.762	26.03	661.162	27.03	686.562	28.03	711.962	29.03	737.362
25.04	636.016	26.04	661.416	27.04	686.816	28.04	712.216	29.04	737.616
25.05	636.270	26.05	661.670	27.05	687.070	28.05	712.470	29.05	737.870
25.06	636.524	26.06	661.924	27.06	687.324	28.06	712.724	29.06	738.124
25.07	636.778	26.07	662.178	27.07	687.578	28.07	712.978	29.07	738.378
25.08	637.032	26.08	662.432	27.08	687.832	28.08	713.232	29.08	738.632
25.09	637.286	26.09	662.686	27.09	688.086	28.09	713.486	29.09	738.886
25.10	637.540	26.10	662.940	27.10	688.340	28.10	713.740	29.10	739.140
25.11	637.794	26.11	663.194	27.11	688.594	28.11	713.994	29.11	739.394
25.12	638.048	26.12	663.448	27.12	688.848	28.12	714.248	29.12	739.648
25.13	638.302	26.13	663.702	27.13	689.102	28.13	714.502	29.13	739.902
25.14	638.556	26.14	663.956	27.14	689.356	28.14	714.756	29.14	740.156
25.15	638.810	26.15	664.210	27.15	689.610	28.15	715.010	29.15	740.410
25.16	639.064	26.16	664.464	27.16	689.864	28.16	715.264	29.16	740.664
25.17	639.318	26.17	664.718	27.17	690.118	28.17	715.518	29.17	740.918
25.18	639.572	26.18	664.972	27.18	690.372	28.18	715.772	29.18	741.172
25.19	639.826	26.19	665.226	27.19	690.626	28.19	716.026	29.19	741.426
25.20	640.080	26.20	665.480	27.20	690.880	28.20	716.280	29.20	741.680
25.21	640.334	26.21	665.734	27.21	691.134	28.21	716.534	29.21	741.934
25.22	640.588	26.22	665.988	27.22	691.388	28.22	716.788	29.22	742.188
25.23	640.842	26.23	666.242	27.23	691.642	28.23	717.042	29.23	742.442
25.24	641.096	26.24	666.496	27.24	691.896	28.24	717.296	29.24	742.696
25.25	641.350	26.25	666.750	27.25	692.150	28.25	717.550	29.25	742.950
25.26	641.604	26.26	667.004	27.26	692.404	28.26	717.804	29.26	743.204
25.27	641.858	26.27	667.258	27.27	692.658	28.27	718.058	29.27	743.458
25.28	642.112	26.28	667.512	27.28	692.912	28.28	718.312	29.28	743.712
25.29	642.366	26.29	667.766	27.29	693.166	28.29	718.566	29.29	743.966
25.30	642.620	26.30	668.020	27.30	693.420	28.30	718.820	29.30	744.220
25.31	642.874	26.31	668.274	27.31	693.674	28.31	719.074	29.31	744.474
25.32	643.128	26.32	668.528	27.32	693.928	28.32	719.328	29.32	744.728
25.33	643.382	26.33	668.782	27.33	694.182	28.33	719.582	29.33	744.982
25.34	643.636	26.34	669.036	27.34	694.436	28.34	719.836	29.34	745.236
25.35	643.890	26.35	669.290	27.35	694.690	28.35	720.090	29.35	745.490
25.36	644.144	26.36	669.544	27.36	694.944	28.36	720.344	29.36	745.744
25.37	644.398	26.37	669.798	27.37	695.198	28.37	720.598	29.37	745.998
25.38	644.652	26.38	670.052	27.38	695.452	28.38	720.852	29.38	746.252
25.39	644.906	26.39	670.306	27.39	695.706	28.39	721.106	29.39	746.506
25.40	645.160	26.40	670.560	27.40	695.960	28.40	721.360	29.40	746.760
25.41	645.414	26.41	670.814	27.41	696.214	28.41	721.614	29.41	747.014
25.42	645.668	26.42	671.068	27.42	696.468	28.42	721.868	29.42	747.268
25.43	645.922	26.43	671.322	27.43	696.722	28.43	722.122	29.43	747.522
25.44	646.176	26.44	671.576	27.44	696.976	28.44	722.376	29.44	747.776
25.45	646.430	26.45	671.830	27.45	697.230	28.45	722.630	29.45	748.030
25.46	646.684	26.46	672.084	27.46	697.484	28.46	722.884	29.46	748.284
25.47	646.938	26.47	672.338	27.47	697.738	28.47	723.138	29.47	748.538
25.48	647.192	26.48	672.592	27.48	697.992	28.48	723.392	29.48	748.792
25.49	647.446	26.49	672.846	27.49	698.246	28.49	723.646	29.49	749.046

Continued - INCHES TO MILLIMETERS 1 INCH = 25.4 MILLIMETERS

INCHES	MILLI-METERS	INCHES	MILLI-METERS	INCHES	MILLI-METERS	INCHES	MILLI-METERS	INCHES	MILLI-METERS
25.50	647.700	26.50	673.100	27.50	698.500	28.50	723.900	29.50	749.300
25.51	647.954	26.51	673.354	27.51	698.754	28.51	724.154	29.51	749.554
25.52	648.208	26.52	673.608	27.52	699.008	28.52	724.408	29.52	749.808
25.53	648.462	26.53	673.862	27.53	699.262	28.53	724.662	29.53	750.062
25.54	648.716	26.54	674.116	27.54	699.516	28.54	724.916	29.54	750.316
25.55	648.970	26.55	674.370	27.55	699.770	28.55	725.170	29.55	750.570
25.56	649.224	26.56	674.624	27.56	700.024	28.56	725.424	29.56	750.824
25.57	649.478	26.57	674.878	27.57	700.278	28.57	725.678	29.57	751.078
25.58	649.732	26.58	675.132	27.58	700.532	28.58	725.932	29.58	751.332
25.59	649.986	26.59	675.386	27.59	700.786	28.59	726.186	29.59	751.586
25.60	650.240	26.60	675.640	27.60	701.040	28.60	726.440	29.60	751.840
25.61	650.494	26.61	675.894	27.61	701.294	28.61	726.694	29.61	752.094
25.62	650.748	26.62	676.148	27.62	701.548	28.62	726.948	29.62	752.348
25.63	651.002	26.63	676.402	27.63	701.802	28.63	727.202	29.63	752.602
25.64	651.256	26.64	676.656	27.64	702.056	28.64	727.456	29.64	752.856
25.65	651.510	26.65	676.910	27.65	702.310	28.65	727.710	29.65	753.110
25.66	651.764	26.66	677.164	27.66	702.564	28.66	727.964	29.66	753.364
25.67	652.018	26.67	677.418	27.67	702.818	28.67	728.218	29.67	753.618
25.68	652.272	26.68	677.672	27.68	703.072	28.68	728.472	29.68	753.872
25.69	652.526	26.69	677.926	27.69	703.326	28.69	728.726	29.69	754.126
25.70	652.780	26.70	678.180	27.70	703.580	28.70	728.980	29.70	754.380
25.71	653.034	26.71	678.434	27.71	703.834	28.71	729.234	29.71	754.634
25.72	653.288	26.72	678.688	27.72	704.088	28.72	729.488	29.72	754.888
25.73	653.542	26.73	678.942	27.73	704.342	28.73	729.742	29.73	755.142
25.74	653.796	26.74	679.196	27.74	704.596	28.74	729.996	29.74	755.396
25.75	654.050	26.75	679.450	27.75	704.850	28.75	730.250	29.75	755.650
25.76	654.304	26.76	679.704	27.76	705.104	28.76	730.504	29.76	755.904
25.77	654.558	26.77	679.958	27.77	705.358	28.77	730.758	29.77	756.158
25.78	654.812	26.78	680.212	27.78	705.612	28.78	731.012	29.78	756.412
25.79	655.066	26.79	680.466	27.79	705.866	28.79	731.266	29.79	756.666
25.80	655.320	26.80	680.720	27.80	706.120	28.80	731.520	29.80	756.920
25.81	655.574	26.81	680.974	27.81	706.374	28.81	731.774	29.81	757.174
25.82	655.828	26.82	681.228	27.82	706.628	28.82	732.028	29.82	757.428
25.83	656.082	26.83	681.482	27.83	706.882	28.83	732.282	29.83	757.682
25.84	656.336	26.84	681.736	27.84	707.136	28.84	732.536	29.84	757.936
25.85	656.590	26.85	681.990	27.85	707.390	28.85	732.790	29.85	758.190
25.86	656.844	26.86	682.244	27.86	707.644	28.86	733.044	29.86	758.444
25.87	657.098	26.87	682.498	27.87	707.898	28.87	733.298	29.87	758.698
25.88	657.352	26.88	682.752	27.88	708.152	28.88	733.552	29.88	758.952
25.89	657.606	26.89	683.006	27.89	708.406	28.89	733.806	29.89	759.206
25.90	657.860	26.90	683.260	27.90	708.660	28.90	734.060	29.90	759.460
25.91	658.114	26.91	683.514	27.91	708.914	28.91	734.314	29.91	759.714
25.92	658.368	26.92	683.768	27.92	709.168	28.92	734.568	29.92	759.968
25.93	658.622	26.93	684.022	27.93	709.422	28.93	734.822	29.93	760.222
25.94	658.876	26.94	684.276	27.94	709.676	28.94	735.076	29.94	760.476
25.95	659.130	26.95	684.530	27.95	709.930	28.95	735.330	29.95	760.730
25.96	659.384	26.96	684.784	27.96	710.184	28.96	735.584	29.96	760.984
25.97	659.638	26.97	685.038	27.97	710.438	28.97	735.838	29.97	761.238
25.98	659.892	26.98	685.292	27.98	710.692	28.98	736.092	29.98	761.492
25.99	660.146	26.99	685.546	27.99	710.946	28.99	736.346	29.99	761.746

Continued - INCHES TO MILLIMETERS 1 INCH = 25.4 MILLIMETERS

INCHES	MILLI-METERS	INCHES	MILLI-METERS	INCHES	MILLI-METERS	INCHES	MILLI-METERS	INCHES	MILLI-METERS
30.00	762.000	31.00	787.400	32.00	812.800	33.00	838.200	34.00	863.600
30.01	762.254	31.01	787.654	32.01	813.054	33.01	838.454	34.01	863.854
30.02	762.508	31.02	787.908	32.02	813.308	33.02	838.708	34.02	864.108
30.03	762.762	31.03	788.162	32.03	813.562	33.03	838.962	34.03	864.362
30.04	763.016	31.04	788.416	32.04	813.816	33.04	839.216	34.04	864.616
30.05	763.270	31.05	788.670	32.05	814.070	33.05	839.470	34.05	864.870
30.06	763.524	31.06	788.924	32.06	814.324	33.06	839.724	34.06	865.124
30.07	763.778	31.07	789.178	32.07	814.578	33.07	839.978	34.07	865.378
30.08	764.032	31.08	789.432	32.08	814.832	33.08	840.232	34.08	865.632
30.09	764.286	31.09	789.686	32.09	815.086	33.09	840.486	34.09	865.886
30.10	764.540	31.10	789.940	32.10	815.340	33.10	840.740	34.10	866.140
30.11	764.794	31.11	790.194	32.11	815.594	33.11	840.994	34.11	866.394
30.12	765.048	31.12	790.448	32.12	815.848	33.12	841.248	34.12	866.648
30.13	765.302	31.13	790.702	32.13	816.102	33.13	841.502	34.13	866.902
30.14	765.556	31.14	790.956	32.14	816.356	33.14	841.756	34.14	867.156
30.15	765.810	31.15	791.210	32.15	816.610	33.15	842.010	34.15	867.410
30.16	766.064	31.16	791.464	32.16	816.864	33.16	842.264	34.16	867.664
30.17	766.318	31.17	791.718	32.17	817.118	33.17	842.518	34.17	867.918
30.18	766.572	31.18	791.972	32.18	817.372	33.18	842.772	34.18	868.172
30.19	766.826	31.19	792.226	32.19	817.626	33.19	843.026	34.19	868.426
30.20	767.080	31.20	792.480	32.20	817.880	33.20	843.280	34.20	868.680
30.21	767.334	31.21	792.734	32.21	818.134	33.21	843.534	34.21	868.934
30.22	767.588	31.22	792.988	32.22	818.388	33.22	843.788	34.22	869.188
30.23	767.842	31.23	793.242	32.23	818.642	33.23	844.042	34.23	869.442
30.24	768.096	31.24	793.496	32.24	818.896	33.24	844.296	34.24	869.696
30.25	768.350	31.25	793.750	32.25	819.150	33.25	844.550	34.25	869.950
30.26	768.604	31.26	794.004	32.26	819.404	33.26	844.804	34.26	870.204
30.27	768.858	31.27	794.258	32.27	819.658	33.27	845.058	34.27	870.458
30.28	769.112	31.28	794.512	32.28	819.912	33.28	845.312	34.28	870.712
30.29	769.366	31.29	794.766	32.29	820.166	33.29	845.566	34.29	870.966
30.30	769.620	31.30	795.020	32.30	820.420	33.30	845.820	34.30	871.220
30.31	769.874	31.31	795.274	32.31	820.674	33.31	846.074	34.31	871.474
30.32	770.128	31.32	795.528	32.32	820.928	33.32	846.328	34.32	871.728
30.33	770.382	31.33	795.782	32.33	821.182	33.33	846.582	34.33	871.982
30.34	770.636	31.34	796.036	32.34	821.436	33.34	846.836	34.34	872.236
30.35	770.890	31.35	796.290	32.35	821.690	33.35	847.090	34.35	872.490
30.36	771.144	31.36	796.544	32.36	821.944	33.36	847.344	34.36	872.744
30.37	771.398	31.37	796.798	32.37	822.198	33.37	847.598	34.37	872.998
30.38	771.652	31.38	797.052	32.38	822.452	33.38	847.852	34.38	873.252
30.39	771.906	31.39	797.306	32.39	822.706	33.39	848.106	34.39	873.506
30.40	772.160	31.40	797.560	32.40	822.960	33.40	848.360	34.40	873.760
30.41	772.414	31.41	797.814	32.41	823.214	33.41	848.614	34.41	874.014
30.42	772.668	31.42	798.068	32.42	823.468	33.42	848.868	34.42	874.268
30.43	772.922	31.43	798.322	32.43	823.722	33.43	849.122	34.43	874.522
30.44	773.176	31.44	798.576	32.44	823.976	33.44	849.376	34.44	874.776
30.45	773.430	31.45	798.830	32.45	824.230	33.45	849.630	34.45	875.030
30.46	773.684	31.46	799.084	32.46	824.484	33.46	849.884	34.46	875.284
30.47	773.938	31.47	799.338	32.47	824.738	33.47	850.138	34.47	875.538
30.48	774.192	31.48	799.592	32.48	824.992	33.48	850.392	34.48	875.792
30.49	774.446	31.49	799.846	32.49	825.246	33.49	850.646	34.49	876.046

Continued - INCHES TO MILLIMETERS 1 INCH = 25.4 MILLIMETERS

INCHES	MILLI-METERS	INCHES	MILLI-METERS	INCHES	MILLI-METERS	INCHES	MILLI-METERS	INCHES	MILLI-METERS
30.50	774.700	31.50	800.100	32.50	825.500	33.50	850.900	34.50	876.300
30.51	774.954	31.51	800.354	32.51	825.754	33.51	851.154	34.51	876.554
30.52	775.208	31.52	800.608	32.52	826.008	33.52	851.408	34.52	876.808
30.53	775.462	31.53	800.862	32.53	826.262	33.53	851.662	34.53	877.062
30.54	775.716	31.54	801.116	32.54	826.516	33.54	851.916	34.54	877.316
30.55	775.970	31.55	801.370	32.55	826.770	33.55	852.170	34.55	877.570
30.56	776.224	31.56	801.624	32.56	827.024	33.56	852.424	34.56	877.824
30.57	776.478	31.57	801.878	32.57	827.278	33.57	852.678	34.57	878.078
30.58	776.732	31.58	802.132	32.58	827.532	33.58	852.932	34.58	878.332
30.59	776.986	31.59	802.386	32.59	827.786	33.59	853.186	34.59	878.586
30.60	777.240	31.60	802.640	32.60	828.040	33.60	853.440	34.60	878.840
30.61	777.494	31.61	802.894	32.61	828.294	33.61	853.694	34.61	879.094
30.62	777.748	31.62	803.148	32.62	828.548	33.62	853.948	34.62	879.348
30.63	778.002	31.63	803.402	32.63	828.802	33.63	854.202	34.63	879.602
30.64	778.256	31.64	803.656	32.64	829.056	33.64	854.456	34.64	879.856
30.65	778.510	31.65	803.910	32.65	829.310	33.65	854.710	34.65	880.110
30.66	778.764	31.66	804.164	32.66	829.564	33.66	854.964	34.66	880.364
30.67	779.018	31.67	804.418	32.67	829.818	33.67	855.218	34.67	880.618
30.68	779.272	31.68	804.672	32.68	830.072	33.68	855.472	34.68	880.872
30.69	779.526	31.69	804.926	32.69	830.326	33.69	855.726	34.69	881.126
30.70	779.780	31.70	805.180	32.70	830.580	33.70	855.980	34.70	881.380
30.71	780.034	31.71	805.434	32.71	830.834	33.71	856.234	34.71	881.634
30.72	780.288	31.72	805.688	32.72	831.088	33.72	856.488	34.72	881.888
30.73	780.542	31.73	805.942	32.73	831.342	33.73	856.742	34.73	882.142
30.74	780.796	31.74	806.196	32.74	831.596	33.74	856.996	34.74	882.396
30.75	781.050	31.75	806.450	32.75	831.850	33.75	857.250	34.75	882.650
30.76	781.304	31.76	806.704	32.76	832.104	33.76	857.504	34.76	882.904
30.77	781.558	31.77	806.958	32.77	832.358	33.77	857.758	34.77	883.158
30.78	781.812	31.78	807.212	32.78	832.612	33.78	858.012	34.78	883.412
30.79	782.066	31.79	807.466	32.79	832.866	33.79	858.266	34.79	883.666
30.80	782.320	31.80	807.720	32.80	833.120	33.80	858.520	34.80	883.920
30.81	782.574	31.81	807.974	32.81	833.374	33.81	858.774	34.81	884.174
30.82	782.828	31.82	808.228	32.82	833.628	33.82	859.028	34.82	884.428
30.83	783.082	31.83	808.482	32.83	833.882	33.83	859.282	34.83	884.682
30.84	783.336	31.84	808.736	32.84	834.136	33.84	859.536	34.84	884.936
30.85	783.590	31.85	808.990	32.85	834.390	33.85	859.790	34.85	885.190
30.86	783.844	31.86	809.244	32.86	834.644	33.86	860.044	34.86	885.444
30.87	784.098	31.87	809.498	32.87	834.898	33.87	860.298	34.87	885.698
30.88	784.352	31.88	809.752	32.88	835.152	33.88	860.552	34.88	885.952
30.89	784.606	31.89	810.006	32.89	835.406	33.89	860.806	34.89	886.206
30.90	784.860	31.90	810.260	32.90	835.660	33.90	861.060	34.90	886.460
30.91	785.114	31.91	810.514	32.91	835.914	33.91	861.314	34.91	886.714
30.92	785.368	31.92	810.768	32.92	836.168	33.92	861.568	34.92	886.968
30.93	785.622	31.93	811.022	32.93	836.422	33.93	861.822	34.93	887.222
30.94	785.876	31.94	811.276	32.94	836.676	33.94	862.076	34.94	887.476
30.95	786.130	31.95	811.530	32.95	836.930	33.95	862.330	34.95	887.730
30.96	786.384	31.96	811.784	32.96	837.184	33.96	862.584	34.96	887.984
30.97	786.638	31.97	812.038	32.97	837.438	33.97	862.838	34.97	888.238
30.98	786.892	31.98	812.292	32.98	837.692	33.98	863.092	34.98	888.492
30.99	787.146	31.99	812.546	32.99	837.946	33.99	863.346	34.99	888.746

Continued - INCHES TO MILLIMETERS 1 INCH = 25.4 MILLIMETERS

INCHES	MILLI-METERS	INCHES	MILLI-METERS	INCHES	MILLI-METERS	INCHES	MILLI-METERS	INCHES	MILLI-METERS
35.00	889.000	36.00	914.400	37.00	939.800	38.00	965.200	39.00	990.600
35.01	889.254	36.01	914.654	37.01	940.054	38.01	965.454	39.01	990.854
35.02	889.508	36.02	914.908	37.02	940.308	38.02	965.708	39.02	991.108
35.03	889.762	36.03	915.162	37.03	940.562	38.03	965.962	39.03	991.362
35.04	890.016	36.04	915.416	37.04	940.816	38.04	966.216	39.04	991.616
35.05	890.270	36.05	915.670	37.05	941.070	38.05	966.470	39.05	991.870
35.06	890.524	36.06	915.924	37.06	941.324	38.06	966.724	39.06	992.124
35.07	890.778	36.07	916.178	37.07	941.578	38.07	966.978	39.07	992.378
35.08	891.032	36.08	916.432	37.08	941.832	38.08	967.232	39.08	992.632
35.09	891.286	36.09	916.686	37.09	942.086	38.09	967.486	39.09	992.886
35.10	891.540	36.10	916.940	37.10	942.340	38.10	967.740	39.10	993.140
35.11	891.794	36.11	917.194	37.11	942.594	38.11	967.994	39.11	993.394
35.12	892.048	36.12	917.448	37.12	942.848	38.12	968.248	39.12	993.648
35.13	892.302	36.13	917.702	37.13	943.102	38.13	968.502	39.13	993.902
35.14	892.556	36.14	917.956	37.14	943.356	38.14	968.756	39.14	994.156
35.15	892.810	36.15	918.210	37.15	943.610	38.15	969.010	39.15	994.410
35.16	893.064	36.16	918.464	37.16	943.864	38.16	969.264	39.16	994.664
35.17	893.318	36.17	918.718	37.17	944.118	38.17	969.518	39.17	994.918
35.18	893.572	36.18	918.972	37.18	944.372	38.18	969.772	39.18	995.172
35.19	893.826	36.19	919.226	37.19	944.626	38.19	970.026	39.19	995.426
35.20	894.080	36.20	919.480	37.20	944.880	38.20	970.280	39.20	995.680
35.21	894.334	36.21	919.734	37.21	945.134	38.21	970.534	39.21	995.934
35.22	894.588	36.22	919.988	37.22	945.388	38.22	970.788	39.22	996.188
35.23	894.842	36.23	920.242	37.23	945.642	38.23	971.042	39.23	996.442
35.24	895.096	36.24	920.496	37.24	945.896	38.24	971.296	39.24	996.696
35.25	895.350	36.25	920.750	37.25	946.150	38.25	971.550	39.25	996.950
35.26	895.604	36.26	921.004	37.26	946.404	38.26	971.804	39.26	997.204
35.27	895.858	36.27	921.258	37.27	946.658	38.27	972.058	39.27	997.458
35.28	896.112	36.28	921.512	37.28	946.912	38.28	972.312	39.28	997.712
35.29	896.366	36.29	921.766	37.29	947.166	38.29	972.566	39.29	997.966
35.30	896.620	36.30	922.020	37.30	947.420	38.30	972.820	39.30	998.220
35.31	896.874	36.31	922.274	37.31	947.674	38.31	973.074	39.31	998.474
35.32	897.128	36.32	922.528	37.32	947.928	38.32	973.328	39.32	998.728
35.33	897.382	36.33	922.782	37.33	948.182	38.33	973.582	39.33	998.982
35.34	897.636	36.34	923.036	37.34	948.436	38.34	973.836	39.34	999.236
35.35	897.890	36.35	923.290	37.35	948.690	38.35	974.090	39.35	999.490
35.36	898.144	36.36	923.544	37.36	948.944	38.36	974.344	39.36	999.744
35.37	898.398	36.37	923.798	37.37	949.198	38.37	974.598	39.37	999.998
35.38	898.652	36.38	924.052	37.38	949.452	38.38	974.852	39.38	1000.252
35.39	898.906	36.39	924.306	37.39	949.706	38.39	975.106	39.39	1000.506
35.40	899.160	36.40	924.560	37.40	949.960	38.40	975.360	39.40	1000.760
35.41	899.414	36.41	924.814	37.41	950.214	38.41	975.614	39.41	1001.014
35.42	899.668	36.42	925.068	37.42	950.468	38.42	975.868	39.42	1001.268
35.43	899.922	36.43	925.322	37.43	950.722	38.43	976.122	39.43	1001.522
35.44	900.176	36.44	925.576	37.44	950.976	38.44	976.376	39.44	1001.776
35.45	900.430	36.45	925.830	37.45	951.230	38.45	976.630	39.45	1002.030
35.46	900.684	36.46	926.084	37.46	951.484	38.46	976.884	39.46	1002.284
35.47	900.938	36.47	926.338	37.47	951.738	38.47	977.138	39.47	1002.538
35.48	901.192	36.48	926.592	37.48	951.992	38.48	977.392	39.48	1002.792
35.49	901.446	36.49	926.846	37.49	952.246	38.49	977.646	39.49	1003.046

Continued - INCHES TO MILLIMETERS 1 INCH = 25.4 MILLIMETERS

INCHES	MILLI-METERS	INCHES	MILLI-METERS	INCHES	MILLI-METERS	INCHES	MILLI-METERS	INCHES	MILLI-METERS
35.50	901.700	36.50	927.100	37.50	952.500	38.50	977.900	39.50	1003.300
35.51	901.954	36.51	927.354	37.51	952.754	38.51	978.154	39.51	1003.554
35.52	902.208	36.52	927.608	37.52	953.008	38.52	978.408	39.52	1003.808
35.53	902.462	36.53	927.862	37.53	953.262	38.53	978.662	39.53	1004.062
35.54	902.716	36.54	928.116	37.54	953.516	38.54	978.916	39.54	1004.316
35.55	902.970	36.55	928.370	37.55	953.770	38.55	979.170	39.55	1004.570
35.56	903.224	36.56	928.624	37.56	954.024	38.56	979.424	39.56	1004.824
35.57	903.478	36.57	928.878	37.57	954.278	38.57	979.678	39.57	1005.078
35.58	903.732	36.58	929.132	37.58	954.532	38.58	979.932	39.58	1005.332
35.59	903.986	36.59	929.386	37.59	954.786	38.59	980.186	39.59	1005.586
35.60	904.240	36.60	929.640	37.60	955.040	38.60	980.440	39.60	1005.840
35.61	904.494	36.61	929.894	37.61	955.294	38.61	980.694	39.61	1006.094
35.62	904.748	36.62	930.148	37.62	955.548	38.62	980.948	39.62	1006.348
35.63	905.002	36.63	930.402	37.63	955.802	38.63	981.202	39.63	1006.602
35.64	905.256	36.64	930.656	37.64	956.056	38.64	981.456	39.64	1006.856
35.65	905.510	36.65	930.910	37.65	956.310	38.65	981.710	39.65	1007.110
35.66	905.764	36.66	931.164	37.66	956.564	38.66	981.964	39.66	1007.364
35.67	906.018	36.67	931.418	37.67	956.818	38.67	982.218	39.67	1007.618
35.68	906.272	36.68	931.672	37.68	957.072	38.68	982.472	39.68	1007.872
35.69	906.526	36.69	931.926	37.69	957.326	38.69	982.726	39.69	1008.126
35.70	906.780	36.70	932.180	37.70	957.580	38.70	982.980	39.70	1008.380
35.71	907.034	36.71	932.434	37.71	957.834	38.71	983.234	39.71	1008.634
35.72	907.288	36.72	932.688	37.72	958.088	38.72	983.488	39.72	1008.888
35.73	907.542	36.73	932.942	37.73	958.342	38.73	983.742	39.73	1009.142
35.74	907.796	36.74	933.196	37.74	958.596	38.74	983.996	39.74	1009.396
35.75	908.050	36.75	933.450	37.75	958.850	38.75	984.250	39.75	1009.650
35.76	908.304	36.76	933.704	37.76	959.104	38.76	984.504	39.76	1009.904
35.77	908.558	36.77	933.958	37.77	959.358	38.77	984.758	39.77	1010.158
35.78	908.812	36.78	934.212	37.78	959.612	38.78	985.012	39.78	1010.412
35.79	909.066	36.79	934.466	37.79	959.866	38.79	985.266	39.79	1010.666
35.80	909.320	36.80	934.720	37.80	960.120	38.80	985.520	39.80	1010.920
35.81	909.574	36.81	934.974	37.81	960.374	38.81	985.774	39.81	1011.174
35.82	909.828	36.82	935.228	37.82	960.628	38.82	986.028	39.82	1011.428
35.83	910.082	36.83	935.482	37.83	960.882	38.83	986.282	39.83	1011.682
35.84	910.336	36.84	935.736	37.84	961.136	38.84	986.536	39.84	1011.936
35.85	910.590	36.85	935.990	37.85	961.390	38.85	986.790	39.85	1012.190
35.86	910.844	36.86	936.244	37.86	961.644	38.86	987.044	39.86	1012.444
35.87	911.098	36.87	936.498	37.87	961.898	38.87	987.298	39.87	1012.698
35.88	911.352	36.88	936.752	37.88	962.152	38.88	987.552	39.88	1012.952
35.89	911.606	36.89	937.006	37.89	962.406	38.89	987.806	39.89	1013.206
35.90	911.860	36.90	937.260	37.90	962.660	38.90	988.060	39.90	1013.460
35.91	912.114	36.91	937.514	37.91	962.914	38.91	988.314	39.91	1013.714
35.92	912.368	36.92	937.768	37.92	963.168	38.92	988.568	39.92	1013.968
35.93	912.622	36.93	938.022	37.93	963.422	38.93	988.822	39.93	1014.222
35.94	912.876	36.94	938.276	37.94	963.676	38.94	989.076	39.94	1014.476
35.95	913.130	36.95	938.530	37.95	963.930	38.95	989.330	39.95	1014.730
35.96	913.384	36.96	938.784	37.96	964.184	38.96	989.584	39.96	1014.984
35.97	913.638	36.97	939.038	37.97	964.438	38.97	989.838	39.97	1015.238
35.98	913.892	36.98	939.292	37.98	964.692	38.98	990.092	39.98	1015.492
35.99	914.146	36.99	939.546	37.99	964.946	38.99	990.346	39.99	1015.746

Continued - INCHES TO MILLIMETERS 1 INCH = 25.4 MILLIMETERS

INCHES	MILLI-METERS	INCHES	MILLI-METERS	INCHES	MILLI-METERS	INCHES	MILLI-METERS	INCHES	MILLI-METERS
40.00	1016.000	41.00	1041.400	42.00	1066.800	43.00	1092.200	44.00	1117.600
40.01	1016.254	41.01	1041.654	42.01	1067.054	43.01	1092.454	44.01	1117.854
40.02	1016.508	41.02	1041.908	42.02	1067.308	43.02	1092.708	44.02	1118.108
40.03	1016.762	41.03	1042.162	42.03	1067.562	43.03	1092.962	44.03	1118.362
40.04	1017.016	41.04	1042.416	42.04	1067.816	43.04	1093.216	44.04	1118.616
40.05	1017.270	41.05	1042.670	42.05	1068.070	43.05	1093.470	44.05	1118.870
40.06	1017.524	41.06	1042.924	42.06	1068.324	43.06	1093.724	44.06	1119.124
40.07	1017.778	41.07	1043.178	42.07	1068.578	43.07	1093.978	44.07	1119.378
40.08	1018.032	41.08	1043.432	42.08	1068.832	43.08	1094.232	44.08	1119.632
40.09	1018.286	41.09	1043.686	42.09	1069.086	43.09	1094.486	44.09	1119.886
40.10	1018.540	41.10	1043.940	42.10	1069.340	43.10	1094.740	44.10	1120.140
40.11	1018.794	41.11	1044.194	42.11	1069.594	43.11	1094.994	44.11	1120.394
40.12	1019.048	41.12	1044.448	42.12	1069.848	43.12	1095.248	44.12	1120.648
40.13	1019.302	41.13	1044.702	42.13	1070.102	43.13	1095.502	44.13	1120.902
40.14	1019.556	41.14	1044.956	42.14	1070.356	43.14	1095.756	44.14	1121.156
40.15	1019.810	41.15	1045.210	42.15	1070.610	43.15	1096.010	44.15	1121.410
40.16	1020.064	41.16	1045.464	42.16	1070.864	43.16	1096.264	44.16	1121.664
40.17	1020.318	41.17	1045.718	42.17	1071.118	43.17	1096.518	44.17	1121.918
40.18	1020.572	41.18	1045.972	42.18	1071.372	43.18	1096.772	44.18	1122.172
40.19	1020.826	41.19	1046.226	42.19	1071.626	43.19	1097.026	44.19	1122.426
40.20	1021.080	41.20	1046.480	42.20	1071.880	43.20	1097.280	44.20	1122.680
40.21	1021.334	41.21	1046.734	42.21	1072.134	43.21	1097.534	44.21	1122.934
40.22	1021.588	41.22	1046.988	42.22	1072.388	43.22	1097.788	44.22	1123.188
40.23	1021.842	41.23	1047.242	42.23	1072.642	43.23	1098.042	44.23	1123.442
40.24	1022.096	41.24	1047.496	42.24	1072.896	43.24	1098.296	44.24	1123.696
40.25	1022.350	41.25	1047.750	42.25	1073.150	43.25	1098.550	44.25	1123.950
40.26	1022.604	41.26	1048.004	42.26	1073.404	43.26	1098.804	44.26	1124.204
40.27	1022.858	41.27	1048.258	42.27	1073.658	43.27	1099.058	44.27	1124.458
40.28	1023.112	41.28	1048.512	42.28	1073.912	43.28	1099.312	44.28	1124.712
40.29	1023.366	41.29	1048.766	42.29	1074.166	43.29	1099.566	44.29	1124.966
40.30	1023.620	41.30	1049.020	42.30	1074.420	43.30	1099.820	44.30	1125.220
40.31	1023.874	41.31	1049.274	42.31	1074.674	43.31	1100.074	44.31	1125.474
40.32	1024.128	41.32	1049.528	42.32	1074.928	43.32	1100.328	44.32	1125.728
40.33	1024.382	41.33	1049.782	42.33	1075.182	43.33	1100.582	44.33	1125.982
40.34	1024.636	41.34	1050.036	42.34	1075.436	43.34	1100.836	44.34	1126.236
40.35	1024.890	41.35	1050.290	42.35	1075.690	43.35	1101.090	44.35	1126.490
40.36	1025.144	41.36	1050.544	42.36	1075.944	43.36	1101.344	44.36	1126.744
40.37	1025.398	41.37	1050.798	42.37	1076.198	43.37	1101.598	44.37	1126.998
40.38	1025.652	41.38	1051.052	42.38	1076.452	43.38	1101.852	44.38	1127.252
40.39	1025.906	41.39	1051.306	42.39	1076.706	43.39	1102.106	44.39	1127.506
40.40	1026.160	41.40	1051.560	42.40	1076.960	43.40	1102.360	44.40	1127.760
40.41	1026.414	41.41	1051.814	42.41	1077.214	43.41	1102.614	44.41	1128.014
40.42	1026.668	41.42	1052.068	42.42	1077.468	43.42	1102.868	44.42	1128.268
40.43	1026.922	41.43	1052.322	42.43	1077.722	43.43	1103.122	44.43	1128.522
40.44	1027.176	41.44	1052.576	42.44	1077.976	43.44	1103.376	44.44	1128.776
40.45	1027.430	41.45	1052.830	42.45	1078.230	43.45	1103.630	44.45	1129.030
40.46	1027.684	41.46	1053.084	42.46	1078.484	43.46	1103.884	44.46	1129.284
40.47	1027.938	41.47	1053.338	42.47	1078.738	43.47	1104.138	44.47	1129.538
40.48	1028.192	41.48	1053.592	42.48	1078.992	43.48	1104.392	44.48	1129.792
40.49	1028.446	41.49	1053.846	42.49	1079.246	43.49	1104.646	44.49	1130.046

Continued - INCHES TO MILLIMETERS 1 INCH = 25.4 MILLIMETERS

INCHES	MILLI-METERS	INCHES	MILLI-METERS	INCHES	MILLI-METERS	INCHES	MILLI-METERS	INCHES	MILLI-METERS
40.50	1028.700	41.50	1054.100	42.50	1079.500	43.50	1104.900	44.50	1130.300
40.51	1028.954	41.51	1054.354	42.51	1079.754	43.51	1105.154	44.51	1130.554
40.52	1029.208	41.52	1054.608	42.52	1080.008	43.52	1105.408	44.52	1130.808
40.53	1029.462	41.53	1054.862	42.53	1080.262	43.53	1105.662	44.53	1131.062
40.54	1029.716	41.54	1055.116	42.54	1080.516	43.54	1105.916	44.54	1131.316
40.55	1029.970	41.55	1055.370	42.55	1080.770	43.55	1106.170	44.55	1131.570
40.56	1030.224	41.56	1055.624	42.56	1081.024	43.56	1106.424	44.56	1131.824
40.57	1030.478	41.57	1055.878	42.57	1081.278	43.57	1106.678	44.57	1132.078
40.58	1030.732	41.58	1056.132	42.58	1081.532	43.58	1106.932	44.58	1132.332
40.59	1030.986	41.59	1056.386	42.59	1081.786	43.59	1107.186	44.59	1132.586
40.60	1031.240	41.60	1056.640	42.60	1082.040	43.60	1107.440	44.60	1132.840
40.61	1031.494	41.61	1056.894	42.61	1082.294	43.61	1107.694	44.61	1133.094
40.62	1031.748	41.62	1057.148	42.62	1082.548	43.62	1107.948	44.62	1133.348
40.63	1032.002	41.63	1057.402	42.63	1082.802	43.63	1108.202	44.63	1133.602
40.64	1032.256	41.64	1057.656	42.64	1083.056	43.64	1108.456	44.64	1133.856
40.65	1032.510	41.65	1057.910	42.65	1083.310	43.65	1108.710	44.65	1134.110
40.66	1032.764	41.66	1058.164	42.66	1083.564	43.66	1108.964	44.66	1134.364
40.67	1033.018	41.67	1058.418	42.67	1083.818	43.67	1109.218	44.67	1134.618
40.68	1033.272	41.68	1058.672	42.68	1084.072	43.68	1109.472	44.68	1134.872
40.69	1033.526	41.69	1058.926	42.69	1084.326	43.69	1109.726	44.69	1135.126
40.70	1033.780	41.70	1059.180	42.70	1084.580	43.70	1109.980	44.70	1135.380
40.71	1034.034	41.71	1059.434	42.71	1084.834	43.71	1110.234	44.71	1135.634
40.72	1034.288	41.72	1059.688	42.72	1085.088	43.72	1110.488	44.72	1135.888
40.73	1034.542	41.73	1059.942	42.73	1085.342	43.73	1110.742	44.73	1136.142
40.74	1034.796	41.74	1060.196	42.74	1085.596	43.74	1110.996	44.74	1136.396
40.75	1035.050	41.75	1060.450	42.75	1085.850	43.75	1111.250	44.75	1136.650
40.76	1035.304	41.76	1060.704	42.76	1086.104	43.76	1111.504	44.76	1136.904
40.77	1035.558	41.77	1060.958	42.77	1086.358	43.77	1111.758	44.77	1137.158
40.78	1035.812	41.78	1061.212	42.78	1086.612	43.78	1112.012	44.78	1137.412
40.79	1036.066	41.79	1061.466	42.79	1086.866	43.79	1112.266	44.79	1137.666
40.80	1036.320	41.80	1061.720	42.80	1087.120	43.80	1112.520	44.80	1137.920
40.81	1036.574	41.81	1061.974	42.81	1087.374	43.81	1112.774	44.81	1138.174
40.82	1036.828	41.82	1062.228	42.82	1087.628	43.82	1113.028	44.82	1138.428
40.83	1037.082	41.83	1062.482	42.83	1087.882	43.83	1113.282	44.83	1138.682
40.84	1037.336	41.84	1062.736	42.84	1088.136	43.84	1113.536	44.84	1138.936
40.85	1037.590	41.85	1062.990	42.85	1088.390	43.85	1113.790	44.85	1139.190
40.86	1037.844	41.86	1063.244	42.86	1088.644	43.86	1114.044	44.86	1139.444
40.87	1038.098	41.87	1063.498	42.87	1088.898	43.87	1114.298	44.87	1139.698
40.88	1038.352	41.88	1063.752	42.88	1089.152	43.88	1114.552	44.88	1139.952
40.89	1038.606	41.89	1064.006	42.89	1089.406	43.89	1114.806	44.89	1140.206
40.90	1038.860	41.90	1064.260	42.90	1089.660	43.90	1115.060	44.90	1140.460
40.91	1039.114	41.91	1064.514	42.91	1089.914	43.91	1115.314	44.91	1140.714
40.92	1039.368	41.92	1064.768	42.92	1090.168	43.92	1115.568	44.92	1140.968
40.93	1039.622	41.93	1065.022	42.93	1090.422	43.93	1115.822	44.93	1141.222
40.94	1039.876	41.94	1065.276	42.94	1090.676	43.94	1116.076	44.94	1141.476
40.95	1040.130	41.95	1065.530	42.95	1090.930	43.95	1116.330	44.95	1141.730
40.96	1040.384	41.96	1065.784	42.96	1091.184	43.96	1116.584	44.96	1141.984
40.97	1040.638	41.97	1066.038	42.97	1091.438	43.97	1116.838	44.97	1142.238
40.98	1040.892	41.98	1066.292	42.98	1091.692	43.98	1117.092	44.98	1142.492
40.99	1041.146	41.99	1066.546	42.99	1091.946	43.99	1117.346	44.99	1142.746

Continued - INCHES TO MILLIMETERS 1 INCH = 25.4 MILLIMETERS

INCHES	MILLI-METERS	INCHES	MILLI-METERS	INCHES	MILLI-METERS	INCHES	MILLI-METERS	INCHES	MILLI-METERS
45.00	1143.000	46.00	1168.400	47.00	1193.800	48.00	1219.200	49.00	1244.600
45.01	1143.254	46.01	1168.654	47.01	1194.054	48.01	1219.454	49.01	1244.854
45.02	1143.508	46.02	1168.908	47.02	1194.308	48.02	1219.708	49.02	1245.108
45.03	1143.762	46.03	1169.162	47.03	1194.562	48.03	1219.962	49.03	1245.362
45.04	1144.016	46.04	1169.416	47.04	1194.816	48.04	1220.216	49.04	1245.616
45.05	1144.270	46.05	1169.670	47.05	1195.070	48.05	1220.470	49.05	1245.870
45.06	1144.524	46.06	1169.924	47.06	1195.324	48.06	1220.724	49.06	1246.124
45.07	1144.778	46.07	1170.178	47.07	1195.578	48.07	1220.978	49.07	1246.378
45.08	1145.032	46.08	1170.432	47.08	1195.832	48.08	1221.232	49.08	1246.632
45.09	1145.286	46.09	1170.686	47.09	1196.086	48.09	1221.486	49.09	1246.886
45.10	1145.540	46.10	1170.940	47.10	1196.340	48.10	1221.740	49.10	1247.140
45.11	1145.794	46.11	1171.194	47.11	1196.594	48.11	1221.994	49.11	1247.394
45.12	1146.048	46.12	1171.448	47.12	1196.848	48.12	1222.248	49.12	1247.648
45.13	1146.302	46.13	1171.702	47.13	1197.102	48.13	1222.502	49.13	1247.902
45.14	1146.556	46.14	1171.956	47.14	1197.356	48.14	1222.756	49.14	1248.156
45.15	1146.810	46.15	1172.210	47.15	1197.610	48.15	1223.010	49.15	1248.410
45.16	1147.064	46.16	1172.464	47.16	1197.864	48.16	1223.264	49.16	1248.664
45.17	1147.318	46.17	1172.718	47.17	1198.118	48.17	1223.518	49.17	1248.918
45.18	1147.572	46.18	1172.972	47.18	1198.372	48.18	1223.772	49.18	1249.172
45.19	1147.826	46.19	1173.226	47.19	1198.626	48.19	1224.026	49.19	1249.426
45.20	1148.080	46.20	1173.480	47.20	1198.880	48.20	1224.280	49.20	1249.680
45.21	1148.334	46.21	1173.734	47.21	1199.134	48.21	1224.534	49.21	1249.934
45.22	1148.588	46.22	1173.988	47.22	1199.388	48.22	1224.788	49.22	1250.188
45.23	1148.842	46.23	1174.242	47.23	1199.642	48.23	1225.042	49.23	1250.442
45.24	1149.096	46.24	1174.496	47.24	1199.896	48.24	1225.296	49.24	1250.696
45.25	1149.350	46.25	1174.750	47.25	1200.150	48.25	1225.550	49.25	1250.950
45.26	1149.604	46.26	1175.004	47.26	1200.404	48.26	1225.804	49.26	1251.204
45.27	1149.858	46.27	1175.258	47.27	1200.658	48.27	1226.058	49.27	1251.458
45.28	1150.112	46.28	1175.512	47.28	1200.912	48.28	1226.312	49.28	1251.712
45.29	1150.366	46.29	1175.766	47.29	1201.166	48.29	1226.566	49.29	1251.966
45.30	1150.620	46.30	1176.020	47.30	1201.420	48.30	1226.820	49.30	1252.220
45.31	1150.874	46.31	1176.274	47.31	1201.674	48.31	1227.074	49.31	1252.474
45.32	1151.128	46.32	1176.528	47.32	1201.928	48.32	1227.328	49.32	1252.728
45.33	1151.382	46.33	1176.782	47.33	1202.182	48.33	1227.582	49.33	1252.982
45.34	1151.636	46.34	1177.036	47.34	1202.436	48.34	1227.836	49.34	1253.236
45.35	1151.890	46.35	1177.290	47.35	1202.690	48.35	1228.090	49.35	1253.490
45.36	1152.144	46.36	1177.544	47.36	1202.944	48.36	1228.344	49.36	1253.744
45.37	1152.398	46.37	1177.798	47.37	1203.198	48.37	1228.598	49.37	1253.998
45.38	1152.652	46.38	1178.052	47.38	1203.452	48.38	1228.852	49.38	1254.252
45.39	1152.906	46.39	1178.306	47.39	1203.706	48.39	1229.106	49.39	1254.506
45.40	1153.160	46.40	1178.560	47.40	1203.960	48.40	1229.360	49.40	1254.760
45.41	1153.414	46.41	1178.814	47.41	1204.214	48.41	1229.614	49.41	1255.014
45.42	1153.668	46.42	1179.068	47.42	1204.468	48.42	1229.868	49.42	1255.268
45.43	1153.922	46.43	1179.322	47.43	1204.722	48.43	1230.122	49.43	1255.522
45.44	1154.176	46.44	1179.576	47.44	1204.976	48.44	1230.376	49.44	1255.776
45.45	1154.430	46.45	1179.830	47.45	1205.230	48.45	1230.630	49.45	1256.030
45.46	1154.684	46.46	1180.084	47.46	1205.484	48.46	1230.884	49.46	1256.284
45.47	1154.938	46.47	1180.338	47.47	1205.738	48.47	1231.138	49.47	1256.538
45.48	1155.192	46.48	1180.592	47.48	1205.992	48.48	1231.392	49.48	1256.792
45.49	1155.446	46.49	1180.846	47.49	1206.246	48.49	1231.646	49.49	1257.046

Continued - INCHES TO MILLIMETERS 1 INCH = 25.4 MILLIMETERS

INCHES	MILLI-METERS	INCHES	MILLI-METERS	INCHES	MILLI-METERS	INCHES	MILLI-METERS	INCHES	MILLI-METERS
45.50	1155.700	46.50	1181.100	47.50	1206.500	48.50	1231.900	49.50	1257.300
45.51	1155.954	46.51	1181.354	47.51	1206.754	48.51	1232.154	49.51	1257.554
45.52	1156.208	46.52	1181.608	47.52	1207.008	48.52	1232.408	49.52	1257.808
45.53	1156.462	46.53	1181.862	47.53	1207.262	48.53	1232.662	49.53	1258.062
45.54	1156.716	46.54	1182.116	47.54	1207.516	48.54	1232.916	49.54	1258.316
45.55	1156.970	46.55	1182.370	47.55	1207.770	48.55	1233.170	49.55	1258.570
45.56	1157.224	46.56	1182.624	47.56	1208.024	48.56	1233.424	49.56	1258.824
45.57	1157.478	46.57	1182.878	47.57	1208.278	48.57	1233.678	49.57	1259.078
45.58	1157.732	46.58	1183.132	47.58	1208.532	48.58	1233.932	49.58	1259.332
45.59	1157.986	46.59	1183.386	47.59	1208.786	48.59	1234.186	49.59	1259.586
45.60	1158.240	46.60	1183.640	47.60	1209.040	48.60	1234.440	49.60	1259.840
45.61	1158.494	46.61	1183.894	47.61	1209.294	48.61	1234.694	49.61	1260.094
45.62	1158.748	46.62	1184.148	47.62	1209.548	48.62	1234.948	49.62	1260.348
45.63	1159.002	46.63	1184.402	47.63	1209.802	48.63	1235.202	49.63	1260.602
45.64	1159.256	46.64	1184.656	47.64	1210.056	48.64	1235.456	49.64	1260.856
45.65	1159.510	46.65	1184.910	47.65	1210.310	48.65	1235.710	49.65	1261.110
45.66	1159.764	46.66	1185.164	47.66	1210.564	48.66	1235.964	49.66	1261.364
45.67	1160.018	46.67	1185.418	47.67	1210.818	48.67	1236.218	49.67	1261.618
45.68	1160.272	46.68	1185.672	47.68	1211.072	48.68	1236.472	49.68	1261.872
45.69	1160.526	46.69	1185.926	47.69	1211.326	48.69	1236.726	49.69	1262.126
45.70	1160.780	46.70	1186.180	47.70	1211.580	48.70	1236.980	49.70	1262.380
45.71	1161.034	46.71	1186.434	47.71	1211.834	48.71	1237.234	49.71	1262.634
45.72	1161.288	46.72	1186.688	47.72	1212.088	48.72	1237.488	49.72	1262.888
45.73	1161.542	46.73	1186.942	47.73	1212.342	48.73	1237.742	49.73	1263.142
45.74	1161.796	46.74	1187.196	47.74	1212.596	48.74	1237.996	49.74	1263.396
45.75	1162.050	46.75	1187.450	47.75	1212.850	48.75	1238.250	49.75	1263.650
45.76	1162.304	46.76	1187.704	47.76	1213.104	48.76	1238.504	49.76	1263.904
45.77	1162.558	46.77	1187.958	47.77	1213.358	48.77	1238.758	49.77	1264.158
45.78	1162.812	46.78	1188.212	47.78	1213.612	48.78	1239.012	49.78	1264.412
45.79	1163.066	46.79	1188.466	47.79	1213.866	48.79	1239.266	49.79	1264.666
45.80	1163.320	46.80	1188.720	47.80	1214.120	48.80	1239.520	49.80	1264.920
45.81	1163.574	46.81	1188.974	47.81	1214.374	48.81	1239.774	49.81	1265.174
45.82	1163.828	46.82	1189.228	47.82	1214.628	48.82	1240.028	49.82	1265.428
45.83	1164.082	46.83	1189.482	47.83	1214.882	48.83	1240.282	49.83	1265.682
45.84	1164.336	46.84	1189.736	47.84	1215.136	48.84	1240.536	49.84	1265.936
45.85	1164.590	46.85	1189.990	47.85	1215.390	48.85	1240.790	49.85	1266.190
45.86	1164.844	46.86	1190.244	47.86	1215.644	48.86	1241.044	49.86	1266.444
45.87	1165.098	46.87	1190.498	47.87	1215.898	48.87	1241.298	49.87	1266.698
45.88	1165.352	46.88	1190.752	47.88	1216.152	48.88	1241.552	49.88	1266.952
45.89	1165.606	46.89	1191.006	47.89	1216.406	48.89	1241.806	49.89	1267.206
45.90	1165.860	46.90	1191.260	47.90	1216.660	48.90	1242.060	49.90	1267.460
45.91	1166.114	46.91	1191.514	47.91	1216.914	48.91	1242.314	49.91	1267.714
45.92	1166.368	46.92	1191.768	47.92	1217.168	48.92	1242.568	49.92	1267.968
45.93	1166.622	46.93	1192.022	47.93	1217.422	48.93	1242.822	49.93	1268.222
45.94	1166.876	46.94	1192.276	47.94	1217.676	48.94	1243.076	49.94	1268.476
45.95	1167.130	46.95	1192.530	47.95	1217.930	48.95	1243.330	49.95	1268.730
45.96	1167.384	46.96	1192.784	47.96	1218.184	48.96	1243.584	49.96	1268.984
45.97	1167.638	46.97	1193.038	47.97	1218.438	48.97	1243.838	49.97	1269.238
45.98	1167.892	46.98	1193.292	47.98	1218.692	48.98	1244.092	49.98	1269.492
45.99	1168.146	46.99	1193.546	47.99	1218.946	48.99	1244.346	49.99	1269.746

Continued - INCHES TO MILLIMETERS 1 INCH = 25.4 MILLIMETERS

INCHES	MILLI-METERS	INCHES	MILLI-METERS	INCHES	MILLI-METERS	INCHES	MILLI-METERS	INCHES	MILLI-METERS
50.00	1270.000	51.00	1295.400	52.00	1320.800	53.00	1346.200	54.00	1371.600
50.01	1270.254	51.01	1295.654	52.01	1321.054	53.01	1346.454	54.01	1371.854
50.02	1270.508	51.02	1295.908	52.02	1321.308	53.02	1346.708	54.02	1372.108
50.03	1270.762	51.03	1296.162	52.03	1321.562	53.03	1346.962	54.03	1372.362
50.04	1271.016	51.04	1296.416	52.04	1321.816	53.04	1347.216	54.04	1372.616
50.05	1271.270	51.05	1296.670	52.05	1322.070	53.05	1347.470	54.05	1372.870
50.06	1271.524	51.06	1296.924	52.06	1322.324	53.06	1347.724	54.06	1373.124
50.07	1271.778	51.07	1297.178	52.07	1322.578	53.07	1347.978	54.07	1373.378
50.08	1272.032	51.08	1297.432	52.08	1322.832	53.08	1348.232	54.08	1373.632
50.09	1272.286	51.09	1297.686	52.09	1323.086	53.09	1348.486	54.09	1373.886
50.10	1272.540	51.10	1297.940	52.10	1323.340	53.10	1348.740	54.10	1374.140
50.11	1272.794	51.11	1298.194	52.11	1323.594	53.11	1348.994	54.11	1374.394
50.12	1273.048	51.12	1298.448	52.12	1323.848	53.12	1349.248	54.12	1374.648
50.13	1273.302	51.13	1298.702	52.13	1324.102	53.13	1349.502	54.13	1374.902
50.14	1273.556	51.14	1298.956	52.14	1324.356	53.14	1349.756	54.14	1375.156
50.15	1273.810	51.15	1299.210	52.15	1324.610	53.15	1350.010	54.15	1375.410
50.16	1274.064	51.16	1299.464	52.16	1324.864	53.16	1350.264	54.16	1375.664
50.17	1274.318	51.17	1299.718	52.17	1325.118	53.17	1350.518	54.17	1375.918
50.18	1274.572	51.18	1299.972	52.18	1325.372	53.18	1350.772	54.18	1376.172
50.19	1274.826	51.19	1300.226	52.19	1325.626	53.19	1351.026	54.19	1376.426
50.20	1275.080	51.20	1300.480	52.20	1325.880	53.20	1351.280	54.20	1376.680
50.21	1275.334	51.21	1300.734	52.21	1326.134	53.21	1351.534	54.21	1376.934
50.22	1275.588	51.22	1300.988	52.22	1326.388	53.22	1351.788	54.22	1377.188
50.23	1275.842	51.23	1301.242	52.23	1326.642	53.23	1352.042	54.23	1377.442
50.24	1276.096	51.24	1301.496	52.24	1326.896	53.24	1352.296	54.24	1377.696
50.25	1276.350	51.25	1301.750	52.25	1327.150	53.25	1352.550	54.25	1377.950
50.26	1276.604	51.26	1302.004	52.26	1327.404	53.26	1352.804	54.26	1378.204
50.27	1276.858	51.27	1302.258	52.27	1327.658	53.27	1353.058	54.27	1378.458
50.28	1277.112	51.28	1302.512	52.28	1327.912	53.28	1353.312	54.28	1378.712
50.29	1277.366	51.29	1302.766	52.29	1328.166	53.29	1353.566	54.29	1378.966
50.30	1277.620	51.30	1303.020	52.30	1328.420	53.30	1353.820	54.30	1379.220
50.31	1277.874	51.31	1303.274	52.31	1328.674	53.31	1354.074	54.31	1379.474
50.32	1278.128	51.32	1303.528	52.32	1328.928	53.32	1354.328	54.32	1379.728
50.33	1278.382	51.33	1303.782	52.33	1329.182	53.33	1354.582	54.33	1379.982
50.34	1278.636	51.34	1304.036	52.34	1329.436	53.34	1354.836	54.34	1380.236
50.35	1278.890	51.35	1304.290	52.35	1329.690	53.35	1355.090	54.35	1380.490
50.36	1279.144	51.36	1304.544	52.36	1329.944	53.36	1355.344	54.36	1380.744
50.37	1279.398	51.37	1304.798	52.37	1330.198	53.37	1355.598	54.37	1380.998
50.38	1279.652	51.38	1305.052	52.38	1330.452	53.38	1355.852	54.38	1381.252
50.39	1279.906	51.39	1305.306	52.39	1330.706	53.39	1356.106	54.39	1381.506
50.40	1280.160	51.40	1305.560	52.40	1330.960	53.40	1356.360	54.40	1381.760
50.41	1280.414	51.41	1305.814	52.41	1331.214	53.41	1356.614	54.41	1382.014
50.42	1280.668	51.42	1306.068	52.42	1331.468	53.42	1356.868	54.42	1382.268
50.43	1280.922	51.43	1306.322	52.43	1331.722	53.43	1357.122	54.43	1382.522
50.44	1281.176	51.44	1306.576	52.44	1331.976	53.44	1357.376	54.44	1382.776
50.45	1281.430	51.45	1306.830	52.45	1332.230	53.45	1357.630	54.45	1383.030
50.46	1281.684	51.46	1307.084	52.46	1332.484	53.46	1357.884	54.46	1383.284
50.47	1281.938	51.47	1307.338	52.47	1332.738	53.47	1358.138	54.47	1383.538
50.48	1282.192	51.48	1307.592	52.48	1332.992	53.48	1358.392	54.48	1383.792
50.49	1282.446	51.49	1307.846	52.49	1333.246	53.49	1358.646	54.49	1384.046

Continued - INCHES TO MILLIMETERS 1 INCH = 25.4 MILLIMETERS

INCHES	MILLI-METERS	INCHES	MILLI-METERS	INCHES	MILLI-METERS	INCHES	MILLI-METERS	INCHES	MILLI-METERS
50.50	1282.700	51.50	1308.100	52.50	1333.500	53.50	1358.900	54.50	1384.300
50.51	1282.954	51.51	1308.354	52.51	1333.754	53.51	1359.154	54.51	1384.554
50.52	1283.208	51.52	1308.608	52.52	1334.008	53.52	1359.408	54.52	1384.808
50.53	1283.462	51.53	1308.862	52.53	1334.262	53.53	1359.662	54.53	1385.062
50.54	1283.716	51.54	1309.116	52.54	1334.516	53.54	1359.916	54.54	1385.316
50.55	1283.970	51.55	1309.370	52.55	1334.770	53.55	1360.170	54.55	1385.570
50.56	1284.224	51.56	1309.624	52.56	1335.024	53.56	1360.424	54.56	1385.824
50.57	1284.478	51.57	1309.878	52.57	1335.278	53.57	1360.678	54.57	1386.078
50.58	1284.732	51.58	1310.132	52.58	1335.532	53.58	1360.932	54.58	1386.332
50.59	1284.986	51.59	1310.386	52.59	1335.786	53.59	1361.186	54.59	1386.586
50.60	1285.240	51.60	1310.640	52.60	1336.040	53.60	1361.440	54.60	1386.840
50.61	1285.494	51.61	1310.894	52.61	1336.294	53.61	1361.694	54.61	1387.094
50.62	1285.748	51.62	1311.148	52.62	1336.548	53.62	1361.948	54.62	1387.348
50.63	1286.002	51.63	1311.402	52.63	1336.802	53.63	1362.202	54.63	1387.602
50.64	1286.256	51.64	1311.656	52.64	1337.056	53.64	1362.456	54.64	1387.856
50.65	1286.510	51.65	1311.910	52.65	1337.310	53.65	1362.710	54.65	1388.110
50.66	1286.764	51.66	1312.164	52.66	1337.564	53.66	1362.964	54.66	1388.364
50.67	1287.018	51.67	1312.418	52.67	1337.818	53.67	1363.218	54.67	1388.618
50.68	1287.272	51.68	1312.672	52.68	1338.072	53.68	1363.472	54.68	1388.872
50.69	1287.526	51.69	1312.926	52.69	1338.326	53.69	1363.726	54.69	1389.126
50.70	1287.780	51.70	1313.180	52.70	1338.580	53.70	1363.980	54.70	1389.380
50.71	1288.034	51.71	1313.434	52.71	1338.834	53.71	1364.234	54.71	1389.634
50.72	1288.288	51.72	1313.688	52.72	1339.088	53.72	1364.488	54.72	1389.888
50.73	1288.542	51.73	1313.942	52.73	1339.342	53.73	1364.742	54.73	1390.142
50.74	1288.796	51.74	1314.196	52.74	1339.596	53.74	1364.996	54.74	1390.396
50.75	1289.050	51.75	1314.450	52.75	1339.850	53.75	1365.250	54.75	1390.650
50.76	1289.304	51.76	1314.704	52.76	1340.104	53.76	1365.504	54.76	1390.904
50.77	1289.558	51.77	1314.958	52.77	1340.358	53.77	1365.758	54.77	1391.158
50.78	1289.812	51.78	1315.212	52.78	1340.612	53.78	1366.012	54.78	1391.412
50.79	1290.066	51.79	1315.466	52.79	1340.866	53.79	1366.266	54.79	1391.666
50.80	1290.320	51.80	1315.720	52.80	1341.120	53.80	1366.520	54.80	1391.920
50.81	1290.574	51.81	1315.974	52.81	1341.374	53.81	1366.774	54.81	1392.174
50.82	1290.828	51.82	1316.228	52.82	1341.628	53.82	1367.028	54.82	1392.428
50.83	1291.082	51.83	1316.482	52.83	1341.882	53.83	1367.282	54.83	1392.682
50.84	1291.336	51.84	1316.736	52.84	1342.136	53.84	1367.536	54.84	1392.936
50.85	1291.590	51.85	1316.990	52.85	1342.390	53.85	1367.790	54.85	1393.190
50.86	1291.844	51.86	1317.244	52.86	1342.644	53.86	1368.044	54.86	1393.444
50.87	1292.098	51.87	1317.498	52.87	1342.898	53.87	1368.298	54.87	1393.698
50.88	1292.352	51.88	1317.752	52.88	1343.152	53.88	1368.552	54.88	1393.952
50.89	1292.606	51.89	1318.006	52.89	1343.406	53.89	1368.806	54.89	1394.206
50.90	1292.860	51.90	1318.260	52.90	1343.660	53.90	1369.060	54.90	1394.460
50.91	1293.114	51.91	1318.514	52.91	1343.914	53.91	1369.314	54.91	1394.714
50.92	1293.368	51.92	1318.768	52.92	1344.168	53.92	1369.568	54.92	1394.968
50.93	1293.622	51.93	1319.022	52.93	1344.422	53.93	1369.822	54.93	1395.222
50.94	1293.876	51.94	1319.276	52.94	1344.676	53.94	1370.076	54.94	1395.476
50.95	1294.130	51.95	1319.530	52.95	1344.930	53.95	1370.330	54.95	1395.730
50.96	1294.384	51.96	1319.784	52.96	1345.184	53.96	1370.584	54.96	1395.984
50.97	1294.638	51.97	1320.038	52.97	1345.438	53.97	1370.838	54.97	1396.238
50.98	1294.892	51.98	1320.292	52.98	1345.692	53.98	1371.092	54.98	1396.492
50.99	1295.146	51.99	1320.546	52.99	1345.946	53.99	1371.346	54.99	1396.746

Continued - INCHES TO MILLIMETERS 1 INCH = 25.4 MILLIMETERS

INCHES	MILLI-METERS	INCHES	MILLI-METERS	INCHES	MILLI-METERS	INCHES	MILLI-METERS	INCHES	MILLI-METERS
55.00	1397.000	56.00	1422.400	57.00	1447.800	58.00	1473.200	59.00	1498.600
55.01	1397.254	56.01	1422.654	57.01	1448.054	58.01	1473.454	59.01	1498.854
55.02	1397.508	56.02	1422.908	57.02	1448.308	58.02	1473.708	59.02	1499.108
55.03	1397.762	56.03	1423.162	57.03	1448.562	58.03	1473.962	59.03	1499.362
55.04	1398.016	56.04	1423.416	57.04	1448.816	58.04	1474.216	59.04	1499.616
55.05	1398.270	56.05	1423.670	57.05	1449.070	58.05	1474.470	59.05	1499.870
55.06	1398.524	56.06	1423.924	57.06	1449.324	58.06	1474.724	59.06	1500.124
55.07	1398.778	56.07	1424.178	57.07	1449.578	58.07	1474.978	59.07	1500.378
55.08	1399.032	56.08	1424.432	57.08	1449.832	58.08	1475.232	59.08	1500.632
55.09	1399.286	56.09	1424.686	57.09	1450.086	58.09	1475.486	59.09	1500.886
55.10	1399.540	56.10	1424.940	57.10	1450.340	58.10	1475.740	59.10	1501.140
55.11	1399.794	56.11	1425.194	57.11	1450.594	58.11	1475.994	59.11	1501.394
55.12	1400.048	56.12	1425.448	57.12	1450.848	58.12	1476.248	59.12	1501.648
55.13	1400.302	56.13	1425.702	57.13	1451.102	58.13	1476.502	59.13	1501.902
55.14	1400.556	56.14	1425.956	57.14	1451.356	58.14	1476.756	59.14	1502.156
55.15	1400.810	56.15	1426.210	57.15	1451.610	58.15	1477.010	59.15	1502.410
55.16	1401.064	56.16	1426.464	57.16	1451.864	58.16	1477.264	59.16	1502.664
55.17	1401.318	56.17	1426.718	57.17	1452.118	58.17	1477.518	59.17	1502.918
55.18	1401.572	56.18	1426.972	57.18	1452.372	58.18	1477.772	59.18	1503.172
55.19	1401.826	56.19	1427.226	57.19	1452.626	58.19	1478.026	59.19	1503.426
55.20	1402.080	56.20	1427.480	57.20	1452.880	58.20	1478.280	59.20	1503.680
55.21	1402.334	56.21	1427.734	57.21	1453.134	58.21	1478.534	59.21	1503.934
55.22	1402.588	56.22	1427.988	57.22	1453.388	58.22	1478.788	59.22	1504.188
55.23	1402.842	56.23	1428.242	57.23	1453.642	58.23	1479.042	59.23	1504.442
55.24	1403.096	56.24	1428.496	57.24	1453.896	58.24	1479.296	59.24	1504.696
55.25	1403.350	56.25	1428.750	57.25	1454.150	58.25	1479.550	59.25	1504.950
55.26	1403.604	56.26	1429.004	57.26	1454.404	58.26	1479.804	59.26	1505.204
55.27	1403.858	56.27	1429.258	57.27	1454.658	58.27	1480.058	59.27	1505.458
55.28	1404.112	56.28	1429.512	57.28	1454.912	58.28	1480.312	59.28	1505.712
55.29	1404.366	56.29	1429.766	57.29	1455.166	58.29	1480.566	59.29	1505.966
55.30	1404.620	56.30	1430.020	57.30	1455.420	58.30	1480.820	59.30	1506.220
55.31	1404.874	56.31	1430.274	57.31	1455.674	58.31	1481.074	59.31	1506.474
55.32	1405.128	56.32	1430.528	57.32	1455.928	58.32	1481.328	59.32	1506.728
55.33	1405.382	56.33	1430.782	57.33	1456.182	58.33	1481.582	59.33	1506.982
55.34	1405.636	56.34	1431.036	57.34	1456.436	58.34	1481.836	59.34	1507.236
55.35	1405.890	56.35	1431.290	57.35	1456.690	58.35	1482.090	59.35	1507.490
55.36	1406.144	56.36	1431.544	57.36	1456.944	58.36	1482.344	59.36	1507.744
55.37	1406.398	56.37	1431.798	57.37	1457.198	58.37	1482.598	59.37	1507.998
55.38	1406.652	56.38	1432.052	57.38	1457.452	58.38	1482.852	59.38	1508.252
55.39	1406.906	56.39	1432.306	57.39	1457.706	58.39	1483.106	59.39	1508.506
55.40	1407.160	56.40	1432.560	57.40	1457.960	58.40	1483.360	59.40	1508.760
55.41	1407.414	56.41	1432.814	57.41	1458.214	58.41	1483.614	59.41	1509.014
55.42	1407.668	56.42	1433.068	57.42	1458.468	58.42	1483.868	59.42	1509.268
55.43	1407.922	56.43	1433.322	57.43	1458.722	58.43	1484.122	59.43	1509.522
55.44	1408.176	56.44	1433.576	57.44	1458.976	58.44	1484.376	59.44	1509.776
55.45	1408.430	56.45	1433.830	57.45	1459.230	58.45	1484.630	59.45	1510.030
55.46	1408.684	56.46	1434.084	57.46	1459.484	58.46	1484.884	59.46	1510.284
55.47	1408.938	56.47	1434.338	57.47	1459.738	58.47	1485.138	59.47	1510.538
55.48	1409.192	56.48	1434.592	57.48	1459.992	58.48	1485.392	59.48	1510.792
55.49	1409.446	56.49	1434.846	57.49	1460.246	58.49	1485.646	59.49	1511.046

Continued - INCHES TO MILLIMETERS 1 INCH = 25.4 MILLIMETERS

INCHES	MILLI-METERS	INCHES	MILLI-METERS	INCHES	MILLI-METERS	INCHES	MILLI-METERS	INCHES	MILLI-METERS
55.50	1409.700	56.50	1435.100	57.50	1460.500	58.50	1485.900	59.50	1511.300
55.51	1409.954	56.51	1435.354	57.51	1460.754	58.51	1486.154	59.51	1511.554
55.52	1410.208	56.52	1435.608	57.52	1461.008	58.52	1486.408	59.52	1511.808
55.53	1410.462	56.53	1435.862	57.53	1461.262	58.53	1486.662	59.53	1512.062
55.54	1410.716	56.54	1436.116	57.54	1461.516	58.54	1486.916	59.54	1512.316
55.55	1410.970	56.55	1436.370	57.55	1461.770	58.55	1487.170	59.55	1512.570
55.56	1411.224	56.56	1436.624	57.56	1462.024	58.56	1487.424	59.56	1512.824
55.57	1411.478	56.57	1436.878	57.57	1462.278	58.57	1487.678	59.57	1513.078
55.58	1411.732	56.58	1437.132	57.58	1462.532	58.58	1487.932	59.58	1513.332
55.59	1411.986	56.59	1437.386	57.59	1462.786	58.59	1488.186	59.59	1513.586
55.60	1412.240	56.60	1437.640	57.60	1463.040	58.60	1488.440	59.60	1513.840
55.61	1412.494	56.61	1437.894	57.61	1463.294	58.61	1488.694	59.61	1514.094
55.62	1412.748	56.62	1438.148	57.62	1463.548	58.62	1488.948	59.62	1514.348
55.63	1413.002	56.63	1438.402	57.63	1463.802	58.63	1489.202	59.63	1514.602
55.64	1413.256	56.64	1438.656	57.64	1464.056	58.64	1489.456	59.64	1514.856
55.65	1413.510	56.65	1438.910	57.65	1464.310	58.65	1489.710	59.65	1515.110
55.66	1413.764	56.66	1439.164	57.66	1464.564	58.66	1489.964	59.66	1515.364
55.67	1414.018	56.67	1439.418	57.67	1464.818	58.67	1490.218	59.67	1515.618
55.68	1414.272	56.68	1439.672	57.68	1465.072	58.68	1490.472	59.68	1515.872
55.69	1414.526	56.69	1439.926	57.69	1465.326	58.69	1490.726	59.69	1516.126
55.70	1414.780	56.70	1440.180	57.70	1465.580	58.70	1490.980	59.70	1516.380
55.71	1415.034	56.71	1440.434	57.71	1465.834	58.71	1491.234	59.71	1516.634
55.72	1415.288	56.72	1440.688	57.72	1466.088	58.72	1491.488	59.72	1516.888
55.73	1415.542	56.73	1440.942	57.73	1466.342	58.73	1491.742	59.73	1517.142
55.74	1415.796	56.74	1441.196	57.74	1466.596	58.74	1491.996	59.74	1517.396
55.75	1416.050	56.75	1441.450	57.75	1466.850	58.75	1492.250	59.75	1517.650
55.76	1416.304	56.76	1441.704	57.76	1467.104	58.76	1492.504	59.76	1517.904
55.77	1416.558	56.77	1441.958	57.77	1467.358	58.77	1492.758	59.77	1518.158
55.78	1416.812	56.78	1442.212	57.78	1467.612	58.78	1493.012	59.78	1518.412
55.79	1417.066	56.79	1442.466	57.79	1467.866	58.79	1493.266	59.79	1518.666
55.80	1417.320	56.80	1442.720	57.80	1468.120	58.80	1493.520	59.80	1518.920
55.81	1417.574	56.81	1442.974	57.81	1468.374	58.81	1493.774	59.81	1519.174
55.82	1417.828	56.82	1443.228	57.82	1468.628	58.82	1494.028	59.82	1519.428
55.83	1418.082	56.83	1443.482	57.83	1468.882	58.83	1494.282	59.83	1519.682
55.84	1418.336	56.84	1443.736	57.84	1469.136	58.84	1494.536	59.84	1519.936
55.85	1418.590	56.85	1443.990	57.85	1469.390	58.85	1494.790	59.85	1520.190
55.86	1418.844	56.86	1444.244	57.86	1469.644	58.86	1495.044	59.86	1520.444
55.87	1419.098	56.87	1444.498	57.87	1469.898	58.87	1495.298	59.87	1520.698
55.88	1419.352	56.88	1444.752	57.88	1470.152	58.88	1495.552	59.88	1520.952
55.89	1419.606	56.89	1445.006	57.89	1470.406	58.89	1495.806	59.89	1521.206
55.90	1419.860	56.90	1445.260	57.90	1470.660	58.90	1496.060	59.90	1521.460
55.91	1420.114	56.91	1445.514	57.91	1470.914	58.91	1496.314	59.91	1521.714
55.92	1420.368	56.92	1445.768	57.92	1471.168	58.92	1496.568	59.92	1521.968
55.93	1420.622	56.93	1446.022	57.93	1471.422	58.93	1496.822	59.93	1522.222
55.94	1420.876	56.94	1446.276	57.94	1471.676	58.94	1497.076	59.94	1522.476
55.95	1421.130	56.95	1446.530	57.95	1471.930	58.95	1497.330	59.95	1522.730
55.96	1421.384	56.96	1446.784	57.96	1472.184	58.96	1497.584	59.96	1522.984
55.97	1421.638	56.97	1447.038	57.97	1472.438	58.97	1497.838	59.97	1523.238
55.98	1421.892	56.98	1447.292	57.98	1472.692	58.98	1498.092	59.98	1523.492
55.99	1422.146	56.99	1447.546	57.99	1472.946	58.99	1498.346	59.99	1523.746

Continued - INCHES TO MILLIMETERS 1 INCH = 25.4 MILLIMETERS

INCHES	MILLI-METERS	INCHES	MILLI-METERS	INCHES	MILLI-METERS	INCHES	MILLI-METERS	INCHES	MILLI-METERS
60.00	1524.000	61.00	1549.400	62.00	1574.800	63.00	1600.200	64.00	1625.600
60.01	1524.254	61.01	1549.654	62.01	1575.054	63.01	1600.454	64.01	1625.854
60.02	1524.508	61.02	1549.908	62.02	1575.308	63.02	1600.708	64.02	1626.108
60.03	1524.762	61.03	1550.162	62.03	1575.562	63.03	1600.962	64.03	1626.362
60.04	1525.016	61.04	1550.416	62.04	1575.816	63.04	1601.216	64.04	1626.616
60.05	1525.270	61.05	1550.670	62.05	1576.070	63.05	1601.470	64.05	1626.870
60.06	1525.524	61.06	1550.924	62.06	1576.324	63.06	1601.724	64.06	1627.124
60.07	1525.778	61.07	1551.178	62.07	1576.578	63.07	1601.978	64.07	1627.378
60.08	1526.032	61.08	1551.432	62.08	1576.832	63.08	1602.232	64.08	1627.632
60.09	1526.286	61.09	1551.686	62.09	1577.086	63.09	1602.486	64.09	1627.886
60.10	1526.540	61.10	1551.940	62.10	1577.340	63.10	1602.740	64.10	1628.140
60.11	1526.794	61.11	1552.194	62.11	1577.594	63.11	1602.994	64.11	1628.394
60.12	1527.048	61.12	1552.448	62.12	1577.848	63.12	1603.248	64.12	1628.648
60.13	1527.302	61.13	1552.702	62.13	1578.102	63.13	1603.502	64.13	1628.902
60.14	1527.556	61.14	1552.956	62.14	1578.356	63.14	1603.756	64.14	1629.156
60.15	1527.810	61.15	1553.210	62.15	1578.610	63.15	1604.010	64.15	1629.410
60.16	1528.064	61.16	1553.464	62.16	1578.864	63.16	1604.264	64.16	1629.664
60.17	1528.318	61.17	1553.718	62.17	1579.118	63.17	1604.518	64.17	1629.918
60.18	1528.572	61.18	1553.972	62.18	1579.372	63.18	1604.772	64.18	1630.172
60.19	1528.826	61.19	1554.226	62.19	1579.626	63.19	1605.026	64.19	1630.426
60.20	1529.080	61.20	1554.480	62.20	1579.880	63.20	1605.280	64.20	1630.680
60.21	1529.334	61.21	1554.734	62.21	1580.134	63.21	1605.534	64.21	1630.934
60.22	1529.588	61.22	1554.988	62.22	1580.388	63.22	1605.788	64.22	1631.188
60.23	1529.842	61.23	1555.242	62.23	1580.642	63.23	1606.042	64.23	1631.442
60.24	1530.096	61.24	1555.496	62.24	1580.896	63.24	1606.296	64.24	1631.696
60.25	1530.350	61.25	1555.750	62.25	1581.150	63.25	1606.550	64.25	1631.950
60.26	1530.604	61.26	1556.004	62.26	1581.404	63.26	1606.804	64.26	1632.204
60.27	1530.858	61.27	1556.258	62.27	1581.658	63.27	1607.058	64.27	1632.458
60.28	1531.112	61.28	1556.512	62.28	1581.912	63.28	1607.312	64.28	1632.712
60.29	1531.366	61.29	1556.766	62.29	1582.166	63.29	1607.566	64.29	1632.966
60.30	1531.620	61.30	1557.020	62.30	1582.420	63.30	1607.820	64.30	1633.220
60.31	1531.874	61.31	1557.274	62.31	1582.674	63.31	1608.074	64.31	1633.474
60.32	1532.128	61.32	1557.528	62.32	1582.928	63.32	1608.328	64.32	1633.728
60.33	1532.382	61.33	1557.782	62.33	1583.182	63.33	1608.582	64.33	1633.982
60.34	1532.636	61.34	1558.036	62.34	1583.436	63.34	1608.836	64.34	1634.236
60.35	1532.890	61.35	1558.290	62.35	1583.690	63.35	1609.090	64.35	1634.490
60.36	1533.144	61.36	1558.544	62.36	1583.944	63.36	1609.344	64.36	1634.744
60.37	1533.398	61.37	1558.798	62.37	1584.198	63.37	1609.598	64.37	1634.998
60.38	1533.652	61.38	1559.052	62.38	1584.452	63.38	1609.852	64.38	1635.252
60.39	1533.906	61.39	1559.306	62.39	1584.706	63.39	1610.106	64.39	1635.506
60.40	1534.160	61.40	1559.560	62.40	1584.960	63.40	1610.360	64.40	1635.760
60.41	1534.414	61.41	1559.814	62.41	1585.214	63.41	1610.614	64.41	1636.014
60.42	1534.668	61.42	1560.068	62.42	1585.468	63.42	1610.868	64.42	1636.268
60.43	1534.922	61.43	1560.322	62.43	1585.722	63.43	1611.122	64.43	1636.522
60.44	1535.176	61.44	1560.576	62.44	1585.976	63.44	1611.376	64.44	1636.776
60.45	1535.430	61.45	1560.830	62.45	1586.230	63.45	1611.630	64.45	1637.030
60.46	1535.684	61.46	1561.084	62.46	1586.484	63.46	1611.884	64.46	1637.284
60.47	1535.938	61.47	1561.338	62.47	1586.738	63.47	1612.138	64.47	1637.538
60.48	1536.192	61.48	1561.592	62.48	1586.992	63.48	1612.392	64.48	1637.792
60.49	1536.446	61.49	1561.846	62.49	1587.246	63.49	1612.646	64.49	1638.046

Continued - INCHES TO MILLIMETERS 1 INCH = 25.4 MILLIMETERS

INCHES	MILLI-METERS	INCHES	MILLI-METERS	INCHES	MILLI-METERS	INCHES	MILLI-METERS	INCHES	MILLI-METERS
60.50	1536.700	61.50	1562.100	62.50	1587.500	63.50	1612.900	64.50	1638.300
60.51	1536.954	61.51	1562.354	62.51	1587.754	63.51	1613.154	64.51	1638.554
60.52	1537.208	61.52	1562.608	62.52	1588.008	63.52	1613.408	64.52	1638.808
60.53	1537.462	61.53	1562.862	62.53	1588.262	63.53	1613.662	64.53	1639.062
60.54	1537.716	61.54	1563.116	62.54	1588.516	63.54	1613.916	64.54	1639.316
60.55	1537.970	61.55	1563.370	62.55	1588.770	63.55	1614.170	64.55	1639.570
60.56	1538.224	61.56	1563.624	62.56	1589.024	63.56	1614.424	64.56	1639.824
60.57	1538.478	61.57	1563.878	62.57	1589.278	63.57	1614.678	64.57	1640.078
60.58	1538.732	61.58	1564.132	62.58	1589.532	63.58	1614.932	64.58	1640.332
60.59	1538.986	61.59	1564.386	62.59	1589.786	63.59	1615.186	64.59	1640.586
60.60	1539.240	61.60	1564.640	62.60	1590.040	63.60	1615.440	64.60	1640.840
60.61	1539.494	61.61	1564.894	62.61	1590.294	63.61	1615.694	64.61	1641.094
60.62	1539.748	61.62	1565.148	62.62	1590.548	63.62	1615.948	64.62	1641.348
60.63	1540.002	61.63	1565.402	62.63	1590.802	63.63	1616.202	64.63	1641.602
60.64	1540.256	61.64	1565.656	62.64	1591.056	63.64	1616.456	64.64	1641.856
60.65	1540.510	61.65	1565.910	62.65	1591.310	63.65	1616.710	64.65	1642.110
60.66	1540.764	61.66	1566.164	62.66	1591.564	63.66	1616.964	64.66	1642.364
60.67	1541.018	61.67	1566.418	62.67	1591.818	63.67	1617.218	64.67	1642.618
60.68	1541.272	61.68	1566.672	62.68	1592.072	63.68	1617.472	64.68	1642.872
60.69	1541.526	61.69	1566.926	62.69	1592.326	63.69	1617.726	64.69	1643.126
60.70	1541.780	61.70	1567.180	62.70	1592.580	63.70	1617.980	64.70	1643.380
60.71	1542.034	61.71	1567.434	62.71	1592.834	63.71	1618.234	64.71	1643.634
60.72	1542.288	61.72	1567.688	62.72	1593.088	63.72	1618.488	64.72	1643.888
60.73	1542.542	61.73	1567.942	62.73	1593.342	63.73	1618.742	64.73	1644.142
60.74	1542.796	61.74	1568.196	62.74	1593.596	63.74	1618.996	64.74	1644.396
60.75	1543.050	61.75	1568.450	62.75	1593.850	63.75	1619.250	64.75	1644.650
60.76	1543.304	61.76	1568.704	62.76	1594.104	63.76	1619.504	64.76	1644.904
60.77	1543.558	61.77	1568.958	62.77	1594.358	63.77	1619.758	64.77	1645.158
60.78	1543.812	61.78	1569.212	62.78	1594.612	63.78	1620.012	64.78	1645.412
60.79	1544.066	61.79	1569.466	62.79	1594.866	63.79	1620.266	64.79	1645.666
60.80	1544.320	61.80	1569.720	62.80	1595.120	63.80	1620.520	64.80	1645.920
60.81	1544.574	61.81	1569.974	62.81	1595.374	63.81	1620.774	64.81	1646.174
60.82	1544.828	61.82	1570.228	62.82	1595.628	63.82	1621.028	64.82	1646.428
60.83	1545.082	61.83	1570.482	62.83	1595.882	63.83	1621.282	64.83	1646.682
60.84	1545.336	61.84	1570.736	62.84	1596.136	63.84	1621.536	64.84	1646.936
60.85	1545.590	61.85	1570.990	62.85	1596.390	63.85	1621.790	64.85	1647.190
60.86	1545.844	61.86	1571.244	62.86	1596.644	63.86	1622.044	64.86	1647.444
60.87	1546.098	61.87	1571.498	62.87	1596.898	63.87	1622.298	64.87	1647.698
60.88	1546.352	61.88	1571.752	62.88	1597.152	63.88	1622.552	64.88	1647.952
60.89	1546.606	61.89	1572.006	62.89	1597.406	63.89	1622.806	64.89	1648.206
60.90	1546.860	61.90	1572.260	62.90	1597.660	63.90	1623.060	64.90	1648.460
60.91	1547.114	61.91	1572.514	62.91	1597.914	63.91	1623.314	64.91	1648.714
60.92	1547.368	61.92	1572.768	62.92	1598.168	63.92	1623.568	64.92	1648.968
60.93	1547.622	61.93	1573.022	62.93	1598.422	63.93	1623.822	64.93	1649.222
60.94	1547.876	61.94	1573.276	62.94	1598.676	63.94	1624.076	64.94	1649.476
60.95	1548.130	61.95	1573.530	62.95	1598.930	63.95	1624.330	64.95	1649.730
60.96	1548.384	61.96	1573.784	62.96	1599.184	63.96	1624.584	64.96	1649.984
60.97	1548.638	61.97	1574.038	62.97	1599.438	63.97	1624.838	64.97	1650.238
60.98	1548.892	61.98	1574.292	62.98	1599.692	63.98	1625.092	64.98	1650.492
60.99	1549.146	61.99	1574.546	62.99	1599.946	63.99	1625.346	64.99	1650.746

Continued - INCHES TO MILLIMETERS 1 INCH = 25.4 MILLIMETERS

INCHES	MILLI-METERS	INCHES	MILLI-METERS	INCHES	MILLI-METERS	INCHES	MILLI-METERS	INCHES	MILLI-METERS
65.00	1651.000	66.00	1676.400	67.00	1701.800	68.00	1727.200	69.00	1752.600
65.01	1651.254	66.01	1676.654	67.01	1702.054	68.01	1727.454	69.01	1752.854
65.02	1651.508	66.02	1676.908	67.02	1702.308	68.02	1727.708	69.02	1753.108
65.03	1651.762	66.03	1677.162	67.03	1702.562	68.03	1727.962	69.03	1753.362
65.04	1652.016	66.04	1677.416	67.04	1702.816	68.04	1728.216	69.04	1753.616
65.05	1652.270	66.05	1677.670	67.05	1703.070	68.05	1728.470	69.05	1753.870
65.06	1652.524	66.06	1677.924	67.06	1703.324	68.06	1728.724	69.06	1754.124
65.07	1652.778	66.07	1678.178	67.07	1703.578	68.07	1728.978	69.07	1754.378
65.08	1653.032	66.08	1678.432	67.08	1703.832	68.08	1729.232	69.08	1754.632
65.09	1653.286	66.09	1678.686	67.09	1704.086	68.09	1729.486	69.09	1754.886
65.10	1653.540	66.10	1678.940	67.10	1704.340	68.10	1729.740	69.10	1755.140
65.11	1653.794	66.11	1679.194	67.11	1704.594	68.11	1729.994	69.11	1755.394
65.12	1654.048	66.12	1679.448	67.12	1704.848	68.12	1730.248	69.12	1755.648
65.13	1654.302	66.13	1679.702	67.13	1705.102	68.13	1730.502	69.13	1755.902
65.14	1654.556	66.14	1679.956	67.14	1705.356	68.14	1730.756	69.14	1756.156
65.15	1654.810	66.15	1680.210	67.15	1705.610	68.15	1731.010	69.15	1756.410
65.16	1655.064	66.16	1680.464	67.16	1705.864	68.16	1731.264	69.16	1756.664
65.17	1655.318	66.17	1680.718	67.17	1706.118	68.17	1731.518	69.17	1756.918
65.18	1655.572	66.18	1680.972	67.18	1706.372	68.18	1731.772	69.18	1757.172
65.19	1655.826	66.19	1681.226	67.19	1706.626	68.19	1732.026	69.19	1757.426
65.20	1656.080	66.20	1681.480	67.20	1706.880	68.20	1732.280	69.20	1757.680
65.21	1656.334	66.21	1681.734	67.21	1707.134	68.21	1732.534	69.21	1757.934
65.22	1656.588	66.22	1681.988	67.22	1707.388	68.22	1732.788	69.22	1758.188
65.23	1656.842	66.23	1682.242	67.23	1707.642	68.23	1733.042	69.23	1758.442
65.24	1657.096	66.24	1682.496	67.24	1707.896	68.24	1733.296	69.24	1758.696
65.25	1657.350	66.25	1682.750	67.25	1708.150	68.25	1733.550	69.25	1758.950
65.26	1657.604	66.26	1683.004	67.26	1708.404	68.26	1733.804	69.26	1759.204
65.27	1657.858	66.27	1683.258	67.27	1708.658	68.27	1734.058	69.27	1759.458
65.28	1658.112	66.28	1683.512	67.28	1708.912	68.28	1734.312	69.28	1759.712
65.29	1658.366	66.29	1683.766	67.29	1709.166	68.29	1734.566	69.29	1759.966
65.30	1658.620	66.30	1684.020	67.30	1709.420	68.30	1734.820	69.30	1760.220
65.31	1658.874	66.31	1684.274	67.31	1709.674	68.31	1735.074	69.31	1760.474
65.32	1659.128	66.32	1684.528	67.32	1709.928	68.32	1735.328	69.32	1760.728
65.33	1659.382	66.33	1684.782	67.33	1710.182	68.33	1735.582	69.33	1760.982
65.34	1659.636	66.34	1685.036	67.34	1710.436	68.34	1735.836	69.34	1761.236
65.35	1659.890	66.35	1685.290	67.35	1710.690	68.35	1736.090	69.35	1761.490
65.36	1660.144	66.36	1685.544	67.36	1710.944	68.36	1736.344	69.36	1761.744
65.37	1660.398	66.37	1685.798	67.37	1711.198	68.37	1736.598	69.37	1761.998
65.38	1660.652	66.38	1686.052	67.38	1711.452	68.38	1736.852	69.38	1762.252
65.39	1660.906	66.39	1686.306	67.39	1711.706	68.39	1737.106	69.39	1762.506
65.40	1661.160	66.40	1686.560	67.40	1711.960	68.40	1737.360	69.40	1762.760
65.41	1661.414	66.41	1686.814	67.41	1712.214	68.41	1737.614	69.41	1763.014
65.42	1661.668	66.42	1687.068	67.42	1712.468	68.42	1737.868	69.42	1763.268
65.43	1661.922	66.43	1687.322	67.43	1712.722	68.43	1738.122	69.43	1763.522
65.44	1662.176	66.44	1687.576	67.44	1712.976	68.44	1738.376	69.44	1763.776
65.45	1662.430	66.45	1687.830	67.45	1713.230	68.45	1738.630	69.45	1764.030
65.46	1662.684	66.46	1688.084	67.46	1713.484	68.46	1738.884	69.46	1764.284
65.47	1662.938	66.47	1688.338	67.47	1713.738	68.47	1739.138	69.47	1764.538
65.48	1663.192	66.48	1688.592	67.48	1713.992	68.48	1739.392	69.48	1764.792
65.49	1663.446	66.49	1688.846	67.49	1714.246	68.49	1739.646	69.49	1765.046

Continued - INCHES TO MILLIMETERS 1 INCH = 25.4 MILLIMETERS

INCHES	MILLI-METERS	INCHES	MILLI-METERS	INCHES	MILLI-METERS	INCHES	MILLI-METERS	INCHES	MILLI-METERS
65.50	1663.700	66.50	1689.100	67.50	1714.500	68.50	1739.900	69.50	1765.300
65.51	1663.954	66.51	1689.354	67.51	1714.754	68.51	1740.154	69.51	1765.554
65.52	1664.208	66.52	1689.608	67.52	1715.008	68.52	1740.408	69.52	1765.808
65.53	1664.462	66.53	1689.862	67.53	1715.262	68.53	1740.662	69.53	1766.062
65.54	1664.716	66.54	1690.116	67.54	1715.516	68.54	1740.916	69.54	1766.316
65.55	1664.970	66.55	1690.370	67.55	1715.770	68.55	1741.170	69.55	1766.570
65.56	1665.224	66.56	1690.624	67.56	1716.024	68.56	1741.424	69.56	1766.824
65.57	1665.478	66.57	1690.878	67.57	1716.278	68.57	1741.678	69.57	1767.078
65.58	1665.732	66.58	1691.132	67.58	1716.532	68.58	1741.932	69.58	1767.332
65.59	1665.986	66.59	1691.386	67.59	1716.786	68.59	1742.186	69.59	1767.586
65.60	1666.240	66.60	1691.640	67.60	1717.040	68.60	1742.440	69.60	1767.840
65.61	1666.494	66.61	1691.894	67.61	1717.294	68.61	1742.694	69.61	1768.094
65.62	1666.748	66.62	1692.148	67.62	1717.548	68.62	1742.948	69.62	1768.348
65.63	1667.002	66.63	1692.402	67.63	1717.802	68.63	1743.202	69.63	1768.602
65.64	1667.256	66.64	1692.656	67.64	1718.056	68.64	1743.456	69.64	1768.856
65.65	1667.510	66.65	1692.910	67.65	1718.310	68.65	1743.710	69.65	1769.110
65.66	1667.764	66.66	1693.164	67.66	1718.564	68.66	1743.964	69.66	1769.364
65.67	1668.018	66.67	1693.418	67.67	1718.818	68.67	1744.218	69.67	1769.618
65.68	1668.272	66.68	1693.672	67.68	1719.072	68.68	1744.472	69.68	1769.872
65.69	1668.526	66.69	1693.926	67.69	1719.326	68.69	1744.726	69.69	1770.126
65.70	1668.780	66.70	1694.180	67.70	1719.580	68.70	1744.980	69.70	1770.380
65.71	1669.034	66.71	1694.434	67.71	1719.834	68.71	1745.234	69.71	1770.634
65.72	1669.288	66.72	1694.688	67.72	1720.088	68.72	1745.488	69.72	1770.888
65.73	1669.542	66.73	1694.942	67.73	1720.342	68.73	1745.742	69.73	1771.142
65.74	1669.796	66.74	1695.196	67.74	1720.596	68.74	1745.996	69.74	1771.396
65.75	1670.050	66.75	1695.450	67.75	1720.850	68.75	1746.250	69.75	1771.650
65.76	1670.304	66.76	1695.704	67.76	1721.104	68.76	1746.504	69.76	1771.904
65.77	1670.558	66.77	1695.958	67.77	1721.358	68.77	1746.758	69.77	1772.158
65.78	1670.812	66.78	1696.212	67.78	1721.612	68.78	1747.012	69.78	1772.412
65.79	1671.066	66.79	1696.466	67.79	1721.866	68.79	1747.266	69.79	1772.666
65.80	1671.320	66.80	1696.720	67.80	1722.120	68.80	1747.520	69.80	1772.920
65.81	1671.574	66.81	1696.974	67.81	1722.374	68.81	1747.774	69.81	1773.174
65.82	1671.828	66.82	1697.228	67.82	1722.628	68.82	1748.028	69.82	1773.428
65.83	1672.082	66.83	1697.482	67.83	1722.882	68.83	1748.282	69.83	1773.682
65.84	1672.336	66.84	1697.736	67.84	1723.136	68.84	1748.536	69.84	1773.936
65.85	1672.590	66.85	1697.990	67.85	1723.390	68.85	1748.790	69.85	1774.190
65.86	1672.844	66.86	1698.244	67.86	1723.644	68.86	1749.044	69.86	1774.444
65.87	1673.098	66.87	1698.498	67.87	1723.898	68.87	1749.298	69.87	1774.698
65.88	1673.352	66.88	1698.752	67.88	1724.152	68.88	1749.552	69.88	1774.952
65.89	1673.606	66.89	1699.006	67.89	1724.406	68.89	1749.806	69.89	1775.206
65.90	1673.860	66.90	1699.260	67.90	1724.660	68.90	1750.060	69.90	1775.460
65.91	1674.114	66.91	1699.514	67.91	1724.914	68.91	1750.314	69.91	1775.714
65.92	1674.368	66.92	1699.768	67.92	1725.168	68.92	1750.568	69.92	1775.968
65.93	1674.622	66.93	1700.022	67.93	1725.422	68.93	1750.822	69.93	1776.222
65.94	1674.876	66.94	1700.276	67.94	1725.676	68.94	1751.076	69.94	1776.476
65.95	1675.130	66.95	1700.530	67.95	1725.930	68.95	1751.330	69.95	1776.730
65.96	1675.384	66.96	1700.784	67.96	1726.184	68.96	1751.584	69.96	1776.984
65.97	1675.638	66.97	1701.038	67.97	1726.438	68.97	1751.838	69.97	1777.238
65.98	1675.892	66.98	1701.292	67.98	1726.692	68.98	1752.092	69.98	1777.492
65.99	1676.146	66.99	1701.546	67.99	1726.946	68.99	1752.346	69.99	1777.746

Continued - INCHES TO MILLIMETERS 1 INCH = 25.4 MILLIMETERS

INCHES	MILLI- METERS	INCHES	MILLI- METERS	INCHES	MILLI- METERS	INCHES	MILLI- METERS	INCHES	MILLI- METERS
70.00	1778.000	71.00	1803.400	72.00	1828.800	73.00	1854.200	74.00	1879.600
70.01	1778.254	71.01	1803.654	72.01	1829.054	73.01	1854.454	74.01	1879.854
70.02	1778.508	71.02	1803.908	72.02	1829.308	73.02	1854.708	74.02	1880.108
70.03	1778.762	71.03	1804.162	72.03	1829.562	73.03	1854.962	74.03	1880.362
70.04	1779.016	71.04	1804.416	72.04	1829.816	73.04	1855.216	74.04	1880.616
70.05	1779.270	71.05	1804.670	72.05	1830.070	73.05	1855.470	74.05	1880.870
70.06	1779.524	71.06	1804.924	72.06	1830.324	73.06	1855.724	74.06	1881.124
70.07	1779.778	71.07	1805.178	72.07	1830.578	73.07	1855.978	74.07	1881.378
70.08	1780.032	71.08	1805.432	72.08	1830.832	73.08	1856.232	74.08	1881.632
70.09	1780.286	71.09	1805.686	72.09	1831.086	73.09	1856.486	74.09	1881.886
70.10	1780.540	71.10	1805.940	72.10	1831.340	73.10	1856.740	74.10	1882.140
70.11	1780.794	71.11	1806.194	72.11	1831.594	73.11	1856.994	74.11	1882.394
70.12	1781.048	71.12	1806.448	72.12	1831.848	73.12	1857.248	74.12	1882.648
70.13	1781.302	71.13	1806.702	72.13	1832.102	73.13	1857.502	74.13	1882.902
70.14	1781.556	71.14	1806.956	72.14	1832.356	73.14	1857.756	74.14	1883.156
70.15	1781.810	71.15	1807.210	72.15	1832.610	73.15	1858.010	74.15	1883.410
70.16	1782.064	71.16	1807.464	72.16	1832.864	73.16	1858.264	74.16	1883.664
70.17	1782.318	71.17	1807.718	72.17	1833.118	73.17	1858.518	74.17	1883.918
70.18	1782.572	71.18	1807.972	72.18	1833.372	73.18	1858.772	74.18	1884.172
70.19	1782.826	71.19	1808.226	72.19	1833.626	73.19	1859.026	74.19	1884.426
70.20	1783.080	71.20	1808.480	72.20	1833.880	73.20	1859.280	74.20	1884.680
70.21	1783.334	71.21	1808.734	72.21	1834.134	73.21	1859.534	74.21	1884.934
70.22	1783.588	71.22	1808.988	72.22	1834.388	73.22	1859.788	74.22	1885.188
70.23	1783.842	71.23	1809.242	72.23	1834.642	73.23	1860.042	74.23	1885.442
70.24	1784.096	71.24	1809.496	72.24	1834.896	73.24	1860.296	74.24	1885.696
70.25	1784.350	71.25	1809.750	72.25	1835.150	73.25	1860.550	74.25	1885.950
70.26	1784.604	71.26	1810.004	72.26	1835.404	73.26	1860.804	74.26	1886.204
70.27	1784.858	71.27	1810.258	72.27	1835.658	73.27	1861.058	74.27	1886.458
70.28	1785.112	71.28	1810.512	72.28	1835.912	73.28	1861.312	74.28	1886.712
70.29	1785.366	71.29	1810.766	72.29	1836.166	73.29	1861.566	74.29	1886.966
70.30	1785.620	71.30	1811.020	72.30	1836.420	73.30	1861.820	74.30	1887.220
70.31	1785.874	71.31	1811.274	72.31	1836.674	73.31	1862.074	74.31	1887.474
70.32	1786.128	71.32	1811.528	72.32	1836.928	73.32	1862.328	74.32	1887.728
70.33	1786.382	71.33	1811.782	72.33	1837.182	73.33	1862.582	74.33	1887.982
70.34	1786.636	71.34	1812.036	72.34	1837.436	73.34	1862.836	74.34	1888.236
70.35	1786.890	71.35	1812.290	72.35	1837.690	73.35	1863.090	74.35	1888.490
70.36	1787.144	71.36	1812.544	72.36	1837.944	73.36	1863.344	74.36	1888.744
70.37	1787.398	71.37	1812.798	72.37	1838.198	73.37	1863.598	74.37	1888.998
70.38	1787.652	71.38	1813.052	72.38	1838.452	73.38	1863.852	74.38	1889.252
70.39	1787.906	71.39	1813.306	72.39	1838.706	73.39	1864.106	74.39	1889.506
70.40	1788.160	71.40	1813.560	72.40	1838.960	73.40	1864.360	74.40	1889.760
70.41	1788.414	71.41	1813.814	72.41	1839.214	73.41	1864.614	74.41	1890.014
70.42	1788.668	71.42	1814.068	72.42	1839.468	73.42	1864.868	74.42	1890.268
70.43	1788.922	71.43	1814.322	72.43	1839.722	73.43	1865.122	74.43	1890.522
70.44	1789.176	71.44	1814.576	72.44	1839.976	73.44	1865.376	74.44	1890.776
70.45	1789.430	71.45	1814.830	72.45	1840.230	73.45	1865.630	74.45	1891.030
70.46	1789.684	71.46	1815.084	72.46	1840.484	73.46	1865.884	74.46	1891.284
70.47	1789.938	71.47	1815.338	72.47	1840.738	73.47	1866.138	74.47	1891.538
70.48	1790.192	71.48	1815.592	72.48	1840.992	73.48	1866.392	74.48	1891.792
70.49	1790.446	71.49	1815.846	72.49	1841.246	73.49	1866.646	74.49	1892.046

Continued - INCHES TO MILLIMETERS 1 INCH = 25.4 MILLIMETERS

INCHES	MILLI-METERS	INCHES	MILLI-METERS	INCHES	MILLI-METERS	INCHES	MILLI-METERS	INCHES	MILLI-METERS
70.50	1790.700	71.50	1816.100	72.50	1841.500	73.50	1866.900	74.50	1892.300
70.51	1790.954	71.51	1816.354	72.51	1841.754	73.51	1867.154	74.51	1892.554
70.52	1791.208	71.52	1816.608	72.52	1842.008	73.52	1867.408	74.52	1892.808
70.53	1791.462	71.53	1816.862	72.53	1842.262	73.53	1867.662	74.53	1893.062
70.54	1791.716	71.54	1817.116	72.54	1842.516	73.54	1867.916	74.54	1893.316
70.55	1791.970	71.55	1817.370	72.55	1842.770	73.55	1868.170	74.55	1893.570
70.56	1792.224	71.56	1817.624	72.56	1843.024	73.56	1868.424	74.56	1893.824
70.57	1792.478	71.57	1817.878	72.57	1843.278	73.57	1868.678	74.57	1894.078
70.58	1792.732	71.58	1818.132	72.58	1843.532	73.58	1868.932	74.58	1894.332
70.59	1792.986	71.59	1818.386	72.59	1843.786	73.59	1869.186	74.59	1894.586
70.60	1793.240	71.60	1818.640	72.60	1844.040	73.60	1869.440	74.60	1894.840
70.61	1793.494	71.61	1818.894	72.61	1844.294	73.61	1869.694	74.61	1895.094
70.62	1793.748	71.62	1819.148	72.62	1844.548	73.62	1869.948	74.62	1895.348
70.63	1794.002	71.63	1819.402	72.63	1844.802	73.63	1870.202	74.63	1895.602
70.64	1794.256	71.64	1819.656	72.64	1845.056	73.64	1870.456	74.64	1895.856
70.65	1794.510	71.65	1819.910	72.65	1845.310	73.65	1870.710	74.65	1896.110
70.66	1794.764	71.66	1820.164	72.66	1845.564	73.66	1870.964	74.66	1896.364
70.67	1795.018	71.67	1820.418	72.67	1845.818	73.67	1871.218	74.67	1896.618
70.68	1795.272	71.68	1820.672	72.68	1846.072	73.68	1871.472	74.68	1896.872
70.69	1795.526	71.69	1820.926	72.69	1846.326	73.69	1871.726	74.69	1897.126
70.70	1795.780	71.70	1821.180	72.70	1846.580	73.70	1871.980	74.70	1897.380
70.71	1796.034	71.71	1821.434	72.71	1846.834	73.71	1872.234	74.71	1897.634
70.72	1796.288	71.72	1821.688	72.72	1847.088	73.72	1872.488	74.72	1897.888
70.73	1796.542	71.73	1821.942	72.73	1847.342	73.73	1872.742	74.73	1898.142
70.74	1796.796	71.74	1822.196	72.74	1847.596	73.74	1872.996	74.74	1898.396
70.75	1797.050	71.75	1822.450	72.75	1847.850	73.75	1873.250	74.75	1898.650
70.76	1797.304	71.76	1822.704	72.76	1848.104	73.76	1873.504	74.76	1898.904
70.77	1797.558	71.77	1822.958	72.77	1848.358	73.77	1873.758	74.77	1899.158
70.78	1797.812	71.78	1823.212	72.78	1848.612	73.78	1874.012	74.78	1899.412
70.79	1798.066	71.79	1823.466	72.79	1848.866	73.79	1874.266	74.79	1899.666
70.80	1798.320	71.80	1823.720	72.80	1849.120	73.80	1874.520	74.80	1899.920
70.81	1798.574	71.81	1823.974	72.81	1849.374	73.81	1874.774	74.81	1900.174
70.82	1798.828	71.82	1824.228	72.82	1849.628	73.82	1875.028	74.82	1900.428
70.83	1799.082	71.83	1824.482	72.83	1849.882	73.83	1875.282	74.83	1900.682
70.84	1799.336	71.84	1824.736	72.84	1850.136	73.84	1875.536	74.84	1900.936
70.85	1799.590	71.85	1824.990	72.85	1850.390	73.85	1875.790	74.85	1901.190
70.86	1799.844	71.86	1825.244	72.86	1850.644	73.86	1876.044	74.86	1901.444
70.87	1800.098	71.87	1825.498	72.87	1850.898	73.87	1876.298	74.87	1901.698
70.88	1800.352	71.88	1825.752	72.88	1851.152	73.88	1876.552	74.88	1901.952
70.89	1800.606	71.89	1826.006	72.89	1851.406	73.89	1876.806	74.89	1902.206
70.90	1800.860	71.90	1826.260	72.90	1851.660	73.90	1877.060	74.90	1902.460
70.91	1801.114	71.91	1826.514	72.91	1851.914	73.91	1877.314	74.91	1902.714
70.92	1801.368	71.92	1826.768	72.92	1852.168	73.92	1877.568	74.92	1902.968
70.93	1801.622	71.93	1827.022	72.93	1852.422	73.93	1877.822	74.93	1903.222
70.94	1801.876	71.94	1827.276	72.94	1852.676	73.94	1878.076	74.94	1903.476
70.95	1802.130	71.95	1827.530	72.95	1852.930	73.95	1878.330	74.95	1903.730
70.96	1802.384	71.96	1827.784	72.96	1853.184	73.96	1878.584	74.96	1903.984
70.97	1802.638	71.97	1828.038	72.97	1853.438	73.97	1878.838	74.97	1904.238
70.98	1802.892	71.98	1828.292	72.98	1853.692	73.98	1879.092	74.98	1904.492
70.99	1803.146	71.99	1828.546	72.99	1853.946	73.99	1879.346	74.99	1904.746

Continued - INCHES TO MILLIMETERS 1 INCH = 25.4 MILLIMETERS

INCHES	MILLI-METERS	INCHES	MILLI-METERS	INCHES	MILLI-METERS	INCHES	MILLI-METERS	INCHES	MILLI-METERS
75.00	1905.000	76.00	1930.400	77.00	1955.800	78.00	1981.200	79.00	2006.600
75.01	1905.254	76.01	1930.654	77.01	1956.054	78.01	1981.454	79.01	2006.854
75.02	1905.508	76.02	1930.908	77.02	1956.308	78.02	1981.708	79.02	2007.108
75.03	1905.762	76.03	1931.162	77.03	1956.562	78.03	1981.962	79.03	2007.362
75.04	1906.016	76.04	1931.416	77.04	1956.816	78.04	1982.216	79.04	2007.616
75.05	1906.270	76.05	1931.670	77.05	1957.070	78.05	1982.470	79.05	2007.870
75.06	1906.524	76.06	1931.924	77.06	1957.324	78.06	1982.724	79.06	2008.124
75.07	1906.778	76.07	1932.178	77.07	1957.578	78.07	1982.978	79.07	2008.378
75.08	1907.032	76.08	1932.432	77.08	1957.832	78.08	1983.232	79.08	2008.632
75.09	1907.286	76.09	1932.686	77.09	1958.086	78.09	1983.486	79.09	2008.886
75.10	1907.540	76.10	1932.940	77.10	1958.340	78.10	1983.740	79.10	2009.140
75.11	1907.794	76.11	1933.194	77.11	1958.594	78.11	1983.994	79.11	2009.394
75.12	1908.048	76.12	1933.448	77.12	1958.848	78.12	1984.248	79.12	2009.648
75.13	1908.302	76.13	1933.702	77.13	1959.102	78.13	1984.502	79.13	2009.902
75.14	1908.556	76.14	1933.956	77.14	1959.356	78.14	1984.756	79.14	2010.156
75.15	1908.810	76.15	1934.210	77.15	1959.610	78.15	1985.010	79.15	2010.410
75.16	1909.064	76.16	1934.464	77.16	1959.864	78.16	1985.264	79.16	2010.664
75.17	1909.318	76.17	1934.718	77.17	1960.118	78.17	1985.518	79.17	2010.918
75.18	1909.572	76.18	1934.972	77.18	1960.372	78.18	1985.772	79.18	2011.172
75.19	1909.826	76.19	1935.226	77.19	1960.626	78.19	1986.026	79.19	2011.426
75.20	1910.080	76.20	1935.480	77.20	1960.880	78.20	1986.280	79.20	2011.680
75.21	1910.334	76.21	1935.734	77.21	1961.134	78.21	1986.534	79.21	2011.934
75.22	1910.588	76.22	1935.988	77.22	1961.388	78.22	1986.788	79.22	2012.188
75.23	1910.842	76.23	1936.242	77.23	1961.642	78.23	1987.042	79.23	2012.442
75.24	1911.096	76.24	1936.496	77.24	1961.896	78.24	1987.296	79.24	2012.696
75.25	1911.350	76.25	1936.750	77.25	1962.150	78.25	1987.550	79.25	2012.950
75.26	1911.604	76.26	1937.004	77.26	1962.404	78.26	1987.804	79.26	2013.204
75.27	1911.858	76.27	1937.258	77.27	1962.658	78.27	1988.058	79.27	2013.458
75.28	1912.112	76.28	1937.512	77.28	1962.912	78.28	1988.312	79.28	2013.712
75.29	1912.366	76.29	1937.766	77.29	1963.166	78.29	1988.566	79.29	2013.966
75.30	1912.620	76.30	1938.020	77.30	1963.420	78.30	1988.820	79.30	2014.220
75.31	1912.874	76.31	1938.274	77.31	1963.674	78.31	1989.074	79.31	2014.474
75.32	1913.128	76.32	1938.528	77.32	1963.928	78.32	1989.328	79.32	2014.728
75.33	1913.382	76.33	1938.782	77.33	1964.182	78.33	1989.582	79.33	2014.982
75.34	1913.636	76.34	1939.036	77.34	1964.436	78.34	1989.836	79.34	2015.236
75.35	1913.890	76.35	1939.290	77.35	1964.690	78.35	1990.090	79.35	2015.490
75.36	1914.144	76.36	1939.544	77.36	1964.944	78.36	1990.344	79.36	2015.744
75.37	1914.398	76.37	1939.798	77.37	1965.198	78.37	1990.598	79.37	2015.998
75.38	1914.652	76.38	1940.052	77.38	1965.452	78.38	1990.852	79.38	2016.252
75.39	1914.906	76.39	1940.306	77.39	1965.706	78.39	1991.106	79.39	2016.506
75.40	1915.160	76.40	1940.560	77.40	1965.960	78.40	1991.360	79.40	2016.760
75.41	1915.414	76.41	1940.814	77.41	1966.214	78.41	1991.614	79.41	2017.014
75.42	1915.668	76.42	1941.068	77.42	1966.468	78.42	1991.868	79.42	2017.268
75.43	1915.922	76.43	1941.322	77.43	1966.722	78.43	1992.122	79.43	2017.522
75.44	1916.176	76.44	1941.576	77.44	1966.976	78.44	1992.376	79.44	2017.776
75.45	1916.430	76.45	1941.830	77.45	1967.230	78.45	1992.630	79.45	2018.030
75.46	1916.684	76.46	1942.084	77.46	1967.484	78.46	1992.884	79.46	2018.284
75.47	1916.938	76.47	1942.338	77.47	1967.738	78.47	1993.138	79.47	2018.538
75.48	1917.192	76.48	1942.592	77.48	1967.992	78.48	1993.392	79.48	2018.792
75.49	1917.446	76.49	1942.846	77.49	1968.246	78.49	1993.646	79.49	2019.046

Continued - INCHES TO MILLIMETERS 1 INCH = 25.4 MILLIMETERS

INCHES	MILLI-METERS	INCHES	MILLI-METERS	INCHES	MILLI-METERS	INCHES	MILLI-METERS	INCHES	MILLI-METERS
75.50	1917.700	76.50	1943.100	77.50	1968.500	78.50	1993.900	79.50	2019.300
75.51	1917.954	76.51	1943.354	77.51	1968.754	78.51	1994.154	79.51	2019.554
75.52	1918.208	76.52	1943.608	77.52	1969.008	78.52	1994.408	79.52	2019.808
75.53	1918.462	76.53	1943.862	77.53	1969.262	78.53	1994.662	79.53	2020.062
75.54	1918.716	76.54	1944.116	77.54	1969.516	78.54	1994.916	79.54	2020.316
75.55	1918.970	76.55	1944.370	77.55	1969.770	78.55	1995.170	79.55	2020.570
75.56	1919.224	76.56	1944.624	77.56	1970.024	78.56	1995.424	79.56	2020.824
75.57	1919.478	76.57	1944.878	77.57	1970.278	78.57	1995.678	79.57	2021.078
75.58	1919.732	76.58	1945.132	77.58	1970.532	78.58	1995.932	79.58	2021.332
75.59	1919.986	76.59	1945.386	77.59	1970.786	78.59	1996.186	79.59	2021.586
75.60	1920.240	76.60	1945.640	77.60	1971.040	78.60	1996.440	79.60	2021.840
75.61	1920.494	76.61	1945.894	77.61	1971.294	78.61	1996.694	79.61	2022.094
75.62	1920.748	76.62	1946.148	77.62	1971.548	78.62	1996.948	79.62	2022.348
75.63	1921.002	76.63	1946.402	77.63	1971.802	78.63	1997.202	79.63	2022.602
75.64	1921.256	76.64	1946.656	77.64	1972.056	78.64	1997.456	79.64	2022.856
75.65	1921.510	76.65	1946.910	77.65	1972.310	78.65	1997.710	79.65	2023.110
75.66	1921.764	76.66	1947.164	77.66	1972.564	78.66	1997.964	79.66	2023.364
75.67	1922.018	76.67	1947.418	77.67	1972.818	78.67	1998.218	79.67	2023.618
75.68	1922.272	76.68	1947.672	77.68	1973.072	78.68	1998.472	79.68	2023.872
75.69	1922.526	76.69	1947.926	77.69	1973.326	78.69	1998.726	79.69	2024.126
75.70	1922.780	76.70	1948.180	77.70	1973.580	78.70	1998.980	79.70	2024.380
75.71	1923.034	76.71	1948.434	77.71	1973.834	78.71	1999.234	79.71	2024.634
75.72	1923.288	76.72	1948.688	77.72	1974.088	78.72	1999.488	79.72	2024.888
75.73	1923.542	76.73	1948.942	77.73	1974.342	78.73	1999.742	79.73	2025.142
75.74	1923.796	76.74	1949.196	77.74	1974.596	78.74	1999.996	79.74	2025.396
75.75	1924.050	76.75	1949.450	77.75	1974.850	78.75	2000.250	79.75	2025.650
75.76	1924.304	76.76	1949.704	77.76	1975.104	78.76	2000.504	79.76	2025.904
75.77	1924.558	76.77	1949.958	77.77	1975.358	78.77	2000.758	79.77	2026.158
75.78	1924.812	76.78	1950.212	77.78	1975.612	78.78	2001.012	79.78	2026.412
75.79	1925.066	76.79	1950.466	77.79	1975.866	78.79	2001.266	79.79	2026.666
75.80	1925.320	76.80	1950.720	77.80	1976.120	78.80	2001.520	79.80	2026.920
75.81	1925.574	76.81	1950.974	77.81	1976.374	78.81	2001.774	79.81	2027.174
75.82	1925.828	76.82	1951.228	77.82	1976.628	78.82	2002.028	79.82	2027.428
75.83	1926.082	76.83	1951.482	77.83	1976.882	78.83	2002.282	79.83	2027.682
75.84	1926.336	76.84	1951.736	77.84	1977.136	78.84	2002.536	79.84	2027.936
75.85	1926.590	76.85	1951.990	77.85	1977.390	78.85	2002.790	79.85	2028.190
75.86	1926.844	76.86	1952.244	77.86	1977.644	78.86	2003.044	79.86	2028.444
75.87	1927.098	76.87	1952.498	77.87	1977.898	78.87	2003.298	79.87	2028.698
75.88	1927.352	76.88	1952.752	77.88	1978.152	78.88	2003.552	79.88	2028.952
75.89	1927.606	76.89	1953.006	77.89	1978.406	78.89	2003.806	79.89	2029.206
75.90	1927.860	76.90	1953.260	77.90	1978.660	78.90	2004.060	79.90	2029.460
75.91	1928.114	76.91	1953.514	77.91	1978.914	78.91	2004.314	79.91	2029.714
75.92	1928.368	76.92	1953.768	77.92	1979.168	78.92	2004.568	79.92	2029.968
75.93	1928.622	76.93	1954.022	77.93	1979.422	78.93	2004.822	79.93	2030.222
75.94	1928.876	76.94	1954.276	77.94	1979.676	78.94	2005.076	79.94	2030.476
75.95	1929.130	76.95	1954.530	77.95	1979.930	78.95	2005.330	79.95	2030.730
75.96	1929.384	76.96	1954.784	77.96	1980.184	78.96	2005.584	79.96	2030.984
75.97	1929.638	76.97	1955.038	77.97	1980.438	78.97	2005.838	79.97	2031.238
75.98	1929.892	76.98	1955.292	77.98	1980.692	78.98	2006.092	79.98	2031.492
75.99	1930.146	76.99	1955.546	77.99	1980.946	78.99	2006.346	79.99	2031.746

Continued - INCHES TO MILLIMETERS 1 INCH = 25.4 MILLIMETERS

INCHES	MILLI-METERS	INCHES	MILLI-METERS	INCHES	MILLI-METERS	INCHES	MILLI-METERS	INCHES	MILLI-METERS
80.00	2032.000	81.00	2057.400	82.00	2082.800	83.00	2108.200	84.00	2133.600
80.01	2032.254	81.01	2057.654	82.01	2083.054	83.01	2108.454	84.01	2133.854
80.02	2032.508	81.02	2057.908	82.02	2083.308	83.02	2108.708	84.02	2134.108
80.03	2032.762	81.03	2058.162	82.03	2083.562	83.03	2108.962	84.03	2134.362
80.04	2033.016	81.04	2058.416	82.04	2083.816	83.04	2109.216	84.04	2134.616
80.05	2033.270	81.05	2058.670	82.05	2084.070	83.05	2109.470	84.05	2134.870
80.06	2033.524	81.06	2058.924	82.06	2084.324	83.06	2109.724	84.06	2135.124
80.07	2033.778	81.07	2059.178	82.07	2084.578	83.07	2109.978	84.07	2135.378
80.08	2034.032	81.08	2059.432	82.08	2084.832	83.08	2110.232	84.08	2135.632
80.09	2034.286	81.09	2059.686	82.09	2085.086	83.09	2110.486	84.09	2135.886
80.10	2034.540	81.10	2059.940	82.10	2085.340	83.10	2110.740	84.10	2136.140
80.11	2034.794	81.11	2060.194	82.11	2085.594	83.11	2110.994	84.11	2136.394
80.12	2035.048	81.12	2060.448	82.12	2085.848	83.12	2111.248	84.12	2136.648
80.13	2035.302	81.13	2060.702	82.13	2086.102	83.13	2111.502	84.13	2136.902
80.14	2035.556	81.14	2060.956	82.14	2086.356	83.14	2111.756	84.14	2137.156
80.15	2035.810	81.15	2061.210	82.15	2086.610	83.15	2112.010	84.15	2137.410
80.16	2036.064	81.16	2061.464	82.16	2086.864	83.16	2112.264	84.16	2137.664
80.17	2036.318	81.17	2061.718	82.17	2087.118	83.17	2112.518	84.17	2137.918
80.18	2036.572	81.18	2061.972	82.18	2087.372	83.18	2112.772	84.18	2138.172
80.19	2036.826	81.19	2062.226	82.19	2087.626	83.19	2113.026	84.19	2138.426
80.20	2037.080	81.20	2062.480	82.20	2087.880	83.20	2113.280	84.20	2138.680
80.21	2037.334	81.21	2062.734	82.21	2088.134	83.21	2113.534	84.21	2138.934
80.22	2037.588	81.22	2062.988	82.22	2088.388	83.22	2113.788	84.22	2139.188
80.23	2037.842	81.23	2063.242	82.23	2088.642	83.23	2114.042	84.23	2139.442
80.24	2038.096	81.24	2063.496	82.24	2088.896	83.24	2114.296	84.24	2139.696
80.25	2038.350	81.25	2063.750	82.25	2089.150	83.25	2114.550	84.25	2139.950
80.26	2038.604	81.26	2064.004	82.26	2089.404	83.26	2114.804	84.26	2140.204
80.27	2038.858	81.27	2064.258	82.27	2089.658	83.27	2115.058	84.27	2140.458
80.28	2039.112	81.28	2064.512	82.28	2089.912	83.28	2115.312	84.28	2140.712
80.29	2039.366	81.29	2064.766	82.29	2090.166	83.29	2115.566	84.29	2140.966
80.30	2039.620	81.30	2065.020	82.30	2090.420	83.30	2115.820	84.30	2141.220
80.31	2039.874	81.31	2065.274	82.31	2090.674	83.31	2116.074	84.31	2141.474
80.32	2040.128	81.32	2065.528	82.32	2090.928	83.32	2116.328	84.32	2141.728
80.33	2040.382	81.33	2065.782	82.33	2091.182	83.33	2116.582	84.33	2141.982
80.34	2040.636	81.34	2066.036	82.34	2091.436	83.34	2116.836	84.34	2142.236
80.35	2040.890	81.35	2066.290	82.35	2091.690	83.35	2117.090	84.35	2142.490
80.36	2041.144	81.36	2066.544	82.36	2091.944	83.36	2117.344	84.36	2142.744
80.37	2041.398	81.37	2066.798	82.37	2092.198	83.37	2117.598	84.37	2142.998
80.38	2041.652	81.38	2067.052	82.38	2092.452	83.38	2117.852	84.38	2143.252
80.39	2041.906	81.39	2067.306	82.39	2092.706	83.39	2118.106	84.39	2143.506
80.40	2042.160	81.40	2067.560	82.40	2092.960	83.40	2118.360	84.40	2143.760
80.41	2042.414	81.41	2067.814	82.41	2093.214	83.41	2118.614	84.41	2144.014
80.42	2042.668	81.42	2068.068	82.42	2093.468	83.42	2118.868	84.42	2144.268
80.43	2042.922	81.43	2068.322	82.43	2093.722	83.43	2119.122	84.43	2144.522
80.44	2043.176	81.44	2068.576	82.44	2093.976	83.44	2119.376	84.44	2144.776
80.45	2043.430	81.45	2068.830	82.45	2094.230	83.45	2119.630	84.45	2145.030
80.46	2043.684	81.46	2069.084	82.46	2094.484	83.46	2119.884	84.46	2145.284
80.47	2043.938	81.47	2069.338	82.47	2094.738	83.47	2120.138	84.47	2145.538
80.48	2044.192	81.48	2069.592	82.48	2094.992	83.48	2120.392	84.48	2145.792
80.49	2044.446	81.49	2069.846	82.49	2095.246	83.49	2120.646	84.49	2146.046

Continued - INCHES TO MILLIMETERS 1 INCH = 25.4 MILLIMETERS

INCHES	MILLI-METERS	INCHES	MILLI-METERS	INCHES	MILLI-METERS	INCHES	MILLI-METERS	INCHES	MILLI-METERS
80.50	2044.700	81.50	2070.100	82.50	2095.500	83.50	2120.900	84.50	2146.300
80.51	2044.954	81.51	2070.354	82.51	2095.754	83.51	2121.154	84.51	2146.554
80.52	2045.208	81.52	2070.608	82.52	2096.008	83.52	2121.408	84.52	2146.808
80.53	2045.462	81.53	2070.862	82.53	2096.262	83.53	2121.662	84.53	2147.062
80.54	2045.716	81.54	2071.116	82.54	2096.516	83.54	2121.916	84.54	2147.316
80.55	2045.970	81.55	2071.370	82.55	2096.770	83.55	2122.170	84.55	2147.570
80.56	2046.224	81.56	2071.624	82.56	2097.024	83.56	2122.424	84.56	2147.824
80.57	2046.478	81.57	2071.878	82.57	2097.278	83.57	2122.678	84.57	2148.078
80.58	2046.732	81.58	2072.132	82.58	2097.532	83.58	2122.932	84.58	2148.332
80.59	2046.986	81.59	2072.386	82.59	2097.786	83.59	2123.186	84.59	2148.586
80.60	2047.240	81.60	2072.640	82.60	2098.040	83.60	2123.440	84.60	2148.840
80.61	2047.494	81.61	2072.894	82.61	2098.294	83.61	2123.694	84.61	2149.094
80.62	2047.748	81.62	2073.148	82.62	2098.548	83.62	2123.948	84.62	2149.348
80.63	2048.002	81.63	2073.402	82.63	2098.802	83.63	2124.202	84.63	2149.602
80.64	2048.256	81.64	2073.656	82.64	2099.056	83.64	2124.456	84.64	2149.856
80.65	2048.510	81.65	2073.910	82.65	2099.310	83.65	2124.710	84.65	2150.110
80.66	2048.764	81.66	2074.164	82.66	2099.564	83.66	2124.964	84.66	2150.364
80.67	2049.018	81.67	2074.418	82.67	2099.818	83.67	2125.218	84.67	2150.618
80.68	2049.272	81.68	2074.672	82.68	2100.072	83.68	2125.472	84.68	2150.872
80.69	2049.526	81.69	2074.926	82.69	2100.326	83.69	2125.726	84.69	2151.126
80.70	2049.780	81.70	2075.180	82.70	2100.580	83.70	2125.980	84.70	2151.380
80.71	2050.034	81.71	2075.434	82.71	2100.834	83.71	2126.234	84.71	2151.634
80.72	2050.288	81.72	2075.688	82.72	2101.088	83.72	2126.488	84.72	2151.888
80.73	2050.542	81.73	2075.942	82.73	2101.342	83.73	2126.742	84.73	2152.142
80.74	2050.796	81.74	2076.196	82.74	2101.596	83.74	2126.996	84.74	2152.396
80.75	2051.050	81.75	2076.450	82.75	2101.850	83.75	2127.250	84.75	2152.650
80.76	2051.304	81.76	2076.704	82.76	2102.104	83.76	2127.504	84.76	2152.904
80.77	2051.558	81.77	2076.958	82.77	2102.358	83.77	2127.758	84.77	2153.158
80.78	2051.812	81.78	2077.212	82.78	2102.612	83.78	2128.012	84.78	2153.412
80.79	2052.066	81.79	2077.466	82.79	2102.866	83.79	2128.266	84.79	2153.666
80.80	2052.320	81.80	2077.720	82.80	2103.120	83.80	2128.520	84.80	2153.920
80.81	2052.574	81.81	2077.974	82.81	2103.374	83.81	2128.774	84.81	2154.174
80.82	2052.828	81.82	2078.228	82.82	2103.628	83.82	2129.028	84.82	2154.428
80.83	2053.082	81.83	2078.482	82.83	2103.882	83.83	2129.282	84.83	2154.682
80.84	2053.336	81.84	2078.736	82.84	2104.136	83.84	2129.536	84.84	2154.936
80.85	2053.590	81.85	2078.990	82.85	2104.390	83.85	2129.790	84.85	2155.190
80.86	2053.844	81.86	2079.244	82.86	2104.644	83.86	2130.044	84.86	2155.444
80.87	2054.098	81.87	2079.498	82.87	2104.898	83.87	2130.298	84.87	2155.698
80.88	2054.352	81.88	2079.752	82.88	2105.152	83.88	2130.552	84.88	2155.952
80.89	2054.606	81.89	2080.006	82.89	2105.406	83.89	2130.806	84.89	2156.206
80.90	2054.860	81.90	2080.260	82.90	2105.660	83.90	2131.060	84.90	2156.460
80.91	2055.114	81.91	2080.514	82.91	2105.914	83.91	2131.314	84.91	2156.714
80.92	2055.368	81.92	2080.768	82.92	2106.168	83.92	2131.568	84.92	2156.968
80.93	2055.622	81.93	2081.022	82.93	2106.422	83.93	2131.822	84.93	2157.222
80.94	2055.876	81.94	2081.276	82.94	2106.676	83.94	2132.076	84.94	2157.476
80.95	2056.130	81.95	2081.530	82.95	2106.930	83.95	2132.330	84.95	2157.730
80.96	2056.384	81.96	2081.784	82.96	2107.184	83.96	2132.584	84.96	2157.984
80.97	2056.638	81.97	2082.038	82.97	2107.438	83.97	2132.838	84.97	2158.238
80.98	2056.892	81.98	2082.292	82.98	2107.692	83.98	2133.092	84.98	2158.492
80.99	2057.146	81.99	2082.546	82.99	2107.946	83.99	2133.346	84.99	2158.746

Continued - INCHES TO MILLIMETERS 1 INCH = 25.4 MILLIMETERS

INCHES	MILLI-METERS	INCHES	MILLI-METERS	INCHES	MILLI-METERS	INCHES	MILLI-METERS	INCHES	MILLI-METERS
85.00	2159.000	86.00	2184.400	87.00	2209.800	88.00	2235.200	89.00	2260.600
85.01	2159.254	86.01	2184.654	87.01	2210.054	88.01	2235.454	89.01	2260.854
85.02	2159.508	86.02	2184.908	87.02	2210.308	88.02	2235.708	89.02	2261.108
85.03	2159.762	86.03	2185.162	87.03	2210.562	88.03	2235.962	89.03	2261.362
85.04	2160.016	86.04	2185.416	87.04	2210.816	88.04	2236.216	89.04	2261.616
85.05	2160.270	86.05	2185.670	87.05	2211.070	88.05	2236.470	89.05	2261.870
85.06	2160.524	86.06	2185.924	87.06	2211.324	88.06	2236.724	89.06	2262.124
85.07	2160.778	86.07	2186.178	87.07	2211.578	88.07	2236.978	89.07	2262.378
85.08	2161.032	86.08	2186.432	87.08	2211.832	88.08	2237.232	89.08	2262.632
85.09	2161.286	86.09	2186.686	87.09	2212.086	88.09	2237.486	89.09	2262.886
85.10	2161.540	86.10	2186.940	87.10	2212.340	88.10	2237.740	89.10	2263.140
85.11	2161.794	86.11	2187.194	87.11	2212.594	88.11	2237.994	89.11	2263.394
85.12	2162.048	86.12	2187.448	87.12	2212.848	88.12	2238.248	89.12	2263.648
85.13	2162.302	86.13	2187.702	87.13	2213.102	88.13	2238.502	89.13	2263.902
85.14	2162.556	86.14	2187.956	87.14	2213.356	88.14	2238.756	89.14	2264.156
85.15	2162.810	86.15	2188.210	87.15	2213.610	88.15	2239.010	89.15	2264.410
85.16	2163.064	86.16	2188.464	87.16	2213.864	88.16	2239.264	89.16	2264.664
85.17	2163.318	86.17	2188.718	87.17	2214.118	88.17	2239.518	89.17	2264.918
85.18	2163.572	86.18	2188.972	87.18	2214.372	88.18	2239.772	89.18	2265.172
85.19	2163.826	86.19	2189.226	87.19	2214.626	88.19	2240.026	89.19	2265.426
85.20	2164.080	86.20	2189.480	87.20	2214.880	88.20	2240.280	89.20	2265.680
85.21	2164.334	86.21	2189.734	87.21	2215.134	88.21	2240.534	89.21	2265.934
85.22	2164.588	86.22	2189.988	87.22	2215.388	88.22	2240.788	89.22	2266.188
85.23	2164.842	86.23	2190.242	87.23	2215.642	88.23	2241.042	89.23	2266.442
85.24	2165.096	86.24	2190.496	87.24	2215.896	88.24	2241.296	89.24	2266.696
85.25	2165.350	86.25	2190.750	87.25	2216.150	88.25	2241.550	89.25	2266.950
85.26	2165.604	86.26	2191.004	87.26	2216.404	88.26	2241.804	89.26	2267.204
85.27	2165.858	86.27	2191.258	87.27	2216.658	88.27	2242.058	89.27	2267.458
85.28	2166.112	86.28	2191.512	87.28	2216.912	88.28	2242.312	89.28	2267.712
85.29	2166.366	86.29	2191.766	87.29	2217.166	88.29	2242.566	89.29	2267.966
85.30	2166.620	86.30	2192.020	87.30	2217.420	88.30	2242.820	89.30	2268.220
85.31	2166.874	86.31	2192.274	87.31	2217.674	88.31	2243.074	89.31	2268.474
85.32	2167.128	86.32	2192.528	87.32	2217.928	88.32	2243.328	89.32	2268.728
85.33	2167.382	86.33	2192.782	87.33	2218.182	88.33	2243.582	89.33	2268.982
85.34	2167.636	86.34	2193.036	87.34	2218.436	88.34	2243.836	89.34	2269.236
85.35	2167.890	86.35	2193.290	87.35	2218.690	88.35	2244.090	89.35	2269.490
85.36	2168.144	86.36	2193.544	87.36	2218.944	88.36	2244.344	89.36	2269.744
85.37	2168.398	86.37	2193.798	87.37	2219.198	88.37	2244.598	89.37	2269.998
85.38	2168.652	86.38	2194.052	87.38	2219.452	88.38	2244.852	89.38	2270.252
85.39	2168.906	86.39	2194.306	87.39	2219.706	88.39	2245.106	89.39	2270.506
85.40	2169.160	86.40	2194.560	87.40	2219.960	88.40	2245.360	89.40	2270.760
85.41	2169.414	86.41	2194.814	87.41	2220.214	88.41	2245.614	89.41	2271.014
85.42	2169.668	86.42	2195.068	87.42	2220.468	88.42	2245.868	89.42	2271.268
85.43	2169.922	86.43	2195.322	87.43	2220.722	88.43	2246.122	89.43	2271.522
85.44	2170.176	86.44	2195.576	87.44	2220.976	88.44	2246.376	89.44	2271.776
85.45	2170.430	86.45	2195.830	87.45	2221.230	88.45	2246.630	89.45	2272.030
85.46	2170.684	86.46	2196.084	87.46	2221.484	88.46	2246.884	89.46	2272.284
85.47	2170.938	86.47	2196.338	87.47	2221.738	88.47	2247.138	89.47	2272.538
85.48	2171.192	86.48	2196.592	87.48	2221.992	88.48	2247.392	89.48	2272.792
85.49	2171.446	86.49	2196.846	87.49	2222.246	88.49	2247.646	89.49	2273.046

Continued - INCHES TO MILLIMETERS 1 INCH = 25.4 MILLIMETERS

INCHES	MILLI-METERS	INCHES	MILLI-METERS	INCHES	MILLI-METERS	INCHES	MILLI-METERS	INCHES	MILLI-METERS
85.50	2171.700	86.50	2197.100	87.50	2222.500	88.50	2247.900	89.50	2273.300
85.51	2171.954	86.51	2197.354	87.51	2222.754	88.51	2248.154	89.51	2273.554
85.52	2172.208	86.52	2197.608	87.52	2223.008	88.52	2248.408	89.52	2273.808
85.53	2172.462	86.53	2197.862	87.53	2223.262	88.53	2248.662	89.53	2274.062
85.54	2172.716	86.54	2198.116	87.54	2223.516	88.54	2248.916	89.54	2274.316
85.55	2172.970	86.55	2198.370	87.55	2223.770	88.55	2249.170	89.55	2274.570
85.56	2173.224	86.56	2198.624	87.56	2224.024	88.56	2249.424	89.56	2274.824
85.57	2173.478	86.57	2198.878	87.57	2224.278	88.57	2249.678	89.57	2275.078
85.58	2173.732	86.58	2199.132	87.58	2224.532	88.58	2249.932	89.58	2275.332
85.59	2173.986	86.59	2199.386	87.59	2224.786	88.59	2250.186	89.59	2275.586
85.60	2174.240	86.60	2199.640	87.60	2225.040	88.60	2250.440	89.60	2275.840
85.61	2174.494	86.61	2199.894	87.61	2225.294	88.61	2250.694	89.61	2276.094
85.62	2174.748	86.62	2200.148	87.62	2225.548	88.62	2250.948	89.62	2276.348
85.63	2175.002	86.63	2200.402	87.63	2225.802	88.63	2251.202	89.63	2276.602
85.64	2175.256	86.64	2200.656	87.64	2226.056	88.64	2251.456	89.64	2276.856
85.65	2175.510	86.65	2200.910	87.65	2226.310	88.65	2251.710	89.65	2277.110
85.66	2175.764	86.66	2201.164	87.66	2226.564	88.66	2251.964	89.66	2277.364
85.67	2176.018	86.67	2201.418	87.67	2226.818	88.67	2252.218	89.67	2277.618
85.68	2176.272	86.68	2201.672	87.68	2227.072	88.68	2252.472	89.68	2277.872
85.69	2176.526	86.69	2201.926	87.69	2227.326	88.69	2252.726	89.69	2278.126
85.70	2176.780	86.70	2202.180	87.70	2227.580	88.70	2252.980	89.70	2278.380
85.71	2177.034	86.71	2202.434	87.71	2227.834	88.71	2253.234	89.71	2278.634
85.72	2177.288	86.72	2202.688	87.72	2228.088	88.72	2253.488	89.72	2278.888
85.73	2177.542	86.73	2202.942	87.73	2228.342	88.73	2253.742	89.73	2279.142
85.74	2177.796	86.74	2203.196	87.74	2228.596	88.74	2253.996	89.74	2279.396
85.75	2178.050	86.75	2203.450	87.75	2228.850	88.75	2254.250	89.75	2279.650
85.76	2178.304	86.76	2203.704	87.76	2229.104	88.76	2254.504	89.76	2279.904
85.77	2178.558	86.77	2203.958	87.77	2229.358	88.77	2254.758	89.77	2280.158
85.78	2178.812	86.78	2204.212	87.78	2229.612	88.78	2255.012	89.78	2280.412
85.79	2179.066	86.79	2204.466	87.79	2229.866	88.79	2255.266	89.79	2280.666
85.80	2179.320	86.80	2204.720	87.80	2230.120	88.80	2255.520	89.80	2280.920
85.81	2179.574	86.81	2204.974	87.81	2230.374	88.81	2255.774	89.81	2281.174
85.82	2179.828	86.82	2205.228	87.82	2230.628	88.82	2256.028	89.82	2281.428
85.83	2180.082	86.83	2205.482	87.83	2230.882	88.83	2256.282	89.83	2281.682
85.84	2180.336	86.84	2205.736	87.84	2231.136	88.84	2256.536	89.84	2281.936
85.85	2180.590	86.85	2205.990	87.85	2231.390	88.85	2256.790	89.85	2282.190
85.86	2180.844	86.86	2206.244	87.86	2231.644	88.86	2257.044	89.86	2282.444
85.87	2181.098	86.87	2206.498	87.87	2231.898	88.87	2257.298	89.87	2282.698
85.88	2181.352	86.88	2206.752	87.88	2232.152	88.88	2257.552	89.88	2282.952
85.89	2181.606	86.89	2207.006	87.89	2232.406	88.89	2257.806	89.89	2283.206
85.90	2181.860	86.90	2207.260	87.90	2232.660	88.90	2258.060	89.90	2283.460
85.91	2182.114	86.91	2207.514	87.91	2232.914	88.91	2258.314	89.91	2283.714
85.92	2182.368	86.92	2207.768	87.92	2233.168	88.92	2258.568	89.92	2283.968
85.93	2182.622	86.93	2208.022	87.93	2233.422	88.93	2258.822	89.93	2284.222
85.94	2182.876	86.94	2208.276	87.94	2233.676	88.94	2259.076	89.94	2284.476
85.95	2183.130	86.95	2208.530	87.95	2233.930	88.95	2259.330	89.95	2284.730
85.96	2183.384	86.96	2208.784	87.96	2234.184	88.96	2259.584	89.96	2284.984
85.97	2183.638	86.97	2209.038	87.97	2234.438	88.97	2259.838	89.97	2285.238
85.98	2183.892	86.98	2209.292	87.98	2234.692	88.98	2260.092	89.98	2285.492
85.99	2184.146	86.99	2209.546	87.99	2234.946	88.99	2260.346	89.99	2285.746

Continued - INCHES TO MILLIMETERS 1 INCH = 25.4 MILLIMETERS

INCHES	MILLI-METERS	INCHES	MILLI-METERS	INCHES	MILLI-METERS	INCHES	MILLI-METERS	INCHES	MILLI-METERS
90.00	2286.000	91.00	2311.400	92.00	2336.800	93.00	2362.200	94.00	2387.600
90.01	2286.254	91.01	2311.654	92.01	2337.054	93.01	2362.454	94.01	2387.854
90.02	2286.508	91.02	2311.908	92.02	2337.308	93.02	2362.708	94.02	2388.108
90.03	2286.762	91.03	2312.162	92.03	2337.562	93.03	2362.962	94.03	2388.362
90.04	2287.016	91.04	2312.416	92.04	2337.816	93.04	2363.216	94.04	2388.616
90.05	2287.270	91.05	2312.670	92.05	2338.070	93.05	2363.470	94.05	2388.870
90.06	2287.524	91.06	2312.924	92.06	2338.324	93.06	2363.724	94.06	2389.124
90.07	2287.778	91.07	2313.178	92.07	2338.578	93.07	2363.978	94.07	2389.378
90.08	2288.032	91.08	2313.432	92.08	2338.832	93.08	2364.232	94.08	2389.632
90.09	2288.286	91.09	2313.686	92.09	2339.086	93.09	2364.486	94.09	2389.886
90.10	2288.540	91.10	2313.940	92.10	2339.340	93.10	2364.740	94.10	2390.140
90.11	2288.794	91.11	2314.194	92.11	2339.594	93.11	2364.994	94.11	2390.394
90.12	2289.048	91.12	2314.448	92.12	2339.848	93.12	2365.248	94.12	2390.648
90.13	2289.302	91.13	2314.702	92.13	2340.102	93.13	2365.502	94.13	2390.902
90.14	2289.556	91.14	2314.956	92.14	2340.356	93.14	2365.756	94.14	2391.156
90.15	2289.810	91.15	2315.210	92.15	2340.610	93.15	2366.010	94.15	2391.410
90.16	2290.064	91.16	2315.464	92.16	2340.864	93.16	2366.264	94.16	2391.664
90.17	2290.318	91.17	2315.718	92.17	2341.118	93.17	2366.518	94.17	2391.918
90.18	2290.572	91.18	2315.972	92.18	2341.372	93.18	2366.772	94.18	2392.172
90.19	2290.826	91.19	2316.226	92.19	2341.626	93.19	2367.026	94.19	2392.426
90.20	2291.080	91.20	2316.480	92.20	2341.880	93.20	2367.280	94.20	2392.680
90.21	2291.334	91.21	2316.734	92.21	2342.134	93.21	2367.534	94.21	2392.934
90.22	2291.588	91.22	2316.988	92.22	2342.388	93.22	2367.788	94.22	2393.188
90.23	2291.842	91.23	2317.242	92.23	2342.642	93.23	2368.042	94.23	2393.442
90.24	2292.096	91.24	2317.496	92,24	2342.896	93.24	2368.296	94.24	2393.696
90.25	2292.350	91.25	2317.750	92.25	2343.150	93.25	2368.550	94.25	2393.950
90.26	2292.604	91.26	2318.004	92.26	2343.404	93.26	2368.804	94.26	2394.204
90.27	2292.858	91.27	2318.258	92.27	2343.658	93.27	2369.058	94.27	2394.458
90.28	2293.112	91.28	2318.512	92.28	2343.912	93.28	2369.312	94.28	2394.712
90.29	2293.366	91.29	2318.766	92.29	2344.166	93.29	2369.566	94.29	2394.966
90.30	2293.620	91.30	2319.020	92.30	2344.420	93.30	2369.820	94.30	2395.220
90.31	2293.874	91.31	2319.274	92.31	2344.674	93.31	2370.074	94.31	2395.474
90.32	2294.128	91.32	2319.528	92.32	2344.928	93.32	2370.328	94.32	2395.728
90.33	2294.382	91.33	2319.782	92.33	2345.182	93.33	2370.582	94.33	2395.982
90.34	2294.636	91.34	2320.036	92.34	2345.436	93.34	2370.836	94.34	2396.236
90.35	2294.890	91.35	2320.290	92.35	2345.690	93.35	2371.090	94.35	2396.490
90.36	2295.144	91.36	2320.544	92.36	2345.944	93.36	2371.344	94.36	2396.744
90.37	2295.398	91.37	2320.798	92.37	2346.198	93.37	2371.598	94.37	2396.998
90.38	2295.652	91.38	2321.052	92.38	2346.452	93.38	2371.852	94.38	2397.252
90.39	2295.906	91.39	2321.306	92.39	2346.706	93.39	2372.106	94.39	2397.506
90.40	2296.160	91.40	2321.560	92.40	2346.960	93.40	2372.360	94.40	2397.760
90.41	2296.414	91.41	2321.814	92.41	2347.214	93.41	2372.614	94.41	2398.014
90.42	2296.668	91.42	2322.068	92.42	2347.468	93.42	2372.868	94.42	2398.268
90.43	2296.922	91.43	2322.322	92.43	2347.722	93.43	2373.122	94.43	2398.522
90.44	2297.176	91.44	2322.576	92.44	2347.976	93.44	2373.376	94.44	2398.776
90.45	2297.430	91.45	2322.830	92.45	2348.230	93.45	2373.630	94.45	2399.030
90.46	2297.684	91.46	2323.084	92.46	2348.484	93.46	2373.884	94.46	2399.284
90.47	2297.938	91.47	2323.338	92.47	2348.738	93.47	2374.138	94.47	2399.538
90.48	2298.192	91.48	2323.592	92.48	2348.992	93.48	2374.392	94.48	2399.792
90.49	2298.446	91.49	2323.846	92.49	2349.246	93.49	2374.646	94.49	2400.046

Continued - INCHES TO MILLIMETERS 1 INCH = 25.4 MILLIMETERS

INCHES	MILLI-METERS	INCHES	MILLI-METERS	INCHES	MILLI-METERS	INCHES	MILLI-METERS	INCHES	MILLI-METERS
90.50	2298.700	91.50	2324.100	92.50	2349.500	93.50	2374.900	94.50	2400.300
90.51	2298.954	91.51	2324.354	92.51	2349.754	93.51	2375.154	94.51	2400.554
90.52	2299.208	91.52	2324.608	92.52	2350.008	93.52	2375.408	94.52	2400.808
90.53	2299.462	91.53	2324.862	92.53	2350.262	93.53	2375.662	94.53	2401.062
90.54	2299.716	91.54	2325.116	92.54	2350.516	93.54	2375.916	94.54	2401.316
90.55	2299.970	91.55	2325.370	92.55	2350.770	93.55	2376.170	94.55	2401.570
90.56	2300.224	91.56	2325.624	92.56	2351.024	93.56	2376.424	94.56	2401.824
90.57	2300.478	91.57	2325.878	92.57	2351.278	93.57	2376.678	94.57	2402.078
90.58	2300.732	91.58	2326.132	92.58	2351.532	93.58	2376.932	94.58	2402.332
90.59	2300.986	91.59	2326.386	92.59	2351.786	93.59	2377.186	94.59	2402.586
90.60	2301.240	91.60	2326.640	92.60	2352.040	93.60	2377.440	94.60	2402.840
90.61	2301.494	91.61	2326.894	92.61	2352.294	93.61	2377.694	94.61	2403.094
90.62	2301.748	91.62	2327.148	92.62	2352.548	93.62	2377.948	94.62	2403.348
90.63	2302.002	91.63	2327.402	92.63	2352.802	93.63	2378.202	94.63	2403.602
90.64	2302.256	91.64	2327.656	92.64	2353.056	93.64	2378.456	94.64	2403.856
90.65	2302.510	91.65	2327.910	92.65	2353.310	93.65	2378.710	94.65	2404.110
90.66	2302.764	91.66	2328.164	92.66	2353.564	93.66	2378.964	94.66	2404.364
90.67	2303.018	91.67	2328.418	92.67	2353.818	93.67	2379.218	94.67	2404.618
90.68	2303.272	91.68	2328.672	92.68	2354.072	93.68	2379.472	94.68	2404.872
90.69	2303.526	91.69	2328.926	92.69	2354.326	93.69	2379.726	94.69	2405.126
90.70	2303.780	91.70	2329.180	92.70	2354.580	93.70	2379.980	94.70	2405.380
90.71	2304.034	91.71	2329.434	92.71	2354.834	93.71	2380.234	94.71	2405.634
90.72	2304.288	91.72	2329.688	92.72	2355.088	93.72	2380.488	94.72	2405.888
90.73	2304.542	91.73	2329.942	92.73	2355.342	93.73	2380.742	94.73	2406.142
90.74	2304.796	91.74	2330.196	92.74	2355.596	93.74	2380.996	94.74	2406.396
90.75	2305.050	91.75	2330.450	92.75	2355.850	93.75	2381.250	94.75	2406.650
90.76	2305.304	91.76	2330.704	92.76	2356.104	93.76	2381.504	94.76	2406.904
90.77	2305.558	91.77	2330.958	92.77	2356.358	93.77	2381.758	94.77	2407.158
90.78	2305.812	91.78	2331.212	92.78	2356.612	93.78	2382.012	94.78	2407.412
90.79	2306.066	91.79	2331.466	92.79	2356.866	93.79	2382.266	94.79	2407.666
90.80	2306.320	91.80	2331.720	92.80	2357.120	93.80	2382.520	94.80	2407.920
90.81	2306.574	91.81	2331.974	92.81	2357.374	93.81	2382.774	94.81	2408.174
90.82	2306.828	91.82	2332.228	92.82	2357.628	93.82	2383.028	94.82	2408.428
90.83	2307.082	91.83	2332.482	92.83	2357.882	93.83	2383.282	94.83	2408.682
90.84	2307.336	91.84	2332.736	92.84	2358.136	93.84	2383.536	94.84	2408.936
90.85	2307.590	91.85	2332.990	92.85	2358.390	93.85	2383.790	94.85	2409.190
90.86	2307.844	91.86	2333.244	92.86	2358.644	93.86	2384.044	94.86	2409.444
90.87	2308.098	91.87	2333.498	92.87	2358.898	93.87	2384.298	94.87	2409.698
90.88	2308.352	91.88	2333.752	92.88	2359.152	93.88	2384.552	94.88	2409.952
90.89	2308.606	91.89	2334.006	92.89	2359.406	93.89	2384.806	94.89	2410.206
90.90	2308.860	91.90	2334.260	92.90	2359.660	93.90	2385.060	94.90	2410.460
90.91	2309.114	91.91	2334.514	92.91	2359.914	93.91	2385.314	94.91	2410.714
90.92	2309.368	91.92	2334.768	92.92	2360.168	93.92	2385.568	94.92	2410.968
90.93	2309.622	91.93	2335.022	92.93	2360.422	93.93	2385.822	94.93	2411.222
90.94	2309.876	91.94	2335.276	92.94	2360.676	93.94	2386.076	94.94	2411.476
90.95	2310.130	91.95	2335.530	92.95	2360.930	93.95	2386.330	94.95	2411.730
90.96	2310.384	91.96	2335.784	92.96	2361.184	93.96	2386.584	94.96	2411.984
90.97	2310.638	91.97	2336.038	92.97	2361.438	93.97	2386.838	94.97	2412.238
90.98	2310.892	91.98	2336.292	92.98	2361.692	93.98	2387.092	94.98	2412.492
90.99	2311.146	91.99	2336.546	92.99	2361.946	93.99	2387.346	94.99	2412.746

Continued - INCHES TO MILLIMETERS 1 INCH = 25.4 MILLIMETERS

INCHES	MILLI-METERS	INCHES	MILLI-METERS	INCHES	MILLI-METERS	INCHES	MILLI-METERS	INCHES	MILLI-METERS
95.00	2413.000	96.00	2438.400	97.00	2463.800	98.00	2489.200	99.00	2514.600
95.01	2413.254	96.01	2438.654	97.01	2464.054	98.01	2489.454	99.01	2514.854
95.02	2413.508	96.02	2438.908	97.02	2464.308	98.02	2489.708	99.02	2515.108
95.03	2413.762	96.03	2439.162	97.03	2464.562	98.03	2489.962	99.03	2515.362
95.04	2414.016	96.04	2439.416	97.04	2464.816	98.04	2490.216	99.04	2515.616
95.05	2414.270	96.05	2439.670	97.05	2465.070	98.05	2490.470	99.05	2515.870
95.06	2414.524	96.06	2439.924	97.06	2465.324	98.06	2490.724	99.06	2516.124
95.07	2414.778	96.07	2440.178	97.07	2465.578	98.07	2490.978	99.07	2516.378
95.08	2415.032	96.08	2440.432	97.08	2465.832	98.08	2491.232	99.08	2516.632
95.09	2415.286	96.09	2440.686	97.09	2466.086	98.09	2491.486	99.09	2516.886
95.10	2415.540	96.10	2440.940	97.10	2466.340	98.10	2491.740	99.10	2517.140
95.11	2415.794	96.11	2441.194	97.11	2466.594	98.11	2491.994	99.11	2517.394
95.12	2416.048	96.12	2441.448	97.12	2466.848	98.12	2492.248	99.12	2517.648
95.13	2416.302	96.13	2441.702	97.13	2467.102	98.13	2492.502	99.13	2517.902
95.14	2416.556	96.14	2441.956	97.14	2467.356	98.14	2492.756	99.14	2518.156
95.15	2416.810	96.15	2442.210	97.15	2467.610	98.15	2493.010	99.15	2518.410
95.16	2417.064	96.16	2442.464	97.16	2467.864	98.16	2493.264	99.16	2518.664
95.17	2417.318	96.17	2442.718	97.17	2468.118	98.17	2493.518	99.17	2518.918
95.18	2417.572	96.18	2442.972	97.18	2468.372	98.18	2493.772	99.18	2519.172
95.19	2417.826	96.19	2443.226	97.19	2468.626	98.19	2494.026	99.19	2519.426
95.20	2418.080	96.20	2443.480	97.20	2468.880	98.20	2494.280	99.20	2519.680
95.21	2418.334	96.21	2443.734	97.21	2469.134	98.21	2494.534	99.21	2519.934
95.22	2418.588	96.22	2443.988	97.22	2469.388	98.22	2494.788	99.22	2520.188
95.23	2418.842	96.23	2444.242	97.23	2469.642	98.23	2495.042	99.23	2520.442
95.24	2419.096	96.24	2444.496	97.24	2469.896	98.24	2495.296	99.24	2520.696
95.25	2419.350	96.25	2444.750	97.25	2470.150	98.25	2495.550	99.25	2520.950
95.26	2419.604	96.26	2445.004	97.26	2470.404	98.26	2495.804	99.26	2521.204
95.27	2419.858	96.27	2445.258	97.27	2470.658	98.27	2496.058	99.27	2521.458
95.28	2420.112	96.28	2445.512	97.28	2470.912	98.28	2496.312	99.28	2521.712
95.29	2420.366	96.29	2445.766	97.29	2471.166	98.29	2496.566	99.29	2521.966
95.30	2420.620	96.30	2446.020	97.30	2471.420	98.30	2496.820	99.30	2522.220
95.31	2420.874	96.31	2446.274	97.31	2471.674	98.31	2497.074	99.31	2522.474
95.32	2421.128	96.32	2446.528	97.32	2471.928	98.32	2497.328	99.32	2522.728
95.33	2421.382	96.33	2446.782	97.33	2472.182	98.33	2497.582	99.33	2522.982
95.34	2421.636	96.34	2447.036	97.34	2472.436	98.34	2497.836	99.34	2523.236
95.35	2421.890	96.35	2447.290	97.35	2472.690	98.35	2498.090	99.35	2523.490
95.36	2422.144	96.36	2447.544	97.36	2472.944	98.36	2498.344	99.36	2523.744
95.37	2422.398	96.37	2447.798	97.37	2473.198	98.37	2498.598	99.37	2523.998
95.38	2422.652	96.38	2448.052	97.38	2473.452	98.38	2498.852	99.38	2524.252
95.39	2422.906	96.39	2448.306	97.39	2473.706	98.39	2499.106	99.39	2524.506
95.40	2423.160	96.40	2448.560	97.40	2473.960	98.40	2499.360	99.40	2524.760
95.41	2423.414	96.41	2448.814	97.41	2474.214	98.41	2499.614	99.41	2525.014
95.42	2423.668	96.42	2449.068	97.42	2474.468	98.42	2499.868	99.42	2525.268
95.43	2423.922	96.43	2449.322	97.43	2474.722	98.43	2500.122	99.43	2525.522
95.44	2424.176	96.44	2449.576	97.44	2474.976	98.44	2500.376	99.44	2525.776
95.45	2424.430	96.45	2449.830	97.45	2475.230	98.45	2500.630	99.45	2526.030
95.46	2424.684	96.46	2450.084	97.46	2475.484	98.46	2500.884	99.46	2526.284
95.47	2424.938	96.47	2450.338	97.47	2475.738	98.47	2501.138	99.47	2526.538
95.48	2425.192	96.48	2450.592	97.48	2475.992	98.48	2501.392	99.48	2526.792
95.49	2425.446	96.49	2450.846	97.49	2476.246	98.49	2501.646	99.49	2527.046

Continued - INCHES TO MILLIMETERS 1 INCH = 25.4 MILLIMETERS

INCHES	MILLI-METERS	INCHES	MILLI-METERS	INCHES	MILLI-METERS	INCHES	MILLI-METERS	INCHES	MILLI-METERS
95.50	2425.700	96.50	2451.100	97.50	2476.500	98.50	2501.900	99.50	2527.300
95.51	2425.954	96.51	2451.354	97.51	2476.754	98.51	2502.154	99.51	2527.554
95.52	2426.208	96.52	2451.608	97.52	2477.008	98.52	2502.408	99.52	2527.808
95.53	2426.462	96.53	2451.862	97.53	2477.262	98.53	2502.662	99.53	2528.062
95.54	2426.716	96.54	2452.116	97.54	2477.516	98.54	2502.916	99.54	2528.316
95.55	2426.970	96.55	2452.370	97.55	2477.770	98.55	2503.170	99.55	2528.570
95.56	2427.224	96.56	2452.624	97.56	2478.024	98.56	2503.424	99.56	2528.824
95.57	2427.478	96.57	2452.878	97.57	2478.278	98.57	2503.678	99.57	2529.078
95.58	2427.732	96.58	2453.132	97.58	2478.532	98.58	2503.932	99.58	2529.332
95.59	2427.986	96.59	2453.386	97.59	2478.786	98.59	2504.186	99.59	2529.586
95.60	2428.240	96.60	2453.640	97.60	2479.040	98.60	2504.440	99.60	2529.840
95.61	2428.494	96.61	2453.894	97.61	2479.294	98.61	2504.694	99.61	2530.094
95.62	2428.748	96.62	2454.148	97.62	2479.548	98.62	2504.948	99.62	2530.348
95.63	2429.002	96.63	2454.402	97.63	2479.802	98.63	2505.202	99.63	2530.602
95.64	2429.256	96.64	2454.656	97.64	2480.056	98.64	2505.456	99.64	2530.856
95.65	2429.510	96.65	2454.910	97.65	2480.310	98.65	2505.710	99.65	2531.110
95.66	2429.764	96.66	2455.164	97.66	2480.564	98.66	2505.964	99.66	2531.364
95.67	2430.018	96.67	2455.418	97.67	2480.818	98.67	2506.218	99.67	2531.618
95.68	2430.272	96.68	2455.672	97.68	2481.072	98.68	2506.472	99.68	2531.872
95.69	2430.526	96.69	2455.926	97.69	2481.326	98.69	2506.726	99.69	2532.126
95.70	2430.780	96.70	2456.180	97.70	2481.580	98.70	2506.980	99.70	2532.380
95.71	2431.034	96.71	2456.434	97.71	2481.834	98.71	2507.234	99.71	2532.634
95.72	2431.288	96.72	2456.688	97.72	2482.088	98.72	2507.488	99.72	2532.888
95.73	2431.542	96.73	2456.942	97.73	2482.342	98.73	2507.742	99.73	2533.142
95.74	2431.796	96.74	2457.196	97.74	2482.596	98.74	2507.996	99.74	2533.396
95.75	2432.050	96.75	2457.450	97.75	2482.850	98.75	2508.250	99.75	2533.650
95.76	2432.304	96.76	2457.704	97.76	2483.104	98.76	2508.504	99.76	2533.904
95.77	2432.558	96.77	2457.958	97.77	2483.358	98.77	2508.758	99.77	2534.158
95.78	2432.812	96.78	2458.212	97.78	2483.612	98.78	2509.012	99.78	2534.412
95.79	2433.066	96.79	2458.466	97.79	2483.866	98.79	2509.266	99.79	2534.666
95.80	2433.320	96.80	2458.720	97.80	2484.120	98.80	2509.520	99.80	2534.920
95.81	2433.574	96.81	2458.974	97.81	2484.374	98.81	2509.774	99.81	2535.174
95.82	2433.828	96.82	2459.228	97.82	2484.628	98.82	2510.028	99.82	2535.428
95.83	2434.082	96.83	2459.482	97.83	2484.882	98.83	2510.282	99.83	2535.682
95.84	2434.336	96.84	2459.736	97.84	2485.136	98.84	2510.536	99.84	2535.936
95.85	2434.590	96.85	2459.990	97.85	2485.390	98.85	2510.790	99.85	2536.190
95.86	2434.844	96.86	2460.244	97.86	2485.644	98.86	2511.044	99.86	2536.444
95.87	2435.098	96.87	2460.498	97.87	2485.898	98.87	2511.298	99.87	2536.698
95.88	2435.352	96.88	2460.752	97.88	2486.152	98.88	2511.552	99.88	2536.952
95.89	2435.606	96.89	2461.006	97.89	2486.406	98.89	2511.806	99.89	2537.206
95.90	2435.860	96.90	2461.260	97.90	2486.660	98.90	2512.060	99.90	2537.460
95.91	2436.114	96.91	2461.514	97.91	2486.914	98.91	2512.314	99.91	2537.714
95.92	2436.368	96.92	2461.768	97.92	2487.168	98.92	2512.568	99.92	2537.968
95.93	2436.622	96.93	2462.022	97.93	2487.422	98.93	2512.822	99.93	2538.222
95.94	2436.876	96.94	2462.276	97.94	2487.676	98.94	2513.076	99.94	2538.476
95.95	2437.130	96.95	2462.530	97.95	2487.930	98.95	2513.330	99.95	2538.730
95.96	2437.384	96.96	2462.784	97.96	2488.184	98.96	2513.584	99.96	2538.984
95.97	2437.638	96.97	2463.038	97.97	2488.438	98.97	2513.838	99.97	2539.238
95.98	2437.892	96.98	2463.292	97.98	2488.692	98.98	2514.092	99.98	2539.492
95.99	2438.146	96.99	2463.546	97.99	2488.946	98.99	2514.346	99.99	2539.746

LENGTH - MILLIMETERS TO INCHES 1 MILLIMETER = 0.039 370 0787 INCH

MILLI-METERS	INCHES	MILLI-METERS	INCHES	MILLI-METERS	INCHES	MILLI-METERS	INCHES	MILLI-METERS	INCHES
0.00	0.00000	1.00	0.03937	2.00	0.07874	3.00	0.11811	4.00	0.15748
0.01	0.00039	1.01	0.03976	2.01	0.07913	3.01	0.11850	4.01	0.15787
0.02	0.00079	1.02	0.04016	2.02	0.07953	3.02	0.11890	4.02	0.15827
0.03	0.00118	1.03	0.04055	2.03	0.07992	3.03	0.11929	4.03	0.15866
0.04	0.00157	1.04	0.04094	2.04	0.08031	3.04	0.11969	4.04	0.15906
0.05	0.00197	1.05	0.04134	2.05	0.08071	3.05	0.12008	4.05	0.15945
0.06	0.00236	1.06	0.04173	2.06	0.08110	3.06	0.12047	4.06	0.15984
0.07	0.00276	1.07	0.04213	2.07	0.08150	3.07	0.12087	4.07	0.16024
0.08	0.00315	1.08	0.04252	2.08	0.08189	3.08	0.12126	4.08	0.16063
0.09	0.00354	1.09	0.04291	2.09	0.08228	3.09	0.12165	4.09	0.16102
0.10	0.00394	1.10	0.04331	2.10	0.08268	3.10	0.12205	4.10	0.16142
0.11	0.00433	1.11	0.04370	2.11	0.08307	3.11	0.12244	4.11	0.16181
0.12	0.00472	1.12	0.04409	2.12	0.08346	3.12	0.12283	4.12	0.16220
0.13	0.00512	1.13	0.04449	2.13	0.08386	3.13	0.12323	4.13	0.16260
0.14	0.00551	1.14	0.04488	2.14	0.08425	3.14	0.12362	4.14	0.16299
0.15	0.00591	1.15	0.04528	2.15	0.08465	3.15	0.12402	4.15	0.16339
0.16	0.00630	1.16	0.04567	2.16	0.08504	3.16	0.12441	4.16	0.16378
0.17	0.00669	1.17	0.04606	2.17	0.08543	3.17	0.12480	4.17	0.16417
0.18	0.00709	1.18	0.04646	2.18	0.08583	3.18	0.12520	4.18	0.16457
0.19	0.00748	1.19	0.04685	2.19	0.08622	3.19	0.12559	4.19	0.16496
0.20	0.00787	1.20	0.04724	2.20	0.08661	3.20	0.12598	4.20	0.16535
0.21	0.00827	1.21	0.04764	2.21	0.08701	3.21	0.12638	4.21	0.16575
0.22	0.00866	1.22	0.04803	2.22	0.08740	3.22	0.12677	4.22	0.16614
0.23	0.00906	1.23	0.04843	2.23	0.08780	3.23	0.12717	4.23	0.16654
0.24	0.00945	1.24	0.04882	2.24	0.08819	3.24	0.12756	4.24	0.16693
0.25	0.00984	1.25	0.04921	2.25	0.08858	3.25	0.12795	4.25	0.16732
0.26	0.01024	1.26	0.04961	2.26	0.08898	3.26	0.12835	4.26	0.16772
0.27	0.01063	1.27	0.05000	2.27	0.08937	3.27	0.12874	4.27	0.16811
0.28	0.01102	1.28	0.05039	2.28	0.08976	3.28	0.12913	4.28	0.16850
0.29	0.01142	1.29	0.05079	2.29	0.09016	3.29	0.12953	4.29	0.16890
0.30	0.01181	1.30	0.05118	2.30	0.09055	3.30	0.12992	4.30	0.16929
0.31	0.01220	1.31	0.05157	2.31	0.09094	3.31	0.13031	4.31	0.16969
0.32	0.01260	1.32	0.05197	2.32	0.09134	3.32	0.13071	4.32	0.17008
0.33	0.01299	1.33	0.05236	2.33	0.09173	3.33	0.13110	4.33	0.17047
0.34	0.01339	1.34	0.05276	2.34	0.09213	3.34	0.13150	4.34	0.17087
0.35	0.01378	1.35	0.05315	2.35	0.09252	3.35	0.13189	4.35	0.17126
0.36	0.01417	1.36	0.05354	2.36	0.09291	3.36	0.13228	4.36	0.17165
0.37	0.01457	1.37	0.05394	2.37	0.09331	3.37	0.13268	4.37	0.17205
0.38	0.01496	1.38	0.05433	2.38	0.09370	3.38	0.13307	4.38	0.17244
0.39	0.01535	1.39	0.05472	2.39	0.09409	3.39	0.13346	4.39	0.17283
0.40	0.01575	1.40	0.05512	2.40	0.09449	3.40	0.13386	4.40	0.17323
0.41	0.01614	1.41	0.05551	2.41	0.09488	3.41	0.13425	4.41	0.17362
0.42	0.01654	1.42	0.05591	2.42	0.09528	3.42	0.13465	4.42	0.17402
0.43	0.01693	1.43	0.05630	2.43	0.09567	3.43	0.13504	4.43	0.17441
0.44	0.01732	1.44	0.05669	2.44	0.09606	3.44	0.13543	4.44	0.17480
0.45	0.01772	1.45	0.05709	2.45	0.09646	3.45	0.13583	4.45	0.17520
0.46	0.01811	1.46	0.05748	2.46	0.09685	3.46	0.13622	4.46	0.17559
0.47	0.01850	1.47	0.05787	2.47	0.09724	3.47	0.13661	4.47	0.17598
0.48	0.01890	1.48	0.05827	2.48	0.09764	3.48	0.13701	4.48	0.17638
0.49	0.01929	1.49	0.05866	2.49	0.09803	3.49	0.13740	4.49	0.17677

Continued - MILLIMETERS TO INCHES 1 MILLIMETER = 0.039 370 0787 INCH

MILLI-METERS	INCHES	MILLI-METERS	INCHES	MILLI-METERS	INCHES	MILLI-METERS	INCHES	MILLI-METERS	INCHES
0.50	0.01969	1.50	0.05906	2.50	0.09843	3.50	0.13780	4.50	0.17717
0.51	0.02008	1.51	0.05945	2.51	0.09882	3.51	0.13819	4.51	0.17756
0.52	0.02047	1.52	0.05984	2.52	0.09921	3.52	0.13858	4.52	0.17795
0.53	0.02087	1.53	0.06024	2.53	0.09961	3.53	0.13898	4.53	0.17835
0.54	0.02126	1.54	0.06063	2.54	0.10000	3.54	0.13937	4.54	0.17874
0.55	0.02165	1.55	0.06102	2.55	0.10039	3.55	0.13976	4.55	0.17913
0.56	0.02205	1.56	0.06142	2.56	0.10079	3.56	0.14016	4.56	0.17953
0.57	0.02244	1.57	0.06181	2.57	0.10118	3.57	0.14055	4.57	0.17992
0.58	0.02283	1.58	0.06220	2.58	0.10157	3.58	0.14094	4.58	0.18031
0.59	0.02323	1.59	0.06260	2.59	0.10197	3.59	0.14134	4.59	0.18071
0.60	0.02362	1.60	0.06299	2.60	0.10236	3.60	0.14173	4.60	0.18110
0.61	0.02402	1.61	0.06339	2.61	0.10276	3.61	0.14213	4.61	0.18150
0.62	0.02441	1.62	0.06378	2.62	0.10315	3.62	0.14252	4.62	0.18189
0.63	0.02480	1.63	0.06417	2.63	0.10354	3.63	0.14291	4.63	0.18228
0.64	0.02520	1.64	0.06457	2.64	0.10394	3.64	0.14331	4.64	0.18268
0.65	0.02559	1.65	0.06496	2.65	0.10433	3.65	0.14370	4.65	0.18307
0.66	0.02598	1.66	0.06535	2.66	0.10472	3.66	0.14409	4.66	0.18346
0.67	0.02638	1.67	0.06575	2.67	0.10512	3.67	0.14449	4.67	0.18386
0.68	0.02677	1.68	0.06614	2.68	0.10551	3.68	0.14488	4.68	0.18425
0.69	0.02717	1.69	0.06654	2.69	0.10591	3.69	0.14528	4.69	0.18465
0.70	0.02756	1.70	0.06693	2.70	0.10630	3.70	0.14567	4.70	0.18504
0.71	0.02795	1.71	0.06732	2.71	0.10669	3.71	0.14606	4.71	0.18543
0.72	0.02835	1.72	0.06772	2.72	0.10709	3.72	0.14646	4.72	0.18583
0.73	0.02874	1.73	0.06811	2.73	0.10748	3.73	0.14685	4.73	0.18622
0.74	0.02913	1.74	0.06850	2.74	0.10787	3.74	0.14724	4.74	0.18661
0.75	0.02953	1.75	0.06890	2.75	0.10827	3.75	0.14764	4.75	0.18701
0.76	0.02992	1.76	0.06929	2.76	0.10866	3.76	0.14803	4.76	0.18740
0.77	0.03031	1.77	0.06969	2.77	0.10906	3.77	0.14843	4.77	0.18780
0.78	0.03071	1.78	0.07008	2.78	0.10945	3.78	0.14882	4.78	0.18819
0.79	0.03110	1.79	0.07047	2.79	0.10984	3.79	0.14921	4.79	0.18858
0.80	0.03150	1.80	0.07087	2.80	0.11024	3.80	0.14961	4.80	0.18898
0.81	0.03189	1.81	0.07126	2.81	0.11063	3.81	0.15000	4.81	0.18937
0.82	0.03228	1.82	0.07165	2.82	0.11102	3.82	0.15039	4.82	0.18976
0.83	0.03268	1.83	0.07205	2.83	0.11142	3.83	0.15079	4.83	0.19016
0.84	0.03307	1.84	0.07244	2.84	0.11181	3.84	0.15118	4.84	0.19055
0.85	0.03346	1.85	0.07283	2.85	0.11220	3.85	0.15157	4.85	0.19094
0.86	0.03386	1.86	0.07323	2.86	0.11260	3.86	0.15197	4.86	0.19134
0.87	0.03425	1.87	0.07362	2.87	0.11299	3.87	0.15236	4.87	0.19173
0.88	0.03465	1.88	0.07402	2.88	0.11339	3.88	0.15276	4.88	0.19213
0.89	0.03504	1.89	0.07441	2.89	0.11378	3.89	0.15315	4.89	0.19252
0.90	0.03543	1.90	0.07480	2.90	0.11417	3.90	0.15354	4.90	0.19291
0.91	0.03583	1.91	0.07520	2.91	0.11457	3.91	0.15394	4.91	0.19331
0.92	0.03622	1.92	0.07559	2.92	0.11496	3.92	0.15433	4.92	0.19370
0.93	0.03661	1.93	0.07598	2.93	0.11535	3.93	0.15472	4.93	0.19409
0.94	0.03701	1.94	0.07638	2.94	0.11575	3.94	0.15512	4.94	0.19449
0.95	0.03740	1.95	0.07677	2.95	0.11614	3.95	0.15551	4.95	0.19488
0.96	0.03780	1.96	0.07717	2.96	0.11654	3.96	0.15591	4.96	0.19528
0.97	0.03819	1.97	0.07756	2.97	0.11693	3.97	0.15630	4.97	0.19567
0.98	0.03858	1.98	0.07795	2.98	0.11732	3.98	0.15669	4.98	0.19606
0.99	0.03898	1.99	0.07835	2.99	0.11772	3.99	0.15709	4.99	0.19646

Continued - MILLIMETERS TO INCHES 1 MILLIMETER = 0.039 370 0787 INCH

MILLI-METERS	INCHES	MILLI-METERS	INCHES	MILLI-METERS	INCHES	MILLI-METERS	INCHES	MILLI-METERS	INCHES
5.00	0.19685	6.00	0.23622	7.00	0.27559	8.00	0.31496	9.00	0.35433
5.01	0.19724	6.01	0.23661	7.01	0.27598	8.01	0.31535	9.01	0.35472
5.02	0.19764	6.02	0.23701	7.02	0.27638	8.02	0.31575	9.02	0.35512
5.03	0.19803	6.03	0.23740	7.03	0.27677	8.03	0.31614	9.03	0.35551
5.04	0.19843	6.04	0.23780	7.04	0.27717	8.04	0.31654	9.04	0.35591
5.05	0.19882	6.05	0.23819	7.05	0.27756	8.05	0.31693	9.05	0.35630
5.06	0.19921	6.06	0.23858	7.06	0.27795	8.06	0.31732	9.06	0.35669
5.07	0.19961	6.07	0.23898	7.07	0.27835	8.07	0.31772	9.07	0.35709
5.08	0.20000	6.08	0.23937	7.08	0.27874	8.08	0.31811	9.08	0.35748
5.09	0.20039	6.09	0.23976	7.09	0.27913	8.09	0.31850	9.09	0.35787
5.10	0.20079	6.10	0.24016	7.10	0.27953	8.10	0.31890	9.10	0.35827
5.11	0.20118	6.11	0.24055	7.11	0.27992	8.11	0.31929	9.11	0.35866
5.12	0.20157	6.12	0.24094	7.12	0.28031	8.12	0.31969	9.12	0.35906
5.13	0.20197	6.13	0.24134	7.13	0.28071	8.13	0.32008	9.13	0.35945
5.14	0.20236	6.14	0.24173	7.14	0.28110	8.14	0.32047	9.14	0.35984
5.15	0.20276	6.15	0.24213	7.15	0.28150	8.15	0.32087	9.15	0.36024
5.16	0.20315	6.16	0.24252	7.16	0.28189	8.16	0.32126	9.16	0.36063
5.17	0.20354	6.17	0.24291	7.17	0.28228	8.17	0.32165	9.17	0.36102
5.18	0.20394	6.18	0.24331	7.18	0.28268	8.18	0.32205	9.18	0.36142
5.19	0.20433	6.19	0.24370	7.19	0.28307	8.19	0.32244	9.19	0.36181
5.20	0.20472	6.20	0.24409	7.20	0.28346	8.20	0.32283	9.20	0.36220
5.21	0.20512	6.21	0.24449	7.21	0.28386	8.21	0.32323	9.21	0.36260
5.22	0.20551	6.22	0.24488	7.22	0.28425	8.22	0.32362	9.22	0.36299
5.23	0.20591	6.23	0.24528	7.23	0.28465	8.23	0.32402	9.23	0.36339
5.24	0.20630	6.24	0.24567	7.24	0.28504	8.24	0.32441	9.24	0.36378
5.25	0.20669	6.25	0.24606	7.25	0.28543	8.25	0.32480	9.25	0.36417
5.26	0.20709	6.26	0.24646	7.26	0.28583	8.26	0.32520	9.26	0.36457
5.27	0.20748	6.27	0.24685	7.27	0.28622	8.27	0.32559	9.27	0.36496
5.28	0.20787	6.28	0.24724	7.28	0.28661	8.28	0.32598	9.28	0.36535
5.29	0.20827	6.29	0.24764	7.29	0.28701	8.29	0.32638	9.29	0.36575
5.30	0.20866	6.30	0.24803	7.30	0.28740	8.30	0.32677	9.30	0.36614
5.31	0.20906	6.31	0.24843	7.31	0.28780	8.31	0.32717	9.31	0.36654
5.32	0.20945	6.32	0.24882	7.32	0.28819	8.32	0.32756	9.32	0.36693
5.33	0.20984	6.33	0.24921	7.33	0.28858	8.33	0.32795	9.33	0.36732
5.34	0.21024	6.34	0.24961	7.34	0.28898	8.34	0.32835	9.34	0.36772
5.35	0.21063	6.35	0.25000	7.35	0.28937	8.35	0.32874	9.35	0.36811
5.36	0.21102	6.36	0.25039	7.36	0.28976	8.36	0.32913	9.36	0.36850
5.37	0.21142	6.37	0.25079	7.37	0.29016	8.37	0.32953	9.37	0.36890
5.38	0.21181	6.38	0.25118	7.38	0.29055	8.38	0.32992	9.38	0.36929
5.39	0.21220	6.39	0.25157	7.39	0.29094	8.39	0.33031	9.39	0.36969
5.40	0.21260	6.40	0.25197	7.40	0.29134	8.40	0.33071	9.40	0.37008
5.41	0.21299	6.41	0.25236	7.41	0.29173	8.41	0.33110	9.41	0.37047
5.42	0.21339	6.42	0.25276	7.42	0.29213	8.42	0.33150	9.42	0.37087
5.43	0.21378	6.43	0.25315	7.43	0.29252	8.43	0.33189	9.43	0.37126
5.44	0.21417	6.44	0.25354	7.44	0.29291	8.44	0.33228	9.44	0.37165
5.45	0.21457	6.45	0.25394	7.45	0.29331	8.45	0.33268	9.45	0.37205
5.46	0.21496	6.46	0.25433	7.46	0.29370	8.46	0.33307	9.46	0.37244
5.47	0.21535	6.47	0.25472	7.47	0.29409	8.47	0.33346	9.47	0.37283
5.48	0.21575	6.48	0.25512	7.48	0.29449	8.48	0.33386	9.48	0.37323
5.49	0.21614	6.49	0.25551	7.49	0.29488	8.49	0.33425	9.49	0.37362

Continued - MILLIMETERS TO INCHES 1 MILLIMETER = 0.039 370 0787 INCH

MILLI-METERS	INCHES	MILLI-METERS	INCHES	MILLI-METERS	INCHES	MILLI-METERS	INCHES	MILLI-METERS	INCHES
5.50	0.21654	6.50	0.25591	7.50	0.29528	8.50	0.33465	9.50	0.37402
5.51	0.21693	6.51	0.25630	7.51	0.29567	8.51	0.33504	9.51	0.37441
5.52	0.21732	6.52	0.25669	7.52	0.29606	8.52	0.33543	9.52	0.37480
5.53	0.21772	6.53	0.25709	7.53	0.29646	8.53	0.33583	9.53	0.37520
5.54	0.21811	6.54	0.25748	7.54	0.29685	8.54	0.33622	9.54	0.37559
5.55	0.21850	6.55	0.25787	7.55	0.29724	8.55	0.33661	9.55	0.37598
5.56	0.21890	6.56	0.25827	7.56	0.29764	8.56	0.33701	9.56	0.37638
5.57	0.21929	6.57	0.25866	7.57	0.29803	8.57	0.33740	9.57	0.37677
5.58	0.21969	6.58	0.25906	7.58	0.29843	8.58	0.33780	9.58	0.37717
5.59	0.22008	6.59	0.25945	7.59	0.29882	8.59	0.33819	9.59	0.37756
5.60	0.22047	6.60	0.25984	7.60	0.29921	8.60	0.33858	9.60	0.37795
5.61	0.22087	6.61	0.26024	7.61	0.29961	8.61	0.33898	9.61	0.37835
5.62	0.22126	6.62	0.26063	7.62	0.30000	8.62	0.33937	9.62	0.37874
5.63	0.22165	6.63	0.26102	7.63	0.30039	8.63	0.33976	9.63	0.37913
5.64	0.22205	6.64	0.26142	7.64	0.30079	8.64	0.34016	9.64	0.37953
5.65	0.22244	6.65	0.26181	7.65	0.30118	8.65	0.34055	9.65	0.37992
5.66	0.22283	6.66	0.26220	7.66	0.30157	8.66	0.34094	9.66	0.38031
5.67	0.22323	6.67	0.26260	7.67	0.30197	8.67	0.34134	9.67	0.38071
5.68	0.22362	6.68	0.26299	7.68	0.30236	8.68	0.34173	9.68	0.38110
5.69	0.22402	6.69	0.26339	7.69	0.30276	8.69	0.34213	9.69	0.38150
5.70	0.22441	6.70	0.26378	7.70	0.30315	8.70	0.34252	9.70	0.38189
5.71	0.22480	6.71	0.26417	7.71	0.30354	8.71	0.34291	9.71	0.38228
5.72	0.22520	6.72	0.26457	7.72	0.30394	8.72	0.34331	9.72	0.38268
5.73	0.22559	6.73	0.26496	7.73	0.30433	8.73	0.34370	9.73	0.38307
5.74	0.22598	6.74	0.26535	7.74	0.30472	8.74	0.34409	9.74	0.38346
5.75	0.22638	6.75	0.26575	7.75	0.30512	8.75	0.34449	9.75	0.38386
5.76	0.22677	6.76	0.26614	7.76	0.30551	8.76	0.34488	9.76	0.38425
5.77	0.22717	6.77	0.26654	7.77	0.30591	8.77	0.34528	9.77	0.38465
5.78	0.22756	6.78	0.26693	7.78	0.30630	8.78	0.34567	9.78	0.38504
5.79	0.22795	6.79	0.26732	7.79	0.30669	8.79	0.34606	9.79	0.38543
5.80	0.22835	6.80	0.26772	7.80	0.30709	8.80	0.34646	9.80	0.38583
5.81	0.22874	6.81	0.26811	7.81	0.30748	8.81	0.34685	9.81	0.38622
5.82	0.22913	6.82	0.26850	7.82	0.30787	8.82	0.34724	9.82	0.38661
5.83	0.22953	6.83	0.26890	7.83	0.30827	8.83	0.34764	9.83	0.38701
5.84	0.22992	6.84	0.26929	7.84	0.30866	8.84	0.34803	9.84	0.38740
5.85	0.23031	6.85	0.26969	7.85	0.30906	8.85	0.34843	9.85	0.38780
5.86	0.23071	6.86	0.27008	7.86	0.30945	8.86	0.34882	9.86	0.38819
5.87	0.23110	6.87	0.27047	7.87	0.30984	8.87	0.34921	9.87	0.38858
5.88	0.23150	6.88	0.27087	7.88	0.31024	8.88	0.34961	9.88	0.38898
5.89	0.23189	6.89	0.27126	7.89	0.31063	8.89	0.35000	9.89	0.38937
5.90	0.23228	6.90	0.27165	7.90	0.31102	8.90	0.35039	9.90	0.38976
5.91	0.23268	6.91	0.27205	7.91	0.31142	8.91	0.35079	9.91	0.39016
5.92	0.23307	6.92	0.27244	7.92	0.31181	8.92	0.35118	9.92	0.39055
5.93	0.23346	6.93	0.27283	7.93	0.31220	8.93	0.35157	9.93	0.39094
5.94	0.23386	6.94	0.27323	7.94	0.31260	8.94	0.35197	9.94	0.39134
5.95	0.23425	6.95	0.27362	7.95	0.31299	8.95	0.35236	9.95	0.39173
5.96	0.23465	6.96	0.27402	7.96	0.31339	8.96	0.35276	9.96	0.39213
5.97	0.23504	6.97	0.27441	7.97	0.31378	8.97	0.35315	9.97	0.39252
5.98	0.23543	6.98	0.27480	7.98	0.31417	8.98	0.35354	9.98	0.39291
5.99	0.23583	6.99	0.27520	7.99	0.31457	8.99	0.35394	9.99	0.39331

Continued - MILLIMETERS TO INCHES 1 MILLIMETER = 0.039 370 0787 INCH

MILLI-METERS	INCHES	MILLI-METERS	INCHES	MILLI-METERS	INCHES	MILLI-METERS	INCHES	MILLI-METERS	INCHES
10.00	0.39370	11.00	0.43307	12.00	0.47244	13.00	0.51181	14.00	0.55118
10.01	0.39409	11.01	0.43346	12.01	0.47283	13.01	0.51220	14.01	0.55157
10.02	0.39449	11.02	0.43386	12.02	0.47323	13.02	0.51260	14.02	0.55197
10.03	0.39488	11.03	0.43425	12.03	0.47362	13.03	0.51299	14.03	0.55236
10.04	0.39528	11.04	0.43465	12.04	0.47402	13.04	0.51339	14.04	0.55276
10.05	0.39567	11.05	0.43504	12.05	0.47441	13.05	0.51378	14.05	0.55315
10.06	0.39606	11.06	0.43543	12.06	0.47480	13.06	0.51417	14.06	0.55354
10.07	0.39646	11.07	0.43583	12.07	0.47520	13.07	0.51457	14.07	0.55394
10.08	0.39685	11.08	0.43622	12.08	0.47559	13.08	0.51496	14.08	0.55433
10.09	0.39724	11.09	0.43661	12.09	0.47598	13.09	0.51535	14.09	0.55472
10.10	0.39764	11.10	0.43701	12.10	0.47638	13.10	0.51575	14.10	0.55512
10.11	0.39803	11.11	0.43740	12.11	0.47677	13.11	0.51614	14.11	0.55551
10.12	0.39843	11.12	0.43780	12.12	0.47717	13.12	0.51654	14.12	0.55591
10.13	0.39882	11.13	0.43819	12.13	0.47756	13.13	0.51693	14.13	0.55630
10.14	0.39921	11.14	0.43858	12.14	0.47795	13.14	0.51732	14.14	0.55669
10.15	0.39961	11.15	0.43898	12.15	0.47835	13.15	0.51772	14.15	0.55709
10.16	0.40000	11.16	0.43937	12.16	0.47874	13.16	0.51811	14.16	0.55748
10.17	0.40039	11.17	0.43976	12.17	0.47913	13.17	0.51850	14.17	0.55787
10.18	0.40079	11.18	0.44016	12.18	0.47953	13.18	0.51890	14.18	0.55827
10.19	0.40118	11.19	0.44055	12.19	0.47992	13.19	0.51929	14.19	0.55866
10.20	0.40157	11.20	0.44094	12.20	0.48031	13.20	0.51969	14.20	0.55906
10.21	0.40197	11.21	0.44134	12.21	0.48071	13.21	0.52008	14.21	0.55945
10.22	0.40236	11.22	0.44173	12.22	0.48110	13.22	0.52047	14.22	0.55984
10.23	0.40276	11.23	0.44213	12.23	0.48150	13.23	0.52087	14.23	0.56024
10.24	0.40315	11.24	0.44252	12.24	0.48189	13.24	0.52126	14.24	0.56063
10.25	0.40354	11.25	0.44291	12.25	0.48228	13.25	0.52165	14.25	0.56102
10.26	0.40394	11.26	0.44331	12.26	0.48268	13.26	0.52205	14.26	0.56142
10.27	0.40433	11.27	0.44370	12.27	0.48307	13.27	0.52244	14.27	0.56181
10.28	0.40472	11.28	0.44409	12.28	0.48346	13.28	0.52283	14.28	0.56220
10.29	0.40512	11.29	0.44449	12.29	0.48386	13.29	0.52323	14.29	0.56260
10.30	0.40551	11.30	0.44488	12.30	0.48425	13.30	0.52362	14.30	0.56299
10.31	0.40591	11.31	0.44528	12.31	0.48465	13.31	0.52402	14.31	0.56339
10.32	0.40630	11.32	0.44567	12.32	0.48504	13.32	0.52441	14.32	0.56378
10.33	0.40669	11.33	0.44606	12.33	0.48543	13.33	0.52480	14.33	0.56417
10.34	0.40709	11.34	0.44646	12.34	0.48583	13.34	0.52520	14.34	0.56457
10.35	0.40748	11.35	0.44685	12.35	0.48622	13.35	0.52559	14.35	0.56496
10.36	0.40787	11.36	0.44724	12.36	0.48661	13.36	0.52598	14.36	0.56535
10.37	0.40827	11.37	0.44764	12.37	0.48701	13.37	0.52638	14.37	0.56575
10.38	0.40866	11.38	0.44803	12.38	0.48740	13.38	0.52677	14.38	0.56614
10.39	0.40906	11.39	0.44843	12.39	0.48780	13.39	0.52717	14.39	0.56654
10.40	0.40945	11.40	0.44882	12.40	0.48819	13.40	0.52756	14.40	0.56693
10.41	0.40984	11.41	0.44921	12.41	0.48858	13.41	0.52795	14.41	0.56732
10.42	0.41024	11.42	0.44961	12.42	0.48898	13.42	0.52835	14.42	0.56772
10.43	0.41063	11.43	0.45000	12.43	0.48937	13.43	0.52874	14.43	0.56811
10.44	0.41102	11.44	0.45039	12.44	0.48976	13.44	0.52913	14.44	0.56850
10.45	0.41142	11.45	0.45079	12.45	0.49016	13.45	0.52953	14.45	0.56890
10.46	0.41181	11.46	0.45118	12.46	0.49055	13.46	0.52992	14.46	0.56929
10.47	0.41220	11.47	0.45157	12.47	0.49094	13.47	0.53031	14.47	0.56969
10.48	0.41260	11.48	0.45197	12.48	0.49134	13.48	0.53071	14.48	0.57008
10.49	0.41299	11.49	0.45236	12.49	0.49173	13.49	0.53110	14.49	0.57047

Continued - MILLIMETERS TO INCHES 1 MILLIMETER = 0.039 370 0787 INCH

MILLI-METERS	INCHES	MILLI-METERS	INCHES	MILLI-METERS	INCHES	MILLI-METERS	INCHES	MILLI-METERS	INCHES
10.50	0.41339	11.50	0.45276	12.50	0.49213	13.50	0.53150	14.50	0.57087
10.51	0.41378	11.51	0.45315	12.51	0.49252	13.51	0.53189	14.51	0.57126
10.52	0.41417	11.52	0.45354	12.52	0.49291	13.52	0.53228	14.52	0.57165
10.53	0.41457	11.53	0.45394	12.53	0.49331	13.53	0.53268	14.53	0.57205
10.54	0.41496	11.54	0.45433	12.54	0.49370	13.54	0.53307	14.54	0.57244
10.55	0.41535	11.55	0.45472	12.55	0.49409	13.55	0.53346	14.55	0.57283
10.56	0.41575	11.56	0.45512	12.56	0.49449	13.56	0.53386	14.56	0.57323
10.57	0.41614	11.57	0.45551	12.57	0.49488	13.57	0.53425	14.57	0.57362
10.58	0.41654	11.58	0.45591	12.58	0.49528	13.58	0.53465	14.58	0.57402
10.59	0.41693	11.59	0.45630	12.59	0.49567	13.59	0.53504	14.59	0.57441
10.60	0.41732	11.60	0.45669	12.60	0.49606	13.60	0.53543	14.60	0.57480
10.61	0.41772	11.61	0.45709	12.61	0.49646	13.61	0.53583	14.61	0.57520
10.62	0.41811	11.62	0.45748	12.62	0.49685	13.62	0.53622	14.62	0.57559
10.63	0.41850	11.63	0.45787	12.63	0.49724	13.63	0.53661	14.63	0.57598
10.64	0.41890	11.64	0.45827	12.64	0.49764	13.64	0.53701	14.64	0.57638
10.65	0.41929	11.65	0.45866	12.65	0.49803	13.65	0.53740	14.65	0.57677
10.66	0.41969	11.66	0.45906	12.66	0.49843	13.66	0.53780	14.66	0.57717
10.67	0.42008	11.67	0.45945	12.67	0.49882	13.67	0.53819	14.67	0.57756
10.68	0.42047	11.68	0.45984	12.68	0.49921	13.68	0.53858	14.68	0.57795
10.69	0.42087	11.69	0.46024	12.69	0.49961	13.69	0.53898	14.69	0.57835
10.70	0.42126	11.70	0.46063	12.70	0.50000	13.70	0.53937	14.70	0.57874
10.71	0.42165	11.71	0.46102	12.71	0.50039	13.71	0.53976	14.71	0.57913
10.72	0.42205	11.72	0.46142	12.72	0.50079	13.72	0.54016	14.72	0.57953
10.73	0.42244	11.73	0.46181	12.73	0.50118	13.73	0.54055	14.73	0.57992
10.74	0.42283	11.74	0.46220	12.74	0.50157	13.74	0.54094	14.74	0.58031
10.75	0.42323	11.75	0.46260	12.75	0.50197	13.75	0.54134	14.75	0.58071
10.76	0.42362	11.76	0.46299	12.76	0.50236	13.76	0.54173	14.76	0.58110
10.77	0.42402	11.77	0.46339	12.77	0.50276	13.77	0.54213	14.77	0.58150
10.78	0.42441	11.78	0.46378	12.78	0.50315	13.78	0.54252	14.78	0.58189
10.79	0.42480	11.79	0.46417	12.79	0.50354	13.79	0.54291	14.79	0.58228
10.80	0.42520	11.80	0.46457	12.80	0.50394	13.80	0.54331	14.80	0.58268
10.81	0.42559	11.81	0.46496	12.81	0.50433	13.81	0.54370	14.81	0.58307
10.82	0.42598	11.82	0.46535	12.82	0.50472	13.82	0.54409	14.82	0.58346
10.83	0.42638	11.83	0.46575	12.83	0.50512	13.83	0.54449	14.83	0.58386
10.84	0.42677	11.84	0.46614	12.84	0.50551	13.84	0.54488	14.84	0.58425
10.85	0.42717	11.85	0.46654	12.85	0.50591	13.85	0.54528	14.85	0.58465
10.86	0.42756	11.86	0.46693	12.86	0.50630	13.86	0.54567	14.86	0.58504
10.87	0.42795	11.87	0.46732	12.87	0.50669	13.87	0.54606	14.87	0.58543
10.88	0.42835	11.88	0.46772	12.88	0.50709	13.88	0.54646	14.88	0.58583
10.89	0.42874	11.89	0.46811	12.89	0.50748	13.89	0.54685	14.89	0.58622
10.90	0.42913	11.90	0.46850	12.90	0.50787	13.90	0.54724	14.90	0.58661
10.91	0.42953	11.91	0.46890	12.91	0.50827	13.91	0.54764	14.91	0.58701
10.92	0.42992	11.92	0.46929	12.92	0.50866	13.92	0.54803	14.92	0.58740
10.93	0.43031	11.93	0.46969	12.93	0.50906	13.93	0.54843	14.93	0.58780
10.94	0.43071	11.94	0.47008	12.94	0.50945	13.94	0.54882	14.94	0.58819
10.95	0.43110	11.95	0.47047	12.95	0.50984	13.95	0.54921	14.95	0.58858
10.96	0.43150	11.96	0.47087	12.96	0.51024	13.96	0.54961	14.96	0.58898
10.97	0.43189	11.97	0.47126	12.97	0.51063	13.97	0.55000	14.97	0.58937
10.98	0.43228	11.98	0.47165	12.98	0.51102	13.98	0.55039	14.98	0.58976
10.99	0.43268	11.99	0.47205	12.99	0.51142	13.99	0.55079	14.99	0.59016

Continued - MILLIMETERS TO INCHES 1 MILLIMETER = 0.039 370 0787 INCH

MILLI-METERS	INCHES	MILLI-METERS	INCHES	MILLI-METERS	INCHES	MILLI-METERS	INCHES	MILLI-METERS	INCHES
15.00	0.59055	16.00	0.62992	17.00	0.66929	18.00	0.70866	19.00	0.74803
15.01	0.59094	16.01	0.63031	17.01	0.66969	18.01	0.70906	19.01	0.74843
15.02	0.59134	16.02	0.63071	17.02	0.67008	18.02	0.70945	19.02	0.74882
15.03	0.59173	16.03	0.63110	17.03	0.67047	18.03	0.70984	19.03	0.74921
15.04	0.59213	16.04	0.63150	17.04	0.67087	18.04	0.71024	19.04	0.74961
15.05	0.59252	16.05	0.63189	17.05	0.67126	18.05	0.71063	19.05	0.75000
15.06	0.59291	16.06	0.63228	17.06	0.67165	18.06	0.71102	19.06	0.75039
15.07	0.59331	16.07	0.63268	17.07	0.67205	18.07	0.71142	19.07	0.75079
15.08	0.59370	16.08	0.63307	17.08	0.67244	18.08	0.71181	19.08	0.75118
15.09	0.59409	16.09	0.63346	17.09	0.67283	18.09	0.71220	19.09	0.75157
15.10	0.59449	16.10	0.63386	17.10	0.67323	18.10	0.71260	19.10	0.75197
15.11	0.59488	16.11	0.63425	17.11	0.67362	18.11	0.71299	19.11	0.75236
15.12	0.59528	16.12	0.63465	17.12	0.67402	18.12	0.71339	19.12	0.75276
15.13	0.59567	16.13	0.63504	17.13	0.67441	18.13	0.71378	19.13	0.75315
15.14	0.59606	16.14	0.63543	17.14	0.67480	18.14	0.71417	19.14	0.75354
15.15	0.59646	16.15	0.63583	17.15	0.67520	18.15	0.71457	19.15	0.75394
15.16	0.59685	16.16	0.63622	17.16	0.67559	18.16	0.71496	19.16	0.75433
15.17	0.59724	16.17	0.63661	17.17	0.67598	18.17	0.71535	19.17	0.75472
15.18	0.59764	16.18	0.63701	17.18	0.67638	18.18	0.71575	19.18	0.75512
15.19	0.59803	16.19	0.63740	17.19	0.67677	18.19	0.71614	19.19	0.75551
15.20	0.59843	16.20	0.63780	17.20	0.67717	18.20	0.71654	19.20	0.75591
15.21	0.59882	16.21	0.63819	17.21	0.67756	18.21	0.71693	19.21	0.75630
15.22	0.59921	16.22	0.63858	17.22	0.67795	18.22	0.71732	19.22	0.75669
15.23	0.59961	16.23	0.63898	17.23	0.67835	18.23	0.71772	19.23	0.75709
15.24	0.60000	16.24	0.63937	17.24	0.67874	18.24	0.71811	19.24	0.75748
15.25	0.60039	16.25	0.63976	17.25	0.67913	18.25	0.71850	19.25	0.75787
15.26	0.60079	16.26	0.64016	17.26	0.67953	18.26	0.71890	19.26	0.75827
15.27	0.60118	16.27	0.64055	17.27	0.67992	18.27	0.71929	19.27	0.75866
15.28	0.60157	16.28	0.64094	17.28	0.68031	18.28	0.71969	19.28	0.75906
15.29	0.60197	16.29	0.64134	17.29	0.68071	18.29	0.72008	19.29	0.75945
15.30	0.60236	16.30	0.64173	17.30	0.68110	18.30	0.72047	19.30	0.75984
15.31	0.60276	16.31	0.64213	17.31	0.68150	18.31	0.72087	19.31	0.76024
15.32	0.60315	16.32	0.64252	17.32	0.68189	18.32	0.72126	19.32	0.76063
15.33	0.60354	16.33	0.64291	17.33	0.68228	18.33	0.72165	19.33	0.76102
15.34	0.60394	16.34	0.64331	17.34	0.68268	18.34	0.72205	19.34	0.76142
15.35	0.60433	16.35	0.64370	17.35	0.68307	18.35	0.72244	19.35	0.76181
15.36	0.60472	16.36	0.64409	17.36	0.68346	18.36	0.72283	19.36	0.76220
15.37	0.60512	16.37	0.64449	17.37	0.68386	18.37	0.72323	19.37	0.76260
15.38	0.60551	16.38	0.64488	17.38	0.68425	18.38	0.72362	19.38	0.76299
15.39	0.60591	16.39	0.64528	17.39	0.68465	18.39	0.72402	19.39	0.76339
15.40	0.60630	16.40	0.64567	17.40	0.68504	18.40	0.72441	19.40	0.76378
15.41	0.60669	16.41	0.64606	17.41	0.68543	18.41	0.72480	19.41	0.76417
15.42	0.60709	16.42	0.64646	17.42	0.68583	18.42	0.72520	19.42	0.76457
15.43	0.60748	16.43	0.64685	17.43	0.68622	18.43	0.72559	19.43	0.76496
15.44	0.60787	16.44	0.64724	17.44	0.68661	18.44	0.72598	19.44	0.76535
15.45	0.60827	16.45	0.64764	17.45	0.68701	18.45	0.72638	19.45	0.76575
15.46	0.60866	16.46	0.64803	17.46	0.68740	18.46	0.72677	19.46	0.76614
15.47	0.60906	16.47	0.64843	17.47	0.68780	18.47	0.72717	19.47	0.76654
15.48	0.60945	16.48	0.64882	17.48	0.68819	18.48	0.72756	19.48	0.76693
15.49	0.60984	16.49	0.64921	17.49	0.68858	18.49	0.72795	19.49	0.76732

Continued - MILLIMETERS TO INCHES 1 MILLIMETER = 0.039 370 0787 INCH

MILLI-METERS	INCHES	MILLI-METERS	INCHES	MILLI-METERS	INCHES	MILLI-METERS	INCHES	MILLI-METERS	INCHES
15.50	0.61024	16.50	0.64961	17.50	0.68898	18.50	0.72835	19.50	0.76772
15.51	0.61063	16.51	0.65000	17.51	0.68937	18.51	0.72874	19.51	0.76811
15.52	0.61102	16.52	0.65039	17.52	0.68976	18.52	0.72913	19.52	0.76850
15.53	0.61142	16.53	0.65079	17.53	0.69016	18.53	0.72953	19.53	0.76890
15.54	0.61181	16.54	0.65118	17.54	0.69055	18.54	0.72992	19.54	0.76929
15.55	0.61220	16.55	0.65157	17.55	0.69094	18.55	0.73031	19.55	0.76969
15.56	0.61260	16.56	0.65197	17.56	0.69134	18.56	0.73071	19.56	0.77008
15.57	0.61299	16.57	0.65236	17.57	0.69173	18.57	0.73110	19.57	0.77047
15.58	0.61339	16.58	0.65276	17.58	0.69213	18.58	0.73150	19.58	0.77087
15.59	0.61378	16.59	0.65315	17.59	0.69252	18.59	0.73189	19.59	0.77126
15.60	0.61417	16.60	0.65354	17.60	0.69291	18.60	0.73228	19.60	0.77165
15.61	0.61457	16.61	0.65394	17.61	0.69331	18.61	0.73268	19.61	0.77205
15.62	0.61496	16.62	0.65433	17.62	0.69370	18.62	0.73307	19.62	0.77244
15.63	0.61535	16.63	0.65472	17.63	0.69409	18.63	0.73346	19.63	0.77283
15.64	0.61575	16.64	0.65512	17.64	0.69449	18.64	0.73386	19.64	0.77323
15.65	0.61614	16.65	0.65551	17.65	0.69488	18.65	0.73425	19.65	0.77362
15.66	0.61654	16.66	0.65591	17.66	0.69528	18.66	0.73465	19.66	0.77402
15.67	0.61693	16.67	0.65630	17.67	0.69567	18.67	0.73504	19.67	0.77441
15.68	0.61732	16.68	0.65669	17.68	0.69606	18.68	0.73543	19.68	0.77480
15.69	0.61772	16.69	0.65709	17.69	0.69646	18.69	0.73583	19.69	0.77520
15.70	0.61811	16.70	0.65748	17.70	0.69685	18.70	0.73622	19.70	0.77559
15.71	0.61850	16.71	0.65787	17.71	0.69724	18.71	0.73661	19.71	0.77598
15.72	0.61890	16.72	0.65827	17.72	0.69764	18.72	0.73701	19.72	0.77638
15.73	0.61929	16.73	0.65866	17.73	0.69803	18.73	0.73740	19.73	0.77677
15.74	0.61969	16.74	0.65906	17.74	0.69843	18.74	0.73780	19.74	0.77717
15.75	0.62008	16.75	0.65945	17.75	0.69882	18.75	0.73819	19.75	0.77756
15.76	0.62047	16.76	0.65984	17.76	0.69921	18.76	0.73858	19.76	0.77795
15.77	0.62087	16.77	0.66024	17.77	0.69961	18.77	0.73898	19.77	0.77835
15.78	0.62126	16.78	0.66063	17.78	0.70000	18.78	0.73937	19.78	0.77874
15.79	0.62165	16.79	0.66102	17.79	0.70039	18.79	0.73976	19.79	0.77913
15.80	0.62205	16.80	0.66142	17.80	0.70079	18.80	0.74016	19.80	0.77953
15.81	0.62244	16.81	0.66181	17.81	0.70118	18.81	0.74055	19.81	0.77992
15.82	0.62283	16.82	0.66220	17.82	0.70157	18.82	0.74094	19.82	0.78031
15.83	0.62323	16.83	0.66260	17.83	0.70197	18.83	0.74134	19.83	0.78071
15.84	0.62362	16.84	0.66299	17.84	0.70236	18.84	0.74173	19.84	0.78110
15.85	0.62402	16.85	0.66339	17.85	0.70276	18.85	0.74213	19.85	0.78150
15.86	0.62441	16.86	0.66378	17.86	0.70315	18.86	0.74252	19.86	0.78189
15.87	0.62480	16.87	0.66417	17.87	0.70354	18.87	0.74291	19.87	0.78228
15.88	0.62520	16.88	0.66457	17.88	0.70394	18.88	0.74331	19.88	0.78268
15.89	0.62559	16.89	0.66496	17.89	0.70433	18.89	0.74370	19.89	0.78307
15.90	0.62598	16.90	0.66535	17.90	0.70472	18.90	0.74409	19.90	0.78346
15.91	0.62638	16.91	0.66575	17.91	0.70512	18.91	0.74449	19.91	0.78386
15.92	0.62677	16.92	0.66614	17.92	0.70551	18.92	0.74488	19.92	0.78425
15.93	0.62717	16.93	0.66654	17.93	0.70591	18.93	0.74528	19.93	0.78465
15.94	0.62756	16.94	0.66693	17.94	0.70630	18.94	0.74567	19.94	0.78504
15.95	0.62795	16.95	0.66732	17.95	0.70669	18.95	0.74606	19.95	0.78543
15.96	0.62835	16.96	0.66772	17.96	0.70709	18.96	0.74646	19.96	0.78583
15.97	0.62874	16.97	0.66811	17.97	0.70748	18.97	0.74685	19.97	0.78622
15.98	0.62913	16.98	0.66850	17.98	0.70787	18.98	0.74724	19.98	0.78661
15.99	0.62953	16.99	0.66890	17.99	0.70827	18.99	0.74764	19.99	0.78701

Continued - MILLIMETERS TO INCHES 1 MILLIMETER = 0.039 370 0787 INCH

MILLI-METERS	INCHES	MILLI-METERS	INCHES	MILLI-METERS	INCHES	MILLI-METERS	INCHES	MILLI-METERS	INCHES
20.00	0.78740	21.00	0.82677	22.00	0.86614	23.00	0.90551	24.00	0.94488
20.01	0.78780	21.01	0.82717	22.01	0.86654	23.01	0.90591	24.01	0.94528
20.02	0.78819	21.02	0.82756	22.02	0.86693	23.02	0.90630	24.02	0.94567
20.03	0.78858	21.03	0.82795	22.03	0.86732	23.03	0.90669	24.03	0.94606
20.04	0.78898	21.04	0.82835	22.04	0.86772	23.04	0.90709	24.04	0.94646
20.05	0.78937	21.05	0.82874	22.05	0.86811	23.05	0.90748	24.05	0.94685
20.06	0.78976	21.06	0.82913	22.06	0.86850	23.06	0.90787	24.06	0.94724
20.07	0.79016	21.07	0.82953	22.07	0.86890	23.07	0.90827	24.07	0.94764
20.08	0.79055	21.08	0.82992	22.08	0.86929	23.08	0.90866	24.08	0.94803
20.09	0.79094	21.09	0.83031	22.09	0.86969	23.09	0.90906	24.09	0.94843
20.10	0.79134	21.10	0.83071	22.10	0.87008	23.10	0.90945	24.10	0.94882
20.11	0.79173	21.11	0.83110	22.11	0.87047	23.11	0.90984	24.11	0.94921
20.12	0.79213	21.12	0.83150	22.12	0.87087	23.12	0.91024	24.12	0.94961
20.13	0.79252	21.13	0.83189	22.13	0.87126	23.13	0.91063	24.13	0.95000
20.14	0.79291	21.14	0.83228	22.14	0.87165	23.14	0.91102	24.14	0.95039
20.15	0.79331	21.15	0.83268	22.15	0.87205	23.15	0.91142	24.15	0.95079
20.16	0.79370	21.16	0.83307	22.16	0.87244	23.16	0.91181	24.16	0.95118
20.17	0.79409	21.17	0.83346	22.17	0.87283	23.17	0.91220	24.17	0.95157
20.18	0.79449	21.18	0.83386	22.18	0.87323	23.18	0.91260	24.18	0.95197
20.19	0.79488	21.19	0.83425	22.19	0.87362	23.19	0.91299	24.19	0.95236
20.20	0.79528	21.20	0.83465	22.20	0.87402	23.20	0.91339	24.20	0.95276
20.21	0.79567	21.21	0.83504	22.21	0.87441	23.21	0.91378	24.21	0.95315
20.22	0.79606	21.22	0.83543	22.22	0.87480	23.22	0.91417	24.22	0.95354
20.23	0.79646	21.23	0.83583	22.23	0.87520	23.23	0.91457	24.23	0.95394
20.24	0.79685	21.24	0.83622	22.24	0.87559	23.24	0.91496	24.24	0.95433
20.25	0.79724	21.25	0.83661	22.25	0.87598	23.25	0.91535	24.25	0.95472
20.26	0.79764	21.26	0.83701	22.26	0.87638	23.26	0.91575	24.26	0.95512
20.27	0.79803	21.27	0.83740	22.27	0.87677	23.27	0.91614	24.27	0.95551
20.28	0.79843	21.28	0.83780	22.28	0.87717	23.28	0.91654	24.28	0.95591
20.29	0.79882	21.29	0.83819	22.29	0.87756	23.29	0.91693	24.29	0.95630
20.30	0.79921	21.30	0.83858	22.30	0.87795	23.30	0.91732	24.30	0.95669
20.31	0.79961	21.31	0.83898	22.31	0.87835	23.31	0.91772	24.31	0.95709
20.32	0.80000	21.32	0.83937	22.32	0.87874	23.32	0.91811	24.32	0.95748
20.33	0.80039	21.33	0.83976	22.33	0.87913	23.33	0.91850	24.33	0.95787
20.34	0.80079	21.34	0.84016	22.34	0.87953	23.34	0.91890	24.34	0.95827
20.35	0.80118	21.35	0.84055	22.35	0.87992	23.35	0.91929	24.35	0.95866
20.36	0.80157	21.36	0.84094	22.36	0.88031	23.36	0.91969	24.36	0.95906
20.37	0.80197	21.37	0.84134	22.37	0.88071	23.37	0.92008	24.37	0.95945
20.38	0.80236	21.38	0.84173	22.38	0.88110	23.38	0.92047	24.38	0.95984
20.39	0.80276	21.39	0.84213	22.39	0.88150	23.39	0.92087	24.39	0.96024
20.40	0.80315	21.40	0.84252	22.40	0.88189	23.40	0.92126	24.40	0.96063
20.41	0.80354	21.41	0.84291	22.41	0.88228	23.41	0.92165	24.41	0.96102
20.42	0.80394	21.42	0.84331	22.42	0.88268	23.42	0.92205	24.42	0.96142
20.43	0.80433	21.43	0.84370	22.43	0.88307	23.43	0.92244	24.43	0.96181
20.44	0.80472	21.44	0.84409	22.44	0.88346	23.44	0.92283	24.44	0.96220
20.45	0.80512	21.45	0.84449	22.45	0.88386	23.45	0.92323	24.45	0.96260
20.46	0.80551	21.46	0.84488	22.46	0.88425	23.46	0.92362	24.46	0.96299
20.47	0.80591	21.47	0.84528	22.47	0.88465	23.47	0.92402	24.47	0.96339
20.48	0.80630	21.48	0.84567	22.48	0.88504	23.48	0.92441	24.48	0.96378
20.49	0.80669	21.49	0.84606	22.49	0.88543	23.49	0.92480	24.49	0.96417

Continued - MILLIMETERS TO INCHES 1 MILLIMETER = 0.039 370 0787 INCH

MILLI-METERS	INCHES	MILLI-METERS	INCHES	MILLI-METERS	INCHES	MILLI-METERS	INCHES	MILLI-METERS	INCHES
20.50	0.80709	21.50	0.84646	22.50	0.88583	23.50	0.92520	24.50	0.96457
20.51	0.80748	21.51	0.84685	22.51	0.88622	23.51	0.92559	24.51	0.96496
20.52	0.80787	21.52	0.84724	22.52	0.88661	23.52	0.92598	24.52	0.96535
20.53	0.80827	21.53	0.84764	22.53	0.88701	23.53	0.92638	24.53	0.96575
20.54	0.80866	21.54	0.84803	22.54	0.88740	23.54	0.92677	24.54	0.96614
20.55	0.80906	21.55	0.84843	22.55	0.88780	23.55	0.92717	24.55	0.96654
20.56	0.80945	21.56	0.84882	22.56	0.88819	23.56	0.92756	24.56	0.96693
20.57	0.80984	21.57	0.84921	22.57	0.88858	23.57	0.92795	24.57	0.96732
20.58	0.81024	21.58	0.84961	22.58	0.88898	23.58	0.92835	24.58	0.96772
20.59	0.81063	21.59	0.85000	22.59	0.88937	23.59	0.92874	24.59	0.96811
20.60	0.81102	21.60	0.85039	22.60	0.88976	23.60	0.92913	24.60	0.96850
20.61	0.81142	21.61	0.85079	22.61	0.89016	23.61	0.92953	24.61	0.96890
20.62	0.81181	21.62	0.85118	22.62	0.89055	23.62	0.92992	24.62	0.96929
20.63	0.81220	21.63	0.85157	22.63	0.89094	23.63	0.93031	24.63	0.96969
20.64	0.81260	21.64	0.85197	22.64	0.89134	23.64	0.93071	24.64	0.97008
20.65	0.81299	21.65	0.85236	22.65	0.89173	23.65	0.93110	24.65	0.97047
20.66	0.81339	21.66	0.85276	22.66	0.89213	23.66	0.93150	24.66	0.97087
20.67	0.81378	21.67	0.85315	22.67	0.89252	23.67	0.93189	24.67	0.97126
20.68	0.81417	21.68	0.85354	22.68	0.89291	23.68	0.93228	24.68	0.97165
20.69	0.81457	21.69	0.85394	22.69	0.89331	23.69	0.93268	24.69	0.97205
20.70	0.81496	21.70	0.85433	22.70	0.89370	23.70	0.93307	24.70	0.97244
20.71	0.81535	21.71	0.85472	22.71	0.89409	23.71	0.93346	24.71	0.97283
20.72	0.81575	21.72	0.85512	22.72	0.89449	23.72	0.93386	24.72	0.97323
20.73	0.81614	21.73	0.85551	22.73	0.89488	23.73	0.93425	24.73	0.97362
20.74	0.81654	21.74	0.85591	22.74	0.89528	23.74	0.93465	24.74	0.97402
20.75	0.81693	21.75	0.85630	22.75	0.89567	23.75	0.93504	24.75	0.97441
20.76	0.81732	21.76	0.85669	22.76	0.89606	23.76	0.93543	24.76	0.97480
20.77	0.81772	21.77	0.85709	22.77	0.89646	23.77	0.93583	24.77	0.97520
20.78	0.81811	21.78	0.85748	22.78	0.89685	23.78	0.93622	24.78	0.97559
20.79	0.81850	21.79	0.85787	22.79	0.89724	23.79	0.93661	24.79	0.97598
20.80	0.81890	21.80	0.85827	22.80	0.89764	23.80	0.93701	24.80	0.97638
20.81	0.81929	21.81	0.85866	22.81	0.89803	23.81	0.93740	24.81	0.97677
20.82	0.81969	21.82	0.85906	22.82	0.89843	23.82	0.93780	24.82	0.97717
20.83	0.82008	21.83	0.85945	22.83	0.89882	23.83	0.93819	24.83	0.97756
20.84	0.82047	21.84	0.85984	22.84	0.89921	23.84	0.93858	24.84	0.97795
20.85	0.82087	21.85	0.86024	22.85	0.89961	23.85	0.93898	24.85	0.97835
20.86	0.82126	21.86	0.86063	22.86	0.90000	23.86	0.93937	24.86	0.97874
20.87	0.82165	21.87	0.86102	22.87	0.90039	23.87	0.93976	24.87	0.97913
20.88	0.82205	21.88	0.86142	22.88	0.90079	23.88	0.94016	24.88	0.97953
20.89	0.82244	21.89	0.86181	22.89	0.90118	23.89	0.94055	24.89	0.97992
20.90	0.82283	21.90	0.86220	22.90	0.90157	23.90	0.94094	24.90	0.98031
20.91	0.82323	21.91	0.86260	22.91	0.90197	23.91	0.94134	24.91	0.98071
20.92	0.82362	21.92	0.86299	22.92	0.90236	23.92	0.94173	24.92	0.98110
20.93	0.82402	21.93	0.86339	22.93	0.90276	23.93	0.94213	24.93	0.98150
20.94	0.82441	21.94	0.86378	22.94	0.90315	23.94	0.94252	24.94	0.98189
20.95	0.82480	21.95	0.86417	22.95	0.90354	23.95	0.94291	24.95	0.98228
20.96	0.82520	21.96	0.86457	22.96	0.90394	23.96	0.94331	24.96	0.98268
20.97	0.82559	21.97	0.86496	22.97	0.90433	23.97	0.94370	24.97	0.98307
20.98	0.82598	21.98	0.86535	22.98	0.90472	23.98	0.94409	24.98	0.98346
20.99	0.82638	21.99	0.86575	22.99	0.90512	23.99	0.94449	24.99	0.98386

Continued - MILLIMETERS TO INCHES 1 MILLIMETER = 0.039 370 0787 INCH

MILLI-METERS	INCHES	MILLI-METERS	INCHES	MILLI-METERS	INCHES	MILLI-METERS	INCHES	MILLI-METERS	INCHES
25.00	0.98425	26.00	1.02362	27.00	1.06299	28.00	1.10236	29.00	1.14173
25.01	0.98465	26.01	1.02402	27.01	1.06339	28.01	1.10276	29.01	1.14213
25.02	0.98504	26.02	1.02441	27.02	1.06378	28.02	1.10315	29.02	1.14252
25.03	0.98543	26.03	1.02480	27.03	1.06417	28.03	1.10354	29.03	1.14291
25.04	0.98583	26.04	1.02520	27.04	1.06457	28.04	1.10394	29.04	1.14331
25.05	0.98622	26.05	1.02559	27.05	1.06496	28.05	1.10433	29.05	1.14370
25.06	0.98661	26.06	1.02598	27.06	1.06535	28.06	1.10472	29.06	1.14409
25.07	0.98701	26.07	1.02638	27.07	1.06575	28.07	1.10512	29.07	1.14449
25.08	0.98740	26.08	1.02677	27.08	1.06614	28.08	1.10551	29.08	1.14488
25.09	0.98780	26.09	1.02717	27.09	1.06654	28.09	1.10591	29.09	1.14528
25.10	0.98819	26.10	1.02756	27.10	1.06693	28.10	1.10630	29.10	1.14567
25.11	0.98858	26.11	1.02795	27.11	1.06732	28.11	1.10669	29.11	1.14606
25.12	0.98898	26.12	1.02835	27.12	1.06772	28.12	1.10709	29.12	1.14646
25.13	0.98937	26.13	1.02874	27.13	1.06811	28.13	1.10748	29.13	1.14685
25.14	0.98976	26.14	1.02913	27.14	1.06850	28.14	1.10787	29.14	1.14724
25.15	0.99016	26.15	1.02953	27.15	1.06890	28.15	1.10827	29.15	1.14764
25.16	0.99055	26.16	1.02992	27.16	1.06929	28.16	1.10866	29.16	1.14803
25.17	0.99094	26.17	1.03031	27.17	1.06969	28.17	1.10906	29.17	1.14843
25.18	0.99134	26.18	1.03071	27.18	1.07008	28.18	1.10945	29.18	1.14882
25.19	0.99173	26.19	1.03110	27.19	1.07047	28.19	1.10984	29.19	1.14921
25.20	0.99213	26.20	1.03150	27.20	1.07087	28.20	1.11024	29.20	1.14961
25.21	0.99252	26.21	1.03189	27.21	1.07126	28.21	1.11063	29.21	1.15000
25.22	0.99291	26.22	1.03228	27.22	1.07165	28.22	1.11102	29.22	1.15039
25.23	0.99331	26.23	1.03268	27.23	1.07205	28.23	1.11142	29.23	1.15079
25.24	0.99370	26.24	1.03307	27.24	1.07244	28.24	1.11181	29.24	1.15118
25.25	0.99409	26.25	1.03346	27.25	1.07283	28.25	1.11220	29.25	1.15157
25.26	0.99449	26.26	1.03386	27.26	1.07323	28.26	1.11260	29.26	1.15197
25.27	0.99488	26.27	1.03425	27.27	1.07362	28.27	1.11299	29.27	1.15236
25.28	0.99528	26.28	1.03465	27.28	1.07402	28.28	1.11339	29.28	1.15276
25.29	0.99567	26.29	1.03504	27.29	1.07441	28.29	1.11378	29.29	1.15315
25.30	0.99606	26.30	1.03543	27.30	1.07480	28.30	1.11417	29.30	1.15354
25.31	0.99646	26.31	1.03583	27.31	1.07520	28.31	1.11457	29.31	1.15394
25.32	0.99685	26.32	1.03622	27.32	1.07559	28.32	1.11496	29.32	1.15433
25.33	0.99724	26.33	1.03661	27.33	1.07598	28.33	1.11535	29.33	1.15472
25.34	0.99764	26.34	1.03701	27.34	1.07638	28.34	1.11575	29.34	1.15512
25.35	0.99803	26.35	1.03740	27.35	1.07677	28.35	1.11614	29.35	1.15551
25.36	0.99843	26.36	1.03780	27.36	1.07717	28.36	1.11654	29.36	1.15591
25.37	0.99882	26.37	1.03819	27.37	1.07756	28.37	1.11693	29.37	1.15630
25.38	0.99921	26.38	1.03858	27.38	1.07795	28.38	1.11732	29.38	1.15669
25.39	0.99961	26.39	1.03898	27.39	1.07835	28.39	1.11772	29.39	1.15709
25.40	1.00000	26.40	1.03937	27.40	1.07874	28.40	1.11811	29.40	1.15748
25.41	1.00039	26.41	1.03976	27.41	1.07913	28.41	1.11850	29.41	1.15787
25.42	1.00079	26.42	1.04016	27.42	1.07953	28.42	1.11890	29.42	1.15827
25.43	1.00118	26.43	1.04055	27.43	1.07992	28.43	1.11929	29.43	1.15866
25.44	1.00157	26.44	1.04094	27.44	1.08031	28.44	1.11969	29.44	1.15906
25.45	1.00197	26.45	1.04134	27.45	1.08071	28.45	1.12008	29.45	1.15945
25.46	1.00236	26.46	1.04173	27.46	1.08110	28.46	1.12047	29.46	1.15984
25.47	1.00276	26.47	1.04213	27.47	1.08150	28.47	1.12087	29.47	1.16024
25.48	1.00315	26.48	1.04252	27.48	1.08189	28.48	1.12126	29.48	1.16063
25.49	1.00354	26.49	1.04291	27.49	1.08228	28.49	1.12165	29.49	1.16102

Continued - MILLIMETERS TO INCHES 1 MILLIMETER = 0.039 370 0787 INCH

MILLI-METERS	INCHES	MILLI-METERS	INCHES	MILLI-METERS	INCHES	MILLI-METERS	INCHES	MILLI-METERS	INCHES
25.50	1.00394	26.50	1.04331	27.50	1.08268	28.50	1.12205	29.50	1.16142
25.51	1.00433	26.51	1.04370	27.51	1.08307	28.51	1.12244	29.51	1.16181
25.52	1.00472	26.52	1.04409	27.52	1.08346	28.52	1.12283	29.52	1.16220
25.53	1.00512	26.53	1.04449	27.53	1.08386	28.53	1.12323	29.53	1.16260
25.54	1.00551	26.54	1.04488	27.54	1.08425	28.54	1.12362	29.54	1.16299
25.55	1.00591	26.55	1.04528	27.55	1.08465	28.55	1.12402	29.55	1.16339
25.56	1.00630	26.56	1.04567	27.56	1.08504	28.56	1.12441	29.56	1.16378
25.57	1.00669	26.57	1.04606	27.57	1.08543	28.57	1.12480	29.57	1.16417
25.58	1.00709	26.58	1.04646	27.58	1.08583	28.58	1.12520	29.58	1.16457
25.59	1.00748	26.59	1.04685	27.59	1.08622	28.59	1.12559	29.59	1.16496
25.60	1.00787	26.60	1.04724	27.60	1.08661	28.60	1.12598	29.60	1.16535
25.61	1.00827	26.61	1.04764	27.61	1.08701	28.61	1.12638	29.61	1.16575
25.62	1.00866	26.62	1.04803	27.62	1.08740	28.62	1.12677	29.62	1.16614
25.63	1.00906	26.63	1.04843	27.63	1.08780	28.63	1.12717	29.63	1.16654
25.64	1.00945	26.64	1.04882	27.64	1.08819	28.64	1.12756	29.64	1.16693
25.65	1.00984	26.65	1.04921	27.65	1.08858	28.65	1.12795	29.65	1.16732
25.66	1.01024	26.66	1.04961	27.66	1.08898	28.66	1.12835	29.66	1.16772
25.67	1.01063	26.67	1.05000	27.67	1.08937	28.67	1.12874	29.67	1.16811
25.68	1.01102	26.68	1.05039	27.68	1.08976	28.68	1.12913	29.68	1.16850
25.69	1.01142	26.69	1.05079	27.69	1.09016	28.69	1.12953	29.69	1.16890
25.70	1.01181	26.70	1.05118	27.70	1.09055	28.70	1.12992	29.70	1.16929
25.71	1.01220	26.71	1.05157	27.71	1.09094	28.71	1.13031	29.71	1.16969
25.72	1.01260	26.72	1.05197	27.72	1.09134	28.72	1.13071	29.72	1.17008
25.73	1.01299	26.73	1.05236	27.73	1.09173	28.73	1.13110	29.73	1.17047
25.74	1.01339	26.74	1.05276	27.74	1.09213	28.74	1.13150	29.74	1.17087
25.75	1.01378	26.75	1.05315	27.75	1.09252	28.75	1.13189	29.75	1.17126
25.76	1.01417	26.76	1.05354	27.76	1.09291	28.76	1.13228	29.76	1.17165
25.77	1.01457	26.77	1.05394	27.77	1.09331	28.77	1.13268	29.77	1.17205
25.78	1.01496	26.78	1.05433	27.78	1.09370	28.78	1.13307	29.78	1.17244
25.79	1.01535	26.79	1.05472	27.79	1.09409	28.79	1.13346	29.79	1.17283
25.80	1.01575	26.80	1.05512	27.80	1.09449	28.80	1.13386	29.80	1.17323
25.81	1.01614	26.81	1.05551	27.81	1.09488	28.81	1.13425	29.81	1.17362
25.82	1.01654	26.82	1.05591	27.82	1.09528	28.82	1.13465	29.82	1.17402
25.83	1.01693	26.83	1.05630	27.83	1.09567	28.83	1.13504	29.83	1.17441
25.84	1.01732	26.84	1.05669	27.84	1.09606	28.84	1.13543	29.84	1.17480
25.85	1.01772	26.85	1.05709	27.85	1.09646	28.85	1.13583	29.85	1.17520
25.86	1.01811	26.86	1.05748	27.86	1.09685	28.86	1.13622	29.86	1.17559
25.87	1.01850	26.87	1.05787	27.87	1.09724	28.87	1.13661	29.87	1.17598
25.88	1.01890	26.88	1.05827	27.88	1.09764	28.88	1.13701	29.88	1.17638
25.89	1.01929	26.89	1.05866	27.89	1.09803	28.89	1.13740	29.89	1.17677
25.90	1.01969	26.90	1.05906	27.90	1.09843	28.90	1.13780	29.90	1.17717
25.91	1.02008	26.91	1.05945	27.91	1.09882	28.91	1.13819	29.91	1.17756
25.92	1.02047	26.92	1.05984	27.92	1.09921	28.92	1.13858	29.92	1.17795
25.93	1.02087	26.93	1.06024	27.93	1.09961	28.93	1.13898	29.93	1.17835
25.94	1.02126	26.94	1.06063	27.94	1.10000	28.94	1.13937	29.94	1.17874
25.95	1.02165	26.95	1.06102	27.95	1.10039	28.95	1.13976	29.95	1.17913
25.96	1.02205	26.96	1.06142	27.96	1.10079	28.96	1.14016	29.96	1.17953
25.97	1.02244	26.97	1.06181	27.97	1.10118	28.97	1.14055	29.97	1.17992
25.98	1.02283	26.98	1.06220	27.98	1.10157	28.98	1.14094	29.98	1.18031
25.99	1.02323	26.99	1.06260	27.99	1.10197	28.99	1.14134	29.99	1.18071

Continued - MILLIMETERS TO INCHES 1 MILLIMETER = 0.039 370 0787 INCH

MILLI-METERS	INCHES	MILLI-METERS	INCHES	MILLI-METERS	INCHES	MILLI-METERS	INCHES	MILLI-METERS	INCHES
30.00	1.18110	31.00	1.22047	32.00	1.25984	33.00	1.29921	34.00	1.33858
30.01	1.18150	31.01	1.22087	32.01	1.26024	33.01	1.29961	34.01	1.33898
30.02	1.18189	31.02	1.22126	32.02	1.26063	33.02	1.30000	34.02	1.33937
30.03	1.18228	31.03	1.22165	32.03	1.26102	33.03	1.30039	34.03	1.33976
30.04	1.18268	31.04	1.22205	32.04	1.26142	33.04	1.30079	34.04	1.34016
30.05	1.18307	31.05	1.22244	32.05	1.26181	33.05	1.30118	34.05	1.34055
30.06	1.18346	31.06	1.22283	32.06	1.26220	33.06	1.30157	34.06	1.34094
30.07	1.18386	31.07	1.22323	32.07	1.26260	33.07	1.30197	34.07	1.34134
30.08	1.18425	31.08	1.22362	32.08	1.26299	33.08	1.30236	34.08	1.34173
30.09	1.18465	31.09	1.22402	32.09	1.26339	33.09	1.30276	34.09	1.34213
30.10	1.18504	31.10	1.22441	32.10	1.26378	33.10	1.30315	34.10	1.34252
30.11	1.18543	31.11	1.22480	32.11	1.26417	33.11	1.30354	34.11	1.34291
30.12	1.18583	31.12	1.22520	32.12	1.26457	33.12	1.30394	34.12	1.34331
30.13	1.18622	31.13	1.22559	32.13	1.26496	33.13	1.30433	34.13	1.34370
30.14	1.18661	31.14	1.22598	32.14	1.26535	33.14	1.30472	34.14	1.34409
30.15	1.18701	31.15	1.22638	32.15	1.26575	33.15	1.30512	34.15	1.34449
30.16	1.18740	31.16	1.22677	32.16	1.26614	33.16	1.30551	34.16	1.34488
30.17	1.18780	31.17	1.22717	32.17	1.26654	33.17	1.30591	34.17	1.34528
30.18	1.18819	31.18	1.22756	32.18	1.26693	33.18	1.30630	34.18	1.34567
30.19	1.18858	31.19	1.22795	32.19	1.26732	33.19	1.30669	34.19	1.34606
30.20	1.18898	31.20	1.22835	32.20	1.26772	33.20	1.30709	34.20	1.34646
30.21	1.18937	31.21	1.22874	32.21	1.26811	33.21	1.30748	34.21	1.34685
30.22	1.18976	31.22	1.22913	32.22	1.26850	33.22	1.30787	34.22	1.34724
30.23	1.19016	31.23	1.22953	32.23	1.26890	33.23	1.30827	34.23	1.34764
30.24	1.19055	31.24	1.22992	32.24	1.26929	33.24	1.30866	34.24	1.34803
30.25	1.19094	31.25	1.23031	32.25	1.26969	33.25	1.30906	34.25	1.34843
30.26	1.19134	31.26	1.23071	32.26	1.27008	33.26	1.30945	34.26	1.34882
30.27	1.19173	31.27	1.23110	32.27	1.27047	33.27	1.30984	34.27	1.34921
30.28	1.19213	31.28	1.23150	32.28	1.27087	33.28	1.31024	34.28	1.34961
30.29	1.19252	31.29	1.23189	32.29	1.27126	33.29	1.31063	34.29	1.35000
30.30	1.19291	31.30	1.23228	32.30	1.27165	33.30	1.31102	34.30	1.35039
30.31	1.19331	31.31	1.23268	32.31	1.27205	33.31	1.31142	34.31	1.35079
30.32	1.19370	31.32	1.23307	32.32	1.27244	33.32	1.31181	34.32	1.35118
30.33	1.19409	31.33	1.23346	32.33	1.27283	33.33	1.31220	34.33	1.35157
30.34	1.19449	31.34	1.23386	32.34	1.27323	33.34	1.31260	34.34	1.35197
30.35	1.19488	31.35	1.23425	32.35	1.27362	33.35	1.31299	34.35	1.35236
30.36	1.19528	31.36	1.23465	32.36	1.27402	33.36	1.31339	34.36	1.35276
30.37	1.19567	31.37	1.23504	32.37	1.27441	33.37	1.31378	34.37	1.35315
30.38	1.19606	31.38	1.23543	32.38	1.27480	33.38	1.31417	34.38	1.35354
30.39	1.19646	31.39	1.23583	32.39	1.27520	33.39	1.31457	34.39	1.35394
30.40	1.19685	31.40	1.23622	32.40	1.27559	33.40	1.31496	34.40	1.35433
30.41	1.19724	31.41	1.23661	32.41	1.27598	33.41	1.31535	34.41	1.35472
30.42	1.19764	31.42	1.23701	32.42	1.27638	33.42	1.31575	34.42	1.35512
30.43	1.19803	31.43	1.23740	32.43	1.27677	33.43	1.31614	34.43	1.35551
30.44	1.19843	31.44	1.23780	32.44	1.27717	33.44	1.31654	34.44	1.35591
30.45	1.19882	31.45	1.23819	32.45	1.27756	33.45	1.31693	34.45	1.35630
30.46	1.19921	31.46	1.23858	32.46	1.27795	33.46	1.31732	34.46	1.35669
30.47	1.19961	31.47	1.23898	32.47	1.27835	33.47	1.31772	34.47	1.35709
30.48	1.20000	31.48	1.23937	32.48	1.27874	33.48	1.31811	34.48	1.35748
30.49	1.20039	31.49	1.23976	32.49	1.27913	33.49	1.31850	34.49	1.35787

Continued - MILLIMETERS TO INCHES 1 MILLIMETER = 0.039 370 0787 INCH

MILLI- METERS	INCHES	MILLI- METERS	INCHES	MILLI- METERS	INCHES	MILLI- METERS	INCHES	MILLI- METERS	INCHES
30.50	1.20079	31.50	1.24016	32.50	1.27953	33.50	1.31890	34.50	1.35827
30.51	1.20118	31.51	1.24055	32.51	1.27992	33.51	1.31929	34.51	1.35866
30.52	1.20157	31.52	1.24094	32.52	1.28031	33.52	1.31969	34.52	1.35906
30.53	1.20197	31.53	1.24134	32.53	1.28071	33.53	1.32008	34.53	1.35945
30.54	1.20236	31.54	1.24173	32.54	1.28110	33.54	1.32047	34.54	1.35984
30.55	1.20276	31.55	1.24213	32.55	1.28150	33.55	1.32087	34.55	1.36024
30.56	1.20315	31.56	1.24252	32.56	1.28189	33.56	1.32126	34.56	1.36063
30.57	1.20354	31.57	1.24291	32.57	1.28228	33.57	1.32165	34.57	1.36102
30.58	1.20394	31.58	1.24331	32.58	1.28268	33.58	1.32205	34.58	1.36142
30.59	1.20433	31.59	1.24370	32.59	1.28307	33.59	1.32244	34.59	1.36181
30.60	1.20472	31.60	1.24409	32.60	1.28346	33.60	1.32283	34.60	1.36220
30.61	1.20512	31.61	1.24449	32.61	1.28386	33.61	1.32323	34.61	1.36260
30.62	1.20551	31.62	1.24488	32.62	1.28425	33.62	1.32362	34.62	1.36299
30.63	1.20591	31.63	1.24528	32.63	1.28465	33.63	1.32402	34.63	1.36339
30.64	1.20630	31.64	1.24567	32.64	1.28504	33.64	1.32441	34.64	1.36378
30.65	1.20669	31.65	1.24606	32.65	1.28543	33.65	1.32480	34.65	1.36417
30.66	1.20709	31.66	1.24646	32.66	1.28583	33.66	1.32520	34.66	1.36457
30.67	1.20748	31.67	1.24685	32.67	1.28622	33.67	1.32559	34.67	1.36496
30.68	1.20787	31.68	1.24724	32.68	1.28661	33.68	1.32598	34.68	1.36535
30.69	1.20827	31.69	1.24764	32.69	1.28701	33.69	1.32638	34.69	1.36575
30.70	1.20866	31.70	1.24803	32.70	1.28740	33.70	1.32677	34.70	1.36614
30.71	1.20906	31.71	1.24843	32.71	1.28780	33.71	1.32717	34.71	1.36654
30.72	1.20945	31.72	1.24882	32.72	1.28819	33.72	1.32756	34.72	1.36693
30.73	1.20984	31.73	1.24921	32.73	1.28858	33.73	1.32795	34.73	1.36732
30.74	1.21024	31.74	1.24961	32.74	1.28898	33.74	1.32835	34.74	1.36772
30.75	1.21063	31.75	1.25000	32.75	1.28937	33.75	1.32874	34.75	1.36811
30.76	1.21102	31.76	1.25039	32.76	1.28976	33.76	1.32913	34.76	1.36850
30.77	1.21142	31.77	1.25079	32.77	1.29016	33.77	1.32953	34.77	1.36890
30.78	1.21181	31.78	1.25118	32.78	1.29055	33.78	1.32992	34.78	1.36929
30.79	1.21220	31.79	1.25157	32.79	1.29094	33.79	1.33031	34.79	1.36969
30.80	1.21260	31.80	1.25197	32.80	1.29134	33.80	1.33071	34.80	1.37008
30.81	1.21299	31.81	1.25236	32.81	1.29173	33.81	1.33110	34.81	1.37047
30.82	1.21339	31.82	1.25276	32.82	1.29213	33.82	1.33150	34.82	1.37087
30.83	1.21378	31.83	1.25315	32.83	1.29252	33.83	1.33189	34.83	1.37126
30.84	1.21417	31.84	1.25354	32.84	1.29291	33.84	1.33228	34.84	1.37165
30.85	1.21457	31.85	1.25394	32.85	1.29331	33.85	1.33268	34.85	1.37205
30.86	1.21496	31.86	1.25433	32.86	1.29370	33.86	1.33307	34.86	1.37244
30.87	1.21535	31.87	1.25472	32.87	1.29409	33.87	1.33346	34.87	1.37283
30.88	1.21575	31.88	1.25512	32.88	1.29449	33.88	1.33386	34.88	1.37323
30.89	1.21614	31.89	1.25551	32.89	1.29488	33.89	1.33425	34.89	1.37362
30.90	1.21654	31.90	1.25591	32.90	1.29528	33.90	1.33465	34.90	1.37402
30.91	1.21693	31.91	1.25630	32.91	1.29567	33.91	1.33504	34.91	1.37441
30.92	1.21732	31.92	1.25669	32.92	1.29606	33.92	1.33543	34.92	1.37480
30.93	1.21772	31.93	1.25709	32.93	1.29646	33.93	1.33583	34.93	1.37520
30.94	1.21811	31.94	1.25748	32.94	1.29685	33.94	1.33622	34.94	1.37559
30.95	1.21850	31.95	1.25787	32.95	1.29724	33.95	1.33661	34.95	1.37598
30.96	1.21890	31.96	1.25827	32.96	1.29764	33.96	1.33701	34.96	1.37638
30.97	1.21929	31.97	1.25866	32.97	1.29803	33.97	1.33740	34.97	1.37677
30.98	1.21969	31.98	1.25906	32.98	1.29843	33.98	1.33780	34.98	1.37717
30.99	1.22008	31.99	1.25945	32.99	1.29882	33.99	1.33819	34.99	1.37756

Continued - MILLIMETERS TO INCHES 1 MILLIMETER = 0.039 370 0787 INCH

MILLI-METERS	INCHES	MILLI-METERS	INCHES	MILLI-METERS	INCHES	MILLI-METERS	INCHES	MILLI-METERS	INCHES
35.00	1.37795	36.00	1.41732	37.00	1.45669	38.00	1.49606	39.00	1.53543
35.01	1.37835	36.01	1.41772	37.01	1.45709	38.01	1.49646	39.01	1.53583
35.02	1.37874	36.02	1.41811	37.02	1.45748	38.02	1.49685	39.02	1.53622
35.03	1.37913	36.03	1.41850	37.03	1.45787	38.03	1.49724	39.03	1.53661
35.04	1.37953	36.04	1.41890	37.04	1.45827	38.04	1.49764	39.04	1.53701
35.05	1.37992	36.05	1.41929	37.05	1.45866	38.05	1.49803	39.05	1.53740
35.06	1.38031	36.06	1.41969	37.06	1.45906	38.06	1.49843	39.06	1.53780
35.07	1.38071	36.07	1.42008	37.07	1.45945	38.07	1.49882	39.07	1.53819
35.08	1.38110	36.08	1.42047	37.08	1.45984	38.08	1.49921	39.08	1.53858
35.09	1.38150	36.09	1.42087	37.09	1.46024	38.09	1.49961	39.09	1.53898
35.10	1.38189	36.10	1.42126	37.10	1.46063	38.10	1.50000	39.10	1.53937
35.11	1.38228	36.11	1.42165	37.11	1.46102	38.11	1.50039	39.11	1.53976
35.12	1.38268	36.12	1.42205	37.12	1.46142	38.12	1.50079	39.12	1.54016
35.13	1.38307	36.13	1.42244	37.13	1.46181	38.13	1.50118	39.13	1.54055
35.14	1.38346	36.14	1.42283	37.14	1.46220	38.14	1.50157	39.14	1.54094
35.15	1.38386	36.15	1.42323	37.15	1.46260	38.15	1.50197	39.15	1.54134
35.16	1.38425	36.16	1.42362	37.16	1.46299	38.16	1.50236	39.16	1.54173
35.17	1.38465	36.17	1.42402	37.17	1.46339	38.17	1.50276	39.17	1.54213
35.18	1.38504	36.18	1.42441	37.18	1.46378	38.18	1.50315	39.18	1.54252
35.19	1.38543	36.19	1.42480	37.19	1.46417	38.19	1.50354	39.19	1.54291
35.20	1.38583	36.20	1.42520	37.20	1.46457	38.20	1.50394	39.20	1.54331
35.21	1.38622	36.21	1.42559	37.21	1.46496	38.21	1.50433	39.21	1.54370
35.22	1.38661	36.22	1.42598	37.22	1.46535	38.22	1.50472	39.22	1.54409
35.23	1.38701	36.23	1.42638	37.23	1.46575	38.23	1.50512	39.23	1.54449
35.24	1.38740	36.24	1.42677	37.24	1.46614	38.24	1.50551	39.24	1.54488
35.25	1.38780	36.25	1.42717	37.25	1.46654	38.25	1.50591	39.25	1.54528
35.26	1.38819	36.26	1.42756	37.26	1.46693	38.26	1.50630	39.26	1.54567
35.27	1.38858	36.27	1.42795	37.27	1.46732	38.27	1.50669	39.27	1.54606
35.28	1.38898	36.28	1.42835	37.28	1.46772	38.28	1.50709	39.28	1.54646
35.29	1.38937	36.29	1.42874	37.29	1.46811	38.29	1.50748	39.29	1.54685
35.30	1.38976	36.30	1.42913	37.30	1.46850	38.30	1.50787	39.30	1.54724
35.31	1.39016	36.31	1.42953	37.31	1.46890	38.31	1.50827	39.31	1.54764
35.32	1.39055	36.32	1.42992	37.32	1.46929	38.32	1.50866	39.32	1.54803
35.33	1.39094	36.33	1.43031	37.33	1.46969	38.33	1.50906	39.33	1.54843
35.34	1.39134	36.34	1.43071	37.34	1.47008	38.34	1.50945	39.34	1.54882
35.35	1.39173	36.35	1.43110	37.35	1.47047	38.35	1.50984	39.35	1.54921
35.36	1.39213	36.36	1.43150	37.36	1.47087	38.36	1.51024	39.36	1.54961
35.37	1.39252	36.37	1.43189	37.37	1.47126	38.37	1.51063	39.37	1.55000
35.38	1.39291	36.38	1.43228	37.38	1.47165	38.38	1.51102	39.38	1.55039
35.39	1.39331	36.39	1.43268	37.39	1.47205	38.39	1.51142	39.39	1.55079
35.40	1.39370	36.40	1.43307	37.40	1.47244	38.40	1.51181	39.40	1.55118
35.41	1.39409	36.41	1.43346	37.41	1.47283	38.41	1.51220	39.41	1.55157
35.42	1.39449	36.42	1.43386	37.42	1.47323	38.42	1.51260	39.42	1.55197
35.43	1.39488	36.43	1.43425	37.43	1.47362	38.43	1.51299	39.43	1.55236
35.44	1.39528	36.44	1.43465	37.44	1.47402	38.44	1.51339	39.44	1.55276
35.45	1.39567	36.45	1.43504	37.45	1.47441	38.45	1.51378	39.45	1.55315
35.46	1.39606	36.46	1.43543	37.46	1.47480	38.46	1.51417	39.46	1.55354
35.47	1.39646	36.47	1.43583	37.47	1.47520	38.47	1.51457	39.47	1.55394
35.48	1.39685	36.48	1.43622	37.48	1.47559	38.48	1.51496	39.48	1.55433
35.49	1.39724	36.49	1.43661	37.49	1.47598	38.49	1.51535	39.49	1.55472

Continued - MILLIMETERS TO INCHES 1 MILLIMETER = 0.039 370 0787 INCH

MILLI-METERS	INCHES	MILLI-METERS	INCHES	MILLI-METERS	INCHES	MILLI-METERS	INCHES	MILLI-METERS	INCHES
35.50	1.39764	36.50	1.43701	37.50	1.47638	38.50	1.51575	39.50	1.55512
35.51	1.39803	36.51	1.43740	37.51	1.47677	38.51	1.51614	39.51	1.55551
35.52	1.39843	36.52	1.43780	37.52	1.47717	38.52	1.51654	39.52	1.55591
35.53	1.39882	36.53	1.43819	37.53	1.47756	38.53	1.51693	39.53	1.55630
35.54	1.39921	36.54	1.43858	37.54	1.47795	38.54	1.51732	39.54	1.55669
35.55	1.39961	36.55	1.43898	37.55	1.47835	38.55	1.51772	39.55	1.55709
35.56	1.40000	36.56	1.43937	37.56	1.47874	38.56	1.51811	39.56	1.55748
35.57	1.40039	36.57	1.43976	37.57	1.47913	38.57	1.51850	39.57	1.55787
35.58	1.40079	36.58	1.44016	37.58	1.47953	38.58	1.51890	39.58	1.55827
35.59	1.40118	36.59	1.44055	37.59	1.47992	38.59	1.51929	39.59	1.55866
35.60	1.40157	36.60	1.44094	37.60	1.48031	38.60	1.51969	39.60	1.55906
35.61	1.40197	36.61	1.44134	37.61	1.48071	38.61	1.52008	39.61	1.55945
35.62	1.40236	36.62	1.44173	37.62	1.48110	38.62	1.52047	39.62	1.55984
35.63	1.40276	36.63	1.44213	37.63	1.48150	38.63	1.52087	39.63	1.56024
35.64	1.40315	36.64	1.44252	37.64	1.48189	38.64	1.52126	39.64	1.56063
35.65	1.40354	36.65	1.44291	37.65	1.48228	38.65	1.52165	39.65	1.56102
35.66	1.40394	36.66	1.44331	37.66	1.48268	38.66	1.52205	39.66	1.56142
35.67	1.40433	36.67	1.44370	37.67	1.48307	38.67	1.52244	39.67	1.56181
35.68	1.40472	36.68	1.44409	37.68	1.48346	38.68	1.52283	39.68	1.56220
35.69	1.40512	36.69	1.44449	37.69	1.48386	38.69	1.52323	39.69	1.56260
35.70	1.40551	36.70	1.44488	37.70	1.48425	38.70	1.52362	39.70	1.56299
35.71	1.40591	36.71	1.44528	37.71	1.48465	38.71	1.52402	39.71	1.56339
35.72	1.40630	36.72	1.44567	37.72	1.48504	38.72	1.52441	39.72	1.56378
35.73	1.40669	36.73	1.44606	37.73	1.48543	38.73	1.52480	39.73	1.56417
35.74	1.40709	36.74	1.44646	37.74	1.48583	38.74	1.52520	39.74	1.56457
35.75	1.40748	36.75	1.44685	37.75	1.48622	38.75	1.52559	39.75	1.56496
35.76	1.40787	36.76	1.44724	37.76	1.48661	38.76	1.52598	39.76	1.56535
35.77	1.40827	36.77	1.44764	37.77	1.48701	38.77	1.52638	39.77	1.56575
35.78	1.40866	36.78	1.44803	37.78	1.48740	38.78	1.52677	39.78	1.56614
35.79	1.40906	36.79	1.44843	37.79	1.48780	38.79	1.52717	39.79	1.56654
35.80	1.40945	36.80	1.44882	37.80	1.48819	38.80	1.52756	39.80	1.56693
35.81	1.40984	36.81	1.44921	37.81	1.48858	38.81	1.52795	39.81	1.56732
35.82	1.41024	36.82	1.44961	37.82	1.48898	38.82	1.52835	39.82	1.56772
35.83	1.41063	36.83	1.45000	37.83	1.48937	38.83	1.52874	39.83	1.56811
35.84	1.41102	36.84	1.45039	37.84	1.48976	38.84	1.52913	39.84	1.56850
35.85	1.41142	36.85	1.45079	37.85	1.49016	38.85	1.52953	39.85	1.56890
35.86	1.41181	36.86	1.45118	37.86	1.49055	38.86	1.52992	39.86	1.56929
35.87	1.41220	36.87	1.45157	37.87	1.49094	38.87	1.53031	39.87	1.56969
35.88	1.41260	36.88	1.45197	37.88	1.49134	38.88	1.53071	39.88	1.57008
35.89	1.41299	36.89	1.45236	37.89	1.49173	38.89	1.53110	39.89	1.57047
35.90	1.41339	36.90	1.45276	37.90	1.49213	38.90	1.53150	39.90	1.57087
35.91	1.41378	36.91	1.45315	37.91	1.49252	38.91	1.53189	39.91	1.57126
35.92	1.41417	36.92	1.45354	37.92	1.49291	38.92	1.53228	39.92	1.57165
35.93	1.41457	36.93	1.45394	37.93	1.49331	38.93	1.53268	39.93	1.57205
35.94	1.41496	36.94	1.45433	37.94	1.49370	38.94	1.53307	39.94	1.57244
35.95	1.41535	36.95	1.45472	37.95	1.49409	38.95	1.53346	39.95	1.57283
35.96	1.41575	36.96	1.45512	37.96	1.49449	38.96	1.53386	39.96	1.57323
35.97	1.41614	36.97	1.45551	37.97	1.49488	38.97	1.53425	39.97	1.57362
35.98	1.41654	36.98	1.45591	37.98	1.49528	38.98	1.53465	39.98	1.57402
35.99	1.41693	36.99	1.45630	37.99	1.49567	38.99	1.53504	39.99	1.57441

Continued - MILLIMETERS TO INCHES 1 MILLIMETER = 0.039 370 0787 INCH

MILLI-METERS	INCHES	MILLI-METERS	INCHES	MILLI-METERS	INCHES	MILLI-METERS	INCHES	MILLI-METERS	INCHES
40.00	1.57480	41.00	1.61417	42.00	1.65354	43.00	1.69291	44.00	1.73228
40.01	1.57520	41.01	1.61457	42.01	1.65394	43.01	1.69331	44.01	1.73268
40.02	1.57559	41.02	1.61496	42.02	1.65433	43.02	1.69370	44.02	1.73307
40.03	1.57598	41.03	1.61535	42.03	1.65472	43.03	1.69409	44.03	1.73346
40.04	1.57638	41.04	1.61575	42.04	1.65512	43.04	1.69449	44.04	1.73386
40.05	1.57677	41.05	1.61614	42.05	1.65551	43.05	1.69488	44.05	1.73425
40.06	1.57717	41.06	1.61654	42.06	1.65591	43.06	1.69528	44.06	1.73465
40.07	1.57756	41.07	1.61693	42.07	1.65630	43.07	1.69567	44.07	1.73504
40.08	1.57795	41.08	1.61732	42.08	1.65669	43.08	1.69606	44.08	1.73543
40.09	1.57835	41.09	1.61772	42.09	1.65709	43.09	1.69646	44.09	1.73583
40.10	1.57874	41.10	1.61811	42.10	1.65748	43.10	1.69685	44.10	1.73622
40.11	1.57913	41.11	1.61850	42.11	1.65787	43.11	1.69724	44.11	1.73661
40.12	1.57953	41.12	1.61890	42.12	1.65827	43.12	1.69764	44.12	1.73701
40.13	1.57992	41.13	1.61929	42.13	1.65866	43.13	1.69803	44.13	1.73740
40.14	1.58031	41.14	1.61969	42.14	1.65906	43.14	1.69843	44.14	1.73780
40.15	1.58071	41.15	1.62008	42.15	1.65945	43.15	1.69882	44.15	1.73819
40.16	1.58110	41.16	1.62047	42.16	1.65984	43.16	1.69921	44.16	1.73858
40.17	1.58150	41.17	1.62087	42.17	1.66024	43.17	1.69961	44.17	1.73898
40.18	1.58189	41.18	1.62126	42.18	1.66063	43.18	1.70000	44.18	1.73937
40.19	1.58228	41.19	1.62165	42.19	1.66102	43.19	1.70039	44.19	1.73976
40.20	1.58268	41.20	1.62205	42.20	1.66142	43.20	1.70079	44.20	1.74016
40.21	1.58307	41.21	1.62244	42.21	1.66181	43.21	1.70118	44.21	1.74055
40.22	1.58346	41.22	1.62283	42.22	1.66220	43.22	1.70157	44.22	1.74094
40.23	1.58386	41.23	1.62323	42.23	1.66260	43.23	1.70197	44.23	1.74134
40.24	1.58425	41.24	1.62362	42.24	1.66299	43.24	1.70236	44.24	1.74173
40.25	1.58465	41.25	1.62402	42.25	1.66339	43.25	1.70276	44.25	1.74213
40.26	1.58504	41.26	1.62441	42.26	1.66378	43.26	1.70315	44.26	1.74252
40.27	1.58543	41.27	1.62480	42.27	1.66417	43.27	1.70354	44.27	1.74291
40.28	1.58583	41.28	1.62520	42.28	1.66457	43.28	1.70394	44.28	1.74331
40.29	1.58622	41.29	1.62559	42.29	1.66496	43.29	1.70433	44.29	1.74370
40.30	1.58661	41.30	1.62598	42.30	1.66535	43.30	1.70472	44.30	1.74409
40.31	1.58701	41.31	1.62638	42.31	1.66575	43.31	1.70512	44.31	1.74449
40.32	1.58740	41.32	1.62677	42.32	1.66614	43.32	1.70551	44.32	1.74488
40.33	1.58780	41.33	1.62717	42.33	1.66654	43.33	1.70591	44.33	1.74528
40.34	1.58819	41.34	1.62756	42.34	1.66693	43.34	1.70630	44.34	1.74567
40.35	1.58858	41.35	1.62795	42.35	1.66732	43.35	1.70669	44.35	1.74606
40.36	1.58898	41.36	1.62835	42.36	1.66772	43.36	1.70709	44.36	1.74646
40.37	1.58937	41.37	1.62874	42.37	1.66811	43.37	1.70748	44.37	1.74685
40.38	1.58976	41.38	1.62913	42.38	1.66850	43.38	1.70787	44.38	1.74724
40.39	1.59016	41.39	1.62953	42.39	1.66890	43.39	1.70827	44.39	1.74764
40.40	1.59055	41.40	1.62992	42.40	1.66929	43.40	1.70866	44.40	1.74803
40.41	1.59094	41.41	1.63031	42.41	1.66969	43.41	1.70906	44.41	1.74843
40.42	1.59134	41.42	1.63071	42.42	1.67008	43.42	1.70945	44.42	1.74882
40.43	1.59173	41.43	1.63110	42.43	1.67047	43.43	1.70984	44.43	1.74921
40.44	1.59213	41.44	1.63150	42.44	1.67087	43.44	1.71024	44.44	1.74961
40.45	1.59252	41.45	1.63189	42.45	1.67126	43.45	1.71063	44.45	1.75000
40.46	1.59291	41.46	1.63228	42.46	1.67165	43.46	1.71102	44.46	1.75039
40.47	1.59331	41.47	1.63268	42.47	1.67205	43.47	1.71142	44.47	1.75079
40.48	1.59370	41.48	1.63307	42.48	1.67244	43.48	1.71181	44.48	1.75118
40.49	1.59409	41.49	1.63346	42.49	1.67283	43.49	1.71220	44.49	1.75157

Continued - MILLIMETERS TO INCHES 1 MILLIMETER = 0.039 370 0787 INCH

MILLI-METERS	INCHES	MILLI-METERS	INCHES	MILLI-METERS	INCHES	MILLI-METERS	INCHES	MILLI-METERS	INCHES
40.50	1.59449	41.50	1.63386	42.50	1.67323	43.50	1.71260	44.50	1.75197
40.51	1.59488	41.51	1.63425	42.51	1.67362	43.51	1.71299	44.51	1.75236
40.52	1.59528	41.52	1.63465	42.52	1.67402	43.52	1.71339	44.52	1.75276
40.53	1.59567	41.53	1.63504	42.53	1.67441	43.53	1.71378	44.53	1.75315
40.54	1.59606	41.54	1.63543	42.54	1.67480	43.54	1.71417	44.54	1.75354
40.55	1.59646	41.55	1.63583	42.55	1.67520	43.55	1.71457	44.55	1.75394
40.56	1.59685	41.56	1.63622	42.56	1.67559	43.56	1.71496	44.56	1.75433
40.57	1.59724	41.57	1.63661	42.57	1.67598	43.57	1.71535	44.57	1.75472
40.58	1.59764	41.58	1.63701	42.58	1.67638	43.58	1.71575	44.58	1.75512
40.59	1.59803	41.59	1.63740	42.59	1.67677	43.59	1.71614	44.59	1.75551
40.60	1.59843	41.60	1.63780	42.60	1.67717	43.60	1.71654	44.60	1.75591
40.61	1.59882	41.61	1.63819	42.61	1.67756	43.61	1.71693	44.61	1.75630
40.62	1.59921	41.62	1.63858	42.62	1.67795	43.62	1.71732	44.62	1.75669
40.63	1.59961	41.63	1.63898	42.63	1.67835	43.63	1.71772	44.63	1.75709
40.64	1.60000	41.64	1.63937	42.64	1.67874	43.64	1.71811	44.64	1.75748
40.65	1.60039	41.65	1.63976	42.65	1.67913	43.65	1.71850	44.65	1.75787
40.66	1.60079	41.66	1.64016	42.66	1.67953	43.66	1.71890	44.66	1.75827
40.67	1.60118	41.67	1.64055	42.67	1.67992	43.67	1.71929	44.67	1.75866
40.68	1.60157	41.68	1.64094	42.68	1.68031	43.68	1.71969	44.68	1.75906
40.69	1.60197	41.69	1.64134	42.69	1.68071	43.69	1.72008	44.69	1.75945
40.70	1.60236	41.70	1.64173	42.70	1.68110	43.70	1.72047	44.70	1.75984
40.71	1.60276	41.71	1.64213	42.71	1.68150	43.71	1.72087	44.71	1.76024
40.72	1.60315	41.72	1.64252	42.72	1.68189	43.72	1.72126	44.72	1.76063
40.73	1.60354	41.73	1.64291	42.73	1.68228	43.73	1.72165	44.73	1.76102
40.74	1.60394	41.74	1.64331	42.74	1.68268	43.74	1.72205	44.74	1.76142
40.75	1.60433	41.75	1.64370	42.75	1.68307	43.75	1.72244	44.75	1.76181
40.76	1.60472	41.76	1.64409	42.76	1.68346	43.76	1.72283	44.76	1.76220
40.77	1.60512	41.77	1.64449	42.77	1.68386	43.77	1.72323	44.77	1.76260
40.78	1.60551	41.78	1.64488	42.78	1.68425	43.78	1.72362	44.78	1.76299
40.79	1.60591	41.79	1.64528	42.79	1.68465	43.79	1.72402	44.79	1.76339
40.80	1.60630	41.80	1.64567	42.80	1.68504	43.80	1.72441	44.80	1.76378
40.81	1.60669	41.81	1.64606	42.81	1.68543	43.81	1.72480	44.81	1.76417
40.82	1.60709	41.82	1.64646	42.82	1.68583	43.82	1.72520	44.82	1.76457
40.83	1.60748	41.83	1.64685	42.83	1.68622	43.83	1.72559	44.83	1.76496
40.84	1.60787	41.84	1.64724	42.84	1.68661	43.84	1.72598	44.84	1.76535
40.85	1.60827	41.85	1.64764	42.85	1.68701	43.85	1.72638	44.85	1.76575
40.86	1.60866	41.86	1.64803	42.86	1.68740	43.86	1.72677	44.86	1.76614
40.87	1.60906	41.87	1.64843	42.87	1.68780	43.87	1.72717	44.87	1.76654
40.88	1.60945	41.88	1.64882	42.88	1.68819	43.88	1.72756	44.88	1.76693
40.89	1.60984	41.89	1.64921	42.89	1.68858	43.89	1.72795	44.89	1.76732
40.90	1.61024	41.90	1.64961	42.90	1.68898	43.90	1.72835	44.90	1.76772
40.91	1.61063	41.91	1.65000	42.91	1.68937	43.91	1.72874	44.91	1.76811
40.92	1.61102	41.92	1.65039	42.92	1.68976	43.92	1.72913	44.92	1.76850
40.93	1.61142	41.93	1.65079	42.93	1.69016	43.93	1.72953	44.93	1.76890
40.94	1.61181	41.94	1.65118	42.94	1.69055	43.94	1.72992	44.94	1.76929
40.95	1.61220	41.95	1.65157	42.95	1.69094	43.95	1.73031	44.95	1.76969
40.96	1.61260	41.96	1.65197	42.96	1.69134	43.96	1.73071	44.96	1.77008
40.97	1.61299	41.97	1.65236	42.97	1.69173	43.97	1.73110	44.97	1.77047
40.98	1.61339	41.98	1.65276	42.98	1.69213	43.98	1.73150	44.98	1.77087
40.99	1.61378	41.99	1.65315	42.99	1.69252	43.99	1.73189	44.99	1.77126

Continued - MILLIMETERS TO INCHES 1 MILLIMETER = 0.039 370 0787 INCH

MILLI-METERS	INCHES	MILLI-METERS	INCHES	MILLI-METERS	INCHES	MILLI-METERS	INCHES	MILLI-METERS	INCHES
45.00	1.77165	46.00	1.81102	47.00	1.85039	48.00	1.88976	49.00	1.92913
45.01	1.77205	46.01	1.81142	47.01	1.85079	48.01	1.89016	49.01	1.92953
45.02	1.77244	46.02	1.81181	47.02	1.85118	48.02	1.89055	49.02	1.92992
45.03	1.77283	46.03	1.81220	47.03	1.85157	48.03	1.89094	49.03	1.93031
45.04	1.77323	46.04	1.81260	47.04	1.85197	48.04	1.89134	49.04	1.93071
45.05	1.77362	46.05	1.81299	47.05	1.85236	48.05	1.89173	49.05	1.93110
45.06	1.77402	46.06	1.81339	47.06	1.85276	48.06	1.89213	49.06	1.93150
45.07	1.77441	46.07	1.81378	47.07	1.85315	48.07	1.89252	49.07	1.93189
45.08	1.77480	46.08	1.81417	47.08	1.85354	48.08	1.89291	49.08	1.93228
45.09	1.77520	46.09	1.81457	47.09	1.85394	48.09	1.89331	49.09	1.93268
45.10	1.77559	46.10	1.81496	47.10	1.85433	48.10	1.89370	49.10	1.93307
45.11	1.77598	46.11	1.81535	47.11	1.85472	48.11	1.89409	49.11	1.93346
45.12	1.77638	46.12	1.81575	47.12	1.85512	48.12	1.89449	49.12	1.93386
45.13	1.77677	46.13	1.81614	47.13	1.85551	48.13	1.89488	49.13	1.93425
45.14	1.77717	46.14	1.81654	47.14	1.85591	48.14	1.89528	49.14	1.93465
45.15	1.77756	46.15	1.81693	47.15	1.85630	48.15	1.89567	49.15	1.93504
45.16	1.77795	46.16	1.81732	47.16	1.85669	48.16	1.89606	49.16	1.93543
45.17	1.77835	46.17	1.81772	47.17	1.85709	48.17	1.89646	49.17	1.93583
45.18	1.77874	46.18	1.81811	47.18	1.85748	48.18	1.89685	49.18	1.93622
45.19	1.77913	46.19	1.81850	47.19	1.85787	48.19	1.89724	49.19	1.93661
45.20	1.77953	46.20	1.81890	47.20	1.85827	48.20	1.89764	49.20	1.93701
45.21	1.77992	46.21	1.81929	47.21	1.85866	48.21	1.89803	49.21	1.93740
45.22	1.78031	46.22	1.81969	47.22	1.85906	48.22	1.89843	49.22	1.93780
45.23	1.78071	46.23	1.82008	47.23	1.85945	48.23	1.89882	49.23	1.93819
45.24	1.78110	46.24	1.82047	47.24	1.85984	48.24	1.89921	49.24	1.93858
45.25	1.78150	46.25	1.82087	47.25	1.86024	48.25	1.89961	49.25	1.93898
45.26	1.78189	46.26	1.82126	47.26	1.86063	48.26	1.90000	49.26	1.93937
45.27	1.78228	46.27	1.82165	47.27	1.86102	48.27	1.90039	49.27	1.93976
45.28	1.78268	46.28	1.82205	47.28	1.86142	48.28	1.90079	49.28	1.94016
45.29	1.78307	46.29	1.82244	47.29	1.86181	48.29	1.90118	49.29	1.94055
45.30	1.78346	46.30	1.82283	47.30	1.86220	48.30	1.90157	49.30	1.94094
45.31	1.78386	46.31	1.82323	47.31	1.86260	48.31	1.90197	49.31	1.94134
45.32	1.78425	46.32	1.82362	47.32	1.86299	48.32	1.90236	49.32	1.94173
45.33	1.78465	46.33	1.82402	47.33	1.86339	48.33	1.90276	49.33	1.94213
45.34	1.78504	46.34	1.82441	47.34	1.86378	48.34	1.90315	49.34	1.94252
45.35	1.78543	46.35	1.82480	47.35	1.86417	48.35	1.90354	49.35	1.94291
45.36	1.78583	46.36	1.82520	47.36	1.86457	48.36	1.90394	49.36	1.94331
45.37	1.78622	46.37	1.82559	47.37	1.86496	48.37	1.90433	49.37	1.94370
45.38	1.78661	46.38	1.82598	47.38	1.86535	48.38	1.90472	49.38	1.94409
45.39	1.78701	46.39	1.82638	47.39	1.86575	48.39	1.90512	49.39	1.94449
45.40	1.78740	46.40	1.82677	47.40	1.86614	48.40	1.90551	49.40	1.94488
45.41	1.78780	46.41	1.82717	47.41	1.86654	48.41	1.90591	49.41	1.94528
45.42	1.78819	46.42	1.82756	47.42	1.86693	48.42	1.90630	49.42	1.94567
45.43	1.78858	46.43	1.82795	47.43	1.86732	48.43	1.90669	49.43	1.94606
45.44	1.78898	46.44	1.82835	47.44	1.86772	48.44	1.90709	49.44	1.94646
45.45	1.78937	46.45	1.82874	47.45	1.86811	48.45	1.90748	49.45	1.94685
45.46	1.78976	46.46	1.82913	47.46	1.86850	48.46	1.90787	49.46	1.94724
45.47	1.79016	46.47	1.82953	47.47	1.86890	48.47	1.90827	49.47	1.94764
45.48	1.79055	46.48	1.82992	47.48	1.86929	48.48	1.90866	49.48	1.94803
45.49	1.79094	46.49	1.83031	47.49	1.86969	48.49	1.90906	49.49	1.94843

Continued - MILLIMETERS TO INCHES 1 MILLIMETER = 0.039 370 0787 INCH

MILLI-METERS	INCHES	MILLI-METERS	INCHES	MILLI-METERS	INCHES	MILLI-METERS	INCHES	MILLI-METERS	INCHES
45.50	1.79134	46.50	1.83071	47.50	1.87008	48.50	1.90945	49.50	1.94882
45.51	1.79173	46.51	1.83110	47.51	1.87047	48.51	1.90984	49.51	1.94921
45.52	1.79213	46.52	1.83150	47.52	1.87087	48.52	1.91024	49.52	1.94961
45.53	1.79252	46.53	1.83189	47.53	1.87126	48.53	1.91063	49.53	1.95000
45.54	1.79291	46.54	1.83228	47.54	1.87165	48.54	1.91102	49.54	1.95039
45.55	1.79331	46.55	1.83268	47.55	1.87205	48.55	1.91142	49.55	1.95079
45.56	1.79370	46.56	1.83307	47.56	1.87244	48.56	1.91181	49.56	1.95118
45.57	1.79409	46.57	1.83346	47.57	1.87283	48.57	1.91220	49.57	1.95157
45.58	1.79449	46.58	1.83386	47.58	1.87323	48.58	1.91260	49.58	1.95197
45.59	1.79488	46.59	1.83425	47.59	1.87362	48.59	1.91299	49.59	1.95236
45.60	1.79528	46.60	1.83465	47.60	1.87402	48.60	1.91339	49.60	1.95276
45.61	1.79567	46.61	1.83504	47.61	1.87441	48.61	1.91378	49.61	1.95315
45.62	1.79606	46.62	1.83543	47.62	1.87480	48.62	1.91417	49.62	1.95354
45.63	1.79646	46.63	1.83583	47.63	1.87520	48.63	1.91457	49.63	1.95394
45.64	1.79685	46.64	1.83622	47.64	1.87559	48.64	1.91496	49.64	1.95433
45.65	1.79724	46.65	1.83661	47.65	1.87598	48.65	1.91535	49.65	1.95472
45.66	1.79764	46.66	1.83701	47.66	1.87638	48.66	1.91575	49.66	1.95512
45.67	1.79803	46.67	1.83740	47.67	1.87677	48.67	1.91614	49.67	1.95551
45.68	1.79843	46.68	1.83780	47.68	1.87717	48.68	1.91654	49.68	1.95591
45.69	1.79882	46.69	1.83819	47.69	1.87756	48.69	1.91693	49.69	1.95630
45.70	1.79921	46.70	1.83858	47.70	1.87795	48.70	1.91732	49.70	1.95669
45.71	1.79961	46.71	1.83898	47.71	1.87835	48.71	1.91772	49.71	1.95709
45.72	1.80000	46.72	1.83937	47.72	1.87874	48.72	1.91811	49.72	1.95748
45.73	1.80039	46.73	1.83976	47.73	1.87913	48.73	1.91850	49.73	1.95787
45.74	1.80079	46.74	1.84016	47.74	1.87953	48.74	1.91890	49.74	1.95827
45.75	1.80118	46.75	1.84055	47.75	1.87992	48.75	1.91929	49.75	1.95866
45.76	1.80157	46.76	1.84094	47.76	1.88031	48.76	1.91969	49.76	1.95906
45.77	1.80197	46.77	1.84134	47.77	1.88071	48.77	1.92008	49.77	1.95945
45.78	1.80236	46.78	1.84173	47.78	1.88110	48.78	1.92047	49.78	1.95984
45.79	1.80276	46.79	1.84213	47.79	1.88150	48.79	1.92087	49.79	1.96024
45.80	1.80315	46.80	1.84252	47.80	1.88189	48.80	1.92126	49.80	1.96063
45.81	1.80354	46.81	1.84291	47.81	1.88228	48.81	1.92165	49.81	1.96102
45.82	1.80394	46.82	1.84331	47.82	1.88268	48.82	1.92205	49.82	1.96142
45.83	1.80433	46.83	1.84370	47.83	1.88307	48.83	1.92244	49.83	1.96181
45.84	1.80472	46.84	1.84409	47.84	1.88346	48.84	1.92283	49.84	1.96220
45.85	1.80512	46.85	1.84449	47.85	1.88386	48.85	1.92323	49.85	1.96260
45.86	1.80551	46.86	1.84488	47.86	1.88425	48.86	1.92362	49.86	1.96299
45.87	1.80591	46.87	1.84528	47.87	1.88465	48.87	1.92402	49.87	1.96339
45.88	1.80630	46.88	1.84567	47.88	1.88504	48.88	1.92441	49.88	1.96378
45.89	1.80669	46.89	1.84606	47.89	1.88543	48.89	1.92480	49.89	1.96417
45.90	1.80709	46.90	1.84646	47.90	1.88583	48.90	1.92520	49.90	1.96457
45.91	1.80748	46.91	1.84685	47.91	1.88622	48.91	1.92559	49.91	1.96496
45.92	1.80787	46.92	1.84724	47.92	1.88661	48.92	1.92598	49.92	1.96535
45.93	1.80827	46.93	1.84764	47.93	1.88701	48.93	1.92638	49.93	1.96575
45.94	1.80866	46.94	1.84803	47.94	1.88740	48.94	1.92677	49.94	1.96614
45.95	1.80906	46.95	1.84843	47.95	1.88780	48.95	1.92717	49.95	1.96654
45.96	1.80945	46.96	1.84882	47.96	1.88819	48.96	1.92756	49.96	1.96693
45.97	1.80984	46.97	1.84921	47.97	1.88858	48.97	1.92795	49.97	1.96732
45.98	1.81024	46.98	1.84961	47.98	1.88898	48.98	1.92835	49.98	1.96772
45.99	1.81063	46.99	1.85000	47.99	1.88937	48.99	1.92874	49.99	1.96811

Continued - MILLIMETERS TO INCHES 1 MILLIMETER = 0.039 370 0787 INCH

MILLI-METERS	INCHES	MILLI-METERS	INCHES	MILLI-METERS	INCHES	MILLI-METERS	INCHES	MILLI-METERS	INCHES
50.00	1.96850	51.00	2.00787	52.00	2.04724	53.00	2.08661	54.00	2.12598
50.01	1.96890	51.01	2.00827	52.01	2.04764	53.01	2.08701	54.01	2.12638
50.02	1.96929	51.02	2.00866	52.02	2.04803	53.02	2.08740	54.02	2.12677
50.03	1.96969	51.03	2.00906	52.03	2.04843	53.03	2.08780	54.03	2.12717
50.04	1.97008	51.04	2.00945	52.04	2.04882	53.04	2.08819	54.04	2.12756
50.05	1.97047	51.05	2.00984	52.05	2.04921	53.05	2.08858	54.05	2.12795
50.06	1.97087	51.06	2.01024	52.06	2.04961	53.06	2.08898	54.06	2.12835
50.07	1.97126	51.07	2.01063	52.07	2.05000	53.07	2.08937	54.07	2.12874
50.08	1.97165	51.08	2.01102	52.08	2.05039	53.08	2.08976	54.08	2.12913
50.09	1.97205	51.09	2.01142	52.09	2.05079	53.09	2.09016	54.09	2.12953
50.10	1.97244	51.10	2.01181	52.10	2.05118	53.10	2.09055	54.10	2.12992
50.11	1.97283	51.11	2.01220	52.11	2.05157	53.11	2.09094	54.11	2.13031
50.12	1.97323	51.12	2.01260	52.12	2.05197	53.12	2.09134	54.12	2.13071
50.13	1.97362	51.13	2.01299	52.13	2.05236	53.13	2.09173	54.13	2.13110
50.14	1.97402	51.14	2.01339	52.14	2.05276	53.14	2.09213	54.14	2.13150
50.15	1.97441	51.15	2.01378	52.15	2.05315	53.15	2.09252	54.15	2.13189
50.16	1.97480	51.16	2.01417	52.16	2.05354	53.16	2.09291	54.16	2.13228
50.17	1.97520	51.17	2.01457	52.17	2.05394	53.17	2.09331	54.17	2.13268
50.18	1.97559	51.18	2.01496	52.18	2.05433	53.18	2.09370	54.18	2.13307
50.19	1.97598	51.19	2.01535	52.19	2.05472	53.19	2.09409	54.19	2.13346
50.20	1.97638	51.20	2.01575	52.20	2.05512	53.20	2.09449	54.20	2.13386
50.21	1.97677	51.21	2.01614	52.21	2.05551	53.21	2.09488	54.21	2.13425
50.22	1.97717	51.22	2.01654	52.22	2.05591	53.22	2.09528	54.22	2.13465
50.23	1.97756	51.23	2.01693	52.23	2.05630	53.23	2.09567	54.23	2.13504
50.24	1.97795	51.24	2.01732	52.24	2.05669	53.24	2.09606	54.24	2.13543
50.25	1.97835	51.25	2.01772	52.25	2.05709	53.25	2.09646	54.25	2.13583
50.26	1.97874	51.26	2.01811	52.26	2.05748	53.26	2.09685	54.26	2.13622
50.27	1.97913	51.27	2.01850	52.27	2.05787	53.27	2.09724	54.27	2.13661
50.28	1.97953	51.28	2.01890	52.28	2.05827	53.28	2.09764	54.28	2.13701
50.29	1.97992	51.29	2.01929	52.29	2.05866	53.29	2.09803	54.29	2.13740
50.30	1.98031	51.30	2.01969	52.30	2.05906	53.30	2.09843	54.30	2.13780
50.31	1.98071	51.31	2.02008	52.31	2.05945	53.31	2.09882	54.31	2.13819
50.32	1.98110	51.32	2.02047	52.32	2.05984	53.32	2.09921	54.32	2.13858
50.33	1.98150	51.33	2.02087	52.33	2.06024	53.33	2.09961	54.33	2.13898
50.34	1.98189	51.34	2.02126	52.34	2.06063	53.34	2.10000	54.34	2.13937
50.35	1.98228	51.35	2.02165	52.35	2.06102	53.35	2.10039	54.35	2.13976
50.36	1.98268	51.36	2.02205	52.36	2.06142	53.36	2.10079	54.36	2.14016
50.37	1.98307	51.37	2.02244	52.37	2.06181	53.37	2.10118	54.37	2.14055
50.38	1.98346	51.38	2.02283	52.38	2.06220	53.38	2.10157	54.38	2.14094
50.39	1.98386	51.39	2.02323	52.39	2.06260	53.39	2.10197	54.39	2.14134
50.40	1.98425	51.40	2.02362	52.40	2.06299	53.40	2.10236	54.40	2.14173
50.41	1.98465	51.41	2.02402	52.41	2.06339	53.41	2.10276	54.41	2.14213
50.42	1.98504	51.42	2.02441	52.42	2.06378	53.42	2.10315	54.42	2.14252
50.43	1.98543	51.43	2.02480	52.43	2.06417	53.43	2.10354	54.43	2.14291
50.44	1.98583	51.44	2.02520	52.44	2.06457	53.44	2.10394	54.44	2.14331
50.45	1.98622	51.45	2.02559	52.45	2.06496	53.45	2.10433	54.45	2.14370
50.46	1.98661	51.46	2.02598	52.46	2.06535	53.46	2.10472	54.46	2.14409
50.47	1.98701	51.47	2.02638	52.47	2.06575	53.47	2.10512	54.47	2.14449
50.48	1.98740	51.48	2.02677	52.48	2.06614	53.48	2.10551	54.48	2.14488
50.49	1.98780	51.49	2.02717	52.49	2.06654	53.49	2.10591	54.49	2.14528

Continued - MILLIMETERS TO INCHES 1 MILLIMETER = 0.039 370 0787 INCH

MILLI-METERS	INCHES	MILLI-METERS	INCHES	MILLI-METERS	INCHES	MILLI-METERS	INCHES	MILLI-METERS	INCHES
50.50	1.98819	51.50	2.02756	52.50	2.06693	53.50	2.10630	54.50	2.14567
50.51	1.98858	51.51	2.02795	52.51	2.06732	53.51	2.10669	54.51	2.14606
50.52	1.98898	51.52	2.02835	52.52	2.06772	53.52	2.10709	54.52	2.14646
50.53	1.98937	51.53	2.02874	52.53	2.06811	53.53	2.10748	54.53	2.14685
50.54	1.98976	51.54	2.02913	52.54	2.06850	53.54	2.10787	54.54	2.14724
50.55	1.99016	51.55	2.02953	52.55	2.06890	53.55	2.10827	54.55	2.14764
50.56	1.99055	51.56	2.02992	52.56	2.06929	53.56	2.10866	54.56	2.14803
50.57	1.99094	51.57	2.03031	52.57	2.06969	53.57	2.10906	54.57	2.14843
50.58	1.99134	51.58	2.03071	52.58	2.07008	53.58	2.10945	54.58	2.14882
50.59	1.99173	51.59	2.03110	52.59	2.07047	53.59	2.10984	54.59	2.14921
50.60	1.99213	51.60	2.03150	52.60	2.07087	53.60	2.11024	54.60	2.14961
50.61	1.99252	51.61	2.03189	52.61	2.07126	53.61	2.11063	54.61	2.15000
50.62	1.99291	51.62	2.03228	52.62	2.07165	53.62	2.11102	54.62	2.15039
50.63	1.99331	51.63	2.03268	52.63	2.07205	53.63	2.11142	54.63	2.15079
50.64	1.99370	51.64	2.03307	52.64	2.07244	53.64	2.11181	54.64	2.15118
50.65	1.99409	51.65	2.03346	52.65	2.07283	53.65	2.11220	54.65	2.15157
50.66	1.99449	51.66	2.03386	52.66	2.07323	53.66	2.11260	54.66	2.15197
50.67	1.99488	51.67	2.03425	52.67	2.07362	53.67	2.11299	54.67	2.15236
50.68	1.99528	51.68	2.03465	52.68	2.07402	53.68	2.11339	54.68	2.15276
50.69	1.99567	51.69	2.03504	52.69	2.07441	53.69	2.11378	54.69	2.15315
50.70	1.99606	51.70	2.03543	52.70	2.07480	53.70	2.11417	54.70	2.15354
50.71	1.99646	51.71	2.03583	52.71	2.07520	53.71	2.11457	54.71	2.15394
50.72	1.99685	51.72	2.03622	52.72	2.07559	53.72	2.11496	54.72	2.15433
50.73	1.99724	51.73	2.03661	52.73	2.07598	53.73	2.11535	54.73	2.15472
50.74	1.99764	51.74	2.03701	52.74	2.07638	53.74	2.11575	54.74	2.15512
50.75	1.99803	51.75	2.03740	52.75	2.07677	53.75	2.11614	54.75	2.15551
50.76	1.99843	51.76	2.03780	52.76	2.07717	53.76	2.11654	54.76	2.15591
50.77	1.99882	51.77	2.03819	52.77	2.07756	53.77	2.11693	54.77	2.15630
50.78	1.99921	51.78	2.03858	52.78	2.07795	53.78	2.11732	54.78	2.15669
50.79	1.99961	51.79	2.03898	52.79	2.07835	53.79	2.11772	54.79	2.15709
50.80	2.00000	51.80	2.03937	52.80	2.07874	53.80	2.11811	54.80	2.15748
50.81	2.00039	51.81	2.03976	52.81	2.07913	53.81	2.11850	54.81	2.15787
50.82	2.00079	51.82	2.04016	52.82	2.07953	53.82	2.11890	54.82	2.15827
50.83	2.00118	51.83	2.04055	52.83	2.07992	53.83	2.11929	54.83	2.15866
50.84	2.00157	51.84	2.04094	52.84	2.08031	53.84	2.11968	54.84	2.15906
50.85	2.00197	51.85	2.04134	52.85	2.08071	53.85	2.12008	54.85	2.15945
50.86	2.00236	51.86	2.04173	52.86	2.08110	53.86	2.12047	54.86	2.15984
50.87	2.00276	51.87	2.04213	52.87	2.08150	53.87	2.12087	54.87	2.16024
50.88	2.00315	51.88	2.04252	52.88	2.08189	53.88	2.12126	54.88	2.16063
50.89	2.00354	51.89	2.04291	52.89	2.08228	53.89	2.12165	54.89	2.16102
50.90	2.00394	51.90	2.04331	52.90	2.08268	53.90	2.12205	54.90	2.16142
50.91	2.00433	51.91	2.04370	52.91	2.08307	53.91	2.12244	54.91	2.16181
50.92	2.00472	51.92	2.04409	52.92	2.08346	53.92	2.12283	54.92	2.16220
50.93	2.00512	51.93	2.04449	52.93	2.08386	53.93	2.12323	54.93	2.16260
50.94	2.00551	51.94	2.04488	52.94	2.08425	53.94	2.12362	54.94	2.16299
50.95	2.00591	51.95	2.04528	52.95	2.08465	53.95	2.12402	54.95	2.16339
50.96	2.00630	51.96	2.04567	52.96	2.08504	53.96	2.12441	54.96	2.16378
50.97	2.00669	51.97	2.04606	52.97	2.08543	53.97	2.12480	54.97	2.16417
50.98	2.00709	51.98	2.04646	52.98	2.08583	53.98	2.12520	54.98	2.16457
50.99	2.00748	51.99	2.04685	52.99	2.08622	53.99	2.12559	54.99	2.16496

Continued - MILLIMETERS TO INCHES 1 MILLIMETER = 0.039 370 0787 INCH

MILLI-METERS	INCHES	MILLI-METERS	INCHES	MILLI-METERS	INCHES	MILLI-METERS	INCHES	MILLI-METERS	INCHES
55.00	2.16535	56.00	2.20472	57.00	2.24409	58.00	2.28346	59.00	2.32283
55.01	2.16575	56.01	2.20512	57.01	2.24449	58.01	2.28386	59.01	2.32323
55.02	2.16614	56.02	2.20551	57.02	2.24488	58.02	2.28425	59.02	2.32362
55.03	2.16654	56.03	2.20591	57.03	2.24528	58.03	2.28465	59.03	2.32402
55.04	2.16693	56.04	2.20630	57.04	2.24567	58.04	2.28504	59.04	2.32441
55.05	2.16732	56.05	2.20669	57.05	2.24606	58.05	2.28543	59.05	2.32480
55.06	2.16772	56.06	2.20709	57.06	2.24646	58.06	2.28583	59.06	2.32520
55.07	2.16811	56.07	2.20748	57.07	2.24685	58.07	2.28622	59.07	2.32559
55.08	2.16850	56.08	2.20787	57.08	2.24724	58.08	2.28661	59.08	2.32598
55.09	2.16890	56.09	2.20827	57.09	2.24764	58.09	2.28701	59.09	2.32638
55.10	2.16929	56.10	2.20866	57.10	2.24803	58.10	2.28740	59.10	2.32677
55.11	2.16969	56.11	2.20906	57.11	2.24843	58.11	2.28780	59.11	2.32717
55.12	2.17008	56.12	2.20945	57.12	2.24882	58.12	2.28819	59.12	2.32756
55.13	2.17047	56.13	2.20984	57.13	2.24921	58.13	2.28858	59.13	2.32795
55.14	2.17087	56.14	2.21024	57.14	2.24961	58.14	2.28898	59.14	2.32835
55.15	2.17126	56.15	2.21063	57.15	2.25000	58.15	2.28937	59.15	2.32874
55.16	2.17165	56.16	2.21102	57.16	2.25039	58.16	2.28976	59.16	2.32913
55.17	2.17205	56.17	2.21142	57.17	2.25079	58.17	2.29016	59.17	2.32953
55.18	2.17244	56.18	2.21181	57.18	2.25118	58.18	2.29055	59.18	2.32992
55.19	2.17283	56.19	2.21220	57.19	2.25157	58.19	2.29094	59.19	2.33031
55.20	2.17323	56.20	2.21260	57.20	2.25197	58.20	2.29134	59.20	2.33071
55.21	2.17362	56.21	2.21299	57.21	2.25236	58.21	2.29173	59.21	2.33110
55.22	2.17402	56.22	2.21339	57.22	2.25276	58.22	2.29213	59.22	2.33150
55.23	2.17441	56.23	2.21378	57.23	2.25315	58.23	2.29252	59.23	2.33189
55.24	2.17480	56.24	2.21417	57.24	2.25354	58.24	2.29291	59.24	2.33228
55.25	2.17520	56.25	2.21457	57.25	2.25394	58.25	2.29331	59.25	2.33268
55.26	2.17559	56.26	2.21496	57.26	2.25433	58.26	2.29370	59.26	2.33307
55.27	2.17598	56.27	2.21535	57.27	2.25472	58.27	2.29409	59.27	2.33346
55.28	2.17638	56.28	2.21575	57.28	2.25512	58.28	2.29449	59.28	2.33386
55.29	2.17677	56.29	2.21614	57.29	2.25551	58.29	2.29488	59.29	2.33425
55.30	2.17717	56.30	2.21654	57.30	2.25591	58.30	2.29528	59.30	2.33465
55.31	2.17756	56.31	2.21693	57.31	2.25630	58.31	2.29567	59.31	2.33504
55.32	2.17795	56.32	2.21732	57.32	2.25669	58.32	2.29606	59.32	2.33543
55.33	2.17835	56.33	2.21772	57.33	2.25709	58.33	2.29646	59.33	2.33583
55.34	2.17874	56.34	2.21811	57.34	2.25748	58.34	2.29685	59.34	2.33622
55.35	2.17913	56.35	2.21850	57.35	2.25787	58.35	2.29724	59.35	2.33661
55.36	2.17953	56.36	2.21890	57.36	2.25827	58.36	2.29764	59.36	2.33701
55.37	2.17992	56.37	2.21929	57.37	2.25866	58.37	2.29803	59.37	2.33740
55.38	2.18031	56.38	2.21968	57.38	2.25906	58.38	2.29843	59.38	2.33780
55.39	2.18071	56.39	2.22008	57.39	2.25945	58.39	2.29882	59.39	2.33819
55.40	2.18110	56.40	2.22047	57.40	2.25984	58.40	2.29921	59.40	2.33858
55.41	2.18150	56.41	2.22087	57.41	2.26024	58.41	2.29961	59.41	2.33898
55.42	2.18189	56.42	2.22126	57.42	2.26063	58.42	2.30000	59.42	2.33937
55.43	2.18228	56.43	2.22165	57.43	2.26102	58.43	2.30039	59.43	2.33976
55.44	2.18268	56.44	2.22205	57.44	2.26142	58.44	2.30079	59.44	2.34016
55.45	2.18307	56.45	2.22244	57.45	2.26181	58.45	2.30118	59.45	2.34055
55.46	2.18346	56.46	2.22283	57.46	2.26220	58.46	2.30157	59.46	2.34094
55.47	2.18386	56.47	2.22323	57.47	2.26260	58.47	2.30197	59.47	2.34134
55.48	2.18425	56.48	2.22362	57.48	2.26299	58.48	2.30236	59.48	2.34173
55.49	2.18465	56.49	2.22402	57.49	2.26339	58.49	2.30276	59.49	2.34213

Continued - MILLIMETERS TO INCHES 1 MILLIMETER = 0.039 370 0787 INCH

MILLI-METERS	INCHES	MILLI-METERS	INCHES	MILLI-METERS	INCHES	MILLI-METERS	INCHES	MILLI-METERS	INCHES
55.50	2.18504	56.50	2.22441	57.50	2.26378	58.50	2.30315	59.50	2.34252
55.51	2.18543	56.51	2.22480	57.51	2.26417	58.51	2.30354	59.51	2.34291
55.52	2.18583	56.52	2.22520	57.52	2.26457	58.52	2.30394	59.52	2.34331
55.53	2.18622	56.53	2.22559	57.53	2.26496	58.53	2.30433	59.53	2.34370
55.54	2.18661	56.54	2.22598	57.54	2.26535	58.54	2.30472	59.54	2.34409
55.55	2.18701	56.55	2.22638	57.55	2.26575	58.55	2.30512	59.55	2.34449
55.56	2.18740	56.56	2.22677	57.56	2.26614	58.56	2.30551	59.56	2.34488
55.57	2.18780	56.57	2.22717	57.57	2.26654	58.57	2.30591	59.57	2.34528
55.58	2.18819	56.58	2.22756	57.58	2.26693	58.58	2.30630	59.58	2.34567
55.59	2.18858	56.59	2.22795	57.59	2.26732	58.59	2.30669	59.59	2.34606
55.60	2.18898	56.60	2.22835	57.60	2.26772	58.60	2.30709	59.60	2.34646
55.61	2.18937	56.61	2.22874	57.61	2.26811	58.61	2.30748	59.61	2.34685
55.62	2.18976	56.62	2.22913	57.62	2.26850	58.62	2.30787	59.62	2.34724
55.63	2.19016	56.63	2.22953	57.63	2.26890	58.63	2.30827	59.63	2.34764
55.64	2.19055	56.64	2.22992	57.64	2.26929	58.64	2.30866	59.64	2.34803
55.65	2.19094	56.65	2.23031	57.65	2.26969	58.65	2.30906	59.65	2.34843
55.66	2.19134	56.66	2.23071	57.66	2.27008	58.66	2.30945	59.66	2.34882
55.67	2.19173	56.67	2.23110	57.67	2.27047	58.67	2.30984	59.67	2.34921
55.68	2.19213	56.68	2.23150	57.68	2.27087	58.68	2.31024	59.68	2.34961
55.69	2.19252	56.69	2.23189	57.69	2.27126	58.69	2.31063	59.69	2.35000
55.70	2.19291	56.70	2.23228	57.70	2.27165	58.70	2.31102	59.70	2.35039
55.71	2.19331	56.71	2.23268	57.71	2.27205	58.71	2.31142	59.71	2.35079
55.72	2.19370	56.72	2.23307	57.72	2.27244	58.72	2.31181	59.72	2.35118
55.73	2.19409	56.73	2.23346	57.73	2.27283	58.73	2.31220	59.73	2.35157
55.74	2.19449	56.74	2.23386	57.74	2.27323	58.74	2.31260	59.74	2.35197
55.75	2.19488	56.75	2.23425	57.75	2.27362	58.75	2.31299	59.75	2.35236
55.76	2.19528	56.76	2.23465	57.76	2.27402	58.76	2.31339	59.76	2.35276
55.77	2.19567	56.77	2.23504	57.77	2.27441	58.77	2.31378	59.77	2.35315
55.78	2.19606	56.78	2.23543	57.78	2.27480	58.78	2.31417	59.78	2.35354
55.79	2.19646	56.79	2.23583	57.79	2.27520	58.79	2.31457	59.79	2.35394
55.80	2.19685	56.80	2.23622	57.80	2.27559	58.80	2.31496	59.80	2.35433
55.81	2.19724	56.81	2.23661	57.81	2.27598	58.81	2.31535	59.81	2.35472
55.82	2.19764	56.82	2.23701	57.82	2.27638	58.82	2.31575	59.82	2.35512
55.83	2.19803	56.83	2.23740	57.83	2.27677	58.83	2.31614	59.83	2.35551
55.84	2.19843	56.84	2.23780	57.84	2.27717	58.84	2.31654	59.84	2.35591
55.85	2.19882	56.85	2.23819	57.85	2.27756	58.85	2.31693	59.85	2.35630
55.86	2.19921	56.86	2.23858	57.86	2.27795	58.86	2.31732	59.86	2.35669
55.87	2.19961	56.87	2.23898	57.87	2.27835	58.87	2.31772	59.87	2.35709
55.88	2.20000	56.88	2.23937	57.88	2.27874	58.88	2.31811	59.88	2.35748
55.89	2.20039	56.89	2.23976	57.89	2.27913	58.89	2.31850	59.89	2.35787
55.90	2.20079	56.90	2.24016	57.90	2.27953	58.90	2.31890	59.90	2.35827
55.91	2.20118	56.91	2.24055	57.91	2.27992	58.91	2.31929	59.91	2.35866
55.92	2.20157	56.92	2.24094	57.92	2.28031	58.92	2.31969	59.92	2.35906
55.93	2.20197	56.93	2.24134	57.93	2.28071	58.93	2.32008	59.93	2.35945
55.94	2.20236	56.94	2.24173	57.94	2.28110	58.94	2.32047	59.94	2.35984
55.95	2.20276	56.95	2.24213	57.95	2.28150	58.95	2.32087	59.95	2.36024
55.96	2.20315	56.96	2.24252	57.96	2.28189	58.96	2.32126	59.96	2.36063
55.97	2.20354	56.97	2.24291	57.97	2.28228	58.97	2.32165	59.97	2.36102
55.98	2.20394	56.98	2.24331	57.98	2.28268	58.98	2.32205	59.98	2.36142
55.99	2.20433	56.99	2.24370	57.99	2.28307	58.99	2.32244	59.99	2.36181

Continued - MILLIMETERS TO INCHES 1 MILLIMETER = 0.039 370 0787 INCH

MILLI-METERS	INCHES	MILLI-METERS	INCHES	MILLI-METERS	INCHES	MILLI-METERS	INCHES	MILLI-METERS	INCHES
60.00	2.36220	61.00	2.40157	62.00	2.44094	63.00	2.48031	64.00	2.51969
60.01	2.36260	61.01	2.40197	62.01	2.44134	63.01	2.48071	64.01	2.52008
60.02	2.36299	61.02	2.40236	62.02	2.44173	63.02	2.48110	64.02	2.52047
60.03	2.36339	61.03	2.40276	62.03	2.44213	63.03	2.48150	64.03	2.52087
60.04	2.36378	61.04	2.40315	62.04	2.44252	63.04	2.48189	64.04	2.52126
60.05	2.36417	61.05	2.40354	62.05	2.44291	63.05	2.48228	64.05	2.52165
60.06	2.36457	61.06	2.40394	62.06	2.44331	63.06	2.48268	64.06	2.52205
60.07	2.36496	61.07	2.40433	62.07	2.44370	63.07	2.48307	64.07	2.52244
60.08	2.36535	61.08	2.40472	62.08	2.44409	63.08	2.48346	64.08	2.52283
60.09	2.36575	61.09	2.40512	62.09	2.44449	63.09	2.48386	64.09	2.52323
60.10	2.36614	61.10	2.40551	62.10	2.44488	63.10	2.48425	64.10	2.52362
60.11	2.36654	61.11	2.40591	62.11	2.44528	63.11	2.48465	64.11	2.52402
60.12	2.36693	61.12	2.40630	62.12	2.44567	63.12	2.48504	64.12	2.52441
60.13	2.36732	61.13	2.40669	62.13	2.44606	63.13	2.48543	64.13	2.52480
60.14	2.36772	61.14	2.40709	62.14	2.44646	63.14	2.48583	64.14	2.52520
60.15	2.36811	61.15	2.40748	62.15	2.44685	63.15	2.48622	64.15	2.52559
60.16	2.36850	61.16	2.40787	62.16	2.44724	63.16	2.48661	64.16	2.52598
60.17	2.36890	61.17	2.40827	62.17	2.44764	63.17	2.48701	64.17	2.52638
60.18	2.36929	61.18	2.40866	62.18	2.44803	63.18	2.48740	64.18	2.52677
60.19	2.36968	61.19	2.40906	62.19	2.44843	63.19	2.48780	64.19	2.52717
60.20	2.37008	61.20	2.40945	62.20	2.44882	63.20	2.48819	64.20	2.52756
60.21	2.37047	61.21	2.40984	62.21	2.44921	63.21	2.48858	64.21	2.52795
60.22	2.37087	61.22	2.41024	62.22	2.44961	63.22	2.48898	64.22	2.52835
60.23	2.37126	61.23	2.41063	62.23	2.45000	63.23	2.48937	64.23	2.52874
60.24	2.37165	61.24	2.41102	62.24	2.45039	63.24	2.48976	64.24	2.52913
60.25	2.37205	61.25	2.41142	62.25	2.45079	63.25	2.49016	64.25	2.52953
60.26	2.37244	61.26	2.41181	62.26	2.45118	63.26	2.49055	64.26	2.52992
60.27	2.37283	61.27	2.41220	62.27	2.45157	63.27	2.49094	64.27	2.53031
60.28	2.37323	61.28	2.41260	62.28	2.45197	63.28	2.49134	64.28	2.53071
60.29	2.37362	61.29	2.41299	62.29	2.45236	63.29	2.49173	64.29	2.53110
60.30	2.37402	61.30	2.41339	62.30	2.45276	63.30	2.49213	64.30	2.53150
60.31	2.37441	61.31	2.41378	62.31	2.45315	63.31	2.49252	64.31	2.53189
60.32	2.37480	61.32	2.41417	62.32	2.45354	63.32	2.49291	64.32	2.53228
60.33	2.37520	61.33	2.41457	62.33	2.45394	63.33	2.49331	64.33	2.53268
60.34	2.37559	61.34	2.41496	62.34	2.45433	63.34	2.49370	64.34	2.53307
60.35	2.37598	61.35	2.41535	62.35	2.45472	63.35	2.49409	64.35	2.53346
60.36	2.37638	61.36	2.41575	62.36	2.45512	63.36	2.49449	64.36	2.53386
60.37	2.37677	61.37	2.41614	62.37	2.45551	63.37	2.49488	64.37	2.53425
60.38	2.37717	61.38	2.41654	62.38	2.45591	63.38	2.49528	64.38	2.53465
60.39	2.37756	61.39	2.41693	62.39	2.45630	63.39	2.49567	64.39	2.53504
60.40	2.37795	61.40	2.41732	62.40	2.45669	63.40	2.49606	64.40	2.53543
60.41	2.37835	61.41	2.41772	62.41	2.45709	63.41	2.49646	64.41	2.53583
60.42	2.37874	61.42	2.41811	62.42	2.45748	63.42	2.49685	64.42	2.53622
60.43	2.37913	61.43	2.41850	62.43	2.45787	63.43	2.49724	64.43	2.53661
60.44	2.37953	61.44	2.41890	62.44	2.45827	63.44	2.49764	64.44	2.53701
60.45	2.37992	61.45	2.41929	62.45	2.45866	63.45	2.49803	64.45	2.53740
60.46	2.38031	61.46	2.41969	62.46	2.45906	63.46	2.49843	64.46	2.53780
60.47	2.38071	61.47	2.42008	62.47	2.45945	63.47	2.49882	64.47	2.53819
60.48	2.38110	61.48	2.42047	62.48	2.45984	63.48	2.49921	64.48	2.53858
60.49	2.38150	61.49	2.42087	62.49	2.46024	63.49	2.49961	64.49	2.53898

Continued - MILLIMETERS TO INCHES 1 MILLIMETER = 0.039 370 0787 INCH

MILLI-METERS	INCHES	MILLI-METERS	INCHES	MILLI-METERS	INCHES	MILLI-METERS	INCHES	MILLI-METERS	INCHES
60.50	2.38189	61.50	2.42126	62.50	2.46063	63.50	2.50000	64.50	2.53937
60.51	2.38228	61.51	2.42165	62.51	2.46102	63.51	2.50039	64.51	2.53976
60.52	2.38268	61.52	2.42205	62.52	2.46142	63.52	2.50079	64.52	2.54016
60.53	2.38307	61.53	2.42244	62.53	2.46181	63.53	2.50118	64.53	2.54055
60.54	2.38346	61.54	2.42283	62.54	2.46220	63.54	2.50157	64.54	2.54094
60.55	2.38386	61.55	2.42323	62.55	2.46260	63.55	2.50197	64.55	2.54134
60.56	2.38425	61.56	2.42362	62.56	2.46299	63.56	2.50236	64.56	2.54173
60.57	2.38465	61.57	2.42402	62.57	2.46339	63.57	2.50276	64.57	2.54213
60.58	2.38504	61.58	2.42441	62.58	2.46378	63.58	2.50315	64.58	2.54252
60.59	2.38543	61.59	2.42480	62.59	2.46417	63.59	2.50354	64.59	2.54291
60.60	2.38583	61.60	2.42520	62.60	2.46457	63.60	2.50394	64.60	2.54331
60.61	2.38622	61.61	2.42559	62.61	2.46496	63.61	2.50433	64.61	2.54370
60.62	2.38661	61.62	2.42598	62.62	2.46535	63.62	2.50472	64.62	2.54409
60.63	2.38701	61.63	2.42638	62.63	2.46575	63.63	2.50512	64.63	2.54449
60.64	2.38740	61.64	2.42677	62.64	2.46614	63.64	2.50551	64.64	2.54488
60.65	2.38780	61.65	2.42717	62.65	2.46654	63.65	2.50591	64.65	2.54528
60.66	2.38819	61.66	2.42756	62.66	2.46693	63.66	2.50630	64.66	2.54567
60.67	2.38858	61.67	2.42795	62.67	2.46732	63.67	2.50669	64.67	2.54606
60.68	2.38898	61.68	2.42835	62.68	2.46772	63.68	2.50709	64.68	2.54646
60.69	2.38937	61.69	2.42874	62.69	2.46811	63.69	2.50748	64.69	2.54685
60.70	2.38976	61.70	2.42913	62.70	2.46850	63.70	2.50787	64.70	2.54724
60.71	2.39016	61.71	2.42953	62.71	2.46890	63.71	2.50827	64.71	2.54764
60.72	2.39055	61.72	2.42992	62.72	2.46929	63.72	2.50866	64.72	2.54803
60.73	2.39094	61.73	2.43031	62.73	2.46968	63.73	2.50906	64.73	2.54843
60.74	2.39134	61.74	2.43071	62.74	2.47008	63.74	2.50945	64.74	2.54882
60.75	2.39173	61.75	2.43110	62.75	2.47047	63.75	2.50984	64.75	2.54921
60.76	2.39213	61.76	2.43150	62.76	2.47087	63.76	2.51024	64.76	2.54961
60.77	2.39252	61.77	2.43189	62.77	2.47126	63.77	2.51063	64.77	2.55000
60.78	2.39291	61.78	2.43228	62.78	2.47165	63.78	2.51102	64.78	2.55039
60.79	2.39331	61.79	2.43268	62.79	2.47205	63.79	2.51142	64.79	2.55079
60.80	2.39370	61.80	2.43307	62.80	2.47244	63.80	2.51181	64.80	2.55118
60.81	2.39409	61.81	2.43346	62.81	2.47283	63.81	2.51220	64.81	2.55157
60.82	2.39449	61.82	2.43386	62.82	2.47323	63.82	2.51260	64.82	2.55197
60.83	2.39488	61.83	2.43425	62.83	2.47362	63.83	2.51299	64.83	2.55236
60.84	2.39528	61.84	2.43465	62.84	2.47402	63.84	2.51339	64.84	2.55276
60.85	2.39567	61.85	2.43504	62.85	2.47441	63.85	2.51378	64.85	2.55315
60.86	2.39606	61.86	2.43543	62.86	2.47480	63.86	2.51417	64.86	2.55354
60.87	2.39646	61.87	2.43583	62.87	2.47520	63.87	2.51457	64.87	2.55394
60.88	2.39685	61.88	2.43622	62.88	2.47559	63.88	2.51496	64.88	2.55433
60.89	2.39724	61.89	2.43661	62.89	2.47598	63.89	2.51535	64.89	2.55472
60.90	2.39764	61.90	2.43701	62.90	2.47638	63.90	2.51575	64.90	2.55512
60.91	2.39803	61.91	2.43740	62.91	2.47677	63.91	2.51614	64.91	2.55551
60.92	2.39843	61.92	2.43780	62.92	2.47717	63.92	2.51654	64.92	2.55591
60.93	2.39882	61.93	2.43819	62.93	2.47756	63.93	2.51693	64.93	2.55630
60.94	2.39921	61.94	2.43858	62.94	2.47795	63.94	2.51732	64.94	2.55669
60.95	2.39961	61.95	2.43898	62.95	2.47835	63.95	2.51772	64.95	2.55709
60.96	2.40000	61.96	2.43937	62.96	2.47874	63.96	2.51811	64.96	2.55748
60.97	2.40039	61.97	2.43976	62.97	2.47913	63.97	2.51850	64.97	2.55787
60.98	2.40079	61.98	2.44016	62.98	2.47953	63.98	2.51890	64.98	2.55827
60.99	2.40118	61.99	2.44055	62.99	2.47992	63.99	2.51929	64.99	2.55866

Continued - MILLIMETERS TO INCHES 1 MILLIMETER = 0.039 370 0787 INCH

MILLI-METERS	INCHES	MILLI-METERS	INCHES	MILLI-METERS	INCHES	MILLI-METERS	INCHES	MILLI-METERS	INCHES
65.00	2.55906	66.00	2.59843	67.00	2.63780	68.00	2.67717	69.00	2.71654
65.01	2.55945	66.01	2.59882	67.01	2.63819	68.01	2.67756	69.01	2.71693
65.02	2.55984	66.02	2.59921	67.02	2.63858	68.02	2.67795	69.02	2.71732
65.03	2.56024	66.03	2.59961	67.03	2.63898	68.03	2.67835	69.03	2.71772
65.04	2.56063	66.04	2.60000	67.04	2.63937	68.04	2.67874	69.04	2.71811
65.05	2.56102	66.05	2.60039	67.05	2.63976	68.05	2.67913	69.05	2.71850
65.06	2.56142	66.06	2.60079	67.06	2.64016	68.06	2.67953	69.06	2.71890
65.07	2.56181	66.07	2.60118	67.07	2.64055	68.07	2.67992	69.07	2.71929
65.08	2.56220	66.08	2.60157	67.08	2.64094	68.08	2.68031	69.08	2.71969
65.09	2.56260	66.09	2.60197	67.09	2.64134	68.09	2.68071	69.09	2.72008
65.10	2.56299	66.10	2.60236	67.10	2.64173	68.10	2.68110	69.10	2.72047
65.11	2.56339	66.11	2.60276	67.11	2.64213	68.11	2.68150	69.11	2.72087
65.12	2.56378	66.12	2.60315	67.12	2.64252	68.12	2.68189	69.12	2.72126
65.13	2.56417	66.13	2.60354	67.13	2.64291	68.13	2.68228	69.13	2.72165
65.14	2.56457	66.14	2.60394	67.14	2.64331	68.14	2.68268	69.14	2.72205
65.15	2.56496	66.15	2.60433	67.15	2.64370	68.15	2.68307	69.15	2.72244
65.16	2.56535	66.16	2.60472	67.16	2.64409	68.16	2.68346	69.16	2.72283
65.17	2.56575	66.17	2.60512	67.17	2.64449	68.17	2.68386	69.17	2.72323
65.18	2.56614	66.18	2.60551	67.18	2.64488	68.18	2.68425	69.18	2.72362
65.19	2.56654	66.19	2.60591	67.19	2.64528	68.19	2.68465	69.19	2.72402
65.20	2.56693	66.20	2.60630	67.20	2.64567	68.20	2.68504	69.20	2.72441
65.21	2.56732	66.21	2.60669	67.21	2.64606	68.21	2.68543	69.21	2.72480
65.22	2.56772	66.22	2.60709	67.22	2.64646	68.22	2.68583	69.22	2.72520
65.23	2.56811	66.23	2.60748	67.23	2.64685	68.23	2.68622	69.23	2.72559
65.24	2.56850	66.24	2.60787	67.24	2.64724	68.24	2.68661	69.24	2.72598
65.25	2.56890	66.25	2.60827	67.25	2.64764	68.25	2.68701	69.25	2.72638
65.26	2.56929	66.26	2.60866	67.26	2.64803	68.26	2.68740	69.26	2.72677
65.27	2.56968	66.27	2.60906	67.27	2.64843	68.27	2.68780	69.27	2.72717
65.28	2.57008	66.28	2.60945	67.28	2.64882	68.28	2.68819	69.28	2.72756
65.29	2.57047	66.29	2.60984	67.29	2.64921	68.29	2.68858	69.29	2.72795
65.30	2.57087	66.30	2.61024	67.30	2.64961	68.30	2.68898	69.30	2.72835
65.31	2.57126	66.31	2.61063	67.31	2.65000	68.31	2.68937	69.31	2.72874
65.32	2.57165	66.32	2.61102	67.32	2.65039	68.32	2.68976	69.32	2.72913
65.33	2.57205	66.33	2.61142	67.33	2.65079	68.33	2.69016	69.33	2.72953
65.34	2.57244	66.34	2.61181	67.34	2.65118	68.34	2.69055	69.34	2.72992
65.35	2.57283	66.35	2.61220	67.35	2.65157	68.35	2.69094	69.35	2.73031
65.36	2.57323	66.36	2.61260	67.36	2.65197	68.36	2.69134	69.36	2.73071
65.37	2.57362	66.37	2.61299	67.37	2.65236	68.37	2.69173	69.37	2.73110
65.38	2.57402	66.38	2.61339	67.38	2.65276	68.38	2.69213	69.38	2.73150
65.39	2.57441	66.39	2.61378	67.39	2.65315	68.39	2.69252	69.39	2.73189
65.40	2.57480	66.40	2.61417	67.40	2.65354	68.40	2.69291	69.40	2.73228
65.41	2.57520	66.41	2.61457	67.41	2.65394	68.41	2.69331	69.41	2.73268
65.42	2.57559	66.42	2.61496	67.42	2.65433	68.42	2.69370	69.42	2.73307
65.43	2.57598	66.43	2.61535	67.43	2.65472	68.43	2.69409	69.43	2.73346
65.44	2.57638	66.44	2.61575	67.44	2.65512	68.44	2.69449	69.44	2.73386
65.45	2.57677	66.45	2.61614	67.45	2.65551	68.45	2.69488	69.45	2.73425
65.46	2.57717	66.46	2.61654	67.46	2.65591	68.46	2.69528	69.46	2.73465
65.47	2.57756	66.47	2.61693	67.47	2.65630	68.47	2.69567	69.47	2.73504
65.48	2.57795	66.48	2.61732	67.48	2.65669	68.48	2.69606	69.48	2.73543
65.49	2.57835	66.49	2.61772	67.49	2.65709	68.49	2.69646	69.49	2.73583

Continued - MILLIMETERS TO INCHES 1 MILLIMETER = 0.039 370 0787 INCH

MILLI-METERS	INCHES	MILLI-METERS	INCHES	MILLI-METERS	INCHES	MILLI-METERS	INCHES	MILLI-METERS	INCHES
65.50	2.57874	66.50	2.61811	67.50	2.65748	68.50	2.69685	69.50	2.73622
65.51	2.57913	66.51	2.61850	67.51	2.65787	68.51	2.69724	69.51	2.73661
65.52	2.57953	66.52	2.61890	67.52	2.65827	68.52	2.69764	69.52	2.73701
65.53	2.57992	66.53	2.61929	67.53	2.65866	68.53	2.69803	69.53	2.73740
65.54	2.58031	66.54	2.61969	67.54	2.65906	68.54	2.69843	69.54	2.73780
65.55	2.58071	66.55	2.62008	67.55	2.65945	68.55	2.69882	69.55	2.73819
65.56	2.58110	66.56	2.62047	67.56	2.65984	68.56	2.69921	69.56	2.73858
65.57	2.58150	66.57	2.62087	67.57	2.66024	68.57	2.69961	69.57	2.73898
65.58	2.58189	66.58	2.62126	67.58	2.66063	68.58	2.70000	69.58	2.73937
65.59	2.58228	66.59	2.62165	67.59	2.66102	68.59	2.70039	69.59	2.73976
65.60	2.58268	66.60	2.62205	67.60	2.66142	68.60	2.70079	69.60	2.74016
65.61	2.58307	66.61	2.62244	67.61	2.66181	68.61	2.70118	69.61	2.74055
65.62	2.58346	66.62	2.62283	67.62	2.66220	68.62	2.70157	69.62	2.74094
65.63	2.58386	66.63	2.62323	67.63	2.66260	68.63	2.70197	69.63	2.74134
65.64	2.58425	66.64	2.62362	67.64	2.66299	68.64	2.70236	69.64	2.74173
65.65	2.58465	66.65	2.62402	67.65	2.66339	68.65	2.70276	69.65	2.74213
65.66	2.58504	66.66	2.62441	67.66	2.66378	68.66	2.70315	69.66	2.74252
65.67	2.58543	66.67	2.62480	67.67	2.66417	68.67	2.70354	69.67	2.74291
65.68	2.58583	66.68	2.62520	67.68	2.66457	68.68	2.70394	69.68	2.74331
65.69	2.58622	66.69	2.62559	67.69	2.66496	68.69	2.70433	69.69	2.74370
65.70	2.58661	66.70	2.62598	67.70	2.66535	68.70	2.70472	69.70	2.74409
65.71	2.58701	66.71	2.62638	67.71	2.66575	68.71	2.70512	69.71	2.74449
65.72	2.58740	66.72	2.62677	67.72	2.66614	68.72	2.70551	69.72	2.74488
65.73	2.58780	66.73	2.62717	67.73	2.66654	68.73	2.70591	69.73	2.74528
65.74	2.58819	66.74	2.62756	67.74	2.66693	68.74	2.70630	69.74	2.74567
65.75	2.58858	66.75	2.62795	67.75	2.66732	68.75	2.70669	69.75	2.74606
65.76	2.58898	66.76	2.62835	67.76	2.66772	68.76	2.70709	69.76	2.74646
65.77	2.58937	66.77	2.62874	67.77	2.66811	68.77	2.70748	69.77	2.74685
65.78	2.58976	66.78	2.62913	67.78	2.66850	68.78	2.70787	69.78	2.74724
65.79	2.59016	66.79	2.62953	67.79	2.66890	68.79	2.70827	69.79	2.74764
65.80	2.59055	66.80	2.62992	67.80	2.66929	68.80	2.70866	69.80	2.74803
65.81	2.59094	66.81	2.63031	67.81	2.66969	68.81	2.70906	69.81	2.74843
65.82	2.59134	66.82	2.63071	67.82	2.67008	68.82	2.70945	69.82	2.74882
65.83	2.59173	66.83	2.63110	67.83	2.67047	68.83	2.70984	69.83	2.74921
65.84	2.59213	66.84	2.63150	67.84	2.67087	68.84	2.71024	69.84	2.74961
65.85	2.59252	66.85	2.63189	67.85	2.67126	68.85	2.71063	69.85	2.75000
65.86	2.59291	66.86	2.63228	67.86	2.67165	68.86	2.71102	69.86	2.75039
65.87	2.59331	66.87	2.63268	67.87	2.67205	68.87	2.71142	69.87	2.75079
65.88	2.59370	66.88	2.63307	67.88	2.67244	68.88	2.71181	69.88	2.75118
65.89	2.59409	66.89	2.63346	67.89	2.67283	68.89	2.71220	69.89	2.75157
65.90	2.59449	66.90	2.63386	67.90	2.67323	68.90	2.71260	69.90	2.75197
65.91	2.59488	66.91	2.63425	67.91	2.67362	68.91	2.71299	69.91	2.75236
65.92	2.59528	66.92	2.63465	67.92	2.67402	68.92	2.71339	69.92	2.75276
65.93	2.59567	66.93	2.63504	67.93	2.67441	68.93	2.71378	69.93	2.75315
65.94	2.59606	66.94	2.63543	67.94	2.67480	68.94	2.71417	69.94	2.75354
65.95	2.59646	66.95	2.63583	67.95	2.67520	68.95	2.71457	69.95	2.75394
65.96	2.59685	66.96	2.63622	67.96	2.67559	68.96	2.71496	69.96	2.75433
65.97	2.59724	66.97	2.63661	67.97	2.67598	68.97	2.71535	69.97	2.75472
65.98	2.59764	66.98	2.63701	67.98	2.67638	68.98	2.71575	69.98	2.75512
65.99	2.59803	66.99	2.63740	67.99	2.67677	68.99	2.71614	69.99	2.75551

Continued - MILLIMETERS TO INCHES 1 MILLIMETER = 0.039 370 0787 INCH

MILLI-METERS	INCHES	MILLI-METERS	INCHES	MILLI-METERS	INCHES	MILLI-METERS	INCHES	MILLI-METERS	INCHES
70.00	2.75591	71.00	2.79528	72.00	2.83465	73.00	2.87402	74.00	2.91339
70.01	2.75630	71.01	2.79567	72.01	2.83504	73.01	2.87441	74.01	2.91378
70.02	2.75669	71.02	2.79606	72.02	2.83543	73.02	2.87480	74.02	2.91417
70.03	2.75709	71.03	2.79646	72.03	2.83583	73.03	2.87520	74.03	2.91457
70.04	2.75748	71.04	2.79685	72.04	2.83622	73.04	2.87559	74.04	2.91496
70.05	2.75787	71.05	2.79724	72.05	2.83661	73.05	2.87598	74.05	2.91535
70.06	2.75827	71.06	2.79764	72.06	2.83701	73.06	2.87638	74.06	2.91575
70.07	2.75866	71.07	2.79803	72.07	2.83740	73.07	2.87677	74.07	2.91614
70.08	2.75906	71.08	2.79843	72.08	2.83780	73.08	2.87717	74.08	2.91654
70.09	2.75945	71.09	2.79882	72.09	2.83819	73.09	2.87756	74.09	2.91693
70.10	2.75984	71.10	2.79921	72.10	2.83858	73.10	2.87795	74.10	2.91732
70.11	2.76024	71.11	2.79961	72.11	2.83898	73.11	2.87835	74.11	2.91772
70.12	2.76063	71.12	2.80000	72.12	2.83937	73.12	2.87874	74.12	2.91811
70.13	2.76102	71.13	2.80039	72.13	2.83976	73.13	2.87913	74.13	2.91850
70.14	2.76142	71.14	2.80079	72.14	2.84016	73.14	2.87953	74.14	2.91890
70.15	2.76181	71.15	2.80118	72.15	2.84055	73.15	2.87992	74.15	2.91929
70.16	2.76220	71.16	2.80157	72.16	2.84094	73.16	2.88031	74.16	2.91969
70.17	2.76260	71.17	2.80197	72.17	2.84134	73.17	2.88071	74.17	2.92008
70.18	2.76299	71.18	2.80236	72.18	2.84173	73.18	2.88110	74.18	2.92047
70.19	2.76339	71.19	2.80276	72.19	2.84213	73.19	2.88150	74.19	2.92087
70.20	2.76378	71.20	2.80315	72.20	2.84252	73.20	2.88189	74.20	2.92126
70.21	2.76417	71.21	2.80354	72.21	2.84291	73.21	2.88228	74.21	2.92165
70.22	2.76457	71.22	2.80394	72.22	2.84331	73.22	2.88268	74.22	2.92205
70.23	2.76496	71.23	2.80433	72.23	2.84370	73.23	2.88307	74.23	2.92244
70.24	2.76535	71.24	2.80472	72.24	2.84409	73.24	2.88346	74.24	2.92283
70.25	2.76575	71.25	2.80512	72.25	2.84449	73.25	2.88386	74.25	2.92323
70.26	2.76614	71.26	2.80551	72.26	2.84488	73.26	2.88425	74.26	2.92362
70.27	2.76654	71.27	2.80591	72.27	2.84528	73.27	2.88465	74.27	2.92402
70.28	2.76693	71.28	2.80630	72.28	2.84567	73.28	2.88504	74.28	2.92441
70.29	2.76732	71.29	2.80669	72.29	2.84606	73.29	2.88543	74.29	2.92480
70.30	2.76772	71.30	2.80709	72.30	2.84646	73.30	2.88583	74.30	2.92520
70.31	2.76811	71.31	2.80748	72.31	2.84685	73.31	2.88622	74.31	2.92559
70.32	2.76850	71.32	2.80787	72.32	2.84724	73.32	2.88661	74.32	2.92598
70.33	2.76890	71.33	2.80827	72.33	2.84764	73.33	2.88701	74.33	2.92638
70.34	2.76929	71.34	2.80866	72.34	2.84803	73.34	2.88740	74.34	2.92677
70.35	2.76969	71.35	2.80906	72.35	2.84843	73.35	2.88780	74.35	2.92717
70.36	2.77008	71.36	2.80945	72.36	2.84882	73.36	2.88819	74.36	2.92756
70.37	2.77047	71.37	2.80984	72.37	2.84921	73.37	2.88858	74.37	2.92795
70.38	2.77087	71.38	2.81024	72.38	2.84961	73.38	2.88898	74.38	2.92835
70.39	2.77126	71.39	2.81063	72.39	2.85000	73.39	2.88937	74.39	2.92874
70.40	2.77165	71.40	2.81102	72.40	2.85039	73.40	2.88976	74.40	2.92913
70.41	2.77205	71.41	2.81142	72.41	2.85079	73.41	2.89016	74.41	2.92953
70.42	2.77244	71.42	2.81181	72.42	2.85118	73.42	2.89055	74.42	2.92992
70.43	2.77283	71.43	2.81220	72.43	2.85157	73.43	2.89094	74.43	2.93031
70.44	2.77323	71.44	2.81260	72.44	2.85197	73.44	2.89134	74.44	2.93071
70.45	2.77362	71.45	2.81299	72.45	2.85236	73.45	2.89173	74.45	2.93110
70.46	2.77402	71.46	2.81339	72.46	2.85276	73.46	2.89213	74.46	2.93150
70.47	2.77441	71.47	2.81378	72.47	2.85315	73.47	2.89252	74.47	2.93189
70.48	2.77480	71.48	2.81417	72.48	2.85354	73.48	2.89291	74.48	2.93228
70.49	2.77520	71.49	2.81457	72.49	2.85394	73.49	2.89331	74.49	2.93268

Continued - MILLIMETERS TO INCHES 1 MILLIMETER = 0.039 370 0787 INCH

MILLI-METERS	INCHES	MILLI-METERS	INCHES	MILLI-METERS	INCHES	MILLI-METERS	INCHES	MILLI-METERS	INCHES
70.50	2.77559	71.50	2.81496	72.50	2.85433	73.50	2.89370	74.50	2.93307
70.51	2.77598	71.51	2.81535	72.51	2.85472	73.51	2.89409	74.51	2.93346
70.52	2.77638	71.52	2.81575	72.52	2.85512	73.52	2.89449	74.52	2.93386
70.53	2.77677	71.53	2.81614	72.53	2.85551	73.53	2.89488	74.53	2.93425
70.54	2.77717	71.54	2.81654	72.54	2.85591	73.54	2.89528	74.54	2.93465
70.55	2.77756	71.55	2.81693	72.55	2.85630	73.55	2.89567	74.55	2.93504
70.56	2.77795	71.56	2.81732	72.56	2.85669	73.56	2.89606	74.56	2.93543
70.57	2.77835	71.57	2.81772	72.57	2.85709	73.57	2.89646	74.57	2.93583
70.58	2.77874	71.58	2.81811	72.58	2.85748	73.58	2.89685	74.58	2.93622
70.59	2.77913	71.59	2.81850	72.59	2.85787	73.59	2.89724	74.59	2.93661
70.60	2.77953	71.60	2.81890	72.60	2.85827	73.60	2.89764	74.60	2.93701
70.61	2.77992	71.61	2.81929	72.61	2.85866	73.61	2.89803	74.61	2.93740
70.62	2.78031	71.62	2.81969	72.62	2.85906	73.62	2.89843	74.62	2.93780
70.63	2.78071	71.63	2.82008	72.63	2.85945	73.63	2.89882	74.63	2.93819
70.64	2.78110	71.64	2.82047	72.64	2.85984	73.64	2.89921	74.64	2.93858
70.65	2.78150	71.65	2.82087	72.65	2.86024	73.65	2.89961	74.65	2.93898
70.66	2.78189	71.66	2.82126	72.66	2.86063	73.66	2.90000	74.66	2.93937
70.67	2.78228	71.67	2.82165	72.67	2.86102	73.67	2.90039	74.67	2.93976
70.68	2.78268	71.68	2.82205	72.68	2.86142	73.68	2.90079	74.68	2.94016
70.69	2.78307	71.69	2.82244	72.69	2.86181	73.69	2.90118	74.69	2.94055
70.70	2.78346	71.70	2.82283	72.70	2.86220	73.70	2.90157	74.70	2.94094
70.71	2.78386	71.71	2.82323	72.71	2.86260	73.71	2.90197	74.71	2.94134
70.72	2.78425	71.72	2.82362	72.72	2.86299	73.72	2.90236	74.72	2.94173
70.73	2.78465	71.73	2.82402	72.73	2.86339	73.73	2.90276	74.73	2.94213
70.74	2.78504	71.74	2.82441	72.74	2.86378	73.74	2.90315	74.74	2.94252
70.75	2.78543	71.75	2.82480	72.75	2.86417	73.75	2.90354	74.75	2.94291
70.76	2.78583	71.76	2.82520	72.76	2.86457	73.76	2.90394	74.76	2.94331
70.77	2.78622	71.77	2.82559	72.77	2.86496	73.77	2.90433	74.77	2.94370
70.78	2.78661	71.78	2.82598	72.78	2.86535	73.78	2.90472	74.78	2.94409
70.79	2.78701	71.79	2.82638	72.79	2.86575	73.79	2.90512	74.79	2.94449
70.80	2.78740	71.80	2.82677	72.80	2.86614	73.80	2.90551	74.80	2.94488
70.81	2.78780	71.81	2.82717	72.81	2.86654	73.81	2.90591	74.81	2.94528
70.82	2.78819	71.82	2.82756	72.82	2.86693	73.82	2.90630	74.82	2.94567
70.83	2.78858	71.83	2.82795	72.83	2.86732	73.83	2.90669	74.83	2.94606
70.84	2.78898	71.84	2.82835	72.84	2.86772	73.84	2.90709	74.84	2.94646
70.85	2.78937	71.85	2.82874	72.85	2.86811	73.85	2.90748	74.85	2.94685
70.86	2.78976	71.86	2.82913	72.86	2.86850	73.86	2.90787	74.86	2.94724
70.87	2.79016	71.87	2.82953	72.87	2.86890	73.87	2.90827	74.87	2.94764
70.88	2.79055	71.88	2.82992	72.88	2.86929	73.88	2.90866	74.88	2.94803
70.89	2.79094	71.89	2.83031	72.89	2.86968	73.89	2.90906	74.89	2.94843
70.90	2.79134	71.90	2.83071	72.90	2.87008	73.90	2.90945	74.90	2.94882
70.91	2.79173	71.91	2.83110	72.91	2.87047	73.91	2.90984	74.91	2.94921
70.92	2.79213	71.92	2.83150	72.92	2.87087	73.92	2.91024	74.92	2.94961
70.93	2.79252	71.93	2.83189	72.93	2.87126	73.93	2.91063	74.93	2.95000
70.94	2.79291	71.94	2.83228	72.94	2.87165	73.94	2.91102	74.94	2.95039
70.95	2.79331	71.95	2.83268	72.95	2.87205	73.95	2.91142	74.95	2.95079
70.96	2.79370	71.96	2.83307	72.96	2.87244	73.96	2.91181	74.96	2.95118
70.97	2.79409	71.97	2.83346	72.97	2.87283	73.97	2.91220	74.97	2.95157
70.98	2.79449	71.98	2.83386	72.98	2.87323	73.98	2.91260	74.98	2.95197
70.99	2.79488	71.99	2.83425	72.99	2.87362	73.99	2.91299	74.99	2.95236

Continued - MILLIMETERS TO INCHES 1 MILLIMETER = 0.039 370 0787 INCH

MILLI-METERS	INCHES	MILLI-METERS	INCHES	MILLI-METERS	INCHES	MILLI-METERS	INCHES	MILLI-METERS	INCHES
75.00	2.95276	76.00	2.99213	77.00	3.03150	78.00	3.07087	79.00	3.11024
75.01	2.95315	76.01	2.99252	77.01	3.03189	78.01	3.07126	79.01	3.11063
75.02	2.95354	76.02	2.99291	77.02	3.03228	78.02	3.07165	79.02	3.11102
75.03	2.95394	76.03	2.99331	77.03	3.03268	78.03	3.07205	79.03	3.11142
75.04	2.95433	76.04	2.99370	77.04	3.03307	78.04	3.07244	79.04	3.11181
75.05	2.95472	76.05	2.99409	77.05	3.03346	78.05	3.07283	79.05	3.11220
75.06	2.95512	76.06	2.99449	77.06	3.03386	78.06	3.07323	79.06	3.11260
75.07	2.95551	76.07	2.99488	77.07	3.03425	78.07	3.07362	79.07	3.11299
75.08	2.95591	76.08	2.99528	77.08	3.03465	78.08	3.07402	79.08	3.11339
75.09	2.95630	76.09	2.99567	77.09	3.03504	78.09	3.07441	79.09	3.11378
75.10	2.95669	76.10	2.99606	77.10	3.03543	78.10	3.07480	79.10	3.11417
75.11	2.95709	76.11	2.99646	77.11	3.03583	78.11	3.07520	79.11	3.11457
75.12	2.95748	76.12	2.99685	77.12	3.03622	78.12	3.07559	79.12	3.11496
75.13	2.95787	76.13	2.99724	77.13	3.03661	78.13	3.07598	79.13	3.11535
75.14	2.95827	76.14	2.99764	77.14	3.03701	78.14	3.07638	79.14	3.11575
75.15	2.95866	76.15	2.99803	77.15	3.03740	78.15	3.07677	79.15	3.11614
75.16	2.95906	76.16	2.99843	77.16	3.03780	78.16	3.07717	79.16	3.11654
75.17	2.95945	76.17	2.99882	77.17	3.03819	78.17	3.07756	79.17	3.11693
75.18	2.95984	76.18	2.99921	77.18	3.03858	78.18	3.07795	79.18	3.11732
75.19	2.96024	76.19	2.99961	77.19	3.03898	78.19	3.07835	79.19	3.11772
75.20	2.96063	76.20	3.00000	77.20	3.03937	78.20	3.07874	79.20	3.11811
75.21	2.96102	76.21	3.00039	77.21	3.03976	78.21	3.07913	79.21	3.11850
75.22	2.96142	76.22	3.00079	77.22	3.04016	78.22	3.07953	79.22	3.11890
75.23	2.96181	76.23	3.00118	77.23	3.04055	78.23	3.07992	79.23	3.11929
75.24	2.96220	76.24	3.00157	77.24	3.04094	78.24	3.08031	79.24	3.11968
75.25	2.96260	76.25	3.00197	77.25	3.04134	78.25	3.08071	79.25	3.12008
75.26	2.96299	76.26	3.00236	77.26	3.04173	78.26	3.08110	79.26	3.12047
75.27	2.96339	76.27	3.00276	77.27	3.04213	78.27	3.08150	79.27	3.12087
75.28	2.96378	76.28	3.00315	77.28	3.04252	78.28	3.08189	79.28	3.12126
75.29	2.96417	76.29	3.00354	77.29	3.04291	78.29	3.08228	79.29	3.12165
75.30	2.96457	76.30	3.00394	77.30	3.04331	78.30	3.08268	79.30	3.12205
75.31	2.96496	76.31	3.00433	77.31	3.04370	78.31	3.08307	79.31	3.12244
75.32	2.96535	76.32	3.00472	77.32	3.04409	78.32	3.08346	79.32	3.12283
75.33	2.96575	76.33	3.00512	77.33	3.04449	78.33	3.08386	79.33	3.12323
75.34	2.96614	76.34	3.00551	77.34	3.04488	78.34	3.08425	79.34	3.12362
75.35	2.96654	76.35	3.00591	77.35	3.04528	78.35	3.08465	79.35	3.12402
75.36	2.96693	76.36	3.00630	77.36	3.04567	78.36	3.08504	79.36	3.12441
75.37	2.96732	76.37	3.00669	77.37	3.04606	78.37	3.08543	79.37	3.12480
75.38	2.96772	76.38	3.00709	77.38	3.04646	78.38	3.08583	79.38	3.12520
75.39	2.96811	76.39	3.00748	77.39	3.04685	78.39	3.08622	79.39	3.12559
75.40	2.96850	76.40	3.00787	77.40	3.04724	78.40	3.08661	79.40	3.12598
75.41	2.96890	76.41	3.00827	77.41	3.04764	78.41	3.08701	79.41	3.12638
75.42	2.96929	76.42	3.00866	77.42	3.04803	78.42	3.08740	79.42	3.12677
75.43	2.96968	76.43	3.00906	77.43	3.04843	78.43	3.08780	79.43	3.12717
75.44	2.97008	76.44	3.00945	77.44	3.04882	78.44	3.08819	79.44	3.12756
75.45	2.97047	76.45	3.00984	77.45	3.04921	78.45	3.08858	79.45	3.12795
75.46	2.97087	76.46	3.01024	77.46	3.04961	78.46	3.08898	79.46	3.12835
75.47	2.97126	76.47	3.01063	77.47	3.05000	78.47	3.08937	79.47	3.12874
75.48	2.97165	76.48	3.01102	77.48	3.05039	78.48	3.08976	79.48	3.12913
75.49	2.97205	76.49	3.01142	77.49	3.05079	78.49	3.09016	79.49	3.12953

Continued - MILLIMETERS TO INCHES 1 MILLIMETER = 0.039 370 0787 INCH

MILLI-METERS	INCHES	MILLI-METERS	INCHES	MILLI-METERS	INCHES	MILLI-METERS	INCHES	MILLI-METERS	INCHES
75.50	2.97244	76.50	3.01181	77.50	3.05118	78.50	3.09055	79.50	3.12992
75.51	2.97283	76.51	3.01220	77.51	3.05157	78.51	3.09094	79.51	3.13031
75.52	2.97323	76.52	3.01260	77.52	3.05197	78.52	3.09134	79.52	3.13071
75.53	2.97362	76.53	3.01299	77.53	3.05236	78.53	3.09173	79.53	3.13110
75.54	2.97402	76.54	3.01339	77.54	3.05276	78.54	3.09213	79.54	3.13150
75.55	2.97441	76.55	3.01378	77.55	3.05315	78.55	3.09252	79.55	3.13189
75.56	2.97480	76.56	3.01417	77.56	3.05354	78.56	3.09291	79.56	3.13228
75.57	2.97520	76.57	3.01457	77.57	3.05394	78.57	3.09331	79.57	3.13268
75.58	2.97559	76.58	3.01496	77.58	3.05433	78.58	3.09370	79.58	3.13307
75.59	2.97598	76.59	3.01535	77.59	3.05472	78.59	3.09409	79.59	3.13346
75.60	2.97638	76.60	3.01575	77.60	3.05512	78.60	3.09449	79.60	3.13386
75.61	2.97677	76.61	3.01614	77.61	3.05551	78.61	3.09488	79.61	3.13425
75.62	2.97717	76.62	3.01654	77.62	3.05591	78.62	3.09528	79.62	3.13465
75.63	2.97756	76.63	3.01693	77.63	3.05630	78.63	3.09567	79.63	3.13504
75.64	2.97795	76.64	3.01732	77.64	3.05669	78.64	3.09606	79.64	3.13543
75.65	2.97835	76.65	3.01772	77.65	3.05709	78.65	3.09646	79.65	3.13583
75.66	2.97874	76.66	3.01811	77.66	3.05748	78.66	3.09685	79.66	3.13622
75.67	2.97913	76.67	3.01850	77.67	3.05787	78.67	3.09724	79.67	3.13661
75.68	2.97953	76.68	3.01890	77.68	3.05827	78.68	3.09764	79.68	3.13701
75.69	2.97992	76.69	3.01929	77.69	3.05866	78.69	3.09803	79.69	3.13740
75.70	2.98031	76.70	3.01969	77.70	3.05906	78.70	3.09843	79.70	3.13780
75.71	2.98071	76.71	3.02008	77.71	3.05945	78.71	3.09882	79.71	3.13819
75.72	2.98110	76.72	3.02047	77.72	3.05984	78.72	3.09921	79.72	3.13858
75.73	2.98150	76.73	3.02087	77.73	3.06024	78.73	3.09961	79.73	3.13898
75.74	2.98189	76.74	3.02126	77.74	3.06063	78.74	3.10000	79.74	3.13937
75.75	2.98228	76.75	3.02165	77.75	3.06102	78.75	3.10039	79.75	3.13976
75.76	2.98268	76.76	3.02205	77.76	3.06142	78.76	3.10079	79.76	3.14016
75.77	2.98307	76.77	3.02244	77.77	3.06181	78.77	3.10118	79.77	3.14055
75.78	2.98346	76.78	3.02283	77.78	3.06220	78.78	3.10157	79.78	3.14094
75.79	2.98386	76.79	3.02323	77.79	3.06260	78.79	3.10197	79.79	3.14134
75.80	2.98425	76.80	3.02362	77.80	3.06299	78.80	3.10236	79.80	3.14173
75.81	2.98465	76.81	3.02402	77.81	3.06339	78.81	3.10276	79.81	3.14213
75.82	2.98504	76.82	3.02441	77.82	3.06378	78.82	3.10315	79.82	3.14252
75.83	2.98543	76.83	3.02480	77.83	3.06417	78.83	3.10354	79.83	3.14291
75.84	2.98583	76.84	3.02520	77.84	3.06457	78.84	3.10394	79.84	3.14331
75.85	2.98622	76.85	3.02559	77.85	3.06496	78.85	3.10433	79.85	3.14370
75.86	2.98661	76.86	3.02598	77.86	3.06535	78.86	3.10472	79.86	3.14409
75.87	2.98701	76.87	3.02638	77.87	3.06575	78.87	3.10512	79.87	3.14449
75.88	2.98740	76.88	3.02677	77.88	3.06614	78.88	3.10551	79.88	3.14488
75.89	2.98780	76.89	3.02717	77.89	3.06654	78.89	3.10591	79.89	3.14528
75.90	2.98819	76.90	3.02756	77.90	3.06693	78.90	3.10630	79.90	3.14567
75.91	2.98858	76.91	3.02795	77.91	3.06732	78.91	3.10669	79.91	3.14606
75.92	2.98898	76.92	3.02835	77.92	3.06772	78.92	3.10709	79.92	3.14646
75.93	2.98937	76.93	3.02874	77.93	3.06811	78.93	3.10748	79.93	3.14685
75.94	2.98976	76.94	3.02913	77.94	3.06850	78.94	3.10787	79.94	3.14724
75.95	2.99016	76.95	3.02953	77.95	3.06890	78.95	3.10827	79.95	3.14764
75.96	2.99055	76.96	3.02992	77.96	3.06929	78.96	3.10866	79.96	3.14803
75.97	2.99094	76.97	3.03031	77.97	3.06968	78.97	3.10906	79.97	3.14843
75.98	2.99134	76.98	3.03071	77.98	3.07008	78.98	3.10945	79.98	3.14882
75.99	2.99173	76.99	3.03110	77.99	3.07047	78.99	3.10984	79.99	3.14921

Continued - MILLIMETERS TO INCHES 1 MILLIMETER = 0.039 370 0787 INCH

MILLI-METERS	INCHES	MILLI-METERS	INCHES	MILLI-METERS	INCHES	MILLI-METERS	INCHES	MILLI-METERS	INCHES
80.00	3.14961	81.00	3.18898	82.00	3.22835	83.00	3.26772	84.00	3.30709
80.01	3.15000	81.01	3.18937	82.01	3.22874	83.01	3.26811	84.01	3.30748
80.02	3.15039	81.02	3.18976	82.02	3.22913	83.02	3.26850	84.02	3.30787
80.03	3.15079	81.03	3.19016	82.03	3.22953	83.03	3.26890	84.03	3.30827
80.04	3.15118	81.04	3.19055	82.04	3.22992	83.04	3.26929	84.04	3.30866
80.05	3.15157	81.05	3.19094	82.05	3.23031	83.05	3.26968	84.05	3.30906
80.06	3.15197	81.06	3.19134	82.06	3.23071	83.06	3.27008	84.06	3.30945
80.07	3.15236	81.07	3.19173	82.07	3.23110	83.07	3.27047	84.07	3.30984
80.08	3.15276	81.08	3.19213	82.08	3.23150	83.08	3.27087	84.08	3.31024
80.09	3.15315	81.09	3.19252	82.09	3.23189	83.09	3.27126	84.09	3.31063
80.10	3.15354	81.10	3.19291	82.10	3.23228	83.10	3.27165	84.10	3.31102
80.11	3.15394	81.11	3.19331	82.11	3.23268	83.11	3.27205	84.11	3.31142
80.12	3.15433	81.12	3.19370	82.12	3.23307	83.12	3.27244	84.12	3.31181
80.13	3.15472	81.13	3.19409	82.13	3.23346	83.13	3.27283	84.13	3.31220
80.14	3.15512	81.14	3.19449	82.14	3.23386	83.14	3.27323	84.14	3.31260
80.15	3.15551	81.15	3.19488	82.15	3.23425	83.15	3.27362	84.15	3.31299
80.16	3.15591	81.16	3.19528	82.16	3.23465	83.16	3.27402	84.16	3.31339
80.17	3.15630	81.17	3.19567	82.17	3.23504	83.17	3.27441	84.17	3.31378
80.18	3.15669	81.18	3.19606	82.18	3.23543	83.18	3.27480	84.18	3.31417
80.19	3.15709	81.19	3.19646	82.19	3.23583	83.19	3.27520	84.19	3.31457
80.20	3.15748	81.20	3.19685	82.20	3.23622	83.20	3.27559	84.20	3.31496
80.21	3.15787	81.21	3.19724	82.21	3.23661	83.21	3.27598	84.21	3.31535
80.22	3.15827	81.22	3.19764	82.22	3.23701	83.22	3.27638	84.22	3.31575
80.23	3.15866	81.23	3.19803	82.23	3.23740	83.23	3.27677	84.23	3.31614
80.24	3.15906	81.24	3.19843	82.24	3.23780	83.24	3.27717	84.24	3.31654
80.25	3.15945	81.25	3.19882	82.25	3.23819	83.25	3.27756	84.25	3.31693
80.26	3.15984	81.26	3.19921	82.26	3.23858	83.26	3.27795	84.26	3.31732
80.27	3.16024	81.27	3.19961	82.27	3.23898	83.27	3.27835	84.27	3.31772
80.28	3.16063	81.28	3.20000	82.28	3.23937	83.28	3.27874	84.28	3.31811
80.29	3.16102	81.29	3.20039	82.29	3.23976	83.29	3.27913	84.29	3.31850
80.30	3.16142	81.30	3.20079	82.30	3.24016	83.30	3.27953	84.30	3.31890
80.31	3.16181	81.31	3.20118	82.31	3.24055	83.31	3.27992	84.31	3.31929
80.32	3.16220	81.32	3.20157	82.32	3.24094	83.32	3.28031	84.32	3.31969
80.33	3.16260	81.33	3.20197	82.33	3.24134	83.33	3.28071	84.33	3.32008
80.34	3.16299	81.34	3.20236	82.34	3.24173	83.34	3.28110	84.34	3.32047
80.35	3.16339	81.35	3.20276	82.35	3.24213	83.35	3.28150	84.35	3.32087
80.36	3.16378	81.36	3.20315	82.36	3.24252	83.36	3.28189	84.36	3.32126
80.37	3.16417	81.37	3.20354	82.37	3.24291	83.37	3.28228	84.37	3.32165
80.38	3.16457	81.38	3.20394	82.38	3.24331	83.38	3.28268	84.38	3.32205
80.39	3.16496	81.39	3.20433	82.39	3.24370	83.39	3.28307	84.39	3.32244
80.40	3.16535	81.40	3.20472	82.40	3.24409	83.40	3.28346	84.40	3.32283
80.41	3.16575	81.41	3.20512	82.41	3.24449	83.41	3.28386	84.41	3.32323
80.42	3.16614	81.42	3.20551	82.42	3.24488	83.42	3.28425	84.42	3.32362
80.43	3.16654	81.43	3.20591	82.43	3.24528	83.43	3.28465	84.43	3.32402
80.44	3.16693	81.44	3.20630	82.44	3.24567	83.44	3.28504	84.44	3.32441
80.45	3.16732	81.45	3.20669	82.45	3.24606	83.45	3.28543	84.45	3.32480
80.46	3.16772	81.46	3.20709	82.46	3.24646	83.46	3.28583	84.46	3.32520
80.47	3.16811	81.47	3.20748	82.47	3.24685	83.47	3.28622	84.47	3.32559
80.48	3.16850	81.48	3.20787	82.48	3.24724	83.48	3.28661	84.48	3.32598
80.49	3.16890	81.49	3.20827	82.49	3.24764	83.49	3.28701	84.49	3.32638

Continued - MILLIMETERS TO INCHES 1 MILLIMETER = 0.039 370 0787 INCH

MILLI-METERS	INCHES	MILLI-METERS	INCHES	MILLI-METERS	INCHES	MILLI-METERS	INCHES	MILLI-METERS	INCHES
80.50	3.16929	81.50	3.20866	82.50	3.24803	83.50	3.28740	84.50	3.32677
80.51	3.16968	81.51	3.20906	82.51	3.24843	83.51	3.28780	84.51	3.32717
80.52	3.17008	81.52	3.20945	82.52	3.24882	83.52	3.28819	84.52	3.32756
80.53	3.17047	81.53	3.20984	82.53	3.24921	83.53	3.28858	84.53	3.32795
80.54	3.17087	81.54	3.21024	82.54	3.24961	83.54	3.28898	84.54	3.32835
80.55	3.17126	81.55	3.21063	82.55	3.25000	83.55	3.28937	84.55	3.32874
80.56	3.17165	81.56	3.21102	82.56	3.25039	83.56	3.28976	84.56	3.32913
80.57	3.17205	81.57	3.21142	82.57	3.25079	83.57	3.29016	84.57	3.32953
80.58	3.17244	81.58	3.21181	82.58	3.25118	83.58	3.29055	84.58	3.32992
80.59	3.17283	81.59	3.21220	82.59	3.25157	83.59	3.29094	84.59	3.33031
80.60	3.17323	81.60	3.21260	82.60	3.25197	83.60	3.29134	84.60	3.33071
80.61	3.17362	81.61	3.21299	82.61	3.25236	83.61	3.29173	84.61	3.33110
80.62	3.17402	81.62	3.21339	82.62	3.25276	83.62	3.29213	84.62	3.33150
80.63	3.17441	81.63	3.21378	82.63	3.25315	83.63	3.29252	84.63	3.33189
80.64	3.17480	81.64	3.21417	82.64	3.25354	83.64	3.29291	84.64	3.33228
80.65	3.17520	81.65	3.21457	82.65	3.25394	83.65	3.29331	84.65	3.33268
80.66	3.17559	81.66	3.21496	82.66	3.25433	83.66	3.29370	84.66	3.33307
80.67	3.17598	81.67	3.21535	82.67	3.25472	83.67	3.29409	84.67	3.33346
80.68	3.17638	81.68	3.21575	82.68	3.25512	83.68	3.29449	84.68	3.33386
80.69	3.17677	81.69	3.21614	82.69	3.25551	83.69	3.29488	84.69	3.33425
80.70	3.17717	81.70	3.21654	82.70	3.25591	83.70	3.29528	84.70	3.33465
80.71	3.17756	81.71	3.21693	82.71	3.25630	83.71	3.29567	84.71	3.33504
80.72	3.17795	81.72	3.21732	82.72	3.25669	83.72	3.29606	84.72	3.33543
80.73	3.17835	81.73	3.21772	82.73	3.25709	83.73	3.29646	84.73	3.33583
80.74	3.17874	81.74	3.21811	82.74	3.25748	83.74	3.29685	84.74	3.33622
80.75	3.17913	81.75	3.21850	82.75	3.25787	83.75	3.29724	84.75	3.33661
80.76	3.17953	81.76	3.21890	82.76	3.25827	83.76	3.29764	84.76	3.33701
80.77	3.17992	81.77	3.21929	82.77	3.25866	83.77	3.29803	84.77	3.33740
80.78	3.18031	81.78	3.21968	82.78	3.25906	83.78	3.29843	84.78	3.33780
80.79	3.18071	81.79	3.22008	82.79	3.25945	83.79	3.29882	84.79	3.33819
80.80	3.18110	81.80	3.22047	82.80	3.25984	83.80	3.29921	84.80	3.33858
80.81	3.18150	81.81	3.22087	82.81	3.26024	83.81	3.29961	84.81	3.33898
80.82	3.18189	81.82	3.22126	82.82	3.26063	83.82	3.30000	84.82	3.33937
80.83	3.18228	81.83	3.22165	82.83	3.26102	83.83	3.30039	84.83	3.33976
80.84	3.18268	81.84	3.22205	82.84	3.26142	83.84	3.30079	84.84	3.34016
80.85	3.18307	81.85	3.22244	82.85	3.26181	83.85	3.30118	84.85	3.34055
80.86	3.18346	81.86	3.22283	82.86	3.26220	83.86	3.30157	84.86	3.34094
80.87	3.18386	81.87	3.22323	82.87	3.26260	83.87	3.30197	84.87	3.34134
80.88	3.18425	81.88	3.22362	82.88	3.26299	83.88	3.30236	84.88	3.34173
80.89	3.18465	81.89	3.22402	82.89	3.26339	83.89	3.30276	84.89	3.34213
80.90	3.18504	81.90	3.22441	82.90	3.26378	83.90	3.30315	84.90	3.34252
80.91	3.18543	81.91	3.22480	82.91	3.26417	83.91	3.30354	84.91	3.34291
80.92	3.18583	81.92	3.22520	82.92	3.26457	83.92	3.30394	84.92	3.34331
80.93	3.18622	81.93	3.22559	82.93	3.26496	83.93	3.30433	84.93	3.34370
80.94	3.18661	81.94	3.22598	82.94	3.26535	83.94	3.30472	84.94	3.34409
80.95	3.18701	81.95	3.22638	82.95	3.26575	83.95	3.30512	84.95	3.34449
80.96	3.18740	81.96	3.22677	82.96	3.26614	83.96	3.30551	84.96	3.34488
80.97	3.18780	81.97	3.22717	82.97	3.26654	83.97	3.30591	84.97	3.34528
80.98	3.18819	81.98	3.22756	82.98	3.26693	83.98	3.30630	84.98	3.34567
80.99	3.18858	81.99	3.22795	82.99	3.26732	83.99	3.30669	84.99	3.34606

Continued - MILLIMETERS TO INCHES 1 MILLIMETER = 0.039 370 0787 INCH

MILLI-METERS	INCHES	MILLI-METERS	INCHES	MILLI-METERS	INCHES	MILLI-METERS	INCHES	MILLI-METERS	INCHES
85.00	3.34646	86.00	3.38583	87.00	3.42520	88.00	3.46457	89.00	3.50394
85.01	3.34685	86.01	3.38622	87.01	3.42559	88.01	3.46496	89.01	3.50433
85.02	3.34724	86.02	3.38661	87.02	3.42598	88.02	3.46535	89.02	3.50472
85.03	3.34764	86.03	3.38701	87.03	3.42638	88.03	3.46575	89.03	3.50512
85.04	3.34803	86.04	3.38740	87.04	3.42677	88.04	3.46614	89.04	3.50551
85.05	3.34843	86.05	3.38780	87.05	3.42717	88.05	3.46654	89.05	3.50591
85.06	3.34882	86.06	3.38819	87.06	3.42756	88.06	3.46693	89.06	3.50630
85.07	3.34921	86.07	3.38858	87.07	3.42795	88.07	3.46732	89.07	3.50669
85.08	3.34961	86.08	3.38898	87.08	3.42835	88.08	3.46772	89.08	3.50709
85.09	3.35000	86.09	3.38937	87.09	3.42874	88.09	3.46811	89.09	3.50748
85.10	3.35039	86.10	3.38976	87.10	3.42913	88.10	3.46850	89.10	3.50787
85.11	3.35079	86.11	3.39016	87.11	3.42953	88.11	3.46890	89.11	3.50827
85.12	3.35118	86.12	3.39055	87.12	3.42992	88.12	3.46929	89.12	3.50866
85.13	3.35157	86.13	3.39094	87.13	3.43031	88.13	3.46968	89.13	3.50906
85.14	3.35197	86.14	3.39134	87.14	3.43071	88.14	3.47008	89.14	3.50945
85.15	3.35236	86.15	3.39173	87.15	3.43110	88.15	3.47047	89.15	3.50984
85.16	3.35276	86.16	3.39213	87.16	3.43150	88.16	3.47087	89.16	3.51024
85.17	3.35315	86.17	3.39252	87.17	3.43189	88.17	3.47126	89.17	3.51063
85.18	3.35354	86.18	3.39291	87.18	3.43228	88.18	3.47165	89.18	3.51102
85.19	3.35394	86.19	3.39331	87.19	3.43268	88.19	3.47205	89.19	3.51142
85.20	3.35433	86.20	3.39370	87.20	3.43307	88.20	3.47244	89.20	3.51181
85.21	3.35472	86.21	3.39409	87.21	3.43346	88.21	3.47283	89.21	3.51220
85.22	3.35512	86.22	3.39449	87.22	3.43386	88.22	3.47323	89.22	3.51260
85.23	3.35551	86.23	3.39488	87.23	3.43425	88.23	3.47362	89.23	3.51299
85.24	3.35591	86.24	3.39528	87.24	3.43465	88.24	3.47402	89.24	3.51339
85.25	3.35630	86.25	3.39567	87.25	3.43504	88.25	3.47441	89.25	3.51378
85.26	3.35669	86.26	3.39606	87.26	3.43543	88.26	3.47480	89.26	3.51417
85.27	3.35709	86.27	3.39646	87.27	3.43583	88.27	3.47520	89.27	3.51457
85.28	3.35748	86.28	3.39685	87.28	3.43622	88.28	3.47559	89.28	3.51496
85.29	3.35787	86.29	3.39724	87.29	3.43661	88.29	3.47598	89.29	3.51535
85.30	3.35827	86.30	3.39764	87.30	3.43701	88.30	3.47638	89.30	3.51575
85.31	3.35866	86.31	3.39803	87.31	3.43740	88.31	3.47677	89.31	3.51614
85.32	3.35906	86.32	3.39843	87.32	3.43780	88.32	3.47717	89.32	3.51654
85.33	3.35945	86.33	3.39882	87.33	3.43819	88.33	3.47756	89.33	3.51693
85.34	3.35984	86.34	3.39921	87.34	3.43858	88.34	3.47795	89.34	3.51732
85.35	3.36024	86.35	3.39961	87.35	3.43898	88.35	3.47835	89.35	3.51772
85.36	3.36063	86.36	3.40000	87.36	3.43937	88.36	3.47874	89.36	3.51811
85.37	3.36102	86.37	3.40039	87.37	3.43976	88.37	3.47913	89.37	3.51850
85.38	3.36142	86.38	3.40079	87.38	3.44016	88.38	3.47953	89.38	3.51890
85.39	3.36181	86.39	3.40118	87.39	3.44055	88.39	3.47992	89.39	3.51929
85.40	3.36220	86.40	3.40157	87.40	3.44094	88.40	3.48031	89.40	3.51969
85.41	3.36260	86.41	3.40197	87.41	3.44134	88.41	3.48071	89.41	3.52008
85.42	3.36299	86.42	3.40236	87.42	3.44173	88.42	3.48110	89.42	3.52047
85.43	3.36339	86.43	3.40276	87.43	3.44213	88.43	3.48150	89.43	3.52087
85.44	3.36378	86.44	3.40315	87.44	3.44252	88.44	3.48189	89.44	3.52126
85.45	3.36417	86.45	3.40354	87.45	3.44291	88.45	3.48228	89.45	3.52165
85.46	3.36457	86.46	3.40394	87.46	3.44331	88.46	3.48268	89.46	3.52205
85.47	3.36496	86.47	3.40433	87.47	3.44370	88.47	3.48307	89.47	3.52244
85.48	3.36535	86.48	3.40472	87.48	3.44409	88.48	3.48346	89.48	3.52283
85.49	3.36575	86.49	3.40512	87.49	3.44449	88.49	3.48386	89.49	3.52323

Continued - MILLIMETERS TO INCHES 1 MILLIMETER = 0.039 370 0787 INCH

MILLI-METERS	INCHES	MILLI-METERS	INCHES	MILLI-METERS	INCHES	MILLI-METERS	INCHES	MILLI-METERS	INCHES
85.50	3.36614	86.50	3.40551	87.50	3.44488	88.50	3.48425	89.50	3.52362
85.51	3.36654	86.51	3.40591	87.51	3.44528	88.51	3.48465	89.51	3.52402
85.52	3.36693	86.52	3.40630	87.52	3.44567	88.52	3.48504	89.52	3.52441
85.53	3.36732	86.53	3.40669	87.53	3.44606	88.53	3.48543	89.53	3.52480
85.54	3.36772	86.54	3.40709	87.54	3.44646	88.54	3.48583	89.54	3.52520
85.55	3.36811	86.55	3.40748	87.55	3.44685	88.55	3.48622	89.55	3.52559
85.56	3.36850	86.56	3.40787	87.56	3.44724	88.56	3.48661	89.56	3.52598
85.57	3.36890	86.57	3.40827	87.57	3.44764	88.57	3.48701	89.57	3.52638
85.58	3.36929	86.58	3.40866	87.58	3.44803	88.58	3.48740	89.58	3.52677
85.59	3.36968	86.59	3.40906	87.59	3.44843	88.59	3.48780	89.59	3.52717
85.60	3.37008	86.60	3.40945	87.60	3.44882	88.60	3.48819	89.60	3.52756
85.61	3.37047	86.61	3.40984	87.61	3.44921	88.61	3.48858	89.61	3.52795
85.62	3.37087	86.62	3.41024	87.62	3.44961	88.62	3.48898	89.62	3.52835
85.63	3.37126	86.63	3.41063	87.63	3.45000	88.63	3.48937	89.63	3.52874
85.64	3.37165	86.64	3.41102	87.64	3.45039	88.64	3.48976	89.64	3.52913
85.65	3.37205	86.65	3.41142	87.65	3.45079	88.65	3.49016	89.65	3.52953
85.66	3.37244	86.66	3.41181	87.66	3.45118	88.66	3.49055	89.66	3.52992
85.67	3.37283	86.67	3.41220	87.67	3.45157	88.67	3.49094	89.67	3.53031
85.68	3.37323	86.68	3.41260	87.68	3.45197	88.68	3.49134	89.68	3.53071
85.69	3.37362	86.69	3.41299	87.69	3.45236	88.69	3.49173	89.69	3.53110
85.70	3.37402	86.70	3.41339	87.70	3.45276	88.70	3.49213	89.70	3.53150
85.71	3.37441	86.71	3.41378	87.71	3.45315	88.71	3.49252	89.71	3.53189
85.72	3.37480	86.72	3.41417	87.72	3.45354	88.72	3.49291	89.72	3.53228
85.73	3.37520	86.73	3.41457	87.73	3.45394	88.73	3.49331	89.73	3.53268
85.74	3.37559	86.74	3.41496	87.74	3.45433	88.74	3.49370	89.74	3.53307
85.75	3.37598	86.75	3.41535	87.75	3.45472	88.75	3.49409	89.75	3.53346
85.76	3.37638	86.76	3.41575	87.76	3.45512	88.76	3.49449	89.76	3.53386
85.77	3.37677	86.77	3.41614	87.77	3.45551	88.77	3.49488	89.77	3.53425
85.78	3.37717	86.78	3.41654	87.78	3.45591	88.78	3.49528	89.78	3.53465
85.79	3.37756	86.79	3.41693	87.79	3.45630	88.79	3.49567	89.79	3.53504
85.80	3.37795	86.80	3.41732	87.80	3.45669	88.80	3.49606	89.80	3.53543
85.81	3.37835	86.81	3.41772	87.81	3.45709	88.81	3.49646	89.81	3.53583
85.82	3.37874	86.82	3.41811	87.82	3.45748	88.82	3.49685	89.82	3.53622
85.83	3.37913	86.83	3.41850	87.83	3.45787	88.83	3.49724	89.83	3.53661
85.84	3.37953	86.84	3.41890	87.84	3.45827	88.84	3.49764	89.84	3.53701
85.85	3.37992	86.85	3.41929	87.85	3.45866	88.85	3.49803	89.85	3.53740
85.86	3.38031	86.86	3.41969	87.86	3.45906	88.86	3.49843	89.86	3.53780
85.87	3.38071	86.87	3.42008	87.87	3.45945	88.87	3.49882	89.87	3.53819
85.88	3.38110	86.88	3.42047	87.88	3.45984	88.88	3.49921	89.88	3.53858
85.89	3.38150	86.89	3.42087	87.89	3.46024	88.89	3.49961	89.89	3.53898
85.90	3.38189	86.90	3.42126	87.90	3.46063	88.90	3.50000	89.90	3.53937
85.91	3.38228	86.91	3.42165	87.91	3.46102	88.91	3.50039	89.91	3.53976
85.92	3.38268	86.92	3.42205	87.92	3.46142	88.92	3.50079	89.92	3.54016
85.93	3.38307	86.93	3.42244	87.93	3.46181	88.93	3.50118	89.93	3.54055
85.94	3.38346	86.94	3.42283	87.94	3.46220	88.94	3.50157	89.94	3.54094
85.95	3.38386	86.95	3.42323	87.95	3.46260	88.95	3.50197	89.95	3.54134
85.96	3.38425	86.96	3.42362	87.96	3.46299	88.96	3.50236	89.96	3.54173
85.97	3.38465	86.97	3.42402	87.97	3.46339	88.97	3.50276	89.97	3.54213
85.98	3.38504	86.98	3.42441	87.98	3.46378	88.98	3.50315	89.98	3.54252
85.99	3.38543	86.99	3.42480	87.99	3.46417	88.99	3.50354	89.99	3.54291

Continued - MILLIMETERS TO INCHES 1 MILLIMETER = 0.039 370 0787 INCH

MILLI-METERS	INCHES	MILLI-METERS	INCHES	MILLI-METERS	INCHES	MILLI-METERS	INCHES	MILLI-METERS	INCHES
90.00	3.54331	91.00	3.58268	92.00	3.62205	93.00	3.66142	94.00	3.70079
90.01	3.54370	91.01	3.58307	92.01	3.62244	93.01	3.66181	94.01	3.70118
90.02	3.54409	91.02	3.58346	92.02	3.62283	93.02	3.66220	94.02	3.70157
90.03	3.54449	91.03	3.58386	92.03	3.62323	93.03	3.66260	94.03	3.70197
90.04	3.54488	91.04	3.58425	92.04	3.62362	93.04	3.66299	94.04	3.70236
90.05	3.54528	91.05	3.58465	92.05	3.62402	93.05	3.66339	94.05	3.70276
90.06	3.54567	91.06	3.58504	92.06	3.62441	93.06	3.66378	94.06	3.70315
90.07	3.54606	91.07	3.58543	92.07	3.62480	93.07	3.66417	94.07	3.70354
90.08	3.54646	91.08	3.58583	92.08	3.62520	93.08	3.66457	94.08	3.70394
90.09	3.54685	91.09	3.58622	92.09	3.62559	93.09	3.66496	94.09	3.70433
90.10	3.54724	91.10	3.58661	92.10	3.62598	93.10	3.66535	94.10	3.70472
90.11	3.54764	91.11	3.58701	92.11	3.62638	93.11	3.66575	94.11	3.70512
90.12	3.54803	91.12	3.58740	92.12	3.62677	93.12	3.66614	94.12	3.70551
90.13	3.54843	91.13	3.58780	92.13	3.62717	93.13	3.66654	94.13	3.70591
90.14	3.54882	91.14	3.58819	92.14	3.62756	93.14	3.66693	94.14	3.70630
90.15	3.54921	91.15	3.58858	92.15	3.62795	93.15	3.66732	94.15	3.70669
90.16	3.54961	91.16	3.58898	92.16	3.62835	93.16	3.66772	94.16	3.70709
90.17	3.55000	91.17	3.58937	92.17	3.62874	93.17	3.66811	94.17	3.70748
90.18	3.55039	91.18	3.58976	92.18	3.62913	93.18	3.66850	94.18	3.70787
90.19	3.55079	91.19	3.59016	92.19	3.62953	93.19	3.66890	94.19	3.70827
90.20	3.55118	91.20	3.59055	92.20	3.62992	93.20	3.66929	94.20	3.70866
90.21	3.55157	91.21	3.59094	92.21	3.63031	93.21	3.66968	94.21	3.70906
90.22	3.55197	91.22	3.59134	92.22	3.63071	93.22	3.67008	94.22	3.70945
90.23	3.55236	91.23	3.59173	92.23	3.63110	93.23	3.67047	94.23	3.70984
90.24	3.55276	91.24	3.59213	92.24	3.63150	93.24	3.67087	94.24	3.71024
90.25	3.55315	91.25	3.59252	92.25	3.63189	93.25	3.67126	94.25	3.71063
90.26	3.55354	91.26	3.59291	92.26	3.63228	93.26	3.67165	94.26	3.71102
90.27	3.55394	91.27	3.59331	92.27	3.63268	93.27	3.67205	94.27	3.71142
90.28	3.55433	91.28	3.59370	92.28	3.63307	93.28	3.67244	94.28	3.71181
90.29	3.55472	91.29	3.59409	92.29	3.63346	93.29	3.67283	94.29	3.71220
90.30	3.55512	91.30	3.59449	92.30	3.63386	93.30	3.67323	94.30	3.71260
90.31	3.55551	91.31	3.59488	92.31	3.63425	93.31	3.67362	94.31	3.71299
90.32	3.55591	91.32	3.59528	92.32	3.63465	93.32	3.67402	94.32	3.71339
90.33	3.55630	91.33	3.59567	92.33	3.63504	93.33	3.67441	94.33	3.71378
90.34	3.55669	91.34	3.59606	92.34	3.63543	93.34	3.67480	94.34	3.71417
90.35	3.55709	91.35	3.59646	92.35	3.63583	93.35	3.67520	94.35	3.71457
90.36	3.55748	91.36	3.59685	92.36	3.63622	93.36	3.67559	94.36	3.71496
90.37	3.55787	91.37	3.59724	92.37	3.63661	93.37	3.67598	94.37	3.71535
90.38	3.55827	91.38	3.59764	92.38	3.63701	93.38	3.67638	94.38	3.71575
90.39	3.55866	91.39	3.59803	92.39	3.63740	93.39	3.67677	94.39	3.71614
90.40	3.55906	91.40	3.59843	92.40	3.63780	93.40	3.67717	94.40	3.71654
90.41	3.55945	91.41	3.59882	92.41	3.63819	93.41	3.67756	94.41	3.71693
90.42	3.55984	91.42	3.59921	92.42	3.63858	93.42	3.67795	94.42	3.71732
90.43	3.56024	91.43	3.59961	92.43	3.63898	93.43	3.67835	94.43	3.71772
90.44	3.56063	91.44	3.60000	92.44	3.63937	93.44	3.67874	94.44	3.71811
90.45	3.56102	91.45	3.60039	92.45	3.63976	93.45	3.67913	94.45	3.71850
90.46	3.56142	91.46	3.60079	92.46	3.64016	93.46	3.67953	94.46	3.71890
90.47	3.56181	91.47	3.60118	92.47	3.64055	93.47	3.67992	94.47	3.71929
90.48	3.56220	91.48	3.60157	92.48	3.64094	93.48	3.68031	94.48	3.71968
90.49	3.56260	91.49	3.60197	92.49	3.64134	93.49	3.68071	94.49	3.72008

Continued - MILLIMETERS TO INCHES 1 MILLIMETER = 0.039 370 0787 INCH

MILLI-METERS	INCHES	MILLI-METERS	INCHES	MILLI-METERS	INCHES	MILLI-METERS	INCHES	MILLI-METERS	INCHES
90.50	3.56299	91.50	3.60236	92.50	3.64173	93.50	3.68110	94.50	3.72047
90.51	3.56339	91.51	3.60276	92.51	3.64213	93.51	3.68150	94.51	3.72087
90.52	3.56378	91.52	3.60315	92.52	3.64252	93.52	3.68189	94.52	3.72126
90.53	3.56417	91.53	3.60354	92.53	3.64291	93.53	3.68228	94.53	3.72165
90.54	3.56457	91.54	3.60394	92.54	3.64331	93.54	3.68268	94.54	3.72205
90.55	3.56496	91.55	3.60433	92.55	3.64370	93.55	3.68307	94.55	3.72244
90.56	3.56535	91.56	3.60472	92.56	3.64409	93.56	3.68346	94.56	3.72283
90.57	3.56575	91.57	3.60512	92.57	3.64449	93.57	3.68386	94.57	3.72323
90.58	3.56614	91.58	3.60551	92.58	3.64488	93.58	3.68425	94.58	3.72362
90.59	3.56654	91.59	3.60591	92.59	3.64528	93.59	3.68465	94.59	3.72402
90.60	3.56693	91.60	3.60630	92.60	3.64567	93.60	3.68504	94.60	3.72441
90.61	3.56732	91.61	3.60669	92.61	3.64606	93.61	3.68543	94.61	3.72480
90.62	3.56772	91.62	3.60709	92.62	3.64646	93.62	3.68583	94.62	3.72520
90.63	3.56811	91.63	3.60748	92.63	3.64685	93.63	3.68622	94.63	3.72559
90.64	3.56850	91.64	3.60787	92.64	3.64724	93.64	3.68661	94.64	3.72598
90.65	3.56890	91.65	3.60827	92.65	3.64764	93.65	3.68701	94.65	3.72638
90.66	3.56929	91.66	3.60866	92.66	3.64803	93.66	3.68740	94.66	3.72677
90.67	3.56968	91.67	3.60906	92.67	3.64843	93.67	3.68780	94.67	3.72717
90.68	3.57008	91.68	3.60945	92.68	3.64882	93.68	3.68819	94.68	3.72756
90.69	3.57047	91.69	3.60984	92.69	3.64921	93.69	3.68858	94.69	3.72795
90.70	3.57087	91.70	3.61024	92.70	3.64961	93.70	3.68898	94.70	3.72835
90.71	3.57126	91.71	3.61063	92.71	3.65000	93.71	3.68937	94.71	3.72874
90.72	3.57165	91.72	3.61102	92.72	3.65039	93.72	3.68976	94.72	3.72913
90.73	3.57205	91.73	3.61142	92.73	3.65079	93.73	3.69016	94.73	3.72953
90.74	3.57244	91.74	3.61181	92.74	3.65118	93.74	3.69055	94.74	3.72992
90.75	3.57283	91.75	3.61220	92.75	3.65157	93.75	3.69094	94.75	3.73031
90.76	3.57323	91.76	3.61260	92.76	3.65197	93.76	3.69134	94.76	3.73071
90.77	3.57362	91.77	3.61299	92.77	3.65236	93.77	3.69173	94.77	3.73110
90.78	3.57402	91.78	3.61339	92.78	3.65276	93.78	3.69213	94.78	3.73150
90.79	3.57441	91.79	3.61378	92.79	3.65315	93.79	3.69252	94.79	3.73189
90.80	3.57480	91.80	3.61417	92.80	3.65354	93.80	3.69291	94.80	3.73228
90.81	3.57520	91.81	3.61457	92.81	3.65394	93.81	3.69331	94.81	3.73268
90.82	3.57559	91.82	3.61496	92.82	3.65433	93.82	3.69370	94.82	3.73307
90.83	3.57598	91.83	3.61535	92.83	3.65472	93.83	3.69409	94.83	3.73346
90.84	3.57638	91.84	3.61575	92.84	3.65512	93.84	3.69449	94.84	3.73386
90.85	3.57677	91.85	3.61614	92.85	3.65551	93.85	3.69488	94.85	3.73425
90.86	3.57717	91.86	3.61654	92.86	3.65591	93.86	3.69528	94.86	3.73465
90.87	3.57756	91.87	3.61693	92.87	3.65630	93.87	3.69567	94.87	3.73504
90.88	3.57795	91.88	3.61732	92.88	3.65669	93.88	3.69606	94.88	3.73543
90.89	3.57835	91.89	3.61772	92.89	3.65709	93.89	3.69646	94.89	3.73583
90.90	3.57874	91.90	3.61811	92.90	3.65748	93.90	3.69685	94.90	3.73622
90.91	3.57913	91.91	3.61850	92.91	3.65787	93.91	3.69724	94.91	3.73661
90.92	3.57953	91.92	3.61890	92.92	3.65827	93.92	3.69764	94.92	3.73701
90.93	3.57992	91.93	3.61929	92.93	3.65866	93.93	3.69803	94.93	3.73740
90.94	3.58031	91.94	3.61968	92.94	3.65906	93.94	3.69843	94.94	3.73780
90.95	3.58071	91.95	3.62008	92.95	3.65945	93.95	3.69882	94.95	3.73819
90.96	3.58110	91.96	3.62047	92.96	3.65984	93.96	3.69921	94.96	3.73858
90.97	3.58150	91.97	3.62087	92.97	3.66024	93.97	3.69961	94.97	3.73898
90.98	3.58189	91.98	3.62126	92.98	3.66063	93.98	3.70000	94.98	3.73937
90.99	3.58228	91.99	3.62165	92.99	3.66102	93.99	3.70039	94.99	3.73976

Continued - MILLIMETERS TO INCHES 1 MILLIMETER = 0.039 370 0787 INCH

MILLI-METERS	INCHES	MILLI-METERS	INCHES	MILLI-METERS	INCHES	MILLI-METERS	INCHES	MILLI-METERS	INCHES
95.00	3.74016	96.00	3.77953	97.00	3.81890	98.00	3.85827	99.00	3.89764
95.01	3.74055	96.01	3.77992	97.01	3.81929	98.01	3.85866	99.01	3.89803
95.02	3.74094	96.02	3.78031	97.02	3.81968	98.02	3.85906	99.02	3.89843
95.03	3.74134	96.03	3.78071	97.03	3.82008	98.03	3.85945	99.03	3.89882
95.04	3.74173	96.04	3.78110	97.04	3.82047	98.04	3.85984	99.04	3.89921
95.05	3.74213	96.05	3.78150	97.05	3.82087	98.05	3.86024	99.05	3.89961
95.06	3.74252	96.06	3.78189	97.06	3.82126	98.06	3.86063	99.06	3.90000
95.07	3.74291	96.07	3.78228	97.07	3.82165	98.07	3.86102	99.07	3.90039
95.08	3.74331	96.08	3.78268	97.08	3.82205	98.08	3.86142	99.08	3.90079
95.09	3.74370	96.09	3.78307	97.09	3.82244	98.09	3.86181	99.09	3.90118
95.10	3.74409	96.10	3.78346	97.10	3.82283	98.10	3.86220	99.10	3.90157
95.11	3.74449	96.11	3.78386	97.11	3.82323	98.11	3.86260	99.11	3.90197
95.12	3.74488	96.12	3.78425	97.12	3.82362	98.12	3.86299	99.12	3.90236
95.13	3.74528	96.13	3.78465	97.13	3.82402	98.13	3.86339	99.13	3.90276
95.14	3.74567	96.14	3.78504	97.14	3.82441	98.14	3.86378	99.14	3.90315
95.15	3.74606	96.15	3.78543	97.15	3.82480	98.15	3.86417	99.15	3.90354
95.16	3.74646	96.16	3.78583	97.16	3.82520	98.16	3.86457	99.16	3.90394
95.17	3.74685	96.17	3.78622	97.17	3.82559	98.17	3.86496	99.17	3.90433
95.18	3.74724	96.18	3.78661	97.18	3.82598	98.18	3.86535	99.18	3.90472
95.19	3.74764	96.19	3.78701	97.19	3.82638	98.19	3.86575	99.19	3.90512
95.20	3.74803	96.20	3.78740	97.20	3.82677	98.20	3.86614	99.20	3.90551
95.21	3.74843	96.21	3.78780	97.21	3.82717	98.21	3.86654	99.21	3.90591
95.22	3.74882	96.22	3.78819	97.22	3.82756	98.22	3.86693	99.22	3.90630
95.23	3.74921	96.23	3.78858	97.23	3.82795	98.23	3.86732	99.23	3.90669
95.24	3.74961	96.24	3.78898	97.24	3.82835	98.24	3.86772	99.24	3.90709
95.25	3.75000	96.25	3.78937	97.25	3.82874	98.25	3.86811	99.25	3.90748
95.26	3.75039	96.26	3.78976	97.26	3.82913	98.26	3.86850	99.26	3.90787
95.27	3.75079	96.27	3.79016	97.27	3.82953	98.27	3.86890	99.27	3.90827
95.28	3.75118	96.28	3.79055	97.28	3.82992	98.28	3.86929	99.28	3.90866
95.29	3.75157	96.29	3.79094	97.29	3.83031	98.29	3.86968	99.29	3.90906
95.30	3.75197	96.30	3.79134	97.30	3.83071	98.30	3.87008	99.30	3.90945
95.31	3.75236	96.31	3.79173	97.31	3.83110	98.31	3.87047	99.31	3.90984
95.32	3.75276	96.32	3.79213	97.32	3.83150	98.32	3.87087	99.32	3.91024
95.33	3.75315	96.33	3.79252	97.33	3.83189	98.33	3.87126	99.33	3.91063
95.34	3.75354	96.34	3.79291	97.34	3.83228	98.34	3.87165	99.34	3.91102
95.35	3.75394	96.35	3.79331	97.35	3.83268	98.35	3.87205	99.35	3.91142
95.36	3.75433	96.36	3.79370	97.36	3.83307	98.36	3.87244	99.36	3.91181
95.37	3.75472	96.37	3.79409	97.37	3.83346	98.37	3.87283	99.37	3.91220
95.38	3.75512	96.38	3.79449	97.38	3.83386	98.38	3.87323	99.38	3.91260
95.39	3.75551	96.39	3.79488	97.39	3.83425	98.39	3.87362	99.39	3.91299
95.40	3.75591	96.40	3.79528	97.40	3.83465	98.40	3.87402	99.40	3.91339
95.41	3.75630	96.41	3.79567	97.41	3.83504	98.41	3.87441	99.41	3.91378
95.42	3.75669	96.42	3.79606	97.42	3.83543	98.42	3.87480	99.42	3.91417
95.43	3.75709	96.43	3.79646	97.43	3.83583	98.43	3.87520	99.43	3.91457
95.44	3.75748	96.44	3.79685	97.44	3.83622	98.44	3.87559	99.44	3.91496
95.45	3.75787	96.45	3.79724	97.45	3.83661	98.45	3.87598	99.45	3.91535
95.46	3.75827	96.46	3.79764	97.46	3.83701	98.46	3.87638	99.46	3.91575
95.47	3.75866	96.47	3.79803	97.47	3.83740	98.47	3.87677	99.47	3.91614
95.48	3.75906	96.48	3.79843	97.48	3.83780	98.48	3.87717	99.48	3.91654
95.49	3.75945	96.49	3.79882	97.49	3.83819	98.49	3.87756	99.49	3.91693

Continued - MILLIMETERS TO INCHES 1 MILLIMETER = 0.039 370 0787 INCH

MILLI-METERS	INCHES	MILLI-METERS	INCHES	MILLI-METERS	INCHES	MILLI-METERS	INCHES	MILLI-METERS	INCHES
95.50	3.75984	96.50	3.79921	97.50	3.83858	98.50	3.87795	99.50	3.91732
95.51	3.76024	96.51	3.79961	97.51	3.83898	98.51	3.87835	99.51	3.91772
95.52	3.76063	96.52	3.80000	97.52	3.83937	98.52	3.87874	99.52	3.91811
95.53	3.76102	96.53	3.80039	97.53	3.83976	98.53	3.87913	99.53	3.91850
95.54	3.76142	96.54	3.80079	97.54	3.84016	98.54	3.87953	99.54	3.91890
95.55	3.76181	96.55	3.80118	97.55	3.84055	98.55	3.87992	99.55	3.91929
95.56	3.76220	96.56	3.80157	97.56	3.84094	98.56	3.88031	99.56	3.91968
95.57	3.76260	96.57	3.80197	97.57	3.84134	98.57	3.88071	99.57	3.92008
95.58	3.76299	96.58	3.80236	97.58	3.84173	98.58	3.88110	99.58	3.92047
95.59	3.76339	96.59	3.80276	97.59	3.84213	98.59	3.88150	99.59	3.92087
95.60	3.76378	96.60	3.80315	97.60	3.84252	98.60	3.88189	99.60	3.92126
95.61	3.76417	96.61	3.80354	97.61	3.84291	98.61	3.88228	99.61	3.92165
95.62	3.76457	96.62	3.80394	97.62	3.84331	98.62	3.88268	99.62	3.92205
95.63	3.76496	96.63	3.80433	97.63	3.84370	98.63	3.88307	99.63	3.92244
95.64	3.76535	96.64	3.80472	97.64	3.84409	98.64	3.88346	99.64	3.92283
95.65	3.76575	96.65	3.80512	97.65	3.84449	98.65	3.88386	99.65	3.92323
95.66	3.76614	96.66	3.80551	97.66	3.84488	98.66	3.88425	99.66	3.92362
95.67	3.76654	96.67	3.80591	97.67	3.84528	98.67	3.88465	99.67	3.92402
95.68	3.76693	96.68	3.80630	97.68	3.84567	98.68	3.88504	99.68	3.92441
95.69	3.76732	96.69	3.80669	97.69	3.84606	98.69	3.88543	99.69	3.92480
95.70	3.76772	96.70	3.80709	97.70	3.84646	98.70	3.88583	99.70	3.92520
95.71	3.76811	96.71	3.80748	97.71	3.84685	98.71	3.88622	99.71	3.92559
95.72	3.76850	96.72	3.80787	97.72	3.84724	98.72	3.88661	99.72	3.92598
95.73	3.76890	96.73	3.80827	97.73	3.84764	98.73	3.88701	99.73	3.92638
95.74	3.76929	96.74	3.80866	97.74	3.84803	98.74	3.88740	99.74	3.92677
95.75	3.76969	96.75	3.80906	97.75	3.84843	98.75	3.88780	99.75	3.92717
95.76	3.77008	96.76	3.80945	97.76	3.84882	98.76	3.88819	99.76	3.92756
95.77	3.77047	96.77	3.80984	97.77	3.84921	98.77	3.88858	99.77	3.92795
95.78	3.77087	96.78	3.81024	97.78	3.84961	98.78	3.88898	99.78	3.92835
95.79	3.77126	96.79	3.81063	97.79	3.85000	98.79	3.88937	99.79	3.92874
95.80	3.77165	96.80	3.81102	97.80	3.85039	98.80	3.88976	99.80	3.92913
95.81	3.77205	96.81	3.81142	97.81	3.85079	98.81	3.89016	99.81	3.92953
95.82	3.77244	96.82	3.81181	97.82	3.85118	98.82	3.89055	99.82	3.92992
95.83	3.77283	96.83	3.81220	97.83	3.85157	98.83	3.89094	99.83	3.93031
95.84	3.77323	96.84	3.81260	97.84	3.85197	98.84	3.89134	99.84	3.93071
95.85	3.77362	96.85	3.81299	97.85	3.85236	98.85	3.89173	99.85	3.93110
95.86	3.77402	96.86	3.81339	97.86	3.85276	98.86	3.89213	99.86	3.93150
95.87	3.77441	96.87	3.81378	97.87	3.85315	98.87	3.89252	99.87	3.93189
95.88	3.77480	96.88	3.81417	97.88	3.85354	98.88	3.89291	99.88	3.93228
95.89	3.77520	96.89	3.81457	97.89	3.85394	98.89	3.89331	99.89	3.93268
95.90	3.77559	96.90	3.81496	97.90	3.85433	98.90	3.89370	99.90	3.93307
95.91	3.77598	96.91	3.81535	97.91	3.85472	98.91	3.89409	99.91	3.93346
95.92	3.77638	96.92	3.81575	97.92	3.85512	98.92	3.89449	99.92	3.93386
95.93	3.77677	96.93	3.81614	97.93	3.85551	98.93	3.89488	99.93	3.93425
95.94	3.77717	96.94	3.81654	97.94	3.85591	98.94	3.89528	99.94	3.93465
95.95	3.77756	96.95	3.81693	97.95	3.85630	98.95	3.89567	99.95	3.93504
95.96	3.77795	96.96	3.81732	97.96	3.85669	98.96	3.89606	99.96	3.93543
95.97	3.77835	96.97	3.81772	97.97	3.85709	98.97	3.89646	99.97	3.93583
95.98	3.77874	96.98	3.81811	97.98	3.85748	98.98	3.89685	99.98	3.93622
95.99	3.77913	96.99	3.81850	97.99	3.85787	98.99	3.89724	99.99	3.93661

LENGTH - FEET TO METERS 1 FOOT = 0.3048 METER

FEET	METERS	FEET	METERS	FEET	METERS	FEET	METERS	FEET	METERS
0	0.0000	100	30.4800	200	60.9600	300	91.4400	400	121.9200
1	0.3048	101	30.7848	201	61.2648	301	91.7448	401	122.2248
2	0.6096	102	31.0896	202	61.5696	302	92.0496	402	122.5296
3	0.9144	103	31.3944	203	61.8744	303	92.3544	403	122.8344
4	1.2192	104	31.6992	204	62.1792	304	92.6592	404	123.1392
5	1.5240	105	32.0040	205	62.4840	305	92.9640	405	123.4440
6	1.8288	106	32.3088	206	62.7888	306	93.2688	406	123.7488
7	2.1336	107	32.6136	207	63.0936	307	93.5736	407	124.0536
8	2.4384	108	32.9184	208	63.3984	308	93.8784	408	124.3584
9	2.7432	109	33.2232	209	63.7032	309	94.1832	409	124.6632
10	3.0480	110	33.5280	210	64.0080	310	94.4880	410	124.9680
11	3.3528	111	33.8328	211	64.3128	311	94.7928	411	125.2728
12	3.6576	112	34.1376	212	64.6176	312	95.0976	412	125.5776
13	3.9624	113	34.4424	213	64.9224	313	95.4024	413	125.8824
14	4.2672	114	34.7472	214	65.2272	314	95.7072	414	126.1872
15	4.5720	115	35.0520	215	65.5320	315	96.0120	415	126.4920
16	4.8768	116	35.3568	216	65.8368	316	96.3168	416	126.7968
17	5.1816	117	35.6616	217	66.1416	317	96.6216	417	127.1016
18	5.4864	118	35.9664	218	66.4464	318	96.9264	418	127.4064
19	5.7912	119	36.2712	219	66.7512	319	97.2312	419	127.7112
20	6.0960	120	36.5760	220	67.0560	320	97.5360	420	128.0160
21	6.4008	121	36.8808	221	67.3608	321	97.8408	421	128.3208
22	6.7056	122	37.1856	222	67.6656	322	98.1456	422	128.6256
23	7.0104	123	37.4904	223	67.9704	323	98.4504	423	128.9304
24	7.3152	124	37.7952	224	68.2752	324	98.7552	424	129.2352
25	7.6200	125	38.1000	225	68.5800	325	99.0600	425	129.5400
26	7.9248	126	38.4048	226	68.8848	326	99.3648	426	129.8448
27	8.2296	127	38.7096	227	69.1896	327	99.6696	427	130.1496
28	8.5344	128	39.0144	228	69.4944	328	99.9744	428	130.4544
29	8.8392	129	39.3192	229	69.7992	329	100.2792	429	130.7592
30	9.1440	130	39.6240	230	70.1040	330	100.5840	430	131.0640
31	9.4488	131	39.9288	231	70.4088	331	100.8888	431	131.3688
32	9.7536	132	40.2336	232	70.7136	332	101.1936	432	131.6736
33	10.0584	133	40.5384	233	71.0184	333	101.4984	433	131.9784
34	10.3632	134	40.8432	234	71.3232	334	101.8032	434	132.2832
35	10.6680	135	41.1480	235	71.6280	335	102.1080	435	132.5880
36	10.9728	136	41.4528	236	71.9328	336	102.4128	436	132.8928
37	11.2776	137	41.7576	237	72.2376	337	102.7176	437	133.1976
38	11.5824	138	42.0624	238	72.5424	338	103.0224	438	133.5024
39	11.8872	139	42.3672	239	72.8472	339	103.3272	439	133.8072
40	12.1920	140	42.6720	240	73.1520	340	103.6320	440	134.1120
41	12.4968	141	42.9768	241	73.4568	341	103.9368	441	134.4168
42	12.8016	142	43.2816	242	73.7616	342	104.2416	442	134.7216
43	13.1064	143	43.5864	243	74.0664	343	104.5464	443	135.0264
44	13.4112	144	43.8912	244	74.3712	344	104.8512	444	135.3312
45	13.7160	145	44.1960	245	74.6760	345	105.1560	445	135.6360
46	14.0208	146	44.5008	246	74.9808	346	105.4608	446	135.9408
47	14.3256	147	44.8056	247	75.2856	347	105.7656	447	136.2456
48	14.6304	148	45.1104	248	75.5904	348	106.0704	448	136.5504
49	14.9352	149	45.4152	249	75.8952	349	106.3752	449	136.8552

Continued - FEET TO METERS 1 FOOT = 0.3048 METER

FEET	METERS	FEET	METERS	FEET	METERS	FEET	METERS	FEET	METERS
50	15.2400	150	45.7200	250	76.2000	350	106.6800	450	137.1600
51	15.5448	151	46.0248	251	76.5048	351	106.9848	451	137.4648
52	15.8496	152	46.3296	252	76.8096	352	107.2896	452	137.7696
53	16.1544	153	46.6344	253	77.1144	353	107.5944	453	138.0744
54	16.4592	154	46.9392	254	77.4192	354	107.8992	454	138.3792
55	16.7640	155	47.2440	255	77.7240	355	108.2040	455	138.6840
56	17.0688	156	47.5488	256	78.0288	356	108.5088	456	138.9888
57	17.3736	157	47.8536	257	78.3336	357	108.8136	457	139.2936
58	17.6784	158	48.1584	258	78.6384	358	109.1184	458	139.5984
59	17.9832	159	48.4632	259	78.9432	359	109.4232	459	139.9032
60	18.2880	160	48.7680	260	79.2480	360	109.7280	460	140.2080
61	18.5928	161	49.0728	261	79.5528	361	110.0328	461	140.5128
62	18.8976	162	49.3776	262	79.8576	362	110.3376	462	140.8176
63	19.2024	163	49.6824	263	80.1624	363	110.6424	463	141.1224
64	19.5072	164	49.9872	264	80.4672	364	110.9472	464	141.4272
65	19.8120	165	50.2920	265	80.7720	365	111.2520	465	141.7320
66	20.1168	166	50.5968	266	81.0768	366	111.5568	466	142.0368
67	20.4216	167	50.9016	267	81.3816	367	111.8616	467	142.3416
68	20.7264	168	51.2064	268	81.6864	368	112.1664	468	142.6464
69	21.0312	169	51.5112	269	81.9912	369	112.4712	469	142.9512
70	21.3360	170	51.8160	270	82.2960	370	112.7760	470	143.2560
71	21.6408	171	52.1208	271	82.6008	371	113.0808	471	143.5608
72	21.9456	172	52.4256	272	82.9056	372	113.3856	472	143.8656
73	22.2504	173	52.7304	273	83.2104	373	113.6904	473	144.1704
74	22.5552	174	53.0352	274	83.5152	374	113.9952	474	144.4752
75	22.8600	175	53.3400	275	83.8200	375	114.3000	475	144.7800
76	23.1648	176	53.6448	276	84.1248	376	114.6048	476	145.0848
77	23.4696	177	53.9496	277	84.4296	377	114.9096	477	145.3896
78	23.7744	178	54.2544	278	84.7344	378	115.2144	478	145.6944
79	24.0792	179	54.5592	279	85.0392	379	115.5192	479	145.9992
80	24.3840	180	54.8640	280	85.3440	380	115.8240	480	146.3040
81	24.6888	181	55.1688	281	85.6488	381	116.1288	481	146.6088
82	24.9936	182	55.4736	282	85.9536	382	116.4336	482	146.9136
83	25.2984	183	55.7784	283	86.2584	383	116.7384	483	147.2184
84	25.6032	184	56.0832	284	86.5632	384	117.0432	484	147.5232
85	25.9080	185	56.3880	285	86.8680	385	117.3480	485	147.8280
86	26.2128	186	56.6928	286	87.1728	386	117.6528	486	148.1328
87	26.5176	187	56.9976	287	87.4776	387	117.9576	487	148.4376
88	26.8224	188	57.3024	288	87.7824	388	118.2624	488	148.7424
89	27.1272	189	57.6072	289	88.0872	389	118.5672	489	149.0472
90	27.4320	190	57.9120	290	88.3920	390	118.8720	490	149.3520
91	27.7368	191	58.2168	291	88.6968	391	119.1768	491	149.6568
92	28.0416	192	58.5216	292	89.0016	392	119.4816	492	149.9616
93	28.3464	193	58.8264	293	89.3064	393	119.7864	493	150.2664
94	28.6512	194	59.1312	294	89.6112	394	120.0912	494	150.5712
95	28.9560	195	59.4360	295	89.9160	395	120.3960	495	150.8760
96	29.2608	196	59.7408	296	90.2208	396	120.7008	496	151.1808
97	29.5656	197	60.0456	297	90.5256	397	121.0056	497	151.4856
98	29.8704	198	60.3504	298	90.8304	398	121.3104	498	151.7904
99	30.1752	199	60.6552	299	91.1352	399	121.6152	499	152.0952

Continued - FEET TO METERS 1 FOOT = 0.3048 METER

FEET	METERS	FEET	METERS	FEET	METERS	FEET	METERS	FEET	METERS
500	152.4000	600	182.8800	700	213.3600	800	243.8400	900	274.3200
501	152.7048	601	183.1848	701	213.6648	801	244.1448	901	274.6248
502	153.0096	602	183.4896	702	213.9696	802	244.4496	902	274.9296
503	153.3144	603	183.7944	703	214.2744	803	244.7544	903	275.2344
504	153.6192	604	184.0992	704	214.5792	804	245.0592	904	275.5392
505	153.9240	605	184.4040	705	214.8840	805	245.3640	905	275.8440
506	154.2288	606	184.7088	706	215.1888	806	245.6688	906	276.1488
507	154.5336	607	185.0136	707	215.4936	807	245.9736	907	276.4536
508	154.8384	608	185.3184	708	215.7984	808	246.2784	908	276.7584
509	155.1432	609	185.6232	709	216.1032	809	246.5832	909	277.0632
510	155.4480	610	185.9280	710	216.4080	810	246.8880	910	277.3680
511	155.7528	611	186.2328	711	216.7128	811	247.1928	911	277.6728
512	156.0576	612	186.5376	712	217.0176	812	247.4976	912	277.9776
513	156.3624	613	186.8424	713	217.3224	813	247.8024	913	278.2824
514	156.6672	614	187.1472	714	217.6272	814	248.1072	914	278.5872
515	156.9720	615	187.4520	715	217.9320	815	248.4120	915	278.8920
516	157.2768	616	187.7568	716	218.2368	816	248.7168	916	279.1968
517	157.5816	617	188.0616	717	218.5416	817	249.0216	917	279.5016
518	157.8864	618	188.3664	718	218.8464	818	249.3264	918	279.8064
519	158.1912	619	188.6712	719	219.1512	819	249.6312	919	280.1112
520	158.4960	620	188.9760	720	219.4560	820	249.9360	920	280.4160
521	158.8008	621	189.2808	721	219.7608	821	250.2408	921	280.7208
522	159.1056	622	189.5856	722	220.0656	822	250.5456	922	281.0256
523	159.4104	623	189.8904	723	220.3704	823	250.8504	923	281.3304
524	159.7152	624	190.1952	724	220.6752	824	251.1552	924	281.6352
525	160.0200	625	190.5000	725	220.9800	825	251.4600	925	281.9400
526	160.3248	626	190.8048	726	221.2848	826	251.7648	926	282.2448
527	160.6296	627	191.1096	727	221.5896	827	252.0696	927	282.5496
528	160.9344	628	191.4144	728	221.8944	828	252.3744	928	282.8544
529	161.2392	629	191.7192	729	222.1992	829	252.6792	929	283.1592
530	161.5440	630	192.0240	730	222.5040	830	252.9840	930	283.4640
531	161.8488	631	192.3288	731	222.8088	831	253.2888	931	283.7688
532	162.1536	632	192.6336	732	223.1136	832	253.5936	932	284.0736
533	162.4584	633	192.9384	733	223.4184	833	253.8984	933	284.3784
534	162.7632	634	193.2432	734	223.7232	834	254.2032	934	284.6832
535	163.0680	635	193.5480	735	224.0280	835	254.5080	935	284.9880
536	163.3728	636	193.8528	736	224.3328	836	254.8128	936	285.2928
537	163.6776	637	194.1576	737	224.6376	837	255.1176	937	285.5976
538	163.9824	638	194.4624	738	224.9424	838	255.4224	938	285.9024
539	164.2872	639	194.7672	739	225.2472	839	255.7272	939	286.2072
540	164.5920	640	195.0720	740	225.5520	840	256.0320	940	286.5120
541	164.8968	641	195.3768	741	225.8568	841	256.3368	941	286.8168
542	165.2016	642	195.6816	742	226.1616	842	256.6416	942	287.1216
543	165.5064	643	195.9864	743	226.4664	843	256.9464	943	287.4264
544	165.8112	644	196.2912	744	226.7712	844	257.2512	944	287.7312
545	166.1160	645	196.5960	745	227.0760	845	257.5560	945	288.0360
546	166.4208	646	196.9008	746	227.3808	846	257.8608	946	288.3408
547	166.7256	647	197.2056	747	227.6856	847	258.1656	947	288.6456
548	167.0304	648	197.5104	748	227.9904	848	258.4704	948	288.9504
549	167.3352	649	197.8152	749	228.2952	849	258.7752	949	289.2552

Continued - FEET TO METERS 1 FOOT = 0.3048 METER

FEET	METERS	FEET	METERS	FEET	METERS	FEET	METERS	FEET	METERS
550	167.6400	650	198.1200	750	228.6000	850	259.0800	950	289.5600
551	167.9448	651	198.4248	751	228.9048	851	259.3848	951	289.8648
552	168.2496	652	198.7296	752	229.2096	852	259.6896	952	290.1696
553	168.5544	653	199.0344	753	229.5144	853	259.9944	953	290.4744
554	168.8592	654	199.3392	754	229.8192	854	260.2992	954	290.7792
555	169.1640	655	199.6440	755	230.1240	855	260.6040	955	291.0840
556	169.4688	656	199.9488	756	230.4288	856	260.9088	956	291.3888
557	169.7736	657	200.2536	757	230.7336	857	261.2136	957	291.6936
558	170.0784	658	200.5584	758	231.0384	858	261.5184	958	291.9984
559	170.3832	659	200.8632	759	231.3432	859	261.8232	959	292.3032
560	170.6880	660	201.1680	760	231.6480	860	262.1280	960	292.6080
561	170.9928	661	201.4728	761	231.9528	861	262.4328	961	292.9128
562	171.2976	662	201.7776	762	232.2576	862	262.7376	962	293.2176
563	171.6024	663	202.0824	763	232.5624	863	263.0424	963	293.5224
564	171.9072	664	202.3872	764	232.8672	864	263.3472	964	293.8272
565	172.2120	665	202.6920	765	233.1720	865	263.6520	965	294.1320
566	172.5168	666	202.9968	766	233.4768	866	263.9568	966	294.4368
567	172.8216	667	203.3016	767	233.7816	867	264.2616	967	294.7416
568	173.1264	668	203.6064	768	234.0864	868	264.5664	968	295.0464
569	173.4312	669	203.9112	769	234.3912	869	264.8712	969	295.3512
570	173.7360	670	204.2160	770	234.6960	870	265.1760	970	295.6560
571	174.0408	671	204.5208	771	235.0008	871	265.4808	971	295.9608
572	174.3456	672	204.8256	772	235.3056	872	265.7856	972	296.2656
573	174.6504	673	205.1304	773	235.6104	873	266.0904	973	296.5704
574	174.9552	674	205.4352	774	235.9152	874	266.3952	974	296.8752
575	175.2600	675	205.7400	775	236.2200	875	266.7000	975	297.1800
576	175.5648	676	206.0448	776	236.5248	876	267.0048	976	297.4848
577	175.8696	677	206.3496	777	236.8296	877	267.3096	977	297.7896
578	176.1744	678	206.6544	778	237.1344	878	267.6144	978	298.0944
579	176.4792	679	206.9592	779	237.4392	879	267.9192	979	298.3992
580	176.7840	680	207.2640	780	237.7440	880	268.2240	980	298.7040
581	177.0888	681	207.5688	781	238.0488	881	268.5288	981	299.0088
582	177.3936	682	207.8736	782	238.3536	882	268.8336	982	299.3136
583	177.6984	683	208.1784	783	238.6584	883	269.1384	983	299.6184
584	178.0032	684	208.4832	784	238.9632	884	269.4432	984	299.9232
585	178.3080	685	208.7880	785	239.2680	885	269.7480	985	300.2280
586	178.6128	686	209.0928	786	239.5728	886	270.0528	986	300.5328
587	178.9176	687	209.3976	787	239.8776	887	270.3576	987	300.8376
588	179.2224	688	209.7024	788	240.1824	888	270.6624	988	301.1424
589	179.5272	689	210.0072	789	240.4872	889	270.9672	989	301.4472
590	179.8320	690	210.3120	790	240.7920	890	271.2720	990	301.7520
591	180.1368	691	210.6168	791	241.0968	891	271.5768	991	302.0568
592	180.4416	692	210.9216	792	241.4016	892	271.8816	992	302.3616
593	180.7464	693	211.2264	793	241.7064	893	272.1864	993	302.6664
594	181.0512	694	211.5312	794	242.0112	894	272.4912	994	302.9712
595	181.3560	695	211.8360	795	242.3160	895	272.7960	995	303.2760
596	181.6608	696	212.1408	796	242.6208	896	273.1008	996	303.5808
597	181.9656	697	212.4456	797	242.9256	897	273.4056	997	303.8856
598	182.2704	698	212.7504	798	243.2304	898	273.7104	998	304.1904
599	182.5752	699	213.0552	799	243.5352	899	274.0152	999	304.4952

LENGTH - METERS TO FEET 1 METER = 3.280 839 895 FEET

METERS	FEET	METERS	FEET	METERS	FEET	METERS	FEET	METERS	FEET
0	0.0000	100	328.0840	200	656.1680	300	984.2520	400	1312.3360
1	3.2808	101	331.3648	201	659.4488	301	987.5328	401	1315.6168
2	6.5617	102	334.6457	202	662.7297	302	990.8136	402	1318.8976
3	9.8425	103	337.9265	203	666.0105	303	994.0945	403	1322.1785
4	13.1234	104	341.2073	204	669.2913	304	997.3753	404	1325.4593
5	16.4042	105	344.4882	205	672.5722	305	1000.6562	405	1328.7401
6	19.6850	106	347.7690	206	675.8530	306	1003.9370	406	1332.0210
7	22.9659	107	351.0499	207	679.1339	307	1007.2178	407	1335.3018
8	26.2467	108	354.3307	208	682.4147	308	1010.4987	408	1338.5827
9	29.5276	109	357.6115	209	685.6955	309	1013.7795	409	1341.8635
10	32.8084	110	360.8924	210	688.9764	310	1017.0604	410	1345.1443
11	36.0892	111	364.1732	211	692.2572	311	1020.3412	411	1348.4252
12	39.3701	112	367.4541	212	695.5381	312	1023.6220	412	1351.7060
13	42.6509	113	370.7349	213	698.8189	313	1026.9029	413	1354.9869
14	45.9318	114	374.0157	214	702.0997	314	1030.1837	414	1358.2677
15	49.2126	115	377.2966	215	705.3806	315	1033.4646	415	1361.5486
16	52.4934	116	380.5774	216	708.6614	316	1036.7454	416	1364.8294
17	55.7743	117	383.8583	217	711.9423	317	1040.0262	417	1368.1102
18	59.0551	118	387.1391	218	715.2231	318	1043.3071	418	1371.3911
19	62.3360	119	390.4199	219	718.5039	319	1046.5879	419	1374.6719
20	65.6168	120	393.7008	220	721.7848	320	1049.8688	420	1377.9527
21	68.8976	121	396.9816	221	725.0656	321	1053.1496	421	1381.2336
22	72.1785	122	400.2625	222	728.3465	322	1056.4304	422	1384.5144
23	75.4593	123	403.5433	223	731.6273	323	1059.7113	423	1387.7953
24	78.7402	124	406.8241	224	734.9081	324	1062.9921	424	1391.0761
25	82.0210	125	410.1050	225	738.1890	325	1066.2729	425	1394.3569
26	85.3018	126	413.3858	226	741.4698	326	1069.5538	426	1397.6378
27	88.5827	127	416.6667	227	744.7506	327	1072.8346	427	1400.9186
28	91.8635	128	419.9475	228	748.0315	328	1076.1155	428	1404.1995
29	95.1444	129	423.2283	229	751.3123	329	1079.3963	429	1407.4803
30	98.4252	130	426.5092	230	754.5932	330	1082.6772	430	1410.7611
31	101.7060	131	429.7900	231	757.8740	331	1085.9580	431	1414.0420
32	104.9869	132	433.0709	232	761.1549	332	1089.2388	432	1417.3228
33	108.2677	133	436.3517	233	764.4357	333	1092.5197	433	1420.6037
34	111.5486	134	439.6325	234	767.7165	334	1095.8005	434	1423.8845
35	114.8294	135	442.9134	235	770.9974	335	1099.0814	435	1427.1653
36	118.1102	136	446.1942	236	774.2782	336	1102.3622	436	1430.4462
37	121.3911	137	449.4751	237	777.5591	337	1105.6430	437	1433.7270
38	124.6719	138	452.7559	238	780.8399	338	1108.9239	438	1437.0079
39	127.9528	139	456.0367	239	784.1207	339	1112.2047	439	1440.2887
40	131.2336	140	459.3176	240	787.4016	340	1115.4855	440	1443.5695
41	134.5144	141	462.5984	241	790.6824	341	1118.7664	441	1446.8504
42	137.7953	142	465.8793	242	793.9632	342	1122.0472	442	1450.1312
43	141.0761	143	469.1601	243	797.2441	343	1125.3281	443	1453.4121
44	144.3570	144	472.4409	244	800.5249	344	1128.6089	444	1456.6929
45	147.6378	145	475.7218	245	803.8058	345	1131.8898	445	1459.9737
46	150.9186	146	479.0026	246	807.0866	346	1135.1706	446	1463.2546
47	154.1995	147	482.2835	247	810.3674	347	1138.4514	447	1466.5354
48	157.4803	148	485.5643	248	813.6483	348	1141.7323	448	1469.8163
49	160.7612	149	488.8451	249	816.9291	349	1145.0131	449	1473.0971

Continued - METERS TC FEET 1 METER = 3.280 839 895 FEET

METERS	FEET	METERS	FEET	METERS	FEET	METERS	FEET	METERS	FEET
50	164.0420	150	492.1260	250	820.2100	350	1148.2940	450	1476.3779
51	167.3228	151	495.4068	251	823.4908	351	1151.5748	451	1479.6588
52	170.6037	152	498.6877	252	826.7717	352	1154.8556	452	1482.9396
53	173.8845	153	501.9685	253	830.0525	353	1158.1365	453	1486.2205
54	177.1654	154	505.2493	254	833.3333	354	1161.4173	454	1489.5013
55	180.4462	155	508.5302	255	836.6142	355	1164.6982	455	1492.7821
56	183.7270	156	511.8110	256	839.8950	356	1167.9790	456	1496.0630
57	187.0079	157	515.0919	257	843.1758	357	1171.2598	457	1499.3438
58	190.2887	158	518.3727	258	846.4567	358	1174.5407	458	1502.6247
59	193.5696	159	521.6535	259	849.7375	359	1177.8215	459	1505.9055
60	196.8504	160	524.9344	260	853.0184	360	1181.1024	460	1509.1863
61	200.1312	161	528.2152	261	856.2992	361	1184.3832	461	1512.4672
62	203.4121	162	531.4961	262	859.5800	362	1187.6640	462	1515.7480
63	206.6929	163	534.7769	263	862.8609	363	1190.9449	463	1519.0289
64	209.9738	164	538.0577	264	866.1417	364	1194.2257	464	1522.3097
65	213.2546	165	541.3386	265	869.4226	365	1197.5065	465	1525.5905
66	216.5354	166	544.6194	266	872.7034	366	1200.7874	466	1528.8714
67	219.8163	167	547.9003	267	875.9842	367	1204.0682	467	1532.1522
68	223.0971	168	551.1811	268	879.2651	368	1207.3491	468	1535.4331
69	226.3780	169	554.4619	269	882.5459	369	1210.6299	469	1538.7139
70	229.6588	170	557.7428	270	885.8268	370	1213.9108	470	1541.9947
71	232.9396	171	561.0236	271	889.1076	371	1217.1916	471	1545.2756
72	236.2205	172	564.3045	272	892.3884	372	1220.4724	472	1548.5564
73	239.5013	173	567.5853	273	895.6693	373	1223.7533	473	1551.8373
74	242.7822	174	570.8661	274	898.9501	374	1227.0341	474	1555.1181
75	246.0630	175	574.1470	275	902.2310	375	1230.3150	475	1558.3989
76	249.3438	176	577.4278	276	905.5118	376	1233.5958	476	1561.6798
77	252.6247	177	580.7087	277	908.7926	377	1236.8766	477	1564.9606
78	255.9055	178	583.9895	278	912.0735	378	1240.1575	478	1568.2415
79	259.1863	179	587.2703	279	915.3543	379	1243.4383	479	1571.5223
80	262.4672	180	590.5512	280	918.6352	380	1246.7191	480	1574.8031
81	265.7480	181	593.8320	281	921.9160	381	1250.0000	481	1578.0840
82	269.0289	182	597.1129	282	925.1968	382	1253.2808	482	1581.3648
83	272.3097	183	600.3937	283	928.4777	383	1256.5617	483	1584.6457
84	275.5905	184	603.6745	284	931.7585	384	1259.8425	484	1587.9265
85	278.8714	185	606.9554	285	935.0394	385	1263.1234	485	1591.2073
86	282.1522	186	610.2362	286	938.3202	386	1266.4042	486	1594.4882
87	285.4331	187	613.5171	287	941.6010	387	1269.6850	487	1597.7690
88	288.7139	188	616.7979	288	944.8819	388	1272.9659	488	1601.0499
89	291.9947	189	620.0787	289	948.1627	389	1276.2467	489	1604.3307
90	295.2756	190	623.3596	290	951.4436	390	1279.5275	490	1607.6115
91	298.5564	191	626.6404	291	954.7244	391	1282.8084	491	1610.8924
92	301.8373	192	629.9213	292	958.0052	392	1286.0892	492	1614.1732
93	305.1181	193	633.2021	293	961.2861	393	1289.3701	493	1617.4541
94	308.3989	194	636.4829	294	964.5669	394	1292.6509	494	1620.7349
95	311.6798	195	639.7638	295	967.8478	395	1295.9317	495	1624.0157
96	314.9606	196	643.0446	296	971.1286	396	1299.2126	496	1627.2966
97	318.2415	197	646.3255	297	974.4094	397	1302.4934	497	1630.5774
98	321.5223	198	649.6063	298	977.6903	398	1305.7743	498	1633.8583
99	324.8031	199	652.8871	299	980.9711	399	1309.0551	499	1637.1391

Continued - METERS TO FEET 1 METER = 3.280 839 895 FEET

METERS	FEET	METERS	FEET	METERS	FEET	METERS	FEET	METERS	FEET
500	1640.4199	600	1968.5039	700	2296.5879	800	2624.6719	900	2952.7559
501	1643.7008	601	1971.7848	701	2299.8687	801	2627.9527	901	2956.0367
502	1646.9816	602	1975.0656	702	2303.1496	802	2631.2336	902	2959.3176
503	1650.2625	603	1978.3465	703	2306.4304	803	2634.5144	903	2962.5984
504	1653.5433	604	1981.6273	704	2309.7113	804	2637.7953	904	2965.8792
505	1656.8241	605	1984.9081	705	2312.9921	805	2641.0761	905	2969.1601
506	1660.1050	606	1988.1890	706	2316.2729	806	2644.3569	906	2972.4409
507	1663.3858	607	1991.4698	707	2319.5538	807	2647.6378	907	2975.7218
508	1666.6667	608	1994.7506	708	2322.8346	808	2650.9186	908	2979.0026
509	1669.9475	609	1998.0315	709	2326.1155	809	2654.1995	909	2982.2834
510	1673.2283	610	2001.3123	710	2329.3963	810	2657.4803	910	2985.5643
511	1676.5092	611	2004.5932	711	2332.6772	811	2660.7611	911	2988.8451
512	1679.7900	612	2007.8740	712	2335.9580	812	2664.0420	912	2992.1260
513	1683.0709	613	2011.1548	713	2339.2388	813	2667.3228	913	2995.4068
514	1686.3517	614	2014.4357	714	2342.5197	814	2670.6037	914	2998.6877
515	1689.6325	615	2017.7165	715	2345.8005	815	2673.8845	915	3001.9685
516	1692.9134	616	2020.9974	716	2349.0814	816	2677.1653	916	3005.2493
517	1696.1942	617	2024.2782	717	2352.3622	817	2680.4462	917	3008.5302
518	1699.4751	618	2027.5591	718	2355.6430	818	2683.7270	918	3011.8110
519	1702.7559	619	2030.8399	719	2358.9239	819	2687.0078	919	3015.0919
520	1706.0367	620	2034.1207	720	2362.2047	820	2690.2887	920	3018.3727
521	1709.3176	621	2037.4016	721	2365.4855	821	2693.5695	921	3021.6535
522	1712.5984	622	2040.6824	722	2368.7664	822	2696.8504	922	3024.9344
523	1715.8793	623	2043.9632	723	2372.0472	823	2700.1312	923	3028.2152
524	1719.1601	624	2047.2441	724	2375.3281	824	2703.4120	924	3031.4960
525	1722.4409	625	2050.5249	725	2378.6089	825	2706.6929	925	3034.7769
526	1725.7218	626	2053.8058	726	2381.8897	826	2709.9737	926	3038.0577
527	1729.0026	627	2057.0866	727	2385.1706	827	2713.2546	927	3041.3386
528	1732.2834	628	2060.3674	728	2388.4514	828	2716.5354	928	3044.6194
529	1735.5643	629	2063.6483	729	2391.7323	829	2719.8163	929	3047.9002
530	1738.8451	630	2066.9291	730	2395.0131	830	2723.0971	930	3051.1811
531	1742.1260	631	2070.2100	731	2398.2939	831	2726.3779	931	3054.4619
532	1745.4068	632	2073.4908	732	2401.5748	832	2729.6588	932	3057.7428
533	1748.6877	633	2076.7716	733	2404.8556	833	2732.9396	933	3061.0236
534	1751.9685	634	2080.0525	734	2408.1365	834	2736.2205	934	3064.3044
535	1755.2493	635	2083.3333	735	2411.4173	835	2739.5013	935	3067.5853
536	1758.5302	636	2086.6142	736	2414.6982	836	2742.7821	936	3070.8661
537	1761.8110	637	2089.8950	737	2417.9790	837	2746.0630	937	3074.1470
538	1765.0919	638	2093.1758	738	2421.2598	838	2749.3438	938	3077.4278
539	1768.3727	639	2096.4567	739	2424.5406	839	2752.6247	939	3080.7086
540	1771.6535	640	2099.7375	740	2427.8215	840	2755.9055	940	3083.9895
541	1774.9344	641	2103.0183	741	2431.1024	841	2759.1863	941	3087.2703
542	1778.2152	642	2106.2992	742	2434.3832	842	2762.4672	942	3090.5511
543	1781.4960	643	2109.5800	743	2437.6640	843	2765.7480	943	3093.8320
544	1784.7769	644	2112.8609	744	2440.9449	844	2769.0288	944	3097.1129
545	1788.0577	645	2116.1417	745	2444.2257	845	2772.3097	945	3100.3937
546	1791.3386	646	2119.4225	746	2447.5065	846	2775.5905	946	3103.6745
547	1794.6194	647	2122.7034	747	2450.7874	847	2778.8714	947	3106.9554
548	1797.9003	648	2125.9842	748	2454.0682	848	2782.1522	948	3110.2362
549	1801.1811	649	2129.2651	749	2457.3491	849	2785.4330	949	3113.5170

Continued - METERS TO FEET 1 METER = 3.280 839 895 FEET

METERS	FEET	METERS	FEET	METERS	FEET	METERS	FEET	METERS	FEET
550	1804.4619	650	2132.5459	750	2460.6299	850	2788.7139	950	3116.7979
551	1807.7428	651	2135.8268	751	2463.9107	851	2791.9947	951	3120.0787
552	1811.0236	652	2139.1076	752	2467.1916	852	2795.2756	952	3123.3596
553	1814.3045	653	2142.3884	753	2470.4724	853	2798.5564	953	3126.6404
554	1817.5853	654	2145.6693	754	2473.7533	854	2801.8372	954	3129.9212
555	1820.8661	655	2148.9501	755	2477.0341	855	2805.1181	955	3133.2021
556	1824.1470	656	2152.2310	756	2480.3149	856	2808.3989	956	3136.4829
557	1827.4278	657	2155.5118	757	2483.5958	857	2811.6798	957	3139.7638
558	1830.7086	658	2158.7926	758	2486.8766	858	2814.9606	958	3143.0446
559	1833.9895	659	2162.0735	759	2490.1575	859	2818.2415	959	3146.3254
560	1837.2703	660	2165.3543	760	2493.4383	860	2821.5223	960	3149.6063
561	1840.5512	661	2168.6352	761	2496.7191	861	2824.8031	961	3152.8871
562	1843.8320	662	2171.9160	762	2500.0000	862	2828.0840	962	3156.1680
563	1847.1129	663	2175.1968	763	2503.2808	863	2831.3648	963	3159.4488
564	1850.3937	664	2178.4777	764	2506.5617	864	2834.6457	964	3162.7296
565	1853.6745	665	2181.7585	765	2509.8425	865	2837.9265	965	3166.0105
566	1856.9554	666	2185.0393	766	2513.1234	866	2841.2073	966	3169.2913
567	1860.2362	667	2188.3202	767	2516.4042	867	2844.4882	967	3172.5721
568	1863.5170	668	2191.6010	768	2519.6850	868	2847.7690	968	3175.8530
569	1866.7979	669	2194.8819	769	2522.9659	869	2851.0498	969	3179.1339
570	1870.0787	670	2198.1627	770	2526.2467	870	2854.3307	970	3182.4147
571	1873.3596	671	2201.4435	771	2529.5275	871	2857.6115	971	3185.6955
572	1876.6404	672	2204.7244	772	2532.8084	872	2860.8924	972	3188.9763
573	1879.9212	673	2208.0052	773	2536.0892	873	2864.1732	973	3192.2572
574	1883.2021	674	2211.2861	774	2539.3701	874	2867.4540	974	3195.5380
575	1886.4829	675	2214.5669	775	2542.6509	875	2870.7349	975	3198.8189
576	1889.7638	676	2217.8477	776	2545.9317	876	2874.0157	976	3202.0997
577	1893.0446	677	2221.1286	777	2549.2126	877	2877.2966	977	3205.3806
578	1896.3255	678	2224.4094	778	2552.4934	878	2880.5774	978	3208.6614
579	1899.6063	679	2227.6903	779	2555.7743	879	2883.8582	979	3211.9422
580	1902.8871	680	2230.9711	780	2559.0551	880	2887.1391	980	3215.2231
581	1906.1680	681	2234.2520	781	2562.3359	881	2890.4199	981	3218.5039
582	1909.4488	682	2237.5328	782	2565.6168	882	2893.7008	982	3221.7848
583	1912.7296	683	2240.8136	783	2568.8976	883	2896.9816	983	3225.0656
584	1916.0105	684	2244.0945	784	2572.1785	884	2900.2625	984	3228.3464
585	1919.2913	685	2247.3753	785	2575.4593	885	2903.5433	985	3231.6273
586	1922.5722	686	2250.6562	786	2578.7401	886	2906.8241	986	3234.9081
587	1925.8530	687	2253.9370	787	2582.0210	887	2910.1050	987	3238.1890
588	1929.1339	688	2257.2178	788	2585.3018	888	2913.3858	988	3241.4698
589	1932.4147	689	2260.4987	789	2588.5827	889	2916.6667	989	3244.7506
590	1935.6955	690	2263.7795	790	2591.8635	890	2919.9475	990	3248.0315
591	1938.9764	691	2267.0604	791	2595.1443	891	2923.2283	991	3251.3123
592	1942.2572	692	2270.3412	792	2598.4252	892	2926.5092	992	3254.5932
593	1945.5380	693	2273.6220	793	2601.7060	893	2929.7900	993	3257.8740
594	1948.8189	694	2276.9029	794	2604.9868	894	2933.0709	994	3261.1548
595	1952.0997	695	2280.1837	795	2608.2677	895	2936.3517	995	3264.4357
596	1955.3806	696	2283.4645	796	2611.5485	896	2939.6325	996	3267.7165
597	1958.6614	697	2286.7454	797	2614.8294	897	2942.9134	997	3270.9973
598	1961.9422	698	2290.0262	798	2618.1102	898	2946.1942	998	3274.2782
599	1965.2231	699	2293.3071	799	2621.3911	899	2949.4750	999	3277.5590

LENGTH - YARDS TO METERS 1 YARD = 0.9144 METER

YARDS	METERS	YARDS	METERS	YARDS	METERS	YARDS	METERS	YARDS	METERS
0	0.0000	100	91.4400	200	182.8800	300	274.3200	400	365.7600
1	0.9144	101	92.3544	201	183.7944	301	275.2344	401	366.6744
2	1.8288	102	93.2688	202	184.7088	302	276.1488	402	367.5888
3	2.7432	103	94.1832	203	185.6232	303	277.0632	403	368.5032
4	3.6576	104	95.0976	204	186.5376	304	277.9776	404	369.4176
5	4.5720	105	96.0120	205	187.4520	305	278.8920	405	370.3320
6	5.4864	106	96.9264	206	188.3664	306	279.8064	406	371.2464
7	6.4008	107	97.8408	207	189.2808	307	280.7208	407	372.1608
8	7.3152	108	98.7552	208	190.1952	308	281.6352	408	373.0752
9	8.2296	109	99.6696	209	191.1096	309	282.5496	409	373.9896
10	9.1440	110	100.5840	210	192.0240	310	283.4640	410	374.9040
11	10.0584	111	101.4984	211	192.9384	311	284.3784	411	375.8184
12	10.9728	112	102.4128	212	193.8528	312	285.2928	412	376.7328
13	11.8872	113	103.3272	213	194.7672	313	286.2072	413	377.6472
14	12.8016	114	104.2416	214	195.6816	314	287.1216	414	378.5616
15	13.7160	115	105.1560	215	196.5960	315	288.0360	415	379.4760
16	14.6304	116	106.0704	216	197.5104	316	288.9504	416	380.3904
17	15.5448	117	106.9848	217	198.4248	317	289.8648	417	381.3048
18	16.4592	118	107.8992	218	199.3392	318	290.7792	418	382.2192
19	17.3736	119	108.8136	219	200.2536	319	291.6936	419	383.1336
20	18.2880	120	109.7280	220	201.1680	320	292.6080	420	384.0480
21	19.2024	121	110.6424	221	202.0824	321	293.5224	421	384.9624
22	20.1168	122	111.5568	222	202.9968	322	294.4368	422	385.8768
23	21.0312	123	112.4712	223	203.9112	323	295.3512	423	386.7912
24	21.9456	124	113.3856	224	204.8256	324	296.2656	424	387.7056
25	22.8600	125	114.3000	225	205.7400	325	297.1800	425	388.6200
26	23.7744	126	115.2144	226	206.6544	326	298.0944	426	389.5344
27	24.6888	127	116.1288	227	207.5688	327	299.0088	427	390.4488
28	25.6032	128	117.0432	228	208.4832	328	299.9232	428	391.3632
29	26.5176	129	117.9576	229	209.3976	329	300.8376	429	392.2776
30	27.4320	130	118.8720	230	210.3120	330	301.7520	430	393.1920
31	28.3464	131	119.7864	231	211.2264	331	302.6664	431	394.1064
32	29.2608	132	120.7008	232	212.1408	332	303.5808	432	395.0208
33	30.1752	133	121.6152	233	213.0552	333	304.4952	433	395.9352
34	31.0896	134	122.5296	234	213.9696	334	305.4096	434	396.8496
35	32.0040	135	123.4440	235	214.8840	335	306.3240	435	397.7640
36	32.9184	136	124.3584	236	215.7984	336	307.2384	436	398.6784
37	33.8328	137	125.2728	237	216.7128	337	308.1528	437	399.5928
38	34.7472	138	126.1872	238	217.6272	338	309.0672	438	400.5072
39	35.6616	139	127.1016	239	218.5416	339	309.9816	439	401.4216
40	36.5760	140	128.0160	240	219.4560	340	310.8960	440	402.3360
41	37.4904	141	128.9304	241	220.3704	341	311.8104	441	403.2504
42	38.4048	142	129.8448	242	221.2848	342	312.7248	442	404.1648
43	39.3192	143	130.7592	243	222.1992	343	313.6392	443	405.0792
44	40.2336	144	131.6736	244	223.1136	344	314.5536	444	405.9936
45	41.1480	145	132.5880	245	224.0280	345	315.4680	445	406.9080
46	42.0624	146	133.5024	246	224.9424	346	316.3824	446	407.8224
47	42.9768	147	134.4168	247	225.8568	347	317.2968	447	408.7368
48	43.8912	148	135.3312	248	226.7712	348	318.2112	448	409.6512
49	44.8056	149	136.2456	249	227.6856	349	319.1256	449	410.5656

Continued - YARDS TO METERS 1 YARD = 0.9144 METER

YARDS	METERS	YARDS	METERS	YARDS	METERS	YARDS	METERS	YARDS	METERS
50	45.7200	150	137.1600	250	228.6000	350	320.0400	450	411.4800
51	46.6344	151	138.0744	251	229.5144	351	320.9544	451	412.3944
52	47.5488	152	138.9888	252	230.4288	352	321.8688	452	413.3088
53	48.4632	153	139.9032	253	231.3432	353	322.7832	453	414.2232
54	49.3776	154	140.8176	254	232.2576	354	323.6976	454	415.1376
55	50.2920	155	141.7320	255	233.1720	355	324.6120	455	416.0520
56	51.2064	156	142.6464	256	234.0864	356	325.5264	456	416.9664
57	52.1208	157	143.5608	257	235.0008	357	326.4408	457	417.8808
58	53.0352	158	144.4752	258	235.9152	358	327.3552	458	418.7952
59	53.9496	159	145.3896	259	236.8296	359	328.2696	459	419.7096
60	54.8640	160	146.3040	260	237.7440	360	329.1840	460	420.6240
61	55.7784	161	147.2184	261	238.6584	361	330.0984	461	421.5384
62	56.6928	162	148.1328	262	239.5728	362	331.0128	462	422.4528
63	57.6072	163	149.0472	263	240.4872	363	331.9272	463	423.3672
64	58.5216	164	149.9616	264	241.4016	364	332.8416	464	424.2816
65	59.4360	165	150.8760	265	242.3160	365	333.7560	465	425.1960
66	60.3504	166	151.7904	266	243.2304	366	334.6704	466	426.1104
67	61.2648	167	152.7048	267	244.1448	367	335.5848	467	427.0248
68	62.1792	168	153.6192	268	245.0592	368	336.4992	468	427.9392
69	63.0936	169	154.5336	269	245.9736	369	337.4136	469	428.8536
70	64.0080	170	155.4480	270	246.8880	370	338.3280	470	429.7680
71	64.9224	171	156.3624	271	247.8024	371	339.2424	471	430.6824
72	65.8368	172	157.2768	272	248.7168	372	340.1568	472	431.5968
73	66.7512	173	158.1912	273	249.6312	373	341.0712	473	432.5112
74	67.6656	174	159.1056	274	250.5456	374	341.9856	474	433.4256
75	68.5800	175	160.0200	275	251.4600	375	342.9000	475	434.3400
76	69.4944	176	160.9344	276	252.3744	376	343.8144	476	435.2544
77	70.4088	177	161.8488	277	253.2888	377	344.7288	477	436.1688
78	71.3232	178	162.7632	278	254.2032	378	345.6432	478	437.0832
79	72.2376	179	163.6776	279	255.1176	379	346.5576	479	437.9976
80	73.1520	180	164.5920	280	256.0320	380	347.4720	480	438.9120
81	74.0664	181	165.5064	281	256.9464	381	348.3864	481	439.8264
82	74.9808	182	166.4208	282	257.8608	382	349.3008	482	440.7408
83	75.8952	183	167.3352	283	258.7752	383	350.2152	483	441.6552
84	76.8096	184	168.2496	284	259.6896	384	351.1296	484	442.5696
85	77.7240	185	169.1640	285	260.6040	385	352.0440	485	443.4840
86	78.6384	186	170.0784	286	261.5184	386	352.9584	486	444.3984
87	79.5528	187	170.9928	287	262.4328	387	353.8728	487	445.3128
88	80.4672	188	171.9072	288	263.3472	388	354.7872	488	446.2272
89	81.3816	189	172.8216	289	264.2616	389	355.7016	489	447.1416
90	82.2960	190	173.7360	290	265.1760	390	356.6160	490	448.0560
91	83.2104	191	174.6504	291	266.0904	391	357.5304	491	448.9704
92	84.1248	192	175.5648	292	267.0048	392	358.4448	492	449.8848
93	85.0392	193	176.4792	293	267.9192	393	359.3592	493	450.7992
94	85.9536	194	177.3936	294	268.8336	394	360.2736	494	451.7136
95	86.8680	195	178.3080	295	269.7480	395	361.1880	495	452.6280
96	87.7824	196	179.2224	296	270.6624	396	362.1024	496	453.5424
97	88.6968	197	180.1368	297	271.5768	397	363.0168	497	454.4568
98	89.6112	198	181.0512	298	272.4912	398	363.9312	498	455.3712
99	90.5256	199	181.9656	299	273.4056	399	364.8456	499	456.2856

Continued - YARDS TO METERS 1 YARD = 0.9144 METER

YARDS	METERS	YARDS	METERS	YARDS	METERS	YARDS	METERS	YARDS	METERS
500	457.2000	600	548.6400	700	640.0800	800	731.5200	900	822.9600
501	458.1144	601	549.5544	701	640.9944	801	732.4344	901	823.8744
502	459.0288	602	550.4688	702	641.9088	802	733.3488	902	824.7888
503	459.9432	603	551.3832	703	642.8232	803	734.2632	903	825.7032
504	460.8576	604	552.2976	704	643.7376	804	735.1776	904	826.6176
505	461.7720	605	553.2120	705	644.6520	805	736.0920	905	827.5320
506	462.6864	606	554.1264	706	645.5664	806	737.0064	906	828.4464
507	463.6008	607	555.0408	707	646.4808	807	737.9208	907	829.3608
508	464.5152	608	555.9552	708	647.3952	808	738.8352	908	830.2752
509	465.4296	609	556.8696	709	648.3096	809	739.7496	909	831.1896
510	466.3440	610	557.7840	710	649.2240	810	740.6640	910	832.1040
511	467.2584	611	558.6984	711	650.1384	811	741.5784	911	833.0184
512	468.1728	612	559.6128	712	651.0528	812	742.4928	912	833.9328
513	469.0872	613	560.5272	713	651.9672	813	743.4072	913	834.8472
514	470.0016	614	561.4416	714	652.8816	814	744.3216	914	835.7616
515	470.9160	615	562.3560	715	653.7960	815	745.2360	915	836.6760
516	471.8304	616	563.2704	716	654.7104	816	746.1504	916	837.5904
517	472.7448	617	564.1848	717	655.6248	817	747.0648	917	838.5048
518	473.6592	618	565.0992	718	656.5392	818	747.9792	918	839.4192
519	474.5736	619	566.0136	719	657.4536	819	748.8936	919	840.3336
520	475.4880	620	566.9280	720	658.3680	820	749.8080	920	841.2480
521	476.4024	621	567.8424	721	659.2824	821	750.7224	921	842.1624
522	477.3168	622	568.7568	722	660.1968	822	751.6368	922	843.0768
523	478.2312	623	569.6712	723	661.1112	823	752.5512	923	843.9912
524	479.1456	624	570.5856	724	662.0256	824	753.4656	924	844.9056
525	480.0600	625	571.5000	725	662.9400	825	754.3800	925	845.8200
526	480.9744	626	572.4144	726	663.8544	826	755.2944	926	846.7344
527	481.8888	627	573.3288	727	664.7688	827	756.2088	927	847.6488
528	482.8032	628	574.2432	728	665.6832	828	757.1232	928	848.5632
529	483.7176	629	575.1576	729	666.5976	829	758.0376	929	849.4776
530	484.6320	630	576.0720	730	667.5120	830	758.9520	930	850.3920
531	485.5464	631	576.9864	731	668.4264	831	759.8664	931	851.3064
532	486.4608	632	577.9008	732	669.3408	832	760.7808	932	852.2208
533	487.3752	633	578.8152	733	670.2552	833	761.6952	933	853.1352
534	488.2896	634	579.7296	734	671.1696	834	762.6096	934	854.0496
535	489.2040	635	580.6440	735	672.0840	835	763.5240	935	854.9640
536	490.1184	636	581.5584	736	672.9984	836	764.4384	936	855.8784
537	491.0328	637	582.4728	737	673.9128	837	765.3528	937	856.7928
538	491.9472	638	583.3872	738	674.8272	838	766.2672	938	857.7072
539	492.8616	639	584.3016	739	675.7416	839	767.1816	939	858.6216
540	493.7760	640	585.2160	740	676.6560	840	768.0960	940	859.5360
541	494.6904	641	586.1304	741	677.5704	841	769.0104	941	860.4504
542	495.6048	642	587.0448	742	678.4848	842	769.9248	942	861.3648
543	496.5192	643	587.9592	743	679.3992	843	770.8392	943	862.2792
544	497.4336	644	588.8736	744	680.3136	844	771.7536	944	863.1936
545	498.3480	645	589.7880	745	681.2280	845	772.6680	945	864.1080
546	499.2624	646	590.7024	746	682.1424	846	773.5824	946	865.0224
547	500.1768	647	591.6168	747	683.0568	847	774.4968	947	865.9368
548	501.0912	648	592.5312	748	683.9712	848	775.4112	948	866.8512
549	502.0056	649	593.4456	749	684.8856	849	776.3256	949	867.7656

Continued - YARDS TO METERS 1 YARD = 0.9144 METER

YARDS	METERS	YARDS	METERS	YARDS	METERS	YARDS	METERS	YARDS	METERS
550	502.9200	650	594.3600	750	685.8000	850	777.2400	950	868.6800
551	503.8344	651	595.2744	751	686.7144	851	778.1544	951	869.5944
552	504.7488	652	596.1888	752	687.6288	852	779.0688	952	870.5088
553	505.6632	653	597.1032	753	688.5432	853	779.9832	953	871.4232
554	506.5776	654	598.0176	754	689.4576	854	780.8976	954	872.3376
555	507.4920	655	598.9320	755	690.3720	855	781.8120	955	873.2520
556	508.4064	656	599.8464	756	691.2864	856	782.7264	956	874.1664
557	509.3208	657	600.7608	757	692.2008	857	783.6408	957	875.0808
558	510.2352	658	601.6752	758	693.1152	858	784.5552	958	875.9952
559	511.1496	659	602.5896	759	694.0296	859	785.4696	959	876.9096
560	512.0640	660	603.5040	760	694.9440	860	786.3840	960	877.8240
561	512.9784	661	604.4184	761	695.8584	861	787.2984	961	878.7384
562	513.8928	662	605.3328	762	696.7728	862	788.2128	962	879.6528
563	514.8072	663	606.2472	763	697.6872	863	789.1272	963	880.5672
564	515.7216	664	607.1616	764	698.6016	864	790.0416	964	881.4816
565	516.6360	665	608.0760	765	699.5160	865	790.9560	965	882.3960
566	517.5504	666	608.9904	766	700.4304	866	791.8704	966	883.3104
567	518.4648	667	609.9048	767	701.3448	867	792.7848	967	884.2248
568	519.3792	668	610.8192	768	702.2592	868	793.6992	968	885.1392
569	520.2936	669	611.7336	769	703.1736	869	794.6136	969	886.0536
570	521.2080	670	612.6480	770	704.0880	870	795.5280	970	886.9680
571	522.1224	671	613.5624	771	705.0024	871	796.4424	971	887.8824
572	523.0368	672	614.4768	772	705.9168	872	797.3568	972	888.7968
573	523.9512	673	615.3912	773	706.8312	873	798.2712	973	889.7112
574	524.8656	674	616.3056	774	707.7456	874	799.1856	974	890.6256
575	525.7800	675	617.2200	775	708.6600	875	800.1000	975	891.5400
576	526.6944	676	618.1344	776	709.5744	876	801.0144	976	892.4544
577	527.6088	677	619.0488	777	710.4888	877	801.9288	977	893.3688
578	528.5232	678	619.9632	778	711.4032	878	802.8432	978	894.2832
579	529.4376	679	620.8776	779	712.3176	879	803.7576	979	895.1976
580	530.3520	680	621.7920	780	713.2320	880	804.6720	980	896.1120
581	531.2664	681	622.7064	781	714.1464	881	805.5864	981	897.0264
582	532.1808	682	623.6208	782	715.0608	882	806.5008	982	897.9408
583	533.0952	683	624.5352	783	715.9752	883	807.4152	983	898.8552
584	534.0096	684	625.4496	784	716.8896	884	808.3296	984	899.7696
585	534.9240	685	626.3640	785	717.8040	885	809.2440	985	900.6840
586	535.8384	686	627.2784	786	718.7184	886	810.1584	986	901.5984
587	536.7528	687	628.1928	787	719.6328	887	811.0728	987	902.5128
588	537.6672	688	629.1072	788	720.5472	888	811.9872	988	903.4272
589	538.5816	689	630.0216	789	721.4616	889	812.9016	989	904.3416
590	539.4960	690	630.9360	790	722.3760	890	813.8160	990	905.2560
591	540.4104	691	631.8504	791	723.2904	891	814.7304	991	906.1704
592	541.3248	692	632.7648	792	724.2048	892	815.6448	992	907.0848
593	542.2392	693	633.6792	793	725.1192	893	816.5592	993	907.9992
594	543.1536	694	634.5936	794	726.0336	894	817.4736	994	908.9136
595	544.0680	695	635.5080	795	726.9480	895	818.3880	995	909.8280
596	544.9824	696	636.4224	796	727.8624	896	819.3024	996	910.7424
597	545.8968	697	637.3368	797	728.7768	897	820.2168	997	911.6568
598	546.8112	698	638.2512	798	729.6912	898	821.1312	998	912.5712
599	547.7256	699	639.1656	799	730.6056	899	822.0456	999	913.4856

LENGTH - METERS TO YARDS 1 METER = 1.093 613 298 YARDS

METERS	YARDS	METERS	YARDS	METERS	YARDS	METERS	YARDS	METERS	YARDS
0	0.0000	100	109.3613	200	218.7227	300	328.0840	400	437.4453
1	1.0936	101	110.4549	201	219.8163	301	329.1776	401	438.5389
2	2.1872	102	111.5486	202	220.9099	302	330.2712	402	439.6325
3	3.2808	103	112.6422	203	222.0035	303	331.3648	403	440.7262
4	4.3745	104	113.7358	204	223.0971	304	332.4584	404	441.8198
5	5.4681	105	114.8294	205	224.1907	305	333.5521	405	442.9134
6	6.5617	106	115.9230	206	225.2843	306	334.6457	406	444.0070
7	7.6553	107	117.0166	207	226.3780	307	335.7393	407	445.1006
8	8.7489	108	118.1102	208	227.4716	308	336.8329	408	446.1942
9	9.8425	109	119.2038	209	228.5652	309	337.9265	409	447.2878
10	10.9361	110	120.2975	210	229.6588	310	339.0201	410	448.3815
11	12.0297	111	121.3911	211	230.7524	311	340.1137	411	449.4751
12	13.1234	112	122.4847	212	231.8460	312	341.2073	412	450.5687
13	14.2170	113	123.5783	213	232.9396	313	342.3010	413	451.6623
14	15.3106	114	124.6719	214	234.0332	314	343.3946	414	452.7559
15	16.4042	115	125.7655	215	235.1269	315	344.4882	415	453.8495
16	17.4978	116	126.8591	216	236.2205	316	345.5818	416	454.9431
17	18.5914	117	127.9528	217	237.3141	317	346.6754	417	456.0367
18	19.6850	118	129.0464	218	238.4077	318	347.7690	418	457.1304
19	20.7787	119	130.1400	219	239.5013	319	348.8626	419	458.2240
20	21.8723	120	131.2336	220	240.5949	320	349.9563	420	459.3176
21	22.9659	121	132.3272	221	241.6885	321	351.0499	421	460.4112
22	24.0595	122	133.4208	222	242.7822	322	352.1435	422	461.5048
23	25.1531	123	134.5144	223	243.8758	323	353.2371	423	462.5984
24	26.2467	124	135.6080	224	244.9694	324	354.3307	424	463.6920
25	27.3403	125	136.7017	225	246.0630	325	355.4243	425	464.7856
26	28.4339	126	137.7953	226	247.1566	326	356.5179	426	465.8793
27	29.5276	127	138.8889	227	248.2502	327	357.6115	427	466.9729
28	30.6212	128	139.9825	228	249.3438	328	358.7052	428	468.0665
29	31.7148	129	141.0761	229	250.4374	329	359.7988	429	469.1601
30	32.8084	130	142.1697	230	251.5311	330	360.8924	430	470.2537
31	33.9020	131	143.2633	231	252.6247	331	361.9860	431	471.3473
32	34.9956	132	144.3570	232	253.7183	332	363.0796	432	472.4409
33	36.0892	133	145.4506	233	254.8119	333	364.1732	433	473.5346
34	37.1829	134	146.5442	234	255.9055	334	365.2668	434	474.6282
35	38.2765	135	147.6378	235	256.9991	335	366.3605	435	475.7218
36	39.3701	136	148.7314	236	258.0927	336	367.4541	436	476.8154
37	40.4637	137	149.8250	237	259.1863	337	368.5477	437	477.9090
38	41.5573	138	150.9186	238	260.2800	338	369.6413	438	479.0026
39	42.6509	139	152.0122	239	261.3736	339	370.7349	439	480.0962
40	43.7445	140	153.1059	240	262.4672	340	371.8285	440	481.1898
41	44.8381	141	154.1995	241	263.5608	341	372.9221	441	482.2835
42	45.9318	142	155.2931	242	264.6544	342	374.0157	442	483.3771
43	47.0254	143	156.3867	243	265.7480	343	375.1094	443	484.4707
44	48.1190	144	157.4803	244	266.8416	344	376.2030	444	485.5643
45	49.2126	145	158.5739	245	267.9353	345	377.2966	445	486.6579
46	50.3062	146	159.6675	246	269.0289	346	378.3902	446	487.7515
47	51.3998	147	160.7612	247	270.1225	347	379.4838	447	488.8451
48	52.4934	148	161.8548	248	271.2161	348	380.5774	448	489.9388
49	53.5871	149	162.9484	249	272.3097	349	381.6710	449	491.0324

Continued - METERS TO YARDS 1 METER = 1.093 613 298 YARDS

METERS	YARDS	METERS	YARDS	METERS	YARDS	METERS	YARDS	METERS	YARDS
50	54.6807	150	164.0420	250	273.4033	350	382.7647	450	492.1260
51	55.7743	151	165.1356	251	274.4969	351	383.8583	451	493.2196
52	56.8679	152	166.2292	252	275.5905	352	384.9519	452	494.3132
53	57.9615	153	167.3228	253	276.6842	353	386.0455	453	495.4068
54	59.0551	154	168.4164	254	277.7778	354	387.1391	454	496.5004
55	60.1487	155	169.5101	255	278.8714	355	388.2327	455	497.5940
56	61.2423	156	170.6037	256	279.9650	356	389.3263	456	498.6877
57	62.3360	157	171.6973	257	281.0586	357	390.4199	457	499.7813
58	63.4296	158	172.7909	258	282.1522	358	391.5136	458	500.8749
59	64.5232	159	173.8845	259	283.2458	359	392.6072	459	501.9685
60	65.6168	160	174.9781	260	284.3395	360	393.7008	460	503.0621
61	66.7104	161	176.0717	261	285.4331	361	394.7944	461	504.1557
62	67.8040	162	177.1654	262	286.5267	362	395.8880	462	505.2493
63	68.8976	163	178.2590	263	287.6203	363	396.9816	463	506.3430
64	69.9913	164	179.3526	264	288.7139	364	398.0752	464	507.4366
65	71.0849	165	180.4462	265	289.8075	365	399.1688	465	508.5302
66	72.1785	166	181.5398	266	290.9011	366	400.2625	466	509.6238
67	73.2721	167	182.6334	267	291.9947	367	401.3561	467	510.7174
68	74.3657	168	183.7270	268	293.0884	368	402.4497	468	511.8110
69	75.4593	169	184.8206	269	294.1820	369	403.5433	469	512.9046
70	76.5529	170	185.9143	270	295.2756	370	404.6369	470	513.9982
71	77.6465	171	187.0079	271	296.3692	371	405.7305	471	515.0919
72	78.7402	172	188.1015	272	297.4628	372	406.8241	472	516.1855
73	79.8338	173	189.1951	273	298.5564	373	407.9178	473	517.2791
74	80.9274	174	190.2887	274	299.6500	374	409.0114	474	518.3727
75	82.0210	175	191.3823	275	300.7437	375	410.1050	475	519.4663
76	83.1146	176	192.4759	276	301.8373	376	411.1986	476	520.5599
77	84.2082	177	193.5696	277	302.9309	377	412.2922	477	521.6535
78	85.3018	178	194.6632	278	304.0245	378	413.3858	478	522.7472
79	86.3954	179	195.7568	279	305.1181	379	414.4794	479	523.8408
80	87.4891	180	196.8504	280	306.2117	380	415.5731	480	524.9344
81	88.5827	181	197.9440	281	307.3053	381	416.6667	481	526.0280
82	89.6763	182	199.0376	282	308.3989	382	417.7603	482	527.1216
83	90.7699	183	200.1312	283	309.4926	383	418.8539	483	528.2152
84	91.8635	184	201.2248	284	310.5862	384	419.9475	484	529.3088
85	92.9571	185	202.3185	285	311.6798	385	421.0411	485	530.4024
86	94.0507	186	203.4121	286	312.7734	386	422.1347	486	531.4961
87	95.1444	187	204.5057	287	313.8670	387	423.2283	487	532.5897
88	96.2380	188	205.5993	288	314.9606	388	424.3220	488	533.6833
89	97.3316	189	206.6929	289	316.0542	389	425.4156	489	534.7769
90	98.4252	190	207.7865	290	317.1479	390	426.5092	490	535.8705
91	99.5188	191	208.8801	291	318.2415	391	427.6028	491	536.9641
92	100.6124	192	209.9738	292	319.3351	392	428.6964	492	538.0577
93	101.7060	193	211.0674	293	320.4287	393	429.7900	493	539.1514
94	102.7996	194	212.1610	294	321.5223	394	430.8836	494	540.2450
95	103.8933	195	213.2546	295	322.6159	395	431.9772	495	541.3386
96	104.9869	196	214.3482	296	323.7095	396	433.0709	496	542.4322
97	106.0805	197	215.4418	297	324.8031	397	434.1645	497	543.5258
98	107.1741	198	216.5354	298	325.8968	398	435.2581	498	544.6194
99	108.2677	199	217.6290	299	326.9904	399	436.3517	499	545.7130

Continued - METERS TO YARDS 1 METER = 1.093 613 298 YARDS

METERS	YARDS	METERS	YARDS	METERS	YARDS	METERS	YARDS	METERS	YARDS
500	546.8066	600	656.1680	700	765.5293	800	874.8906	900	984.2520
501	547.9003	601	657.2616	701	766.6229	801	875.9842	901	985.3456
502	548.9939	602	658.3552	702	767.7165	802	877.0779	902	986.4392
503	550.0875	603	659.4488	703	768.8101	803	878.1715	903	987.5328
504	551.1811	604	660.5424	704	769.9038	804	879.2651	904	988.6264
505	552.2747	605	661.6360	705	770.9974	805	880.3587	905	989.7200
506	553.3683	606	662.7297	706	772.0910	806	881.4523	906	990.8136
507	554.4619	607	663.8233	707	773.1846	807	882.5459	907	991.9073
508	555.5555	608	664.9169	708	774.2782	808	883.6395	908	993.0009
509	556.6492	609	666.0105	709	775.3718	809	884.7332	909	994.0945
510	557.7428	610	667.1041	710	776.4654	810	885.8268	910	995.1881
511	558.8364	611	668.1977	711	777.5591	811	886.9204	911	996.2817
512	559.9300	612	669.2913	712	778.6527	812	888.0140	912	997.3753
513	561.0236	613	670.3849	713	779.7463	813	889.1076	913	998.4689
514	562.1172	614	671.4786	714	780.8399	814	890.2012	914	999.5625
515	563.2108	615	672.5722	715	781.9335	815	891.2948	915	1000.6562
516	564.3045	616	673.6658	716	783.0271	816	892.3884	916	1001.7498
517	565.3981	617	674.7594	717	784.1207	817	893.4821	917	1002.8434
518	566.4917	618	675.8530	718	785.2143	818	894.5757	918	1003.9370
519	567.5853	619	676.9466	719	786.3080	819	895.6693	919	1005.0306
520	568.6789	620	678.0402	720	787.4016	820	896.7629	920	1006.1242
521	569.7725	621	679.1339	721	788.4952	821	897.8565	921	1007.2178
522	570.8661	622	680.2275	722	789.5888	822	898.9501	922	1008.3115
523	571.9597	623	681.3211	723	790.6824	823	900.0437	923	1009.4051
524	573.0534	624	682.4147	724	791.7760	824	901.1374	924	1010.4987
525	574.1470	625	683.5083	725	792.8696	825	902.2310	925	1011.5923
526	575.2406	626	684.6019	726	793.9632	826	903.3246	926	1012.6859
527	576.3342	627	685.6955	727	795.0569	827	904.4182	927	1013.7795
528	577.4278	628	686.7891	728	796.1505	828	905.5118	928	1014.8731
529	578.5214	629	687.8828	729	797.2441	829	906.6054	929	1015.9668
530	579.6150	630	688.9764	730	798.3377	830	907.6990	930	1017.0604
531	580.7087	631	690.0700	731	799.4313	831	908.7926	931	1018.1540
532	581.8023	632	691.1636	732	800.5249	832	909.8863	932	1019.2476
533	582.8959	633	692.2572	733	801.6185	833	910.9799	933	1020.3412
534	583.9895	634	693.3508	734	802.7122	834	912.0735	934	1021.4348
535	585.0831	635	694.4444	735	803.8058	835	913.1671	935	1022.5284
536	586.1767	636	695.5381	736	804.8994	836	914.2607	936	1023.6220
537	587.2703	637	696.6317	737	805.9930	837	915.3543	937	1024.7157
538	588.3640	638	697.7253	738	807.0866	838	916.4479	938	1025.8093
539	589.4576	639	698.8189	739	808.1802	839	917.5415	939	1026.9029
540	590.5512	640	699.9125	740	809.2738	840	918.6352	940	1027.9965
541	591.6448	641	701.0061	741	810.3674	841	919.7288	941	1029.0901
542	592.7384	642	702.0997	742	811.4611	842	920.8224	942	1030.1837
543	593.8320	643	703.1933	743	812.5547	843	921.9160	943	1031.2773
544	594.9256	644	704.2870	744	813.6483	844	923.0096	944	1032.3709
545	596.0192	645	705.3806	745	814.7419	845	924.1032	945	1033.4646
546	597.1129	646	706.4742	746	815.8355	846	925.1968	946	1034.5582
547	598.2065	647	707.5678	747	816.9291	847	926.2905	947	1035.6518
548	599.3001	648	708.6614	748	818.0227	848	927.3841	948	1036.7454
549	600.3937	649	709.7550	749	819.1164	849	928.4777	949	1037.8390

Continued - METERS TO YARDS 1 METER = 1.093 613 298 YARDS

METERS	YARDS	METERS	YARDS	METERS	YARDS	METERS	YARDS	METERS	YARDS
550	601.4873	650	710.8486	750	820.2100	850	929.5713	950	1038.9326
551	602.5809	651	711.9423	751	821.3036	851	930.6649	951	1040.0262
552	603.6745	652	713.0359	752	822.3972	852	931.7585	952	1041.1199
553	604.7682	653	714.1295	753	823.4908	853	932.8521	953	1042.2135
554	605.8618	654	715.2231	754	824.5844	854	933.9458	954	1043.3071
555	606.9554	655	716.3167	755	825.6780	855	935.0394	955	1044.4007
556	608.0490	656	717.4103	756	826.7717	856	936.1330	956	1045.4943
557	609.1426	657	718.5039	757	827.8653	857	937.2266	957	1046.5879
558	610.2362	658	719.5975	758	828.9589	858	938.3202	958	1047.6815
559	611.3298	659	720.6912	759	830.0525	859	939.4138	959	1048.7751
560	612.4234	660	721.7848	760	831.1461	860	940.5074	960	1049.8688
561	613.5171	661	722.8784	761	832.2397	861	941.6010	961	1050.9624
562	614.6107	662	723.9720	762	833.3333	862	942.6947	962	1052.0560
563	615.7043	663	725.0656	763	834.4269	863	943.7883	963	1053.1496
564	616.7979	664	726.1592	764	835.5206	864	944.8819	964	1054.2432
565	617.8915	665	727.2528	765	836.6142	865	945.9755	965	1055.3368
566	618.9851	666	728.3465	766	837.7078	866	947.0691	966	1056.4304
567	620.0787	667	729.4401	767	838.8014	867	948.1627	967	1057.5240
568	621.1723	668	730.5337	768	839.8950	868	949.2563	968	1058.6177
569	622.2660	669	731.6273	769	840.9886	869	950.3500	969	1059.7113
570	623.3596	670	732.7209	770	842.0822	870	951.4436	970	1060.8049
571	624.4532	671	733.8145	771	843.1758	871	952.5372	971	1061.8985
572	625.5468	672	734.9081	772	844.2695	872	953.6308	972	1062.9921
573	626.6404	673	736.0017	773	845.3631	873	954.7244	973	1064.0857
574	627.7340	674	737.0954	774	846.4567	874	955.8180	974	1065.1793
575	628.8276	675	738.1890	775	847.5503	875	956.9116	975	1066.2729
576	629.9213	676	739.2826	776	848.6439	876	958.0052	976	1067.3666
577	631.0149	677	740.3762	777	849.7375	877	959.0989	977	1068.4602
578	632.1085	678	741.4698	778	850.8311	878	960.1925	978	1069.5538
579	633.2021	679	742.5634	779	851.9248	879	961.2861	979	1070.6474
580	634.2957	680	743.6570	780	853.0184	880	962.3797	980	1071.7410
581	635.3893	681	744.7506	781	854.1120	881	963.4733	981	1072.8346
582	636.4829	682	745.8443	782	855.2056	882	964.5669	982	1073.9283
583	637.5765	683	746.9379	783	856.2992	883	965.6605	983	1075.0219
584	638.6702	684	748.0315	784	857.3928	884	966.7542	984	1076.1155
585	639.7638	685	749.1251	785	858.4864	885	967.8478	985	1077.2091
586	640.8574	686	750.2187	786	859.5800	886	968.9414	986	1078.3027
587	641.9510	687	751.3123	787	860.6737	887	970.0350	987	1079.3963
588	643.0446	688	752.4059	788	861.7673	888	971.1286	988	1080.4899
589	644.1382	689	753.4996	789	862.8609	889	972.2222	989	1081.5835
590	645.2318	690	754.5932	790	863.9545	890	973.3158	990	1082.6772
591	646.3255	691	755.6868	791	865.0481	891	974.4094	991	1083.7708
592	647.4191	692	756.7804	792	866.1417	892	975.5031	992	1084.8644
593	648.5127	693	757.8740	793	867.2353	893	976.5967	993	1085.9580
594	649.6063	694	758.9676	794	868.3290	894	977.6903	994	1087.0516
595	650.6999	695	760.0612	795	869.4226	895	978.7839	995	1088.1452
596	651.7935	696	761.1549	796	870.5162	896	979.8775	996	1089.2388
597	652.8871	697	762.2485	797	871.6098	897	980.9711	997	1090.3324
598	653.9808	698	763.3421	798	872.7034	898	982.0647	998	1091.4261
599	655.0744	699	764.4357	799	873.7970	899	983.1583	999	1092.5197

LENGTH - MILES TO KILOMETERS 1 MILE = 1.609 344 KILOMETERS

MILES	KILO-METERS	MILES	KILO-METERS	MILES	KILO-METERS	MILES	KILO-METERS	MILES	KILO-METERS
0	0.0000	100	160.9344	200	321.8688	300	482.8032	400	643.7376
1	1.6093	101	162.5437	201	323.4781	301	484.4125	401	645.3469
2	3.2187	102	164.1531	202	325.0875	302	486.0219	402	646.9563
3	4.8280	103	165.7624	203	326.6968	303	487.6312	403	648.5656
4	6.4374	104	167.3718	204	328.3062	304	489.2406	404	650.1750
5	8.0467	105	168.9811	205	329.9155	305	490.8499	405	651.7843
6	9.6561	106	170.5905	206	331.5249	306	492.4593	406	653.3937
7	11.2654	107	172.1998	207	333.1342	307	494.0686	407	655.0030
8	12.8748	108	173.8092	208	334.7436	308	495.6780	408	656.6124
9	14.4841	109	175.4185	209	336.3529	309	497.2873	409	658.2217
10	16.0934	110	177.0278	210	337.9622	310	498.8966	410	659.8310
11	17.7028	111	178.6372	211	339.5716	311	500.5060	411	661.4404
12	19.3121	112	180.2465	212	341.1809	312	502.1153	412	663.0497
13	20.9215	113	181.8559	213	342.7903	313	503.7247	413	664.6591
14	22.5308	114	183.4652	214	344.3996	314	505.3340	414	666.2684
15	24.1402	115	185.0746	215	346.0090	315	506.9434	415	667.8778
16	25.7495	116	186.6839	216	347.6183	316	508.5527	416	669.4871
17	27.3588	117	188.2932	217	349.2276	317	510.1620	417	671.0964
18	28.9682	118	189.9026	218	350.8370	318	511.7714	418	672.7058
19	30.5775	119	191.5119	219	352.4463	319	513.3807	419	674.3151
20	32.1869	120	193.1213	220	354.0557	320	514.9901	420	675.9245
21	33.7962	121	194.7306	221	355.6650	321	516.5994	421	677.5338
22	35.4056	122	196.3400	222	357.2744	322	518.2088	422	679.1432
23	37.0149	123	197.9493	223	358.8837	323	519.8181	423	680.7525
24	38.6243	124	199.5587	224	360.4931	324	521.4275	424	682.3619
25	40.2336	125	201.1680	225	362.1024	325	523.0368	425	683.9712
26	41.8429	126	202.7773	226	363.7117	326	524.6461	426	685.5805
27	43.4523	127	204.3867	227	365.3211	327	526.2555	427	687.1899
28	45.0616	128	205.9960	228	366.9304	328	527.8648	428	688.7992
29	46.6710	129	207.6054	229	368.5398	329	529.4742	429	690.4086
30	48.2803	130	209.2147	230	370.1491	330	531.0835	430	692.0179
31	49.8897	131	210.8241	231	371.7585	331	532.6929	431	693.6273
32	51.4990	132	212.4334	232	373.3678	332	534.3022	432	695.2366
33	53.1084	133	214.0428	233	374.9771	333	535.9116	433	696.8459
34	54.7177	134	215.6521	234	376.5865	334	537.5209	434	698.4553
35	56.3270	135	217.2614	235	378.1958	335	539.1302	435	700.0646
36	57.9364	136	218.8708	236	379.8052	336	540.7396	436	701.6740
37	59.5457	137	220.4801	237	381.4145	337	542.3489	437	703.2833
38	61.1551	138	222.0895	238	383.0239	338	543.9583	438	704.8927
39	62.7644	139	223.6988	239	384.6332	339	545.5676	439	706.5020
40	64.3738	140	225.3082	240	386.2426	340	547.1770	440	708.1114
41	65.9831	141	226.9175	241	387.8519	341	548.7863	441	709.7207
42	67.5924	142	228.5268	242	389.4612	342	550.3956	442	711.3300
43	69.2018	143	230.1362	243	391.0706	343	552.0050	443	712.9394
44	70.8111	144	231.7455	244	392.6799	344	553.6143	444	714.5487
45	72.4205	145	233.3549	245	394.2893	345	555.2237	445	716.1581
46	74.0298	146	234.9642	246	395.8986	346	556.8330	446	717.7674
47	75.6392	147	236.5736	247	397.5080	347	558.4424	447	719.3768
48	77.2485	148	238.1829	248	399.1173	348	560.0517	448	720.9861
49	78.8579	149	239.7923	249	400.7267	349	561.6611	449	722.5955

Continued - MILES TO KILOMETERS 1 MILE = 1.609 344 KILOMETERS

MILES	KILO-METERS	MILES	KILO-METERS	MILES	KILO-METERS	MILES	KILO-METERS	MILES	KILO-METERS
50	80.4672	150	241.4016	250	402.3360	350	563.2704	450	724.2048
51	82.0765	151	243.0109	251	403.9453	351	564.8797	451	725.8141
52	83.6859	152	244.6203	252	405.5547	352	566.4891	452	727.4235
53	85.2952	153	246.2296	253	407.1640	353	568.0984	453	729.0328
54	86.9046	154	247.8390	254	408.7734	354	569.7078	454	730.6422
55	88.5139	155	249.4483	255	410.3827	355	571.3171	455	732.2515
56	90.1233	156	251.0577	256	411.9921	356	572.9265	456	733.8609
57	91.7326	157	252.6670	257	413.6014	357	574.5358	457	735.4702
58	93.3420	158	254.2764	258	415.2108	358	576.1451	458	737.0796
59	94.9513	159	255.8857	259	416.8201	359	577.7545	459	738.6889
60	96.5606	160	257.4950	260	418.4294	360	579.3638	460	740.2982
61	98.1700	161	259.1044	261	420.0388	361	580.9732	461	741.9076
62	99.7793	162	260.7137	262	421.6481	362	582.5825	462	743.5169
63	101.3887	163	262.3231	263	423.2575	363	584.1919	463	745.1263
64	102.9980	164	263.9324	264	424.8668	364	585.8012	464	746.7356
65	104.6074	165	265.5418	265	426.4762	365	587.4106	465	748.3450
66	106.2167	166	267.1511	266	428.0855	366	589.0199	466	749.9543
67	107.8260	167	268.7604	267	429.6948	367	590.6292	467	751.5636
68	109.4354	168	270.3698	268	431.3042	368	592.2386	468	753.1730
69	111.0447	169	271.9791	269	432.9135	369	593.8479	469	754.7823
70	112.6541	170	273.5885	270	434.5229	370	595.4573	470	756.3917
71	114.2634	171	275.1978	271	436.1322	371	597.0666	471	758.0010
72	115.8728	172	276.8072	272	437.7416	372	598.6760	472	759.6104
73	117.4821	173	278.4165	273	439.3509	373	600.2853	473	761.2197
74	119.0915	174	280.0259	274	440.9603	374	601.8947	474	762.8291
75	120.7008	175	281.6352	275	442.5696	375	603.5040	475	764.4384
76	122.3101	176	283.2445	276	444.1789	376	605.1133	476	766.0477
77	123.9195	177	284.8539	277	445.7883	377	606.7227	477	767.6571
78	125.5288	178	286.4632	278	447.3976	378	608.3320	478	769.2664
79	127.1382	179	288.0726	279	449.0070	379	609.9414	479	770.8758
80	128.7475	180	289.6819	280	450.6163	380	611.5507	480	772.4851
81	130.3569	181	291.2913	281	452.2257	381	613.1601	481	774.0945
82	131.9662	182	292.9006	282	453.8350	382	614.7694	482	775.7038
83	133.5756	183	294.5100	283	455.4444	383	616.3788	483	777.3131
84	135.1849	184	296.1193	284	457.0537	384	617.9881	484	778.9225
85	136.7942	185	297.7286	285	458.6630	385	619.5974	485	780.5318
86	138.4036	186	299.3380	286	460.2724	386	621.2068	486	782.1412
87	140.0129	187	300.9473	287	461.8817	387	622.8161	487	783.7505
88	141.6223	188	302.5567	288	463.4911	388	624.4255	488	785.3599
89	143.2316	189	304.1660	289	465.1004	389	626.0348	489	786.9692
90	144.8410	190	305.7754	290	466.7098	390	627.6442	490	788.5786
91	146.4503	191	307.3847	291	468.3191	391	629.2535	491	790.1879
92	148.0596	192	308.9940	292	469.9284	392	630.8628	492	791.7972
93	149.6690	193	310.6034	293	471.5378	393	632.4722	493	793.4066
94	151.2783	194	312.2127	294	473.1471	394	634.0815	494	795.0159
95	152.8877	195	313.8221	295	474.7565	395	635.6909	495	796.6253
96	154.4970	196	315.4314	296	476.3658	396	637.3002	496	798.2346
97	156.1064	197	317.0408	297	477.9752	397	638.9096	497	799.8440
98	157.7157	198	318.6501	298	479.5845	398	640.5189	498	801.4533
99	159.3251	199	320.2595	299	481.1939	399	642.1283	499	803.0627

Continued - MILES TO KILOMETERS 1 MILE = 1.609 344 KILOMETERS

MILES	KILO-METERS	MILES	KILO-METERS	MILES	KILO-METERS	MILES	KILO-METERS	MILES	KILO-METERS
500	804.6720	600	965.6064	700	1126.5408	800	1287.4752	900	1448.4096
501	806.2813	601	967.2157	701	1128.1501	801	1289.0845	901	1450.0189
502	807.8907	602	968.8251	702	1129.7595	802	1290.6939	902	1451.6283
503	809.5000	603	970.4344	703	1131.3688	803	1292.3032	903	1453.2376
504	811.1094	604	972.0438	704	1132.9782	804	1293.9126	904	1454.8470
505	812.7187	605	973.6531	705	1134.5875	805	1295.5219	905	1456.4563
506	814.3281	606	975.2625	706	1136.1969	806	1297.1313	906	1458.0657
507	815.9374	607	976.8718	707	1137.8062	807	1298.7406	907	1459.6750
508	817.5468	608	978.4812	708	1139.4155	808	1300.3499	908	1461.2843
509	819.1561	609	980.0905	709	1141.0249	809	1301.9593	909	1462.8937
510	820.7654	610	981.6998	710	1142.6342	810	1303.5686	910	1464.5030
511	822.3748	611	983.3092	711	1144.2436	811	1305.1780	911	1466.1124
512	823.9841	612	984.9185	712	1145.8529	812	1306.7873	912	1467.7217
513	825.5935	613	986.5279	713	1147.4623	813	1308.3967	913	1469.3311
514	827.2028	614	988.1372	714	1149.0716	814	1310.0060	914	1470.9404
515	828.8122	615	989.7466	715	1150.6810	815	1311.6154	915	1472.5498
516	830.4215	616	991.3559	716	1152.2903	816	1313.2247	916	1474.1591
517	832.0308	617	992.9652	717	1153.8996	817	1314.8340	917	1475.7684
518	833.6402	618	994.5746	718	1155.5090	818	1316.4434	918	1477.3778
519	835.2495	619	996.1839	719	1157.1183	819	1318.0527	919	1478.9871
520	836.8589	620	997.7933	720	1158.7277	820	1319.6621	920	1480.5965
521	838.4682	621	999.4026	721	1160.3370	821	1321.2714	921	1482.2058
522	840.0776	622	1001.0120	722	1161.9464	822	1322.8808	922	1483.8152
523	841.6869	623	1002.6213	723	1163.5557	823	1324.4901	923	1485.4245
524	843.2963	624	1004.2307	724	1165.1651	824	1326.0995	924	1487.0339
525	844.9056	625	1005.8400	725	1166.7744	825	1327.7088	925	1488.6432
526	846.5149	626	1007.4493	726	1168.3837	826	1329.3181	926	1490.2525
527	848.1243	627	1009.0587	727	1169.9931	827	1330.9275	927	1491.8619
528	849.7336	628	1010.6680	728	1171.6024	828	1332.5368	928	1493.4712
529	851.3430	629	1012.2774	729	1173.2118	829	1334.1462	929	1495.0806
530	852.9523	630	1013.8867	730	1174.8211	830	1335.7555	930	1496.6899
531	854.5617	631	1015.4961	731	1176.4305	831	1337.3649	931	1498.2993
532	856.1710	632	1017.1054	732	1178.0398	832	1338.9742	932	1499.9086
533	857.7803	633	1018.7148	733	1179.6492	833	1340.5835	933	1501.5179
534	859.3897	634	1020.3241	734	1181.2585	834	1342.1929	934	1503.1273
535	860.9990	635	1021.9334	735	1182.8678	835	1343.8022	935	1504.7366
536	862.6084	636	1023.5428	736	1184.4772	836	1345.4116	936	1506.3460
537	864.2177	637	1025.1521	737	1186.0865	837	1347.0209	937	1507.9553
538	865.8271	638	1026.7615	738	1187.6959	838	1348.6303	938	1509.5647
539	867.4364	639	1028.3708	739	1189.3052	839	1350.2396	939	1511.1740
540	869.0458	640	1029.9802	740	1190.9146	840	1351.8490	940	1512.7834
541	870.6551	641	1031.5895	741	1192.5239	841	1353.4583	941	1514.3927
542	872.2645	642	1033.1988	742	1194.1332	842	1355.0676	942	1516.0020
543	873.8738	643	1034.8082	743	1195.7426	843	1356.6770	943	1517.6114
544	875.4831	644	1036.4175	744	1197.3519	844	1358.2863	944	1519.2207
545	877.0925	645	1038.0269	745	1198.9613	845	1359.8957	945	1520.8301
546	878.7018	646	1039.6362	746	1200.5706	846	1361.5050	946	1522.4394
547	880.3112	647	1041.2456	747	1202.1800	847	1363.1144	947	1524.0488
548	881.9205	648	1042.8549	748	1203.7893	848	1364.7237	948	1525.6581
549	883.5299	649	1044.4642	749	1205.3987	849	1366.3331	949	1527.2675

Continued - MILES TO KILOMETERS 1 MILE = 1.609 344 KILOMETERS

MILES	KILO-METERS	MILES	KILO-METERS	MILES	KILO-METERS	MILES	KILO-METERS	MILES	KILO-METERS
550	885.1392	650	1046.0736	750	1207.0080	850	1367.9424	950	1528.8768
551	886.7485	651	1047.6829	751	1208.6173	851	1369.5517	951	1530.4861
552	888.3579	652	1049.2923	752	1210.2267	852	1371.1611	952	1532.0955
553	889.9672	653	1050.9016	753	1211.8360	853	1372.7704	953	1533.7048
554	891.5766	654	1052.5110	754	1213.4454	854	1374.3798	954	1535.3142
555	893.1859	655	1054.1203	755	1215.0547	855	1375.9891	955	1536.9235
556	894.7953	656	1055.7297	756	1216.6641	856	1377.5985	956	1538.5329
557	896.4046	657	1057.3390	757	1218.2734	857	1379.2078	957	1540.1422
558	898.0140	658	1058.9483	758	1219.8828	858	1380.8172	958	1541.7516
559	899.6233	659	1060.5577	759	1221.4921	859	1382.4265	959	1543.3609
560	901.2326	660	1062.1670	760	1223.1014	860	1384.0358	960	1544.9702
561	902.8420	661	1063.7764	761	1224.7108	861	1385.6452	961	1546.5796
562	904.4513	662	1065.3857	762	1226.3201	862	1387.2545	962	1548.1889
563	906.0607	663	1066.9951	763	1227.9295	863	1388.8639	963	1549.7983
564	907.6700	664	1068.6044	764	1229.5388	864	1390.4732	964	1551.4076
565	909.2794	665	1070.2138	765	1231.1482	865	1392.0826	965	1553.0170
566	910.8887	666	1071.8231	766	1232.7575	866	1393.6919	966	1554.6263
567	912.4980	667	1073.4324	767	1234.3669	867	1395.3012	967	1556.2356
568	914.1074	668	1075.0418	768	1235.9762	868	1396.9106	968	1557.8450
569	915.7167	669	1076.6511	769	1237.5855	869	1398.5199	969	1559.4543
570	917.3261	670	1078.2605	770	1239.1949	870	1400.1293	970	1561.0637
571	918.9354	671	1079.8698	771	1240.8042	871	1401.7386	971	1562.6730
572	920.5448	672	1081.4792	772	1242.4136	872	1403.3480	972	1564.2824
573	922.1541	673	1083.0885	773	1244.0229	873	1404.9573	973	1565.8917
574	923.7635	674	1084.6978	774	1245.6322	874	1406.5667	974	1567.5011
575	925.3728	675	1086.3072	775	1247.2416	875	1408.1760	975	1569.1104
576	926.9821	676	1087.9165	776	1248.8509	876	1409.7853	976	1570.7197
577	928.5915	677	1089.5259	777	1250.4603	877	1411.3947	977	1572.3291
578	930.2008	678	1091.1352	778	1252.0696	878	1413.0040	978	1573.9384
579	931.8102	679	1092.7446	779	1253.6790	879	1414.6134	979	1575.5478
580	933.4195	680	1094.3539	780	1255.2883	880	1416.2227	980	1577.1571
581	935.0289	681	1095.9633	781	1256.8977	881	1417.8321	981	1578.7665
582	936.6382	682	1097.5726	782	1258.5070	882	1419.4414	982	1580.3758
583	938.2476	683	1099.1819	783	1260.1163	883	1421.0508	983	1581.9852
584	939.8569	684	1100.7913	784	1261.7257	884	1422.6601	984	1583.5945
585	941.4662	685	1102.4006	785	1263.3350	885	1424.2694	985	1585.2038
586	943.0756	686	1104.0100	786	1264.9444	886	1425.8788	986	1586.8132
587	944.6849	687	1105.6193	787	1266.5537	887	1427.4881	987	1588.4225
588	946.2943	688	1107.2287	788	1268.1631	888	1429.0975	988	1590.0319
589	947.9036	689	1108.8380	789	1269.7724	889	1430.7068	989	1591.6412
590	949.5130	690	1110.4474	790	1271.3818	890	1432.3162	990	1593.2506
591	951.1223	691	1112.0567	791	1272.9911	891	1433.9255	991	1594.8599
592	952.7317	692	1113.6660	792	1274.6004	892	1435.5349	992	1596.4693
593	954.3410	693	1115.2754	793	1276.2098	893	1437.1442	993	1598.0786
594	955.9503	694	1116.8847	794	1277.8191	894	1438.7535	994	1599.6879
595	957.5597	695	1118.4941	795	1279.4285	895	1440.3629	995	1601.2973
596	959.1690	696	1120.1034	796	1281.0378	896	1441.9722	996	1602.9066
597	960.7784	697	1121.7128	797	1282.6472	897	1443.5816	997	1604.5160
598	962.3877	698	1123.3221	798	1284.2565	898	1445.1909	998	1606.1253
599	963.9971	699	1124.9315	799	1285.8659	899	1446.8002	999	1607.7346

LENGTH - KILOMETERS TO MILES 1 KILOMETER = 0.621 371 192 MILE

KILO-METERS	MILES	KILO-METERS	MILES	KILO-METERS	MILES	KILO-METERS	MILES	KILO-METERS	MILES
0	0.00000	100	62.13712	200	124.27424	300	186.41136	400	248.54848
1	0.62137	101	62.75849	201	124.89561	301	187.03273	401	249.16985
2	1.24274	102	63.37986	202	125.51698	302	187.65410	402	249.79122
3	1.86411	103	64.00123	203	126.13835	303	188.27547	403	250.41259
4	2.48548	104	64.62260	204	126.75972	304	188.89684	404	251.03396
5	3.10686	105	65.24397	205	127.38109	305	189.51821	405	251.65533
6	3.72823	106	65.86535	206	128.00246	306	190.13959	406	252.27670
7	4.34960	107	66.48672	207	128.62384	307	190.76096	407	252.89808
8	4.97097	108	67.10809	208	129.24521	308	191.38233	408	253.51945
9	5.59234	109	67.72946	209	129.86658	309	192.00370	409	254.14082
10	6.21371	110	68.35083	210	130.48795	310	192.62507	410	254.76219
11	6.83508	111	68.97220	211	131.10932	311	193.24644	411	255.38356
12	7.45645	112	69.59357	212	131.73069	312	193.86781	412	256.00493
13	8.07783	113	70.21494	213	132.35206	313	194.48918	413	256.62630
14	8.69920	114	70.83632	214	132.97343	314	195.11055	414	257.24767
15	9.32057	115	71.45769	215	133.59481	315	195.73193	415	257.86905
16	9.94194	116	72.07906	216	134.21618	316	196.35330	416	258.49041
17	10.56331	117	72.70043	217	134.83755	317	196.97467	417	259.11179
18	11.18468	118	73.32180	218	135.45892	318	197.59604	418	259.73316
19	11.80605	119	73.94317	219	136.08029	319	198.21741	419	260.35453
20	12.42742	120	74.56454	220	136.70166	320	198.83878	420	260.97590
21	13.04879	121	75.18591	221	137.32303	321	199.46015	421	261.59727
22	13.67017	122	75.80729	222	137.94440	322	200.08152	422	262.21864
23	14.29154	123	76.42866	223	138.56577	323	200.70289	423	262.84001
24	14.91291	124	77.05003	224	139.18715	324	201.32427	424	263.46138
25	15.53428	125	77.67140	225	139.80852	325	201.94564	425	264.08276
26	16.15565	126	78.29277	226	140.42989	326	202.56701	426	264.70413
27	16.77702	127	78.91414	227	141.05126	327	203.18838	427	265.32550
28	17.39839	128	79.53551	228	141.67263	328	203.80975	428	265.94687
29	18.01976	129	80.15688	229	142.29400	329	204.43112	429	266.56824
30	18.64114	130	80.77825	230	142.91537	330	205.05249	430	267.18961
31	19.26251	131	81.39963	231	143.53675	331	205.67386	431	267.81098
32	19.88388	132	82.02100	232	144.15812	332	206.29524	432	268.43235
33	20.50525	133	82.64237	233	144.77949	333	206.91661	433	269.05373
34	21.12662	134	83.26374	234	145.40086	334	207.53798	434	269.67510
35	21.74799	135	83.88511	235	146.02223	335	208.15935	435	270.29647
36	22.36936	136	84.50648	236	146.64360	336	208.78072	436	270.91784
37	22.99073	137	85.12785	237	147.26497	337	209.40209	437	271.53921
38	23.61211	138	85.74922	238	147.88634	338	210.02346	438	272.16058
39	24.23348	139	86.37060	239	148.50772	339	210.64483	439	272.78195
40	24.85485	140	86.99197	240	149.12909	340	211.26620	440	273.40332
41	25.47622	141	87.61334	241	149.75046	341	211.88758	441	274.02470
42	26.09759	142	88.23471	242	150.37183	342	212.50895	442	274.64606
43	26.71896	143	88.85608	243	150.99320	343	213.13032	443	275.26744
44	27.34033	144	89.47745	244	151.61457	344	213.75169	444	275.88881
45	27.96170	145	90.09882	245	152.23594	345	214.37306	445	276.51018
46	28.58307	146	90.72019	246	152.85731	346	214.99443	446	277.13155
47	29.20445	147	91.34157	247	153.47868	347	215.61580	447	277.75292
48	29.82582	148	91.96294	248	154.10006	348	216.23717	448	278.37429
49	30.44719	149	92.58431	249	154.72143	349	216.85855	449	278.99566

Continued - KILOMETERS TO MILES 1 KILOMETER = 0.621 371 192 MILE

KILO-METERS	MILES	KILO-METERS	MILES	KILO-METERS	MILES	KILO-METERS	MILES	KILO-METERS	MILES
50	31.06856	150	93.20568	250	155.34280	350	217.47992	450	279.61703
51	31.68993	151	93.82705	251	155.96417	351	218.10129	451	280.23841
52	32.31130	152	94.44842	252	156.58554	352	218.72266	452	280.85978
53	32.93267	153	95.06979	253	157.20691	353	219.34403	453	281.48115
54	33.55404	154	95.69116	254	157.82828	354	219.96540	454	282.10252
55	34.17542	155	96.31253	255	158.44965	355	220.58677	455	282.72389
56	34.79679	156	96.93391	256	159.07103	356	221.20815	456	283.34526
57	35.41816	157	97.55528	257	159.69240	357	221.82952	457	283.96663
58	36.03953	158	98.17665	258	160.31377	358	222.45089	458	284.58801
59	36.66090	159	98.79802	259	160.93514	359	223.07226	459	285.20938
60	37.28227	160	99.41939	260	161.55651	360	223.69363	460	285.83075
61	37.90364	161	100.04076	261	162.17788	361	224.31500	461	286.45212
62	38.52501	162	100.66213	262	162.79925	362	224.93637	462	287.07349
63	39.14639	163	101.28350	263	163.42062	363	225.55774	463	287.69486
64	39.76776	164	101.90488	264	164.04199	364	226.17911	464	288.31623
65	40.38913	165	102.52625	265	164.66337	365	226.80049	465	288.93760
66	41.01050	166	103.14762	266	165.28474	366	227.42186	466	289.55898
67	41.63187	167	103.76899	267	165.90611	367	228.04323	467	290.18035
68	42.25324	168	104.39036	268	166.52748	368	228.66460	468	290.80172
69	42.87461	169	105.01173	269	167.14885	369	229.28597	469	291.42309
70	43.49598	170	105.63310	270	167.77022	370	229.90734	470	292.04446
71	44.11735	171	106.25447	271	168.39159	371	230.52871	471	292.66583
72	44.73873	172	106.87584	272	169.01296	372	231.15008	472	293.28720
73	45.36010	173	107.49722	273	169.63433	373	231.77145	473	293.90857
74	45.98147	174	108.11859	274	170.25571	374	232.39283	474	294.52995
75	46.60284	175	108.73996	275	170.87708	375	233.01420	475	295.15131
76	47.22421	176	109.36133	276	171.49845	376	233.63557	476	295.77269
77	47.84558	177	109.98270	277	172.11982	377	234.25694	477	296.39406
78	48.46695	178	110.60407	278	172.74119	378	234.87831	478	297.01543
79	49.08832	179	111.22544	279	173.36256	379	235.49968	479	297.63680
80	49.70970	180	111.84681	280	173.98393	380	236.12105	480	298.25817
81	50.33107	181	112.46819	281	174.60530	381	236.74242	481	298.87954
82	50.95244	182	113.08956	282	175.22668	382	237.36380	482	299.50092
83	51.57381	183	113.71093	283	175.84805	383	237.98517	483	300.12228
84	52.19518	184	114.33230	284	176.46942	384	238.60654	484	300.74366
85	52.81655	185	114.95367	285	177.09079	385	239.22791	485	301.36503
86	53.43792	186	115.57504	286	177.71216	386	239.84928	486	301.98640
87	54.05929	187	116.19641	287	178.33353	387	240.47065	487	302.60777
88	54.68067	188	116.81778	288	178.95490	388	241.09202	488	303.22914
89	55.30204	189	117.43916	289	179.57627	389	241.71339	489	303.85051
90	55.92341	190	118.06053	290	180.19765	390	242.33476	490	304.47188
91	56.54478	191	118.68190	291	180.81902	391	242.95614	491	305.09325
92	57.16615	192	119.30327	292	181.44039	392	243.57751	492	305.71463
93	57.78752	193	119.92464	293	182.06176	393	244.19888	493	306.33600
94	58.40889	194	120.54601	294	182.68313	394	244.82025	494	306.95737
95	59.03026	195	121.16738	295	183.30450	395	245.44162	495	307.57874
96	59.65163	196	121.78875	296	183.92587	396	246.06299	496	308.20011
97	60.27301	197	122.41012	297	184.54724	397	246.68436	497	308.82148
98	60.89438	198	123.03150	298	185.16862	398	247.30573	498	309.44285
99	61.51575	199	123.65287	299	185.78999	399	247.92710	499	310.06422

Continued - KILOMETERS TO MILES 1 KILOMETER = 0.621 371 192 MILE

KILO-METERS	MILES	KILO-METERS	MILES	KILO-METERS	MILES	KILO-METERS	MILES	KILO-METERS	MILES
500	310.68560	600	372.82272	700	434.95984	800	497.09695	900	559.23407
501	311.30696	601	373.44409	701	435.58121	801	497.71833	901	559.85545
502	311.92834	602	374.06546	702	436.20258	802	498.33969	902	560.47681
503	312.54971	603	374.68683	703	436.82395	803	498.96107	903	561.09818
504	313.17108	604	375.30820	704	437.44532	804	499.58244	904	561.71956
505	313.79245	605	375.92957	705	438.06669	805	500.20381	905	562.34093
506	314.41382	606	376.55094	706	438.68806	806	500.82518	906	562.96230
507	315.03519	607	377.17231	707	439.30943	807	501.44655	907	563.58367
508	315.65657	608	377.79369	708	439.93081	808	502.06792	908	564.20504
509	316.27794	609	378.41505	709	440.55217	809	502.68929	909	564.82641
510	316.89931	610	379.03643	710	441.17355	810	503.31067	910	565.44778
511	317.52068	611	379.65780	711	441.79492	811	503.93204	911	566.06915
512	318.14205	612	380.27917	712	442.41629	812	504.55341	912	566.69053
513	318.76342	613	380.90054	713	443.03766	813	505.17478	913	567.31190
514	319.38479	614	381.52191	714	443.65903	814	505.79615	914	567.93327
515	320.00616	615	382.14328	715	444.28040	815	506.41752	915	568.55464
516	320.62753	616	382.76465	716	444.90178	816	507.03889	916	569.17601
517	321.24891	617	383.38602	717	445.52314	817	507.66026	917	569.79738
518	321.87028	618	384.00740	718	446.14452	818	508.28164	918	570.41875
519	322.49165	619	384.62877	719	446.76589	819	508.90301	919	571.04012
520	323.11302	620	385.25014	720	447.38726	820	509.52438	920	571.66150
521	323.73439	621	385.87151	721	448.00863	821	510.14575	921	572.28287
522	324.35576	622	386.49288	722	448.63000	822	510.76712	922	572.90424
523	324.97713	623	387.11425	723	449.25137	823	511.38849	923	573.52561
524	325.59850	624	387.73562	724	449.87274	824	512.00986	924	574.14698
525	326.21988	625	388.35699	725	450.49411	825	512.63123	925	574.76835
526	326.84125	626	388.97837	726	451.11549	826	513.25260	926	575.38972
527	327.46262	627	389.59974	727	451.73686	827	513.87398	927	576.01109
528	328.08399	628	390.22111	728	452.35823	828	514.49535	928	576.63246
529	328.70536	629	390.84248	729	452.97960	829	515.11671	929	577.25384
530	329.32673	630	391.46385	730	453.60097	830	515.73809	930	577.87521
531	329.94810	631	392.08522	731	454.22234	831	516.35946	931	578.49658
532	330.56947	632	392.70659	732	454.84371	832	516.98083	932	579.11795
533	331.19085	633	393.32796	733	455.46508	833	517.60220	933	579.73932
534	331.81222	634	393.94934	734	456.08646	834	518.22357	934	580.36069
535	332.43359	635	394.57071	735	456.70782	835	518.84494	935	580.98206
536	333.05496	636	395.19208	736	457.32920	836	519.46632	936	581.60343
537	333.67633	637	395.81345	737	457.95057	837	520.08768	937	582.22481
538	334.29770	638	396.43482	738	458.57194	838	520.70906	938	582.84618
539	334.91907	639	397.05619	739	459.19331	839	521.33043	939	583.46754
540	335.54044	640	397.67756	740	459.81468	840	521.95180	940	584.08892
541	336.16182	641	398.29893	741	460.43605	841	522.57317	941	584.71029
542	336.78318	642	398.92030	742	461.05743	842	523.19454	942	585.33167
543	337.40456	643	399.54168	743	461.67879	843	523.81591	943	585.95303
544	338.02593	644	400.16305	744	462.30017	844	524.43729	944	586.57440
545	338.64730	645	400.78442	745	462.92154	845	525.05865	945	587.19578
546	339.26867	646	401.40579	746	463.54291	846	525.68002	946	587.81715
547	339.89004	647	402.02716	747	464.16428	847	526.30140	947	588.43851
548	340.51141	648	402.64853	748	464.78565	848	526.92277	948	589.05989
549	341.13279	649	403.26991	749	465.40702	849	527.54414	949	589.68126

Continued - KILOMETERS TO MILES 1 KILOMETER = 0.621 371 192 MILE

KILO-METERS	MILES	KILO-METERS	MILES	KILO-METERS	MILES	KILO-METERS	MILES	KILO-METERS	MILES
550	341.75415	650	403.89127	750	466.02839	850	528.16551	950	590.30263
551	342.37553	651	404.51265	751	466.64977	851	528.78688	951	590.92400
552	342.99690	652	405.13402	752	467.27114	852	529.40826	952	591.54537
553	343.61827	653	405.75539	753	467.89251	853	530.02962	953	592.16675
554	344.23964	654	406.37676	754	468.51388	854	530.65099	954	592.78812
555	344.86101	655	406.99813	755	469.13525	855	531.27237	955	593.40948
556	345.48238	656	407.61950	756	469.75662	856	531.89374	956	594.03086
557	346.10375	657	408.24087	757	470.37799	857	532.51511	957	594.65223
558	346.72512	658	408.86224	758	470.99936	858	533.13648	958	595.27360
559	347.34650	659	409.48362	759	471.62074	859	533.75785	959	595.89497
560	347.96787	660	410.10499	760	472.24211	860	534.37923	960	596.51634
561	348.58924	661	410.72636	761	472.86348	861	535.00060	961	597.13771
562	349.21061	662	411.34773	762	473.48485	862	535.62196	962	597.75909
563	349.83198	663	411.96910	763	474.10622	863	536.24334	963	598.38046
564	350.45335	664	412.59047	764	474.72759	864	536.86471	964	599.00183
565	351.07472	665	413.21184	765	475.34896	865	537.48608	965	599.62320
566	351.69609	666	413.83321	766	475.97033	866	538.10745	966	600.24457
567	352.31747	667	414.45459	767	476.59171	867	538.72882	967	600.86594
568	352.93884	668	415.07595	768	477.21308	868	539.35020	968	601.48731
569	353.56021	669	415.69733	769	477.83445	869	539.97157	969	602.10868
570	354.18158	670	416.31870	770	478.45582	870	540.59293	970	602.73006
571	354.80295	671	416.94007	771	479.07719	871	541.21431	971	603.35143
572	355.42432	672	417.56144	772	479.69856	872	541.83568	972	603.97280
573	356.04569	673	418.18281	773	480.31993	873	542.45705	973	604.59417
574	356.66706	674	418.80418	774	480.94130	874	543.07842	974	605.21554
575	357.28844	675	419.42556	775	481.56268	875	543.69979	975	605.83691
576	357.90981	676	420.04692	776	482.18404	876	544.32116	976	606.45828
577	358.53118	677	420.66830	777	482.80542	877	544.94254	977	607.07965
578	359.15255	678	421.28967	778	483.42679	878	545.56390	978	607.70103
579	359.77392	679	421.91104	779	484.04816	879	546.18528	979	608.32240
580	360.39529	680	422.53241	780	484.66953	880	546.80665	980	608.94376
581	361.01666	681	423.15378	781	485.29090	881	547.42802	981	609.56514
582	361.63803	682	423.77515	782	485.91227	882	548.04939	982	610.18651
583	362.25940	683	424.39652	783	486.53364	883	548.67076	983	610.80788
584	362.88078	684	425.01789	784	487.15501	884	549.29213	984	611.42925
585	363.50215	685	425.63927	785	487.77639	885	549.91351	985	612.05062
586	364.12352	686	426.26064	786	488.39776	886	550.53487	986	612.67200
587	364.74489	687	426.88201	787	489.01913	887	551.15624	987	613.29337
588	365.36626	688	427.50338	788	489.64050	888	551.77762	988	613.91473
589	365.98763	689	428.12475	789	490.26187	889	552.39899	989	614.53611
590	366.60900	690	428.74612	790	490.88324	890	553.02036	990	615.15748
591	367.23037	691	429.36749	791	491.50461	891	553.64173	991	615.77885
592	367.85175	692	429.98886	792	492.12598	892	554.26310	992	616.40022
593	368.47312	693	430.61024	793	492.74736	893	554.88448	993	617.02159
594	369.09449	694	431.23161	794	493.36872	894	555.50584	994	617.64297
595	369.71586	695	431.85298	795	493.99010	895	556.12721	995	618.26434
596	370.33723	696	432.47435	796	494.61147	896	556.74859	996	618.88570
597	370.95860	697	433.09572	797	495.23284	897	557.36996	997	619.50708
598	371.57997	698	433.71709	798	495.85421	898	557.99133	998	620.12845
599	372.20134	699	434.33846	799	496.47558	899	558.61270	999	620.74982

MASS - AVDP POUNDS TO KILOGRAMS 1 AVDP POUND = 0.453 592 37 KILOGRAM

AVDP POUNDS	KILO-GRAMS	AVDP POUNDS	KILO-GRAMS	AVDP POUNDS	KILO-GRAMS	AVDP POUNDS	KILO-GRAMS	AVDP POUNDS	KILO-GRAMS
0	0.00000	100	45.35924	200	90.71847	300	136.07771	400	181.43695
1	0.45359	101	45.81283	201	91.17207	301	136.53130	401	181.89054
2	0.90718	102	46.26642	202	91.62566	302	136.98490	402	182.34413
3	1.36078	103	46.72001	203	92.07925	303	137.43849	403	182.79772
4	1.81437	104	47.17361	204	92.53284	304	137.89208	404	183.25132
5	2.26796	105	47.62720	205	92.98644	305	138.34567	405	183.70491
6	2.72155	106	48.08079	206	93.44003	306	138.79926	406	184.15850
7	3.17515	107	48.53438	207	93.89362	307	139.25286	407	184.61209
8	3.62874	108	48.98798	208	94.34721	308	139.70645	408	185.06569
9	4.08233	109	49.44157	209	94.80081	309	140.16004	409	185.51928
10	4.53592	110	49.89516	210	95.25440	310	140.61363	410	185.97287
11	4.98952	111	50.34875	211	95.70799	311	141.06723	411	186.42646
12	5.44311	112	50.80235	212	96.16158	312	141.52082	412	186.88006
13	5.89670	113	51.25594	213	96.61517	313	141.97441	413	187.33365
14	6.35029	114	51.70953	214	97.06877	314	142.42800	414	187.78724
15	6.80389	115	52.16312	215	97.52236	315	142.88160	415	188.24083
16	7.25748	116	52.61671	216	97.97595	316	143.33519	416	188.69443
17	7.71107	117	53.07031	217	98.42954	317	143.78878	417	189.14802
18	8.16466	118	53.52390	218	98.88314	318	144.24237	418	189.60161
19	8.61826	119	53.97749	219	99.33673	319	144.69596	419	190.05520
20	9.07185	120	54.43108	220	99.79032	320	145.14956	420	190.50879
21	9.52544	121	54.88468	221	100.24391	321	145.60315	421	190.96239
22	9.97903	122	55.33827	222	100.69751	322	146.05674	422	191.41598
23	10.43262	123	55.79186	223	101.15110	323	146.51033	423	191.86957
24	10.88622	124	56.24545	224	101.60469	324	146.96393	424	192.32316
25	11.33981	125	56.69905	225	102.05828	325	147.41752	425	192.77676
26	11.79340	126	57.15264	226	102.51188	326	147.87111	426	193.23035
27	12.24699	127	57.60623	227	102.96547	327	148.32471	427	193.68394
28	12.70059	128	58.05982	228	103.41906	328	148.77830	428	194.13753
29	13.15418	129	58.51342	229	103.87265	329	149.23189	429	194.59113
30	13.60777	130	58.96701	230	104.32625	330	149.68548	430	195.04472
31	14.06136	131	59.42060	231	104.77984	331	150.13907	431	195.49831
32	14.51496	132	59.87419	232	105.23343	332	150.59267	432	195.95190
33	14.96855	133	60.32779	233	105.68702	333	151.04626	433	196.40550
34	15.42214	134	60.78138	234	106.14061	334	151.49985	434	196.85909
35	15.87573	135	61.23497	235	106.59421	335	151.95344	435	197.31268
36	16.32933	136	61.68856	236	107.04780	336	152.40704	436	197.76627
37	16.78292	137	62.14215	237	107.50139	337	152.86063	437	198.21987
38	17.23651	138	62.59575	238	107.95498	338	153.31422	438	198.67346
39	17.69010	139	63.04934	239	108.40858	339	153.76781	439	199.12705
40	18.14369	140	63.50293	240	108.86217	340	154.22141	440	199.58064
41	18.59729	141	63.95652	241	109.31576	341	154.67500	441	200.03424
42	19.05088	142	64.41012	242	109.76935	342	155.12859	442	200.48783
43	19.50447	143	64.86371	243	110.22295	343	155.58218	443	200.94142
44	19.95806	144	65.31730	244	110.67654	344	156.03577	444	201.39501
45	20.41166	145	65.77089	245	111.13013	345	156.48937	445	201.84860
46	20.86525	146	66.22449	246	111.58372	346	156.94296	446	202.30220
47	21.31884	147	66.67808	247	112.03732	347	157.39655	447	202.75579
48	21.77243	148	67.13167	248	112.49091	348	157.85014	448	203.20938
49	22.22603	149	67.58526	249	112.94450	349	158.30374	449	203.66297

Continued - AVDP POUNDS TO KILOGRAMS 1 AVDP POUND = 0.453 592 37 KILOGRAM

AVDP POUNDS	KILO-GRAMS	AVDP POUNDS	KILO-GRAMS	AVDP POUNDS	KILO-GRAMS	AVDP POUNDS	KILO-GRAMS	AVDP POUNDS	KILO-GRAMS
50	22.67962	150	68.03886	250	113.39809	350	158.75733	450	204.11657
51	23.13321	151	68.49245	251	113.85168	351	159.21092	451	204.57016
52	23.58680	152	68.94604	252	114.30528	352	159.66451	452	205.02375
53	24.04040	153	69.39963	253	114.75887	353	160.11811	453	205.47734
54	24.49399	154	69.85322	254	115.21246	354	160.57170	454	205.93093
55	24.94758	155	70.30682	255	115.66605	355	161.02529	455	206.38453
56	25.40117	156	70.76041	256	116.11965	356	161.47888	456	206.83812
57	25.85476	157	71.21400	257	116.57324	357	161.93248	457	207.29171
58	26.30836	158	71.66759	258	117.02683	358	162.38607	458	207.74530
59	26.76195	159	72.12119	259	117.48042	359	162.83966	459	208.19890
60	27.21554	160	72.57478	260	117.93402	360	163.29325	460	208.65249
61	27.66913	161	73.02837	261	118.38761	361	163.74685	461	209.10608
62	28.12273	162	73.48196	262	118.84120	362	164.20044	462	209.55968
63	28.57632	163	73.93556	263	119.29479	363	164.65403	463	210.01327
64	29.02991	164	74.38915	264	119.74839	364	165.10762	464	210.46686
65	29.48350	165	74.84274	265	120.20198	365	165.56121	465	210.92045
66	29.93710	166	75.29633	266	120.65557	366	166.01481	466	211.37404
67	30.39069	167	75.74993	267	121.10916	367	166.46840	467	211.82764
68	30.84428	168	76.20352	268	121.56275	368	166.92199	468	212.28123
69	31.29787	169	76.65711	269	122.01635	369	167.37558	469	212.73482
70	31.75147	170	77.11070	270	122.46994	370	167.82918	470	213.18841
71	32.20506	171	77.56429	271	122.92353	371	168.28277	471	213.64201
72	32.65865	172	78.01789	272	123.37712	372	168.73636	472	214.09560
73	33.11224	173	78.47148	273	123.83072	373	169.18995	473	214.54919
74	33.56583	174	78.92507	274	124.28431	374	169.64355	474	215.00278
75	34.01943	175	79.37866	275	124.73790	375	170.09714	475	215.45638
76	34.47302	176	79.83226	276	125.19149	376	170.55073	476	215.90997
77	34.92661	177	80.28585	277	125.64509	377	171.00432	477	216.36356
78	35.38020	178	80.73944	278	126.09868	378	171.45792	478	216.81715
79	35.83380	179	81.19303	279	126.55227	379	171.91151	479	217.27074
80	36.28739	180	81.64663	280	127.00586	380	172.36510	480	217.72434
81	36.74098	181	82.10022	281	127.45946	381	172.81869	481	218.17793
82	37.19457	182	82.55381	282	127.91305	382	173.27229	482	218.63152
83	37.64817	183	83.00740	283	128.36664	383	173.72588	483	219.08511
84	38.10176	184	83.46100	284	128.82023	384	174.17947	484	219.53871
85	38.55535	185	83.91459	285	129.27382	385	174.63306	485	219.99230
86	39.00894	186	84.36818	286	129.72742	386	175.08665	486	220.44589
87	39.46254	187	84.82177	287	130.18101	387	175.54025	487	220.89948
88	39.91613	188	85.27536	288	130.63460	388	175.99384	488	221.35308
89	40.36972	189	85.72896	289	131.08819	389	176.44743	489	221.80667
90	40.82331	190	86.18255	290	131.54179	390	176.90102	490	222.26026
91	41.27691	191	86.63614	291	131.99538	391	177.35462	491	222.71385
92	41.73050	192	87.08974	292	132.44897	392	177.80821	492	223.16745
93	42.18409	193	87.54333	293	132.90256	393	178.26180	493	223.62104
94	42.63768	194	87.99692	294	133.35616	394	178.71539	494	224.07463
95	43.09128	195	88.45051	295	133.80975	395	179.16899	495	224.52822
96	43.54487	196	88.90410	296	134.26334	396	179.62258	496	224.98182
97	43.99846	197	89.35770	297	134.71693	397	180.07617	497	225.43541
98	44.45205	198	89.81129	298	135.17053	398	180.52976	498	225.88900
99	44.90564	199	90.26488	299	135.62412	399	180.98335	499	226.34259

Continued - AVDP POUNDS TO KILOGRAMS 1 AVDP POUND = 0.453 592 37 KILOGRAM

AVDP POUNDS	KILO-GRAMS	AVDP POUNDS	KILO-GRAMS	AVDP POUNDS	KILO-GRAMS	AVDP POUNDS	KILO-GRAMS	AVDP POUNDS	KILO-GRAMS
500	226.79618	600	272.15542	700	317.51466	800	362.87389	900	408.23313
501	227.24978	601	272.60901	701	317.96825	801	363.32749	901	408.68673
502	227.70337	602	273.06261	702	318.42184	802	363.78108	902	409.14032
503	228.15696	603	273.51620	703	318.87543	803	364.23467	903	409.59391
504	228.61055	604	273.96979	704	319.32903	804	364.68826	904	410.04750
505	229.06415	605	274.42338	705	319.78262	805	365.14186	905	410.50109
506	229.51774	606	274.87698	706	320.23621	806	365.59545	906	410.95469
507	229.97133	607	275.33057	707	320.68980	807	366.04904	907	411.40828
508	230.42492	608	275.78416	708	321.14340	808	366.50263	908	411.86187
509	230.87852	609	276.23775	709	321.59699	809	366.95623	909	412.31546
510	231.33211	610	276.69135	710	322.05058	810	367.40982	910	412.76905
511	231.78570	611	277.14494	711	322.50417	811	367.86341	911	413.22265
512	232.23929	612	277.59853	712	322.95777	812	368.31701	912	413.67624
513	232.69289	613	278.05212	713	323.41136	813	368.77060	913	414.12983
514	233.14648	614	278.50571	714	323.86495	814	369.22419	914	414.58342
515	233.60007	615	278.95930	715	324.31854	815	369.67778	915	415.03702
516	234.05366	616	279.41290	716	324.77214	816	370.13137	916	415.49061
517	234.50726	617	279.86649	717	325.22573	817	370.58496	917	415.94420
518	234.96085	618	280.32008	718	325.67932	818	371.03856	918	416.39780
519	235.41444	619	280.77367	719	326.13291	819	371.49215	919	416.85139
520	235.86803	620	281.22727	720	326.58651	820	371.94574	920	417.30498
521	236.32162	621	281.68086	721	327.04010	821	372.39933	921	417.75857
522	236.77522	622	282.13445	722	327.49369	822	372.85293	922	418.21217
523	237.22881	623	282.58805	723	327.94728	823	373.30652	923	418.66576
524	237.68240	624	283.04164	724	328.40088	824	373.76011	924	419.11935
525	238.13599	625	283.49523	725	328.85447	825	374.21370	925	419.57294
526	238.58959	626	283.94882	726	329.30806	826	374.66730	926	420.02654
527	239.04318	627	284.40242	727	329.76165	827	375.12089	927	420.48013
528	239.49677	628	284.85601	728	330.21524	828	375.57448	928	420.93372
529	239.95036	629	285.30960	729	330.66884	829	376.02807	929	421.38731
530	240.40396	630	285.76319	730	331.12243	830	376.48167	930	421.84090
531	240.85755	631	286.21679	731	331.57602	831	376.93526	931	422.29449
532	241.31114	632	286.67038	732	332.02961	832	377.38885	932	422.74809
533	241.76473	633	287.12397	733	332.48321	833	377.84244	933	423.20168
534	242.21832	634	287.57756	734	332.93680	834	378.29604	934	423.65527
535	242.67192	635	288.03115	735	333.39039	835	378.74963	935	424.10886
536	243.12551	636	288.48475	736	333.84398	836	379.20322	936	424.56246
537	243.57910	637	288.93834	737	334.29758	837	379.65681	937	425.01605
538	244.03269	638	289.39193	738	334.75117	838	380.11040	938	425.46964
539	244.48629	639	289.84552	739	335.20476	839	380.56400	939	425.92323
540	244.93988	640	290.29911	740	335.65835	840	381.01759	940	426.37683
541	245.39347	641	290.75271	741	336.11195	841	381.47118	941	426.83042
542	245.84706	642	291.20630	742	336.56554	842	381.92477	942	427.28401
543	246.30066	643	291.65989	743	337.01913	843	382.37837	943	427.73760
544	246.75425	644	292.11348	744	337.47272	844	382.83196	944	428.19120
545	247.20784	645	292.56708	745	337.92632	845	383.28555	945	428.64479
546	247.66143	646	293.02067	746	338.37991	846	383.73914	946	429.09838
547	248.11503	647	293.47426	747	338.83350	847	384.19274	947	429.55198
548	248.56862	648	293.92786	748	339.28709	848	384.64633	948	430.00557
549	249.02221	649	294.38145	749	339.74068	849	385.09992	949	430.45916

Continued - AVDP POUNDS TO KILOGRAMS 1 AVDP POUND = 0.453 592 37 KILOGRAM

AVDP POUNDS	KILO-GRAMS	AVDP POUNDS	KILO-GRAMS	AVDP POUNDS	KILO-GRAMS	AVDP POUNDS	KILO-GRAMS	AVDP POUNDS	KILO-GRAMS
550	249.47580	650	294.83504	750	340.19427	850	385.55351	950	430.91275
551	249.92940	651	295.28863	751	340.64787	851	386.00711	951	431.36634
552	250.38299	652	295.74223	752	341.10146	852	386.46070	952	431.81993
553	250.83658	653	296.19582	753	341.55505	853	386.91429	953	432.27353
554	251.29017	654	296.64941	754	342.00864	854	387.36788	954	432.72712
555	251.74376	655	297.10300	755	342.46224	855	387.82148	955	433.18071
556	252.19736	656	297.55659	756	342.91583	856	388.27507	956	433.63430
557	252.65095	657	298.01019	757	343.36942	857	388.72866	957	434.08790
558	253.10454	658	298.46378	758	343.82302	858	389.18225	958	434.54149
559	253.55813	659	298.91737	759	344.27661	859	389.63585	959	434.99508
560	254.01173	660	299.37096	760	344.73020	860	390.08944	960	435.44867
561	254.46532	661	299.82455	761	345.18379	861	390.54303	961	435.90227
562	254.91891	662	300.27815	762	345.63739	862	390.99662	962	436.35586
563	255.37250	663	300.73174	763	346.09098	863	391.45021	963	436.80945
564	255.82610	664	301.18533	764	346.54457	864	391.90381	964	437.26304
565	256.27969	665	301.63892	765	346.99816	865	392.35740	965	437.71664
566	256.73328	666	302.09252	766	347.45176	866	392.81099	966	438.17023
567	257.18687	667	302.54611	767	347.90535	867	393.26458	967	438.62382
568	257.64046	668	302.99970	768	348.35894	868	393.71818	968	439.07741
569	258.09406	669	303.45329	769	348.81253	869	394.17177	969	439.53101
570	258.54765	670	303.90689	770	349.26612	870	394.62536	970	439.98460
571	259.00124	671	304.36048	771	349.71972	871	395.07895	971	440.43819
572	259.45483	672	304.81407	772	350.17331	872	395.53255	972	440.89178
573	259.90843	673	305.26766	773	350.62690	873	395.98614	973	441.34538
574	260.36202	674	305.72126	774	351.08049	874	396.43973	974	441.79897
575	260.81561	675	306.17485	775	351.53408	875	396.89332	975	442.25256
576	261.26920	676	306.62844	776	351.98768	876	397.34692	976	442.70615
577	261.72280	677	307.08204	777	352.44127	877	397.80051	977	443.15974
578	262.17639	678	307.53563	778	352.89486	878	398.25410	978	443.61334
579	262.62998	679	307.98922	779	353.34845	879	398.70769	979	444.06693
580	263.08357	680	308.44281	780	353.80205	880	399.16129	980	444.52052
581	263.53717	681	308.89640	781	354.25564	881	399.61488	981	444.97411
582	263.99076	682	309.34999	782	354.70923	882	400.06847	982	445.42771
583	264.44435	683	309.80359	783	355.16283	883	400.52206	983	445.88130
584	264.89794	684	310.25718	784	355.61642	884	400.97565	984	446.33489
585	265.35154	685	310.71077	785	356.07001	885	401.42924	985	446.78848
586	265.80513	686	311.16436	786	356.52360	886	401.88284	986	447.24208
587	266.25872	687	311.61796	787	356.97720	887	402.33643	987	447.69567
588	266.71231	688	312.07155	788	357.43079	888	402.79002	988	448.14926
589	267.16590	689	312.52514	789	357.88438	889	403.24361	989	448.60285
590	267.61950	690	312.97873	790	358.33797	890	403.69721	990	449.05645
591	268.07309	691	313.43233	791	358.79156	891	404.15080	991	449.51004
592	268.52668	692	313.88592	792	359.24516	892	404.60439	992	449.96363
593	268.98027	693	314.33951	793	359.69875	893	405.05799	993	450.41722
594	269.43387	694	314.79310	794	360.15234	894	405.51158	994	450.87082
595	269.88746	695	315.24670	795	360.60593	895	405.96517	995	451.32441
596	270.34105	696	315.70029	796	361.05952	896	406.41876	996	451.77800
597	270.79464	697	316.15388	797	361.51312	897	406.87236	997	452.23159
598	271.24824	698	316.60747	798	361.96671	898	407.32595	998	452.68518
599	271.70183	699	317.06107	799	362.42030	899	407.77954	999	453.13878

MASS – KILOGRAMS TO AVDP POUNDS 1 KILOGRAM = 2.204 622 622 AVDP POUNDS

KILO-GRAMS	AVDP POUNDS	KILO-GRAMS	AVDP POUNDS	KILO-GRAMS	AVDP POUNDS	KILO-GRAMS	AVDP POUNDS	KILO-GRAMS	AVDP POUNDS
0	0.0000	100	220.4623	200	440.9245	300	661.3868	400	881.8490
1	2.2046	101	222.6669	201	443.1291	301	663.5914	401	884.0537
2	4.4092	102	224.8715	202	445.3338	302	665.7960	402	886.2583
3	6.6139	103	227.0761	203	447.5384	303	668.0006	403	888.4629
4	8.8185	104	229.2808	204	449.7430	304	670.2053	404	890.6675
5	11.0231	105	231.4854	205	451.9476	305	672.4099	405	892.8722
6	13.2277	106	233.6900	206	454.1523	306	674.6145	406	895.0768
7	15.4324	107	235.8946	207	456.3569	307	676.8191	407	897.2814
8	17.6370	108	238.0992	208	458.5615	308	679.0238	408	899.4860
9	19.8416	109	240.3039	209	460.7661	309	681.2284	409	901.6907
10	22.0462	110	242.5085	210	462.9707	310	683.4330	410	903.8953
11	24.2508	111	244.7131	211	465.1754	311	685.6376	411	906.0999
12	26.4555	112	246.9177	212	467.3800	312	687.8423	412	908.3045
13	28.6601	113	249.1224	213	469.5846	313	690.0469	413	910.5091
14	30.8647	114	251.3270	214	471.7892	314	692.2515	414	912.7138
15	33.0693	115	253.5316	215	473.9939	315	694.4561	415	914.9184
16	35.2740	116	255.7362	216	476.1985	316	696.6607	416	917.1230
17	37.4786	117	257.9408	217	478.4031	317	698.8654	417	919.3276
18	39.6832	118	260.1455	218	480.6077	318	701.0700	418	921.5323
19	41.8878	119	262.3501	219	482.8124	319	703.2746	419	923.7369
20	44.0925	120	264.5547	220	485.0170	320	705.4792	420	925.9415
21	46.2971	121	266.7593	221	487.2216	321	707.6839	421	928.1461
22	48.5017	122	268.9640	222	489.4262	322	709.8885	422	930.3507
23	50.7063	123	271.1686	223	491.6308	323	712.0931	423	932.5554
24	52.9109	124	273.3732	224	493.8355	324	714.2977	424	934.7600
25	55.1156	125	275.5778	225	496.0401	325	716.5023	425	936.9646
26	57.3202	126	277.7824	226	498.2447	326	718.7070	426	939.1692
27	59.5248	127	279.9871	227	500.4493	327	720.9116	427	941.3739
28	61.7294	128	282.1917	228	502.6540	328	723.1162	428	943.5785
29	63.9341	129	284.3963	229	504.8586	329	725.3208	429	945.7831
30	66.1387	130	286.6009	230	507.0632	330	727.5255	430	947.9877
31	68.3433	131	288.8056	231	509.2678	331	729.7301	431	950.1923
32	70.5479	132	291.0102	232	511.4724	332	731.9347	432	952.3970
33	72.7525	133	293.2148	233	513.6771	333	734.1393	433	954.6016
34	74.9572	134	295.4194	234	515.8817	334	736.3440	434	956.8062
35	77.1618	135	297.6241	235	518.0863	335	738.5486	435	959.0108
36	79.3664	136	299.8287	236	520.2909	336	740.7532	436	961.2155
37	81.5710	137	302.0333	237	522.4956	337	742.9578	437	963.4201
38	83.7757	138	304.2379	238	524.7002	338	745.1624	438	965.6247
39	85.9803	139	306.4425	239	526.9048	339	747.3671	439	967.8293
40	88.1849	140	308.6472	240	529.1094	340	749.5717	440	970.0340
41	90.3895	141	310.8518	241	531.3140	341	751.7763	441	972.2386
42	92.5941	142	313.0564	242	533.5187	342	753.9809	442	974.4432
43	94.7988	143	315.2610	243	535.7233	343	756.1856	443	976.6478
44	97.0034	144	317.4657	244	537.9279	344	758.3902	444	978.8524
45	99.2080	145	319.6703	245	540.1325	345	760.5948	445	981.0571
46	101.4126	146	321.8749	246	542.3372	346	762.7994	446	983.2617
47	103.6173	147	324.0795	247	544.5418	347	765.0041	447	985.4663
48	105.8219	148	326.2841	248	546.7464	348	767.2087	448	987.6709
49	108.0265	149	328.4888	249	548.9510	349	769.4133	449	989.8756

Continued - KILOGRAMS TO AVDP POUNDS 1 KILOGRAM = 2.204 622 622 AVDP POUNDS

KILO-GRAMS	AVDP POUNDS	KILO-GRAMS	AVDP POUNDS	KILO-GRAMS	AVDP POUNDS	KILO-GRAMS	AVDP POUNDS	KILO-GRAMS	AVDP POUNDS
50	110.2311	150	330.6934	250	551.1557	350	771.6179	450	992.0802
51	112.4358	151	332.8980	251	553.3603	351	773.8225	451	994.2848
52	114.6404	152	335.1026	252	555.5649	352	776.0272	452	996.4894
53	116.8450	153	337.3073	253	557.7695	353	778.2318	453	998.6940
54	119.0496	154	339.5119	254	559.9741	354	780.4364	454	1000.8987
55	121.2542	155	341.7165	255	562.1788	355	782.6410	455	1003.1033
56	123.4589	156	343.9211	256	564.3834	356	784.8456	456	1005.3079
57	125.6635	157	346.1258	257	566.5880	357	787.0503	457	1007.5125
58	127.8681	158	348.3304	258	568.7926	358	789.2549	458	1009.7172
59	130.0727	159	350.5350	259	570.9973	359	791.4595	459	1011.9218
60	132.2774	160	352.7396	260	573.2019	360	793.6641	460	1014.1264
61	134.4820	161	354.9442	261	575.4065	361	795.8688	461	1016.3310
62	136.6866	162	357.1489	262	577.6111	362	798.0734	462	1018.5357
63	138.8912	163	359.3535	263	579.8158	363	800.2780	463	1020.7403
64	141.0958	164	361.5581	264	582.0204	364	802.4826	464	1022.9449
65	143.3005	165	363.7627	265	584.2250	365	804.6873	465	1025.1495
66	145.5051	166	365.9674	266	586.4296	366	806.8919	466	1027.3541
67	147.7097	167	368.1720	267	588.6342	367	809.0965	467	1029.5588
68	149.9143	168	370.3766	268	590.8389	368	811.3011	468	1031.7634
69	152.1190	169	372.5812	269	593.0435	369	813.5057	469	1033.9680
70	154.3236	170	374.7858	270	595.2481	370	815.7104	470	1036.1726
71	156.5282	171	376.9905	271	597.4527	371	817.9150	471	1038.3772
72	158.7328	172	379.1951	272	599.6573	372	820.1196	472	1040.5819
73	160.9375	173	381.3997	273	601.8620	373	822.3242	473	1042.7865
74	163.1421	174	383.6043	274	604.0666	374	824.5289	474	1044.9911
75	165.3467	175	385.8090	275	606.2712	375	826.7335	475	1047.1957
76	167.5513	176	388.0136	276	608.4758	376	828.9381	476	1049.4004
77	169.7559	177	390.2182	277	610.6805	377	831.1427	477	1051.6050
78	171.9606	178	392.4228	278	612.8851	378	833.3474	478	1053.8096
79	174.1652	179	394.6274	279	615.0897	379	835.5520	479	1056.0142
80	176.3698	180	396.8321	280	617.2943	380	837.7566	480	1058.2189
81	178.5744	181	399.0367	281	619.4990	381	839.9612	481	1060.4235
82	180.7791	182	401.2413	282	621.7036	382	842.1658	482	1062.6281
83	182.9837	183	403.4459	283	623.9082	383	844.3705	483	1064.8327
84	185.1883	184	405.6506	284	626.1128	384	846.5751	484	1067.0373
85	187.3929	185	407.8552	285	628.3174	385	848.7797	485	1069.2420
86	189.5975	186	410.0598	286	630.5221	386	850.9843	486	1071.4466
87	191.8022	187	412.2644	287	632.7267	387	853.1889	487	1073.6512
88	194.0068	188	414.4691	288	634.9313	388	855.3936	488	1075.8558
89	196.2114	189	416.6737	289	637.1359	389	857.5982	489	1078.0605
90	198.4160	190	418.8783	290	639.3406	390	859.8028	490	1080.2651
91	200.6207	191	421.0829	291	641.5452	391	862.0074	491	1082.4697
92	202.8253	192	423.2875	292	643.7498	392	864.2121	492	1084.6743
93	205.0299	193	425.4922	293	645.9544	393	866.4167	493	1086.8790
94	207.2345	194	427.6968	294	648.1590	394	868.6213	494	1089.0836
95	209.4391	195	429.9014	295	650.3637	395	870.8259	495	1091.2882
96	211.6438	196	432.1060	296	652.5683	396	873.0306	496	1093.4928
97	213.8484	197	434.3107	297	654.7729	397	875.2352	497	1095.6974
98	216.0530	198	436.5153	298	656.9775	398	877.4398	498	1097.9021
99	218.2576	199	438.7199	299	659.1822	399	879.6444	499	1100.1067

Continued - KILOGRAMS TO AVDP POUNDS 1 KILOGRAM = 2.204 622 622 AVDP POUNDS

KILO-GRAMS	AVDP POUNDS	KILO-GRAMS	AVDP POUNDS	KILO-GRAMS	AVDP POUNDS	KILO-GRAMS	AVDP POUNDS	KILO-GRAMS	AVDP POUNDS
500	1102.3113	600	1322.7736	700	1543.2358	800	1763.6981	900	1984.1604
501	1104.5159	601	1324.9782	701	1545.4405	801	1765.9027	901	1986.3650
502	1106.7206	602	1327.1828	702	1547.6451	802	1768.1073	902	1988.5696
503	1108.9252	603	1329.3874	703	1549.8497	803	1770.3120	903	1990.7742
504	1111.1298	604	1331.5921	704	1552.0543	804	1772.5166	904	1992.9789
505	1113.3344	605	1333.7967	705	1554.2589	805	1774.7212	905	1995.1835
506	1115.5390	606	1336.0013	706	1556.4636	806	1776.9258	906	1997.3881
507	1117.7437	607	1338.2059	707	1558.6682	807	1779.1304	907	1999.5927
508	1119.9483	608	1340.4106	708	1560.8728	808	1781.3351	908	2001.7973
509	1122.1529	609	1342.6152	709	1563.0774	809	1783.5397	909	2004.0020
510	1124.3575	610	1344.8198	710	1565.2821	810	1785.7443	910	2006.2066
511	1126.5621	611	1347.0244	711	1567.4867	811	1787.9489	911	2008.4112
512	1128.7668	612	1349.2290	712	1569.6913	812	1790.1536	912	2010.6158
513	1130.9714	613	1351.4337	713	1571.8959	813	1792.3582	913	2012.8204
514	1133.1760	614	1353.6383	714	1574.1005	814	1794.5628	914	2015.0251
515	1135.3806	615	1355.8429	715	1576.3052	815	1796.7674	915	2017.2297
516	1137.5853	616	1358.0475	716	1578.5098	816	1798.9721	916	2019.4343
517	1139.7899	617	1360.2522	717	1580.7144	817	1801.1767	917	2021.6389
518	1141.9945	618	1362.4568	718	1582.9190	818	1803.3813	918	2023.8436
519	1144.1991	619	1364.6614	719	1585.1237	819	1805.5859	919	2026.0482
520	1146.4038	620	1366.8660	720	1587.3283	820	1807.7905	920	2028.2528
521	1148.6084	621	1369.0706	721	1589.5329	821	1809.9952	921	2030.4574
522	1150.8130	622	1371.2753	722	1591.7375	822	1812.1998	922	2032.6620
523	1153.0176	623	1373.4799	723	1593.9422	823	1814.4044	923	2034.8667
524	1155.2222	624	1375.6845	724	1596.1468	824	1816.6090	924	2037.0713
525	1157.4269	625	1377.8891	725	1598.3514	825	1818.8137	925	2039.2759
526	1159.6315	626	1380.0938	726	1600.5560	826	1821.0183	926	2041.4805
527	1161.8361	627	1382.2984	727	1602.7606	827	1823.2229	927	2043.6852
528	1164.0407	628	1384.5030	728	1604.9653	828	1825.4275	928	2045.8898
529	1166.2454	629	1386.7076	729	1607.1699	829	1827.6322	929	2048.0944
530	1168.4500	630	1388.9122	730	1609.3745	830	1829.8368	930	2050.2990
531	1170.6546	631	1391.1169	731	1611.5791	831	1832.0414	931	2052.5037
532	1172.8592	632	1393.3215	732	1613.7838	832	1834.2460	932	2054.7083
533	1175.0639	633	1395.5261	733	1615.9884	833	1836.4506	933	2056.9129
534	1177.2685	634	1397.7307	734	1618.1930	834	1838.6553	934	2059.1175
535	1179.4731	635	1399.9354	735	1620.3976	835	1840.8599	935	2061.3221
536	1181.6777	636	1402.1400	736	1622.6022	836	1843.0645	936	2063.5268
537	1183.8823	637	1404.3446	737	1624.8069	837	1845.2691	937	2065.7314
538	1186.0870	638	1406.5492	738	1627.0115	838	1847.4738	938	2067.9360
539	1188.2916	639	1408.7538	739	1629.2161	839	1849.6784	939	2070.1406
540	1190.4962	640	1410.9585	740	1631.4207	840	1851.8830	940	2072.3452
541	1192.7008	641	1413.1631	741	1633.6254	841	1854.0876	941	2074.5499
542	1194.9055	642	1415.3677	742	1635.8300	842	1856.2922	942	2076.7545
543	1197.1101	643	1417.5723	743	1638.0346	843	1858.4969	943	2078.9591
544	1199.3147	644	1419.7770	744	1640.2392	844	1860.7015	944	2081.1638
545	1201.5193	645	1421.9816	745	1642.4438	845	1862.9061	945	2083.3684
546	1203.7240	646	1424.1862	746	1644.6485	846	1865.1107	946	2085.5730
547	1205.9286	647	1426.3908	747	1646.8531	847	1867.3154	947	2087.7776
548	1208.1332	648	1428.5955	748	1649.0577	848	1869.5200	948	2089.9822
549	1210.3378	649	1430.8001	749	1651.2623	849	1871.7246	949	2092.1869

Continued - KILOGRAMS TO AVDP POUNDS 1 KILOGRAM = 2.204 622 622 AVDP POUNDS

KILO-GRAMS	AVDP POUNDS	KILO-GRAMS	AVDP POUNDS	KILO-GRAMS	AVDP POUNDS	KILO-GRAMS	AVDP POUNDS	KILO-GRAMS	AVDP POUNDS
550	1212.5424	650	1433.0047	750	1653.4670	850	1873.9292	950	2094.3915
551	1214.7471	651	1435.2093	751	1655.6716	851	1876.1339	951	2096.5961
552	1216.9517	652	1437.4139	752	1657.8762	852	1878.3385	952	2098.8007
553	1219.1563	653	1439.6186	753	1660.0808	853	1880.5431	953	2101.0053
554	1221.3609	654	1441.8232	754	1662.2854	854	1882.7477	954	2103.2100
555	1223.5656	655	1444.0278	755	1664.4901	855	1884.9523	955	2105.4146
556	1225.7702	656	1446.2324	756	1666.6947	856	1887.1570	956	2107.6192
557	1227.9748	657	1448.4371	757	1668.8993	857	1889.3616	957	2109.8239
558	1230.1794	658	1450.6417	758	1671.1039	858	1891.5662	958	2112.0285
559	1232.3840	659	1452.8463	759	1673.3086	859	1893.7708	959	2114.2331
560	1234.5887	660	1455.0509	760	1675.5132	860	1895.9754	960	2116.4377
561	1236.7933	661	1457.2556	761	1677.7178	861	1898.1801	961	2118.6423
562	1238.9979	662	1459.4602	762	1679.9224	862	1900.3847	962	2120.8470
563	1241.2025	663	1461.6648	763	1682.1271	863	1902.5893	963	2123.0516
564	1243.4072	664	1463.8694	764	1684.3317	864	1904.7939	964	2125.2562
565	1245.6118	665	1466.0740	765	1686.5363	865	1906.9986	965	2127.4608
566	1247.8164	666	1468.2787	766	1688.7409	866	1909.2032	966	2129.6654
567	1250.0210	667	1470.4833	767	1690.9455	867	1911.4078	967	2131.8701
568	1252.2256	668	1472.6879	768	1693.1502	868	1913.6124	968	2134.0747
569	1254.4303	669	1474.8925	769	1695.3548	869	1915.8170	969	2136.2793
570	1256.6349	670	1477.0972	770	1697.5594	870	1918.0217	970	2138.4839
571	1258.8395	671	1479.3018	771	1699.7640	871	1920.2263	971	2140.6886
572	1261.0441	672	1481.5064	772	1701.9687	872	1922.4309	972	2142.8932
573	1263.2488	673	1483.7110	773	1704.1733	873	1924.6355	973	2145.0978
574	1265.4534	674	1485.9156	774	1706.3779	874	1926.8402	974	2147.3024
575	1267.6580	675	1488.1203	775	1708.5825	875	1929.0448	975	2149.5070
576	1269.8626	676	1490.3249	776	1710.7872	876	1931.2494	976	2151.7117
577	1272.0672	677	1492.5295	777	1712.9918	877	1933.4540	977	2153.9163
578	1274.2719	678	1494.7341	778	1715.1964	878	1935.6587	978	2156.1209
579	1276.4765	679	1496.9388	779	1717.4010	879	1937.8633	979	2158.3255
580	1278.6811	680	1499.1434	780	1719.6056	880	1940.0679	980	2160.5302
581	1280.8857	681	1501.3480	781	1721.8103	881	1942.2725	981	2162.7348
582	1283.0904	682	1503.5526	782	1724.0149	882	1944.4771	982	2164.9394
583	1285.2950	683	1505.7572	783	1726.2195	883	1946.6818	983	2167.1440
584	1287.4996	684	1507.9619	784	1728.4241	884	1948.8864	984	2169.3487
585	1289.7042	685	1510.1665	785	1730.6288	885	1951.0910	985	2171.5533
586	1291.9088	686	1512.3711	786	1732.8334	886	1953.2956	986	2173.7579
587	1294.1135	687	1514.5757	787	1735.0380	887	1955.5003	987	2175.9625
588	1296.3181	688	1516.7804	788	1737.2426	888	1957.7049	988	2178.1671
589	1298.5227	689	1518.9850	789	1739.4473	889	1959.9095	989	2180.3718
590	1300.7273	690	1521.1896	790	1741.6519	890	1962.1141	990	2182.5764
591	1302.9320	691	1523.3942	791	1743.8565	891	1964.3188	991	2184.7810
592	1305.1366	692	1525.5988	792	1746.0611	892	1966.5234	992	2186.9856
593	1307.3412	693	1527.8035	793	1748.2657	893	1968.7280	993	2189.1902
594	1309.5458	694	1530.0081	794	1750.4704	894	1970.9326	994	2191.3949
595	1311.7505	695	1532.2127	795	1752.6750	895	1973.1372	995	2193.5995
596	1313.9551	696	1534.4173	796	1754.8796	896	1975.3419	996	2195.8041
597	1316.1597	697	1536.6220	797	1757.0842	897	1977.5465	997	2198.0087
598	1318.3643	698	1538.8266	798	1759.2888	898	1979.7511	998	2200.2134
599	1320.5689	699	1541.0312	799	1761.4935	899	1981.9557	999	2202.4180

Capacity, or Volume

VOLUME-CAPACITY - CUBIC FEET TO CUBIC METERS 1 CUBIC FOOT = 0.028 316 846 592 CUBIC METER

CUBIC FEET	CUBIC METERS	CUBIC FEET	CUBIC METERS	CUBIC FEET	CUBIC METERS	CUBIC FEET	CUBIC METERS	CUBIC FEET	CUBIC METERS
0	0.000000	100	2.831685	200	5.663369	300	8.495054	400	11.326739
1	0.028317	101	2.860002	201	5.691686	301	8.523371	401	11.355055
2	0.056634	102	2.888318	202	5.720003	302	8.551688	402	11.383372
3	0.084951	103	2.916635	203	5.748320	303	8.580004	403	11.411689
4	0.113267	104	2.944952	204	5.776637	304	8.608321	404	11.440006
5	0.141584	105	2.973269	205	5.804954	305	8.636638	405	11.468323
6	0.169901	106	3.001586	206	5.833270	306	8.664955	406	11.496640
7	0.198218	107	3.029903	207	5.861587	307	8.693272	407	11.524956
8	0.226535	108	3.058219	208	5.889904	308	8.721589	408	11.553273
9	0.254852	109	3.086536	209	5.918221	309	8.749906	409	11.581590
10	0.283168	110	3.114853	210	5.946538	310	8.778222	410	11.609907
11	0.311485	111	3.143170	211	5.974855	311	8.806539	411	11.638224
12	0.339802	112	3.171487	212	6.003171	312	8.834856	412	11.666541
13	0.368119	113	3.199804	213	6.031488	313	8.863173	413	11.694858
14	0.396436	114	3.228121	214	6.059805	314	8.891490	414	11.723174
15	0.424753	115	3.256437	215	6.088122	315	8.919807	415	11.751491
16	0.453070	116	3.284754	216	6.116439	316	8.948123	416	11.779808
17	0.481386	117	3.313071	217	6.144756	317	8.976440	417	11.808125
18	0.509703	118	3.341388	218	6.173073	318	9.004757	418	11.836442
19	0.538020	119	3.369705	219	6.201389	319	9.033074	419	11.864759
20	0.566337	120	3.398022	220	6.229706	320	9.061391	420	11.893075
21	0.594654	121	3.426338	221	6.258023	321	9.089708	421	11.921392
22	0.622971	122	3.454655	222	6.286340	322	9.118025	422	11.949709
23	0.651287	123	3.482972	223	6.314657	323	9.146341	423	11.978026
24	0.679604	124	3.511289	224	6.342974	324	9.174658	424	12.006343
25	0.707921	125	3.539606	225	6.371290	325	9.202975	425	12.034660
26	0.736238	126	3.567923	226	6.399607	326	9.231292	426	12.062977
27	0.764555	127	3.596240	227	6.427924	327	9.259609	427	12.091293
28	0.792872	128	3.624556	228	6.456241	328	9.287926	428	12.119610
29	0.821189	129	3.652873	229	6.484558	329	9.316242	429	12.147927
30	0.849505	130	3.681190	230	6.512875	330	9.344559	430	12.176244
31	0.877822	131	3.709507	231	6.541192	331	9.372876	431	12.204561
32	0.906139	132	3.737824	232	6.569508	332	9.401193	432	12.232878
33	0.934456	133	3.766141	233	6.597825	333	9.429510	433	12.261194
34	0.962773	134	3.794457	234	6.626142	334	9.457827	434	12.289511
35	0.991090	135	3.822774	235	6.654459	335	9.486144	435	12.317828
36	1.019406	136	3.851091	236	6.682776	336	9.514460	436	12.346145
37	1.047723	137	3.879408	237	6.711093	337	9.542777	437	12.374462
38	1.076040	138	3.907725	238	6.739409	338	9.571094	438	12.402779
39	1.104357	139	3.936042	239	6.767726	339	9.599411	439	12.431096
40	1.132674	140	3.964359	240	6.796043	340	9.627728	440	12.459412
41	1.160991	141	3.992675	241	6.824360	341	9.656045	441	12.487729
42	1.189308	142	4.020992	242	6.852677	342	9.684361	442	12.516046
43	1.217624	143	4.049309	243	6.880994	343	9.712678	443	12.544363
44	1.245941	144	4.077626	244	6.909311	344	9.740995	444	12.572680
45	1.274258	145	4.105943	245	6.937627	345	9.769312	445	12.600997
46	1.302575	146	4.134260	246	6.965944	346	9.797629	446	12.629313
47	1.330892	147	4.162576	247	6.994261	347	9.825946	447	12.657630
48	1.359209	148	4.190893	248	7.022578	348	9.854263	448	12.685947
49	1.387525	149	4.219210	249	7.050895	349	9.882579	449	12.714264

Continued - CUBIC FEET TO CUBIC METERS 1 CUBIC FOOT = 0.028 316 846 592 CUBIC METER

CUBIC FEET	CUBIC METERS	CUBIC FEET	CUBIC METERS	CUBIC FEET	CUBIC METERS	CUBIC FEET	CUBIC METERS	CUBIC FEET	CUBIC METERS
50	1.415842	150	4.247527	250	7.079212	350	9.910896	450	12.742581
51	1.444159	151	4.275844	251	7.107528	351	9.939213	451	12.770898
52	1.472476	152	4.304161	252	7.135845	352	9.967530	452	12.799215
53	1.500793	153	4.332478	253	7.164162	353	9.995847	453	12.827531
54	1.529110	154	4.360794	254	7.192479	354	10.024164	454	12.855848
55	1.557427	155	4.389111	255	7.220796	355	10.052480	455	12.884165
56	1.585743	156	4.417428	256	7.249113	356	10.080797	456	12.912482
57	1.614060	157	4.445745	257	7.277430	357	10.109114	457	12.940799
58	1.642377	158	4.474062	258	7.305746	358	10.137431	458	12.969116
59	1.670694	159	4.502379	259	7.334063	359	10.165748	459	12.997432
60	1.699011	160	4.530695	260	7.362380	360	10.194065	460	13.025749
61	1.727328	161	4.559012	261	7.390697	361	10.222382	461	13.054066
62	1.755644	162	4.587329	262	7.419014	362	10.250698	462	13.082383
63	1.783961	163	4.615646	263	7.447331	363	10.279015	463	13.110700
64	1.812278	164	4.643963	264	7.475647	364	10.307332	464	13.139017
65	1.840595	165	4.672280	265	7.503964	365	10.335649	465	13.167334
66	1.868912	166	4.700597	266	7.532281	366	10.363966	466	13.195650
67	1.897229	167	4.728913	267	7.560598	367	10.392283	467	13.223967
68	1.925546	168	4.757230	268	7.588915	368	10.420599	468	13.252284
69	1.953862	169	4.785547	269	7.617232	369	10.448916	469	13.280601
70	1.982179	170	4.813864	270	7.645549	370	10.477233	470	13.308918
71	2.010496	171	4.842181	271	7.673865	371	10.505550	471	13.337235
72	2.038813	172	4.870498	272	7.702182	372	10.533867	472	13.365551
73	2.067130	173	4.898814	273	7.730499	373	10.562184	473	13.393868
74	2.095447	174	4.927131	274	7.758816	374	10.590501	474	13.422185
75	2.123763	175	4.955448	275	7.787133	375	10.618817	475	13.450502
76	2.152080	176	4.983765	276	7.815450	376	10.647134	476	13.478819
77	2.180397	177	5.012082	277	7.843766	377	10.675451	477	13.507136
78	2.208714	178	5.040399	278	7.872083	378	10.703768	478	13.535453
79	2.237031	179	5.068716	279	7.900400	379	10.732085	479	13.563769
80	2.265348	180	5.097032	280	7.928717	380	10.760402	480	13.592086
81	2.293665	181	5.125349	281	7.957034	381	10.788718	481	13.620403
82	2.321981	182	5.153666	282	7.985351	382	10.817035	482	13.648720
83	2.350298	183	5.181983	283	8.013667	383	10.845352	483	13.677037
84	2.378615	184	5.210300	284	8.041984	384	10.873669	484	13.705354
85	2.406932	185	5.238617	285	8.070301	385	10.901986	485	13.733671
86	2.435249	186	5.266933	286	8.098618	386	10.930303	486	13.761987
87	2.463566	187	5.295250	287	8.126935	387	10.958620	487	13.790304
88	2.491882	188	5.323567	288	8.155252	388	10.986936	488	13.818621
89	2.520199	189	5.351884	289	8.183569	389	11.015253	489	13.846938
90	2.548516	190	5.380201	290	8.211885	390	11.043570	490	13.875255
91	2.576833	191	5.408518	291	8.240202	391	11.071887	491	13.903572
92	2.605150	192	5.436835	292	8.268519	392	11.100204	492	13.931888
93	2.633467	193	5.465151	293	8.296836	393	11.128521	493	13.960205
94	2.661784	194	5.493468	294	8.325153	394	11.156837	494	13.988522
95	2.690100	195	5.521785	295	8.353470	395	11.185154	495	14.016839
96	2.718417	196	5.550102	296	8.381787	396	11.213471	496	14.045156
97	2.746734	197	5.578419	297	8.410103	397	11.241788	497	14.073473
98	2.775051	198	5.606736	298	8.438420	398	11.270105	498	14.101790
99	2.803368	199	5.635052	299	8.466737	399	11.298422	499	14.130106

Continued - CUBIC FEET TO CUBIC METERS 1 CUBIC FOOT = 0.028 316 846 592 CUBIC METER

CUBIC FEET	CUBIC METERS	CUBIC FEET	CUBIC METERS	CUBIC FEET	CUBIC METERS	CUBIC FEET	CUBIC METERS	CUBIC FEET	CUBIC METERS
500	14.158423	600	16.990108	700	19.821793	800	22.653477	900	25.485162
501	14.186740	601	17.018425	701	19.850109	801	22.681794	901	25.513479
502	14.215057	602	17.046741	702	19.878426	802	22.710111	902	25.541795
503	14.243374	603	17.075058	703	19.906743	803	22.738428	903	25.570112
504	14.271691	604	17.103375	704	19.935060	804	22.766745	904	25.598429
505	14.300007	605	17.131692	705	19.963377	805	22.795061	905	25.626746
506	14.328324	606	17.160009	706	19.991693	806	22.823378	906	25.655063
507	14.356641	607	17.188326	707	20.020010	807	22.851695	907	25.683380
508	14.384958	608	17.216643	708	20.048327	808	22.880012	908	25.711697
509	14.413275	609	17.244959	709	20.076644	809	22.908329	909	25.740013
510	14.441592	610	17.273276	710	20.104961	810	22.936646	910	25.768330
511	14.469909	611	17.301593	711	20.133278	811	22.964962	911	25.796647
512	14.498225	612	17.329910	712	20.161595	812	22.993279	912	25.824964
513	14.526542	613	17.358227	713	20.189912	813	23.021596	913	25.853281
514	14.554859	614	17.386544	714	20.218228	814	23.049913	914	25.881598
515	14.583176	615	17.414860	715	20.246545	815	23.078230	915	25.909914
516	14.611493	616	17.443177	716	20.274862	816	23.106547	916	25.938231
517	14.639810	617	17.471494	717	20.303179	817	23.134864	917	25.966548
518	14.668126	618	17.499811	718	20.331496	818	23.163180	918	25.994865
519	14.696443	619	17.528128	719	20.359812	819	23.191497	919	26.023182
520	14.724760	620	17.556445	720	20.388129	820	23.219814	920	26.051499
521	14.753077	621	17.584762	721	20.416446	821	23.248131	921	26.079816
522	14.781394	622	17.613078	722	20.444763	822	23.276448	922	26.108132
523	14.809711	623	17.641395	723	20.473080	823	23.304765	923	26.136449
524	14.838028	624	17.669712	724	20.501397	824	23.333081	924	26.164766
525	14.866344	625	17.698029	725	20.529714	825	23.361398	925	26.193083
526	14.894661	626	17.726346	726	20.558031	826	23.389715	926	26.221400
527	14.922978	627	17.754663	727	20.586347	827	23.418032	927	26.249717
528	14.951295	628	17.782979	728	20.614664	828	23.446349	928	26.278033
529	14.979612	629	17.811296	729	20.642981	829	23.474666	929	26.306350
530	15.007929	630	17.839613	730	20.671298	830	23.502983	930	26.334667
531	15.036245	631	17.867930	731	20.699615	831	23.531299	931	26.362984
532	15.064562	632	17.896247	732	20.727931	832	23.559616	932	26.391301
533	15.092879	633	17.924564	733	20.756248	833	23.587933	933	26.419618
534	15.121196	634	17.952881	734	20.784565	834	23.616250	934	26.447935
535	15.149513	635	17.981197	735	20.812882	835	23.644567	935	26.476251
536	15.177830	636	18.009514	736	20.841199	836	23.672884	936	26.504568
537	15.206147	637	18.037831	737	20.869516	837	23.701200	937	26.532885
538	15.234463	638	18.066148	738	20.897833	838	23.729517	938	26.561202
539	15.262780	639	18.094465	739	20.926150	839	23.757834	939	26.589519
540	15.291097	640	18.122782	740	20.954466	840	23.786151	940	26.617836
541	15.319414	641	18.151098	741	20.982783	841	23.814468	941	26.646152
542	15.347731	642	18.179415	742	21.011100	842	23.842785	942	26.674469
543	15.376048	643	18.207732	743	21.039417	843	23.871102	943	26.702786
544	15.404364	644	18.236049	744	21.067734	844	23.899418	944	26.731103
545	15.432681	645	18.264366	745	21.096051	845	23.927735	945	26.759420
546	15.460998	646	18.292683	746	21.124367	846	23.956052	946	26.787737
547	15.489315	647	18.321000	747	21.152684	847	23.984369	947	26.816054
548	15.517632	648	18.349316	748	21.181001	848	24.012686	948	26.844370
549	15.545949	649	18.377633	749	21.209318	849	24.041003	949	26.872687

Continued - CUBIC FEET TO CUBIC METERS 1 CUBIC FOOT = 0.028 316 846 592 CUBIC METER

CUBIC FEET	CUBIC METERS	CUBIC FEET	CUBIC METERS	CUBIC FEET	CUBIC METERS	CUBIC FEET	CUBIC METERS	CUBIC FEET	CUBIC METERS
550	15.574266	650	18.405950	750	21.237635	850	24.069319	950	26.901004
551	15.602582	651	18.434267	751	21.265952	851	24.097636	951	26.929321
552	15.630899	652	18.462584	752	21.294269	852	24.125953	952	26.957638
553	15.659216	653	18.490901	753	21.322585	853	24.154270	953	26.985955
554	15.687533	654	18.519217	754	21.350902	854	24.182587	954	27.014271
555	15.715850	655	18.547534	755	21.379219	855	24.210904	955	27.042588
556	15.744167	656	18.575851	756	21.407536	856	24.239221	956	27.070905
557	15.772483	657	18.604168	757	21.435853	857	24.267537	957	27.099222
558	15.800800	658	18.632485	758	21.464170	858	24.295854	958	27.127539
559	15.829117	659	18.660802	759	21.492486	859	24.324171	959	27.155856
560	15.857434	660	18.689119	760	21.520803	860	24.352488	960	27.184173
561	15.885751	661	18.717435	761	21.549120	861	24.380805	961	27.212489
562	15.914068	662	18.745752	762	21.577437	862	24.409122	962	27.240806
563	15.942385	663	18.774069	763	21.605754	863	24.437438	963	27.269123
564	15.970701	664	18.802386	764	21.634071	864	24.465755	964	27.297440
565	15.999018	665	18.830703	765	21.662388	865	24.494072	965	27.325757
566	16.027335	666	18.859020	766	21.690704	866	24.522389	966	27.354074
567	16.055652	667	18.887336	767	21.719021	867	24.550706	967	27.382390
568	16.083969	668	18.915653	768	21.747338	868	24.579023	968	27.410707
569	16.112286	669	18.943970	769	21.775655	869	24.607340	969	27.439024
570	16.140602	670	18.972287	770	21.803972	870	24.635656	970	27.467341
571	16.168919	671	19.000604	771	21.832289	871	24.663973	971	27.495658
572	16.197236	672	19.028921	772	21.860605	872	24.692290	972	27.523975
573	16.225553	673	19.057238	773	21.888922	873	24.720607	973	27.552292
574	16.253870	674	19.085554	774	21.917239	874	24.748924	974	27.580608
575	16.282187	675	19.113871	775	21.945556	875	24.777241	975	27.608925
576	16.310503	676	19.142188	776	21.973873	876	24.805557	976	27.637242
577	16.338820	677	19.170505	777	22.002190	877	24.833874	977	27.665559
578	16.367137	678	19.198822	778	22.030507	878	24.862191	978	27.693876
579	16.395454	679	19.227139	779	22.058823	879	24.890508	979	27.722193
580	16.423771	680	19.255455	780	22.087140	880	24.918825	980	27.750510
581	16.452088	681	19.283772	781	22.115457	881	24.947142	981	27.778826
582	16.480405	682	19.312089	782	22.143774	882	24.975459	982	27.807143
583	16.508721	683	19.340406	783	22.172091	883	25.003775	983	27.835460
584	16.537038	684	19.368723	784	22.200408	884	25.032092	984	27.863777
585	16.565355	685	19.397040	785	22.228724	885	25.060409	985	27.892094
586	16.593672	686	19.425357	786	22.257041	886	25.088726	986	27.920411
587	16.621989	687	19.453674	787	22.285358	887	25.117043	987	27.948727
588	16.650306	688	19.481990	788	22.313675	888	25.145360	988	27.977044
589	16.678622	689	19.510307	789	22.341992	889	25.173676	989	28.005361
590	16.706939	690	19.538624	790	22.370309	890	25.201993	990	28.033678
591	16.735256	691	19.566941	791	22.398626	891	25.230310	991	28.061995
592	16.763573	692	19.595258	792	22.426942	892	25.258627	992	28.090312
593	16.791890	693	19.623574	793	22.455259	893	25.286944	993	28.118629
594	16.820207	694	19.651891	794	22.483576	894	25.315261	994	28.146945
595	16.848524	695	19.680208	795	22.511893	895	25.343578	995	28.175262
596	16.876840	696	19.708525	796	22.540210	896	25.371894	996	28.203579
597	16.905157	697	19.736842	797	22.568527	897	25.400211	997	28.231896
598	16.933474	698	19.765159	798	22.596843	898	25.428528	998	28.260213
599	16.961791	699	19.793476	799	22.625160	899	25.456845	999	28.288530

260-260 O - 67 - 12

VOLUME-CAPACITY - CUBIC METERS TO CUBIC FEET 1 CUBIC METER = 35.314 6667 CUBIC FEET

CUBIC METERS	CUBIC FEET	CUBIC METERS	CUBIC FEET	CUBIC METERS	CUBIC FEET	CUBIC METERS	CUBIC FEET	CUBIC METERS	CUBIC FEET
0	0.000	100	3531.467	200	7062.933	300	10594.400	400	14125.867
1	35.315	101	3566.781	201	7098.248	301	10629.715	401	14161.181
2	70.629	102	3602.096	202	7133.563	302	10665.029	402	14196.496
3	105.944	103	3637.411	203	7168.877	303	10700.344	403	14231.811
4	141.259	104	3672.725	204	7204.192	304	10735.659	404	14267.125
5	176.573	105	3708.040	205	7239.507	305	10770.973	405	14302.440
6	211.888	106	3743.355	206	7274.821	306	10806.288	406	14337.755
7	247.203	107	3778.669	207	7310.136	307	10841.603	407	14373.069
8	282.517	108	3813.984	208	7345.451	308	10876.917	408	14408.384
9	317.832	109	3849.299	209	7380.765	309	10912.232	409	14443.699
10	353.147	110	3884.613	210	7416.080	310	10947.547	410	14479.013
11	388.461	111	3919.928	211	7451.395	311	10982.861	411	14514.328
12	423.776	112	3955.243	212	7486.709	312	11018.176	412	14549.643
13	459.091	113	3990.557	213	7522.024	313	11053.491	413	14584.957
14	494.405	114	4025.872	214	7557.339	314	11088.805	414	14620.272
15	529.720	115	4061.187	215	7592.653	315	11124.120	415	14655.587
16	565.035	116	4096.501	216	7627.968	316	11159.435	416	14690.901
17	600.349	117	4131.816	217	7663.283	317	11194.749	417	14726.216
18	635.664	118	4167.131	218	7698.597	318	11230.064	418	14761.531
19	670.979	119	4202.445	219	7733.912	319	11265.379	419	14796.845
20	706.293	120	4237.760	220	7769.227	320	11300.693	420	14832.160
21	741.608	121	4273.075	221	7804.541	321	11336.008	421	14867.475
22	776.923	122	4308.389	222	7839.856	322	11371.323	422	14902.789
23	812.237	123	4343.704	223	7875.171	323	11406.637	423	14938.104
24	847.552	124	4379.019	224	7910.485	324	11441.952	424	14973.419
25	882.867	125	4414.333	225	7945.800	325	11477.267	425	15008.733
26	918.181	126	4449.648	226	7981.115	326	11512.581	426	15044.048
27	953.496	127	4484.963	227	8016.429	327	11547.896	427	15079.363
28	988.811	128	4520.277	228	8051.744	328	11583.211	428	15114.677
29	1024.125	129	4555.592	229	8087.059	329	11618.525	429	15149.992
30	1059.440	130	4590.907	230	8122.373	330	11653.840	430	15185.307
31	1094.755	131	4626.221	231	8157.688	331	11689.155	431	15220.621
32	1130.069	132	4661.536	232	8193.003	332	11724.469	432	15255.936
33	1165.384	133	4696.851	233	8228.317	333	11759.784	433	15291.251
34	1200.699	134	4732.165	234	8263.632	334	11795.099	434	15326.565
35	1236.013	135	4767.480	235	8298.947	335	11830.413	435	15361.880
36	1271.328	136	4802.795	236	8334.261	336	11865.728	436	15397.195
37	1306.643	137	4838.109	237	8369.576	337	11901.043	437	15432.509
38	1341.957	138	4873.424	238	8404.891	338	11936.357	438	15467.824
39	1377.272	139	4908.739	239	8440.205	339	11971.672	439	15503.139
40	1412.587	140	4944.053	240	8475.520	340	12006.987	440	15538.453
41	1447.901	141	4979.368	241	8510.835	341	12042.301	441	15573.768
42	1483.216	142	5014.683	242	8546.149	342	12077.616	442	15609.083
43	1518.531	143	5049.997	243	8581.464	343	12112.931	443	15644.397
44	1553.845	144	5085.312	244	8616.779	344	12148.245	444	15679.712
45	1589.160	145	5120.627	245	8652.093	345	12183.560	445	15715.027
46	1624.475	146	5155.941	246	8687.408	346	12218.875	446	15750.341
47	1659.789	147	5191.256	247	8722.723	347	12254.189	447	15785.656
48	1695.104	148	5226.571	248	8758.037	348	12289.504	448	15820.971
49	1730.419	149	5261.885	249	8793.352	349	12324.819	449	15856.285

Continued - CUBIC METERS TO CUBIC FEET 1 CUBIC METER = 35.314 6667 CUBIC FEET

CUBIC METERS	CUBIC FEET	CUBIC METERS	CUBIC FEET	CUBIC METERS	CUBIC FEET	CUBIC METERS	CUBIC FEET	CUBIC METERS	CUBIC FEET
50	1765.733	150	5297.200	250	8828.667	350	12360.133	450	15891.600
51	1801.048	151	5332.515	251	8863.981	351	12395.448	451	15926.915
52	1836.363	152	5367.829	252	8899.296	352	12430.763	452	15962.229
53	1871.677	153	5403.144	253	8934.611	353	12466.077	453	15997.544
54	1906.992	154	5438.459	254	8969.925	354	12501.392	454	16032.859
55	1942.307	155	5473.773	255	9005.240	355	12536.707	455	16068.173
56	1977.621	156	5509.088	256	9040.555	356	12572.021	456	16103.488
57	2012.936	157	5544.403	257	9075.869	357	12607.336	457	16138.803
58	2048.251	158	5579.717	258	9111.184	358	12642.651	458	16174.117
59	2083.565	159	5615.032	259	9146.499	359	12677.965	459	16209.432
60	2118.880	160	5650.347	260	9181.813	360	12713.280	460	16244.747
61	2154.195	161	5685.661	261	9217.128	361	12748.595	461	16280.061
62	2189.509	162	5720.976	262	9252.443	362	12783.909	462	16315.376
63	2224.824	163	5756.291	263	9287.757	363	12819.224	463	16350.691
64	2260.139	164	5791.605	264	9323.072	364	12854.539	464	16386.005
65	2295.453	165	5826.920	265	9358.387	365	12889.853	465	16421.320
66	2330.768	166	5862.235	266	9393.701	366	12925.168	466	16456.635
67	2366.083	167	5897.549	267	9429.016	367	12960.483	467	16491.949
68	2401.397	168	5932.864	268	9464.331	368	12995.797	468	16527.264
69	2436.712	169	5968.179	269	9499.645	369	13031.112	469	16562.579
70	2472.027	170	6003.493	270	9534.960	370	13066.427	470	16597.893
71	2507.341	171	6038.808	271	9570.275	371	13101.741	471	16633.208
72	2542.656	172	6074.123	272	9605.589	372	13137.056	472	16668.523
73	2577.971	173	6109.437	273	9640.904	373	13172.371	473	16703.837
74	2613.285	174	6144.752	274	9676.219	374	13207.685	474	16739.152
75	2648.600	175	6180.067	275	9711.533	375	13243.000	475	16774.467
76	2683.915	176	6215.381	276	9746.848	376	13278.315	476	16809.781
77	2719.229	177	6250.696	277	9782.163	377	13313.629	477	16845.096
78	2754.544	178	6286.011	278	9817.477	378	13348.944	478	16880.411
79	2789.859	179	6321.325	279	9852.792	379	13384.259	479	16915.725
80	2825.173	180	6356.640	280	9888.107	380	13419.573	480	16951.040
81	2860.488	181	6391.955	281	9923.421	381	13454.888	481	16986.354
82	2895.803	182	6427.269	282	9958.736	382	13490.203	482	17021.669
83	2931.117	183	6462.584	283	9994.051	383	13525.517	483	17056.984
84	2966.432	184	6497.899	284	10029.365	384	13560.832	484	17092.299
85	3001.747	185	6533.213	285	10064.680	385	13596.147	485	17127.613
86	3037.061	186	6568.528	286	10099.995	386	13631.461	486	17162.928
87	3072.376	187	6603.843	287	10135.309	387	13666.776	487	17198.243
88	3107.691	188	6639.157	288	10170.624	388	13702.091	488	17233.557
89	3143.005	189	6674.472	289	10205.939	389	13737.405	489	17268.872
90	3178.320	190	6709.787	290	10241.253	390	13772.720	490	17304.187
91	3213.635	191	6745.101	291	10276.568	391	13808.035	491	17339.501
92	3248.949	192	6780.416	292	10311.883	392	13843.349	492	17374.816
93	3284.264	193	6815.731	293	10347.197	393	13878.664	493	17410.131
94	3319.579	194	6851.045	294	10382.512	394	13913.979	494	17445.445
95	3354.893	195	6886.360	295	10417.827	395	13949.293	495	17480.760
96	3390.208	196	6921.675	296	10453.141	396	13984.608	496	17516.075
97	3425.523	197	6956.989	297	10488.456	397	14019.923	497	17551.389
98	3460.837	198	6992.304	298	10523.771	398	14055.237	498	17586.704
99	3496.152	199	7027.619	299	10559.085	399	14090.552	499	17622.019

Continued - CUBIC METERS TO CUBIC FEET 1 CUBIC METER = 35.314 6667 CUBIC FEET

CUBIC METERS	CUBIC FEET	CUBIC METERS	CUBIC FEET	CUBIC METERS	CUBIC FEET	CUBIC METERS	CUBIC FEET	CUBIC METERS	CUBIC FEET
500	17657.333	600	21188.800	700	24720.267	800	28251.733	900	31783.200
501	17692.648	601	21224.115	701	24755.581	801	28287.048	901	31818.515
502	17727.963	602	21259.429	702	24790.896	802	28322.363	902	31853.829
503	17763.277	603	21294.744	703	24826.211	803	28357.677	903	31889.144
504	17798.592	604	21330.059	704	24861.525	804	28392.992	904	31924.459
505	17833.906	605	21365.373	705	24896.840	805	28428.307	905	31959.773
506	17869.221	606	21400.688	706	24932.155	806	28463.621	906	31995.088
507	17904.536	607	21436.003	707	24967.469	807	28498.936	907	32030.403
508	17939.851	608	21471.317	708	25002.784	808	28534.251	908	32065.717
509	17975.165	609	21506.632	709	25038.099	809	28569.565	909	32101.032
510	18010.480	610	21541.947	710	25073.413	810	28604.880	910	32136.347
511	18045.795	611	21577.261	711	25108.728	811	28640.195	911	32171.661
512	18081.109	612	21612.576	712	25144.043	812	28675.509	912	32206.976
513	18116.424	613	21647.891	713	25179.357	813	28710.824	913	32242.291
514	18151.739	614	21683.205	714	25214.672	814	28746.139	914	32277.605
515	18187.053	615	21718.520	715	25249.987	815	28781.453	915	32312.920
516	18222.368	616	21753.835	716	25285.301	816	28816.768	916	32348.235
517	18257.683	617	21789.149	717	25320.616	817	28852.083	917	32383.549
518	18292.997	618	21824.464	718	25355.931	818	28887.397	918	32418.864
519	18328.312	619	21859.779	719	25391.245	819	28922.712	919	32454.179
520	18363.627	620	21895.093	720	25426.560	820	28958.027	920	32489.493
521	18398.941	621	21930.408	721	25461.875	821	28993.341	921	32524.808
522	18434.256	622	21965.723	722	25497.189	822	29028.656	922	32560.123
523	18469.571	623	22001.037	723	25532.504	823	29063.971	923	32595.437
524	18504.885	624	22036.352	724	25567.819	824	29099.285	924	32630.752
525	18540.200	625	22071.667	725	25603.133	825	29134.600	925	32666.067
526	18575.515	626	22106.981	726	25638.448	826	29169.915	926	32701.381
527	18610.829	627	22142.296	727	25673.763	827	29205.229	927	32736.696
528	18646.144	628	22177.611	728	25709.077	828	29240.544	928	32772.011
529	18681.458	629	22212.925	729	25744.392	829	29275.859	929	32807.325
530	18716.773	630	22248.240	730	25779.707	830	29311.173	930	32842.640
531	18752.088	631	22283.555	731	25815.021	831	29346.488	931	32877.955
532	18787.403	632	22318.869	732	25850.336	832	29381.803	932	32913.269
533	18822.717	633	22354.184	733	25885.651	833	29417.117	933	32948.584
534	18858.032	634	22389.499	734	25920.965	834	29452.432	934	32983.898
535	18893.347	635	22424.813	735	25956.280	835	29487.747	935	33019.213
536	18928.661	636	22460.128	736	25991.595	836	29523.061	936	33054.528
537	18963.976	637	22495.443	737	26026.909	837	29558.376	937	33089.842
538	18999.291	638	22530.757	738	26062.224	838	29593.691	938	33125.157
539	19034.605	639	22566.072	739	26097.539	839	29629.005	939	33160.472
540	19069.920	640	22601.387	740	26132.853	840	29664.320	940	33195.787
541	19105.235	641	22636.701	741	26168.168	841	29699.635	941	33231.101
542	19140.549	642	22672.016	742	26203.483	842	29734.949	942	33266.416
543	19175.864	643	22707.331	743	26238.797	843	29770.264	943	33301.730
544	19211.179	644	22742.645	744	26274.112	844	29805.579	944	33337.045
545	19246.493	645	22777.960	745	26309.427	845	29840.893	945	33372.360
546	19281.808	646	22813.275	746	26344.741	846	29876.208	946	33407.674
547	19317.123	647	22848.589	747	26380.056	847	29911.523	947	33442.989
548	19352.437	648	22883.904	748	26415.371	848	29946.837	948	33478.304
549	19387.752	649	22919.219	749	26450.685	849	29982.152	949	33513.619

Continued - CUBIC METERS TO CUBIC FEET 1 CUBIC METER = 35.314 6667 CUBIC FEET

CUBIC METERS	CUBIC FEET	CUBIC METERS	CUBIC FEET	CUBIC METERS	CUBIC FEET	CUBIC METERS	CUBIC FEET	CUBIC METERS	CUBIC FEET
550	19423.067	650	22954.533	750	26486.000	850	30017.467	950	33548.933
551	19458.381	651	22989.848	751	26521.315	851	30052.781	951	33584.248
552	19493.696	652	23025.163	752	26556.629	852	30088.096	952	33619.563
553	19529.010	653	23060.477	753	26591.944	853	30123.411	953	33654.877
554	19564.325	654	23095.792	754	26627.259	854	30158.725	954	33690.192
555	19599.640	655	23131.107	755	26662.573	855	30194.040	955	33725.506
556	19634.955	656	23166.421	756	26697.888	856	30229.355	956	33760.821
557	19670.269	657	23201.736	757	26733.203	857	30264.669	957	33796.136
558	19705.584	658	23237.051	758	26768.517	858	30299.984	958	33831.451
559	19740.899	659	23272.365	759	26803.832	859	30335.299	959	33866.765
560	19776.213	660	23307.680	760	26839.147	860	30370.613	960	33902.080
561	19811.528	661	23342.995	761	26874.461	861	30405.928	961	33937.395
562	19846.843	662	23378.309	762	26909.776	862	30441.243	962	33972.709
563	19882.157	663	23413.624	763	26945.091	863	30476.557	963	34008.024
564	19917.472	664	23448.939	764	26980.405	864	30511.872	964	34043.338
565	19952.787	665	23484.253	765	27015.720	865	30547.187	965	34078.653
566	19988.101	666	23519.568	766	27051.035	866	30582.501	966	34113.968
567	20023.416	667	23554.883	767	27086.349	867	30617.816	967	34149.283
568	20058.731	668	23590.197	768	27121.664	868	30653.131	968	34184.597
569	20094.045	669	23625.512	769	27156.979	869	30688.445	969	34219.912
570	20129.360	670	23660.827	770	27192.293	870	30723.760	970	34255.227
571	20164.675	671	23696.141	771	27227.608	871	30759.075	971	34290.541
572	20199.989	672	23731.456	772	27262.923	872	30794.389	972	34325.856
573	20235.304	673	23766.771	773	27298.237	873	30829.704	973	34361.170
574	20270.619	674	23802.085	774	27333.552	874	30865.019	974	34396.485
575	20305.933	675	23837.400	775	27368.867	875	30900.333	975	34431.800
576	20341.248	676	23872.715	776	27404.181	876	30935.648	976	34467.115
577	20376.563	677	23908.029	777	27439.496	877	30970.963	977	34502.429
578	20411.877	678	23943.344	778	27474.811	878	31006.277	978	34537.744
579	20447.192	679	23978.659	779	27510.125	879	31041.592	979	34573.059
580	20482.507	680	24013.973	780	27545.440	880	31076.907	980	34608.373
581	20517.821	681	24049.288	781	27580.755	881	31112.221	981	34643.688
582	20553.136	682	24084.603	782	27616.069	882	31147.536	982	34679.002
583	20588.451	683	24119.917	783	27651.384	883	31182.851	983	34714.317
584	20623.765	684	24155.232	784	27686.699	884	31218.165	984	34749.632
585	20659.080	685	24190.547	785	27722.013	885	31253.480	985	34784.946
586	20694.395	686	24225.861	786	27757.328	886	31288.795	986	34820.261
587	20729.709	687	24261.176	787	27792.643	887	31324.109	987	34855.576
588	20765.024	688	24296.491	788	27827.957	888	31359.424	988	34890.891
589	20800.339	689	24331.805	789	27863.272	889	31394.739	989	34926.205
590	20835.653	690	24367.120	790	27898.587	890	31430.053	990	34961.520
591	20870.968	691	24402.435	791	27933.901	891	31465.368	991	34996.834
592	20906.283	692	24437.749	792	27969.216	892	31500.683	992	35032.149
593	20941.597	693	24473.064	793	28004.531	893	31535.997	993	35067.464
594	20976.912	694	24508.379	794	28039.845	894	31571.312	994	35102.778
595	21012.227	695	24543.693	795	28075.160	895	31606.627	995	35138.093
596	21047.541	696	24579.008	796	28110.475	896	31641.941	996	35173.408
597	21082.856	697	24614.323	797	28145.789	897	31677.256	997	35208.723
598	21118.171	698	24649.637	798	28181.104	898	31712.571	998	35244.037
599	21153.485	699	24684.952	799	28216.419	899	31747.885	999	35279.352

VOLUME-CAPACITY - CUBIC YARDS TO CUBIC METERS 1 CUBIC YARD = 0.764 554 857 984 CUBIC METER

CUBIC YARDS	CUBIC METERS	CUBIC YARDS	CUBIC METERS	CUBIC YARDS	CUBIC METERS	CUBIC YARDS	CUBIC METERS	CUBIC YARDS	CUBIC METERS
0	0.00000	100	76.45549	200	152.91097	300	229.36646	400	305.82194
1	0.76455	101	77.22004	201	153.67553	301	230.13101	401	306.58649
2	1.52911	102	77.98460	202	154.44008	302	230.89557	402	307.35105
3	2.29366	103	78.74915	203	155.20464	303	231.66012	403	308.11560
4	3.05822	104	79.51371	204	155.96919	304	232.42468	404	308.88016
5	3.82277	105	80.27826	205	156.73375	305	233.18923	405	309.64471
6	4.58733	106	81.04281	206	157.49830	306	233.95378	406	310.40927
7	5.35188	107	81.80737	207	158.26286	307	234.71834	407	311.17382
8	6.11644	108	82.57192	208	159.02741	308	235.48289	408	311.93838
9	6.88099	109	83.33648	209	159.79196	309	236.24745	409	312.70293
10	7.64555	110	84.10103	210	160.55652	310	237.01200	410	313.46749
11	8.41010	111	84.86559	211	161.32107	311	237.77656	411	314.23204
12	9.17466	112	85.63014	212	162.08563	312	238.54111	412	314.99660
13	9.93921	113	86.39470	213	162.85018	313	239.30567	413	315.76115
14	10.70377	114	87.15925	214	163.61474	314	240.07022	414	316.52571
15	11.46832	115	87.92381	215	164.37929	315	240.83478	415	317.29026
16	12.23288	116	88.68836	216	165.14385	316	241.59933	416	318.05482
17	12.99743	117	89.45292	217	165.90840	317	242.36389	417	318.81937
18	13.76199	118	90.21747	218	166.67296	318	243.12844	418	319.58393
19	14.52654	119	90.98203	219	167.43751	319	243.89300	419	320.34848
20	15.29110	120	91.74658	220	168.20207	320	244.65755	420	321.11304
21	16.05565	121	92.51114	221	168.96662	321	245.42211	421	321.87759
22	16.82021	122	93.27569	222	169.73118	322	246.18666	422	322.64215
23	17.58476	123	94.04025	223	170.49573	323	246.95122	423	323.40670
24	18.34932	124	94.80480	224	171.26029	324	247.71577	424	324.17126
25	19.11387	125	95.56936	225	172.02484	325	248.48033	425	324.93581
26	19.87843	126	96.33391	226	172.78940	326	249.24488	426	325.70037
27	20.64298	127	97.09847	227	173.55395	327	250.00944	427	326.46492
28	21.40754	128	97.86302	228	174.31851	328	250.77399	428	327.22948
29	22.17209	129	98.62758	229	175.08306	329	251.53855	429	327.99403
30	22.93665	130	99.39213	230	175.84762	330	252.30310	430	328.75859
31	23.70120	131	100.15669	231	176.61217	331	253.06766	431	329.52314
32	24.46576	132	100.92124	232	177.37673	332	253.83221	432	330.28770
33	25.23031	133	101.68580	233	178.14128	333	254.59677	433	331.05225
34	25.99487	134	102.45035	234	178.90584	334	255.36132	434	331.81681
35	26.75942	135	103.21491	235	179.67039	335	256.12588	435	332.58136
36	27.52397	136	103.97946	236	180.43495	336	256.89043	436	333.34592
37	28.28853	137	104.74401	237	181.19950	337	257.65498	437	334.11047
38	29.05308	138	105.50857	238	181.96406	338	258.41954	438	334.87503
39	29.81764	139	106.27312	239	182.72861	339	259.18409	439	335.63958
40	30.58219	140	107.03768	240	183.49317	340	259.94865	440	336.40414
41	31.34675	141	107.80223	241	184.25772	341	260.71320	441	337.16869
42	32.11130	142	108.56679	242	185.02227	342	261.47776	442	337.93325
43	32.87586	143	109.33134	243	185.78683	343	262.24231	443	338.69780
44	33.64041	144	110.09590	244	186.55138	344	263.00687	444	339.46236
45	34.40497	145	110.86045	245	187.31594	345	263.77142	445	340.22691
46	35.16952	146	111.62501	246	188.08049	346	264.53598	446	340.99147
47	35.93408	147	112.38956	247	188.84505	347	265.30053	447	341.75602
48	36.69863	148	113.15412	248	189.60960	348	266.06509	448	342.52058
49	37.46319	149	113.91867	249	190.37416	349	266.82964	449	343.28513

Continued - CUBIC YARDS TO CUBIC METERS 1 CUBIC YARD = 0.764 554 857 984 CUBIC METER

CUBIC YARDS	CUBIC METERS	CUBIC YARDS	CUBIC METERS	CUBIC YARDS	CUBIC METERS	CUBIC YARDS	CUBIC METERS	CUBIC YARDS	CUBIC METERS
50	38.22774	150	114.68323	250	191.13871	350	267.59420	450	344.04968
51	38.99230	151	115.44778	251	191.90327	351	268.35875	451	344.81424
52	39.75685	152	116.21234	252	192.66782	352	269.12331	452	345.57879
53	40.52141	153	116.97689	253	193.43238	353	269.88786	453	346.34335
54	41.28596	154	117.74145	254	194.19693	354	270.65242	454	347.10790
55	42.05052	155	118.50600	255	194.96149	355	271.41697	455	347.87246
56	42.81507	156	119.27056	256	195.72604	356	272.18153	456	348.63701
57	43.57963	157	120.03511	257	196.49060	357	272.94608	457	349.40157
58	44.34418	158	120.79967	258	197.25515	358	273.71064	458	350.16612
59	45.10874	159	121.56422	259	198.01971	359	274.47519	459	350.93068
60	45.87329	160	122.32878	260	198.78426	360	275.23975	460	351.69523
61	46.63785	161	123.09333	261	199.54882	361	276.00430	461	352.45979
62	47.40240	162	123.85789	262	200.31337	362	276.76886	462	353.22434
63	48.16696	163	124.62244	263	201.07793	363	277.53341	463	353.98890
64	48.93151	164	125.38700	264	201.84248	364	278.29797	464	354.75345
65	49.69607	165	126.15155	265	202.60704	365	279.06252	465	355.51801
66	50.46062	166	126.91611	266	203.37159	366	279.82708	466	356.28256
67	51.22518	167	127.68066	267	204.13615	367	280.59163	467	357.04712
68	51.98973	168	128.44522	268	204.90070	368	281.35619	468	357.81167
69	52.75428	169	129.20977	269	205.66526	369	282.12074	469	358.57623
70	53.51884	170	129.97433	270	206.42981	370	282.88530	470	359.34078
71	54.28339	171	130.73888	271	207.19437	371	283.64985	471	360.10534
72	55.04795	172	131.50344	272	207.95892	372	284.41441	472	360.86989
73	55.81250	173	132.26799	273	208.72347	373	285.17896	473	361.63445
74	56.57706	174	133.03255	274	209.48803	374	285.94352	474	362.39900
75	57.34161	175	133.79710	275	210.25258	375	286.70807	475	363.16356
76	58.10617	176	134.56166	276	211.01714	376	287.47263	476	363.92811
77	58.87072	177	135.32621	277	211.78169	377	288.23718	477	364.69267
78	59.63528	178	136.09076	278	212.54625	378	289.00174	478	365.45722
79	60.39983	179	136.85532	279	213.31080	379	289.76629	479	366.22178
80	61.16439	180	137.61987	280	214.07536	380	290.53085	480	366.98633
81	61.92894	181	138.38443	281	214.83991	381	291.29540	481	367.75089
82	62.69350	182	139.14898	282	215.60447	382	292.05996	482	368.51544
83	63.45805	183	139.91354	283	216.36902	383	292.82451	483	369.27999
84	64.22261	184	140.67809	284	217.13358	384	293.58907	484	370.04455
85	64.98716	185	141.44265	285	217.89813	385	294.35362	485	370.80910
86	65.75172	186	142.20720	286	218.66269	386	295.11817	486	371.57366
87	66.51627	187	142.97176	287	219.42724	387	295.88273	487	372.33821
88	67.28083	188	143.73631	288	220.19180	388	296.64728	488	373.10277
89	68.04538	189	144.50087	289	220.95635	389	297.41184	489	373.86732
90	68.80994	190	145.26542	290	221.72091	390	298.17639	490	374.63188
91	69.57449	191	146.02998	291	222.48546	391	298.94095	491	375.39643
92	70.33905	192	146.79453	292	223.25002	392	299.70550	492	376.16099
93	71.10360	193	147.55909	293	224.01457	393	300.47006	493	376.92554
94	71.86816	194	148.32364	294	224.77913	394	301.23461	494	377.69010
95	72.63271	195	149.08820	295	225.54368	395	301.99917	495	378.45465
96	73.39727	196	149.85275	296	226.30824	396	302.76372	496	379.21921
97	74.16182	197	150.61731	297	227.07279	397	303.52828	497	379.98376
98	74.92638	198	151.38186	298	227.83735	398	304.29283	498	380.74832
99	75.69093	199	152.14642	299	228.60190	399	305.05739	499	381.51287

Continued - CUBIC YARDS TO CUBIC METERS 1 CUBIC YARD = 0.764 554 857 984 CUBIC METER

CUBIC YARDS	CUBIC METERS	CUBIC YARDS	CUBIC METERS	CUBIC YARDS	CUBIC METERS	CUBIC YARDS	CUBIC METERS	CUBIC YARDS	CUBIC METERS
500	382.27743	600	458.73291	700	535.18840	800	611.64388	900	688.09937
501	383.04198	601	459.49747	701	535.95295	801	612.40844	901	688.86392
502	383.80654	602	460.26202	702	536.71751	802	613.17299	902	689.62848
503	384.57109	603	461.02658	703	537.48206	803	613.93755	903	690.39304
504	385.33565	604	461.79113	704	538.24662	804	614.70210	904	691.15759
505	386.10020	605	462.55569	705	539.01117	805	615.46666	905	691.92214
506	386.86476	606	463.32024	706	539.77573	806	616.23121	906	692.68670
507	387.62931	607	464.08480	707	540.54028	807	616.99577	907	693.45126
508	388.39387	608	464.84935	708	541.30483	808	617.76032	908	694.21581
509	389.15842	609	465.61391	709	542.06939	809	618.52488	909	694.98036
510	389.92298	610	466.37846	710	542.83395	810	619.28943	910	695.74492
511	390.68753	611	467.14302	711	543.59850	811	620.05399	911	696.50948
512	391.45209	612	467.90757	712	544.36305	812	620.81854	912	697.27402
513	392.21664	613	468.67213	713	545.12761	813	621.58310	913	698.03858
514	392.98119	614	469.43668	714	545.89217	814	622.34765	914	698.80314
515	393.74575	615	470.20124	715	546.65672	815	623.11221	915	699.56769
516	394.51030	616	470.96579	716	547.42127	816	623.87676	916	700.33224
517	395.27486	617	471.73035	717	548.18583	817	624.64131	917	701.09680
518	396.03941	618	472.49490	718	548.95039	818	625.40587	918	701.86136
519	396.80397	619	473.25946	719	549.71494	819	626.17043	919	702.62591
520	397.56852	620	474.02401	720	550.47949	820	626.93498	920	703.39046
521	398.33308	621	474.78857	721	551.24405	821	627.69953	921	704.15502
522	399.09763	622	475.55312	722	552.00861	822	628.46409	922	704.91958
523	399.86219	623	476.31768	723	552.77316	823	629.22865	923	705.68413
524	400.62674	624	477.08223	724	553.53771	824	629.99320	924	706.44868
525	401.39130	625	477.84678	725	554.30227	825	630.75775	925	707.21324
526	402.15585	626	478.61134	726	555.06683	826	631.52231	926	707.97780
527	402.92041	627	479.37589	727	555.83138	827	632.28687	927	708.74235
528	403.68496	628	480.14045	728	556.59593	828	633.05142	928	709.50690
529	404.44952	629	480.90500	729	557.36049	829	633.81597	929	710.27146
530	405.21407	630	481.66956	730	558.12505	830	634.58053	930	711.03601
531	405.97863	631	482.43411	731	558.88960	831	635.34509	931	711.80057
532	406.74318	632	483.19867	732	559.65415	832	636.10964	932	712.56512
533	407.50774	633	483.96322	733	560.41871	833	636.87419	933	713.32968
534	408.27229	634	484.72778	734	561.18327	834	637.63875	934	714.09423
535	409.03685	635	485.49233	735	561.94781	835	638.40331	935	714.85879
536	409.80140	636	486.25689	736	562.71237	836	639.16785	936	715.62334
537	410.56596	637	487.02144	737	563.47693	837	639.93241	937	716.38790
538	411.33051	638	487.78600	738	564.24148	838	640.69697	938	717.15245
539	412.09507	639	488.55055	739	565.00603	839	641.46152	939	717.91701
540	412.85962	640	489.31511	740	565.77059	840	642.22607	940	718.68156
541	413.62418	641	490.07966	741	566.53515	841	642.99063	941	719.44612
542	414.38873	642	490.84422	742	567.29970	842	643.75519	942	720.21067
543	415.15329	643	491.60877	743	568.06425	843	644.51974	943	720.97523
544	415.91784	644	492.37333	744	568.82881	844	645.28429	944	721.73978
545	416.68240	645	493.13788	745	569.59337	845	646.04885	945	722.50433
546	417.44695	646	493.90244	746	570.35792	846	646.81341	946	723.26889
547	418.21151	647	494.66699	747	571.12247	847	647.57796	947	724.03345
548	418.97606	648	495.43155	748	571.88703	848	648.34251	948	724.79800
549	419.74062	649	496.19610	749	572.65159	849	649.10707	949	725.56255

Continued - CUBIC YARDS TO CUBIC METERS 1 CUBIC YARD = 0.764 554 857 984 CUBIC METER

CUBIC YARDS	CUBIC METERS	CUBIC YARDS	CUBIC METERS	CUBIC YARDS	CUBIC METERS	CUBIC YARDS	CUBIC METERS	CUBIC YARDS	CUBIC METERS
550	420.50517	650	496.96066	750	573.41614	850	649.87163	950	726.32711
551	421.26973	651	497.72521	751	574.18069	851	650.63618	951	727.09167
552	422.03428	652	498.48977	752	574.94525	852	651.40073	952	727.85622
553	422.79884	653	499.25432	753	575.70980	853	652.16529	953	728.62077
554	423.56339	654	500.01888	754	576.47436	854	652.92985	954	729.38533
555	424.32795	655	500.78343	755	577.23891	855	653.69440	955	730.14989
556	425.09250	656	501.54799	756	578.00347	856	654.45895	956	730.91444
557	425.85706	657	502.31254	757	578.76802	857	655.22351	957	731.67899
558	426.62161	658	503.07710	758	579.53258	858	655.98807	958	732.44355
559	427.38617	659	503.84165	759	580.29713	859	656.75262	959	733.20811
560	428.15072	660	504.60620	760	581.06169	860	657.51717	960	733.97266
561	428.91527	661	505.37076	761	581.82624	861	658.28173	961	734.73721
562	429.67983	662	506.13531	762	582.59080	862	659.04629	962	735.50177
563	430.44438	663	506.89987	763	583.35535	863	659.81084	963	736.26633
564	431.20894	664	507.66442	764	584.11991	864	660.57539	964	737.03088
565	431.97349	665	508.42898	765	584.88446	865	661.33995	965	737.79543
566	432.73805	666	509.19353	766	585.64902	866	662.10450	966	738.55999
567	433.50260	667	509.95809	767	586.41357	867	662.86906	967	739.32455
568	434.26716	668	510.72264	768	587.17813	868	663.63361	968	740.08910
569	435.03171	669	511.48720	769	587.94268	869	664.39817	969	740.85365
570	435.79627	670	512.25175	770	588.70724	870	665.16272	970	741.61821
571	436.56082	671	513.01630	771	589.47179	871	665.92728	971	742.38277
572	437.32538	672	513.78086	772	590.23634	872	666.69183	972	743.14732
573	438.08993	673	514.54542	773	591.00090	873	667.45639	973	743.91187
574	438.85449	674	515.30997	774	591.76546	874	668.22094	974	744.67643
575	439.61904	675	516.07452	775	592.53001	875	668.98550	975	745.44099
576	440.38360	676	516.83908	776	593.29456	876	669.75005	976	746.20554
577	441.14815	677	517.60364	777	594.05912	877	670.51461	977	746.97009
578	441.91270	678	518.36819	778	594.82368	878	671.27916	978	747.73465
579	442.67726	679	519.13274	779	595.58823	879	672.04372	979	748.49920
580	443.44181	680	519.89730	780	596.35278	880	672.80827	980	749.26376
581	444.20637	681	520.66186	781	597.11734	881	673.57282	981	750.02831
582	444.97092	682	521.42641	782	597.88190	882	674.33738	982	750.79287
583	445.73548	683	522.19096	783	598.64645	883	675.10194	983	751.55742
584	446.50003	684	522.95552	784	599.41100	884	675.86649	984	752.32198
585	447.26459	685	523.72008	785	600.17556	885	676.63104	985	753.08653
586	448.02914	686	524.48463	786	600.94012	886	677.39560	986	753.85109
587	448.79370	687	525.24918	787	601.70467	887	678.16016	987	754.61564
588	449.55825	688	526.01374	788	602.46922	888	678.92471	988	755.38020
589	450.32281	689	526.77829	789	603.23378	889	679.68926	989	756.14475
590	451.08736	690	527.54285	790	603.99834	890	680.45382	990	756.90931
591	451.85192	691	528.30740	791	604.76289	891	681.21838	991	757.67386
592	452.61647	692	529.07196	792	605.52744	892	681.98293	992	758.43842
593	453.38103	693	529.83651	793	606.29200	893	682.74748	993	759.20297
594	454.14558	694	530.60107	794	607.05656	894	683.51204	994	759.96752
595	454.91014	695	531.36562	795	607.82111	895	684.27660	995	760.73208
596	455.67469	696	532.13018	796	608.58566	896	685.04115	996	761.49664
597	456.43925	697	532.89473	797	609.35022	897	685.80570	997	762.26119
598	457.20380	698	533.65929	798	610.11478	898	686.57026	998	763.02574
599	457.96836	699	534.42384	799	610.87933	899	687.33482	999	763.79030

VOLUME-CAPACITY - CUBIC METERS TO CUBIC YARDS 1 CUBIC METER = 1.307 950 619 CUBIC YARDS

CUBIC METERS	CUBIC YARDS	CUBIC METERS	CUBIC YARDS	CUBIC METERS	CUBIC YARDS	CUBIC METERS	CUBIC YARDS	CUBIC METERS	CUBIC YARDS
0	0.0000	100	130.7951	200	261.5901	300	392.3852	400	523.1802
1	1.3080	101	132.1030	201	262.8981	301	393.6931	401	524.4882
2	2.6159	102	133.4110	202	264.2060	302	395.0011	402	525.7961
3	3.9239	103	134.7189	203	265.5140	303	396.3090	403	527.1041
4	5.2318	104	136.0269	204	266.8219	304	397.6170	404	528.4120
5	6.5398	105	137.3348	205	268.1299	305	398.9249	405	529.7200
6	7.8477	106	138.6428	206	269.4378	306	400.2329	406	531.0279
7	9.1557	107	139.9507	207	270.7458	307	401.5408	407	532.3359
8	10.4636	108	141.2587	208	272.0537	308	402.8488	408	533.6438
9	11.7716	109	142.5666	209	273.3617	309	404.1567	409	534.9518
10	13.0795	110	143.8746	210	274.6696	310	405.4647	410	536.2598
11	14.3875	111	145.1825	211	275.9776	311	406.7726	411	537.5677
12	15.6954	112	146.4905	212	277.2855	312	408.0806	412	538.8756
13	17.0034	113	147.7984	213	278.5935	313	409.3885	413	540.1836
14	18.3113	114	149.1064	214	279.9014	314	410.6965	414	541.4916
15	19.6193	115	150.4143	215	281.2094	315	412.0044	415	542.7995
16	20.9272	116	151.7223	216	282.5173	316	413.3124	416	544.1075
17	22.2352	117	153.0302	217	283.8253	317	414.6203	417	545.4154
18	23.5431	118	154.3382	218	285.1332	318	415.9283	418	546.7234
19	24.8511	119	155.6461	219	286.4412	319	417.2362	419	548.0313
20	26.1590	120	156.9541	220	287.7491	320	418.5442	420	549.3393
21	27.4670	121	158.2620	221	289.0571	321	419.8521	421	550.6472
22	28.7749	122	159.5700	222	290.3650	322	421.1601	422	551.9552
23	30.0829	123	160.8779	223	291.6730	323	422.4680	423	553.2631
24	31.3908	124	162.1859	224	292.9809	324	423.7760	424	554.5711
25	32.6988	125	163.4938	225	294.2889	325	425.0840	425	555.8790
26	34.0067	126	164.8018	226	295.5968	326	426.3919	426	557.1870
27	35.3147	127	166.1097	227	296.9048	327	427.6998	427	558.4949
28	36.6226	128	167.4177	228	298.2127	328	429.0078	428	559.8029
29	37.9306	129	168.7256	229	299.5207	329	430.3158	429	561.1108
30	39.2385	130	170.0336	230	300.8286	330	431.6237	430	562.4188
31	40.5465	131	171.3415	231	302.1366	331	432.9317	431	563.7267
32	41.8544	132	172.6495	232	303.4445	332	434.2396	432	565.0347
33	43.1624	133	173.9574	233	304.7525	333	435.5476	433	566.3426
34	44.4703	134	175.2654	234	306.0604	334	436.8555	434	567.6506
35	45.7783	135	176.5733	235	307.3684	335	438.1635	435	568.9585
36	47.0862	136	177.8813	236	308.6763	336	439.4714	436	570.2665
37	48.3942	137	179.1892	237	309.9843	337	440.7794	437	571.5744
38	49.7021	138	180.4972	238	311.2922	338	442.0873	438	572.8824
39	51.0101	139	181.8051	239	312.6002	339	443.3953	439	574.1903
40	52.3180	140	183.1131	240	313.9081	340	444.7032	440	575.4983
41	53.6260	141	184.4210	241	315.2161	341	446.0112	441	576.8062
42	54.9339	142	185.7290	242	316.5240	342	447.3191	442	578.1142
43	56.2419	143	187.0369	243	317.8320	343	448.6271	443	579.4221
44	57.5498	144	188.3449	244	319.1399	344	449.9350	444	580.7301
45	58.8578	145	189.6528	245	320.4479	345	451.2430	445	582.0380
46	60.1657	146	190.9608	246	321.7558	346	452.5509	446	583.3460
47	61.4737	147	192.2687	247	323.0638	347	453.8589	447	584.6539
48	62.7816	148	193.5767	248	324.3717	348	455.1668	448	585.9619
49	64.0896	149	194.8846	249	325.6797	349	456.4748	449	587.2698

Continued - CUBIC METERS TO CUBIC YARDS 1 CUBIC METER = 1.307 950 619 CUBIC YARDS

CUBIC METERS	CUBIC YARDS	CUBIC METERS	CUBIC YARDS	CUBIC METERS	CUBIC YARDS	CUBIC METERS	CUBIC YARDS	CUBIC METERS	CUBIC YARDS
50	65.3975	150	196.1926	250	326.9877	350	457.7827	450	588.5778
51	66.7055	151	197.5005	251	328.2956	351	459.0907	451	589.8857
52	68.0134	152	198.8085	252	329.6036	352	460.3986	452	591.1937
53	69.3214	153	200.1164	253	330.9115	353	461.7066	453	592.5016
54	70.6293	154	201.4244	254	332.2195	354	463.0145	454	593.8096
55	71.9373	155	202.7323	255	333.5274	355	464.3225	455	595.1175
56	73.2452	156	204.0403	256	334.8354	356	465.6304	456	596.4255
57	74.5532	157	205.3482	257	336.1433	357	466.9384	457	597.7334
58	75.8611	158	206.6562	258	337.4513	358	468.2463	458	599.0414
59	77.1691	159	207.9641	259	338.7592	359	469.5543	459	600.3493
60	78.4770	160	209.2721	260	340.0672	360	470.8622	460	601.6573
61	79.7850	161	210.5800	261	341.3751	361	472.1702	461	602.9652
62	81.0929	162	211.8880	262	342.6831	362	473.4781	462	604.2732
63	82.4009	163	213.1959	263	343.9910	363	474.7861	463	605.5811
64	83.7088	164	214.5039	264	345.2990	364	476.0940	464	606.8891
65	85.0168	165	215.8119	265	346.6069	365	477.4020	465	608.1970
66	86.3247	166	217.1198	266	347.9149	366	478.7099	466	609.5050
67	87.6327	167	218.4278	267	349.2228	367	480.0179	467	610.8129
68	88.9406	168	219.7357	268	350.5308	368	481.3258	468	612.1209
69	90.2486	169	221.0437	269	351.8387	369	482.6338	469	613.4288
70	91.5565	170	222.3516	270	353.1467	370	483.9417	470	614.7368
71	92.8645	171	223.6596	271	354.4546	371	485.2497	471	616.0447
72	94.1724	172	224.9675	272	355.7626	372	486.5576	472	617.3527
73	95.4804	173	226.2755	273	357.0705	373	487.8656	473	618.6606
74	96.7883	174	227.5834	274	358.3785	374	489.1735	474	619.9686
75	98.0963	175	228.8914	275	359.6864	375	490.4815	475	621.2765
76	99.4042	176	230.1993	276	360.9944	376	491.7894	476	622.5845
77	100.7122	177	231.5073	277	362.3023	377	493.0974	477	623.8924
78	102.0201	178	232.8152	278	363.6103	378	494.4053	478	625.2004
79	103.3281	179	234.1232	279	364.9182	379	495.7133	479	626.5083
80	104.6360	180	235.4311	280	366.2262	380	497.0212	480	627.8163
81	105.9440	181	236.7391	281	367.5341	381	498.3292	481	629.1242
82	107.2520	182	238.0470	282	368.8421	382	499.6371	482	630.4322
83	108.5599	183	239.3550	283	370.1500	383	500.9451	483	631.7401
84	109.8679	184	240.6629	284	371.4580	384	502.2530	484	633.0481
85	111.1758	185	241.9709	285	372.7659	385	503.5610	485	634.3560
86	112.4838	186	243.2788	286	374.0739	386	504.8689	486	635.6640
87	113.7917	187	244.5868	287	375.3818	387	506.1769	487	636.9719
88	115.0997	188	245.8947	288	376.6898	388	507.4848	488	638.2799
89	116.4076	189	247.2027	289	377.9977	389	508.7928	489	639.5878
90	117.7156	190	248.5106	290	379.3057	390	510.1007	490	640.8958
91	119.0235	191	249.8186	291	380.6136	391	511.4087	491	642.2038
92	120.3315	192	251.1265	292	381.9216	392	512.7166	492	643.5117
93	121.6394	193	252.4345	293	383.2295	393	514.0246	493	644.8196
94	122.9474	194	253.7424	294	384.5375	394	515.3325	494	646.1276
95	124.2553	195	255.0504	295	385.8454	395	516.6405	495	647.4356
96	125.5633	196	256.3583	296	387.1534	396	517.9484	496	648.7435
97	126.8712	197	257.6663	297	388.4613	397	519.2564	497	650.0515
98	128.1792	198	258.9742	298	389.7693	398	520.5643	498	651.3594
99	129.4871	199	260.2822	299	391.0772	399	521.8723	499	652.6674

Continued - CUBIC METERS TO CUBIC YARDS 1 CUBIC METER = 1.307 950 619 CUBIC YARDS

CUBIC METERS	CUBIC YARDS	CUBIC METERS	CUBIC YARDS	CUBIC METERS	CUBIC YARDS	CUBIC METERS	CUBIC YARDS	CUBIC METERS	CUBIC YARDS
500	653.9753	600	784.7704	700	915.5654	800	1046.3605	900	1177.1555
501	655.2833	601	786.0783	701	916.8734	801	1047.6684	901	1178.4635
502	656.5912	602	787.3863	702	918.1813	802	1048.9764	902	1179.7715
503	657.8992	603	788.6942	703	919.4893	803	1050.2843	903	1181.0794
504	659.2071	604	790.0022	704	920.7972	804	1051.5923	904	1182.3873
505	660.5151	605	791.3101	705	922.1052	805	1052.9002	905	1183.6953
506	661.8230	606	792.6181	706	923.4131	806	1054.2082	906	1185.0033
507	663.1310	607	793.9260	707	924.7211	807	1055.5161	907	1186.3112
508	664.4389	608	795.2340	708	926.0290	808	1056.8241	908	1187.6192
509	665.7469	609	796.5419	709	927.3370	809	1058.1320	909	1188.9271
510	667.0548	610	797.8499	710	928.6449	810	1059.4400	910	1190.2350
511	668.3628	611	799.1578	711	929.9529	811	1060.7479	911	1191.5430
512	669.6707	612	800.4658	712	931.2608	812	1062.0559	912	1192.8510
513	670.9787	613	801.7737	713	932.5688	813	1063.3638	913	1194.1589
514	672.2866	614	803.0817	714	933.8767	814	1064.6718	914	1195.4669
515	673.5946	615	804.3896	715	935.1847	815	1065.9798	915	1196.7748
516	674.9025	616	805.6976	716	936.4926	816	1067.2877	916	1198.0828
517	676.2105	617	807.0055	717	937.8006	817	1068.5956	917	1199.3907
518	677.5184	618	808.3135	718	939.1085	818	1069.9036	918	1200.6987
519	678.8264	619	809.6214	719	940.4165	819	1071.2115	919	1202.0066
520	680.1343	620	810.9294	720	941.7244	820	1072.5195	920	1203.3146
521	681.4423	621	812.2373	721	943.0324	821	1073.8275	921	1204.6225
522	682.7502	622	813.5453	722	944.3403	822	1075.1354	922	1205.9305
523	684.0582	623	814.8532	723	945.6483	823	1076.4433	923	1207.2384
524	685.3661	624	816.1612	724	946.9562	824	1077.7513	924	1208.5464
525	686.6741	625	817.4691	725	948.2642	825	1079.0592	925	1209.8543
526	687.9820	626	818.7771	726	949.5721	826	1080.3672	926	1211.1623
527	689.2900	627	820.0850	727	950.8801	827	1081.6752	927	1212.4702
528	690.5979	628	821.3930	728	952.1880	828	1082.9831	928	1213.7782
529	691.9059	629	822.7009	729	953.4960	829	1084.2910	929	1215.0861
530	693.2138	630	824.0089	730	954.8039	830	1085.5990	930	1216.3941
531	694.5218	631	825.3168	731	956.1119	831	1086.9070	931	1217.7020
532	695.8297	632	826.6248	732	957.4198	832	1088.2149	932	1219.0100
533	697.1377	633	827.9327	733	958.7278	833	1089.5229	933	1220.3179
534	698.4456	634	829.2407	734	960.0358	834	1090.8308	934	1221.6259
535	699.7536	635	830.5486	735	961.3437	835	1092.1388	935	1222.9338
536	701.0615	636	831.8566	736	962.6516	836	1093.4467	936	1224.2418
537	702.3695	637	833.1645	737	963.9596	837	1094.7547	937	1225.5497
538	703.6774	638	834.4725	738	965.2675	838	1096.0626	938	1226.8577
539	704.9854	639	835.7804	739	966.5755	839	1097.3706	939	1228.1656
540	706.2933	640	837.0884	740	967.8835	840	1098.6785	940	1229.4736
541	707.6013	641	838.3963	741	969.1914	841	1099.9865	941	1230.7815
542	708.9092	642	839.7043	742	970.4994	842	1101.2944	942	1232.0895
543	710.2172	643	841.0122	743	971.8073	843	1102.6024	943	1233.3974
544	711.5251	644	842.3202	744	973.1153	844	1103.9103	944	1234.7054
545	712.8331	645	843.6281	745	974.4232	845	1105.2183	945	1236.0133
546	714.1410	646	844.9361	746	975.7312	846	1106.5262	946	1237.3213
547	715.4490	647	846.2440	747	977.0391	847	1107.8342	947	1238.6292
548	716.7569	648	847.5520	748	978.3471	848	1109.1421	948	1239.9372
549	718.0649	649	848.8599	749	979.6550	849	1110.4501	949	1241.2451

Continued - CUBIC METERS TO CUBIC YARDS 1 CUBIC METER = 1.307 950 619 CUBIC YARDS

CUBIC METERS	CUBIC YARDS	CUBIC METERS	CUBIC YARDS	CUBIC METERS	CUBIC YARDS	CUBIC METERS	CUBIC YARDS	CUBIC METERS	CUBIC YARDS
550	719.3728	650	850.1679	750	980.9630	850	1111.7580	950	1242.5531
551	720.6808	651	851.4758	751	982.2709	851	1113.0660	951	1243.8610
552	721.9887	652	852.7838	752	983.5789	852	1114.3739	952	1245.1690
553	723.2967	653	854.0918	753	984.8868	853	1115.6819	953	1246.4769
554	724.6046	654	855.3997	754	986.1948	854	1116.9898	954	1247.7849
555	725.9126	655	856.7076	755	987.5027	855	1118.2978	955	1249.0928
556	727.2205	656	858.0156	756	988.8107	856	1119.6057	956	1250.4008
557	728.5285	657	859.3235	757	990.1186	857	1120.9137	957	1251.7087
558	729.8364	658	860.6315	758	991.4266	858	1122.2216	958	1253.0167
559	731.1444	659	861.9395	759	992.7345	859	1123.5296	959	1254.3246
560	732.4523	660	863.2474	760	994.0425	860	1124.8375	960	1255.6326
561	733.7603	661	864.5554	761	995.3504	861	1126.1455	961	1256.9405
562	735.0682	662	865.8633	762	996.6584	862	1127.4534	962	1258.2485
563	736.3762	663	867.1713	763	997.9663	863	1128.7614	963	1259.5564
564	737.6841	664	868.4792	764	999.2743	864	1130.0693	964	1260.8644
565	738.9921	665	869.7872	765	1000.5822	865	1131.3773	965	1262.1723
566	740.3000	666	871.0951	766	1001.8902	866	1132.6852	966	1263.4803
567	741.6080	667	872.4031	767	1003.1981	867	1133.9932	967	1264.7882
568	742.9159	668	873.7110	768	1004.5061	868	1135.3011	968	1266.0962
569	744.2239	669	875.0190	769	1005.8140	869	1136.6091	969	1267.4041
570	745.5318	670	876.3269	770	1007.1220	870	1137.9170	970	1268.7121
571	746.8398	671	877.6349	771	1008.4299	871	1139.2250	971	1270.0200
572	748.1478	672	878.9428	772	1009.7379	872	1140.5329	972	1271.3280
573	749.4557	673	880.2508	773	1011.0458	873	1141.8409	973	1272.6359
574	750.7636	674	881.5587	774	1012.3538	874	1143.1488	974	1273.9439
575	752.0716	675	882.8667	775	1013.6617	875	1144.4568	975	1275.2518
576	753.3796	676	884.1746	776	1014.9697	876	1145.7647	976	1276.5598
577	754.6875	677	885.4826	777	1016.2776	877	1147.0727	977	1277.8677
578	755.9955	678	886.7905	778	1017.5856	878	1148.3806	978	1279.1757
579	757.3034	679	888.0985	779	1018.8935	879	1149.6886	979	1280.4836
580	758.6114	680	889.4064	780	1020.2015	880	1150.9965	980	1281.7916
581	759.9193	681	890.7144	781	1021.5094	881	1152.3045	981	1283.0995
582	761.2273	682	892.0223	782	1022.8174	882	1153.6124	982	1284.4075
583	762.5352	683	893.3303	783	1024.1253	883	1154.9204	983	1285.7155
584	763.8432	684	894.6382	784	1025.4333	884	1156.2283	984	1287.0234
585	765.1511	685	895.9462	785	1026.7412	885	1157.5363	985	1288.3313
586	766.4591	686	897.2541	786	1028.0492	886	1158.8442	986	1289.6393
587	767.7670	687	898.5621	787	1029.3571	887	1160.1522	987	1290.9473
588	769.0750	688	899.8700	788	1030.6651	888	1161.4601	988	1292.2552
589	770.3829	689	901.1780	789	1031.9730	889	1162.7681	989	1293.5632
590	771.6909	690	902.4859	790	1033.2810	890	1164.0760	990	1294.8711
591	772.9988	691	903.7939	791	1034.5889	891	1165.3840	991	1296.1790
592	774.3068	692	905.1018	792	1035.8969	892	1166.6919	992	1297.4870
593	775.6147	693	906.4098	793	1037.2048	893	1167.9999	993	1298.7950
594	776.9227	694	907.7177	794	1038.5128	894	1169.3078	994	1300.1029
595	778.2306	695	909.0257	795	1039.8207	895	1170.6158	995	1301.4109
596	779.5386	696	910.3336	796	1041.1287	896	1171.9238	996	1302.7188
597	780.8465	697	911.6416	797	1042.4366	897	1173.2317	997	1304.0268
598	782.1545	698	912.9495	798	1043.7446	898	1174.5396	998	1305.3347
599	783.4624	699	914.2575	799	1045.0525	899	1175.8476	999	1306.6427

VOLUME-CAPACITY - LIQUID QUARTS TO LITERS 1 LIQUID QUART = 0.946 352 946 LITER

LIQUID QUARTS	LITERS	LIQUID QUARTS	LITERS	LIQUID QUARTS	LITERS	LIQUID QUARTS	LITERS	LIQUID QUARTS	LITERS
0	0.00000	100	94.63529	200	189.27059	300	283.90588	400	378.54118
1	0.94635	101	95.58165	201	190.21694	301	284.85223	401	379.48753
2	1.89271	102	96.52800	202	191.16329	302	285.79859	402	380.43388
3	2.83906	103	97.47435	203	192.10965	303	286.74494	403	381.38023
4	3.78541	104	98.42071	204	193.05600	304	287.69129	404	382.32659
5	4.73176	105	99.36706	205	194.00235	305	288.63765	405	383.27294
6	5.67812	106	100.31341	206	194.94871	306	289.58400	406	384.21929
7	6.62447	107	101.25976	207	195.89506	307	290.53035	407	385.16565
8	7.57082	108	102.20612	208	196.84141	308	291.47670	408	386.11200
9	8.51718	109	103.15247	209	197.78776	309	292.42306	409	387.05835
10	9.46353	110	104.09882	210	198.73412	310	293.36941	410	388.00470
11	10.40988	111	105.04518	211	199.68047	311	294.31577	411	388.95106
12	11.35624	112	105.99153	212	200.62682	312	295.26212	412	389.89741
13	12.30259	113	106.93788	213	201.57318	313	296.20847	413	390.84377
14	13.24894	114	107.88424	214	202.51953	314	297.15482	414	391.79012
15	14.19529	115	108.83059	215	203.46588	315	298.10118	415	392.73647
16	15.14165	116	109.77694	216	204.41224	316	299.04753	416	393.68282
17	16.08800	117	110.72329	217	205.35859	317	299.99388	417	394.62918
18	17.03435	118	111.66965	218	206.30494	318	300.94024	418	395.57553
19	17.98071	119	112.61600	219	207.25129	319	301.88659	419	396.52188
20	18.92706	120	113.56235	220	208.19765	320	302.83294	420	397.46824
21	19.87341	121	114.50871	221	209.14400	321	303.77929	421	398.41459
22	20.81976	122	115.45506	222	210.09035	322	304.72565	422	399.36094
23	21.76612	123	116.40141	223	211.03671	323	305.67200	423	400.30729
24	22.71247	124	117.34776	224	211.98306	324	306.61835	424	401.25365
25	23.65882	125	118.29412	225	212.92941	325	307.56470	425	402.20000
26	24.60518	126	119.24047	226	213.87576	326	308.51106	426	403.14635
27	25.55153	127	120.18682	227	214.82212	327	309.45741	427	404.09270
28	26.49788	128	121.13318	228	215.76847	328	310.40376	428	405.03906
29	27.44424	129	122.07953	229	216.71482	329	311.35012	429	405.98541
30	28.39059	130	123.02588	230	217.66118	330	312.29647	430	406.93176
31	29.33694	131	123.97223	231	218.60753	331	313.24282	431	407.87812
32	30.28329	132	124.91859	232	219.55388	332	314.18917	432	408.82447
33	31.22965	133	125.86494	233	220.50023	333	315.13553	433	409.77082
34	32.17600	134	126.81129	234	221.44659	334	316.08188	434	410.71717
35	33.12235	135	127.75765	235	222.39294	335	317.02823	435	411.66353
36	34.06871	136	128.70400	236	223.33929	336	317.97459	436	412.60988
37	35.01506	137	129.65035	237	224.28565	337	318.92094	437	413.55624
38	35.96141	138	130.59670	238	225.23200	338	319.86729	438	414.50259
39	36.90776	139	131.54306	239	226.17835	339	320.81364	439	415.44894
40	37.85412	140	132.48941	240	227.12471	340	321.76000	440	416.39529
41	38.80047	141	133.43576	241	228.07106	341	322.70635	441	417.34165
42	39.74682	142	134.38212	242	229.01741	342	323.65271	442	418.28800
43	40.69318	143	135.32847	243	229.96376	343	324.59906	443	419.23435
44	41.63953	144	136.27482	244	230.91012	344	325.54541	444	420.18071
45	42.58588	145	137.22118	245	231.85647	345	326.49176	445	421.12706
46	43.53224	146	138.16753	246	232.80282	346	327.43812	446	422.07341
47	44.47859	147	139.11388	247	233.74918	347	328.38447	447	423.01976
48	45.42494	148	140.06023	248	234.69553	348	329.33082	448	423.96612
49	46.37129	149	141.00659	249	235.64188	349	330.27718	449	424.91247

Continued - LIQUID QUARTS TO LITERS 1 LIQUID QUART = 0.946 352 946 LITER

LIQUID QUARTS	LITERS	LIQUID QUARTS	LITERS	LIQUID QUARTS	LITERS	LIQUID QUARTS	LITERS	LIQUID QUARTS	LITERS
50	47.31765	150	141.95294	250	236.58824	350	331.22353	450	425.85882
51	48.26400	151	142.89929	251	237.53459	351	332.16988	451	426.80518
52	49.21035	152	143.84565	252	238.48094	352	333.11623	452	427.75153
53	50.15671	153	144.79200	253	239.42729	353	334.06259	453	428.69788
54	51.10306	154	145.73835	254	240.37365	354	335.00894	454	429.64423
55	52.04941	155	146.68471	255	241.32000	355	335.95529	455	430.59059
56	52.99576	156	147.63106	256	242.26635	356	336.90165	456	431.53694
57	53.94212	157	148.57741	257	243.21271	357	337.84800	457	432.48329
58	54.88847	158	149.52376	258	244.15906	358	338.79435	458	433.42965
59	55.83482	159	150.47012	259	245.10541	359	339.74070	459	434.37600
60	56.78118	160	151.41647	260	246.05176	360	340.68706	460	435.32235
61	57.72753	161	152.36282	261	246.99812	361	341.63341	461	436.26870
62	58.67388	162	153.30918	262	247.94447	362	342.57977	462	437.21506
63	59.62024	163	154.25553	263	248.89082	363	343.52612	463	438.16141
64	60.56659	164	155.20188	264	249.83718	364	344.47247	464	439.10777
65	61.51294	165	156.14824	265	250.78353	365	345.41882	465	440.05412
66	62.45929	166	157.09459	266	251.72988	366	346.36518	466	441.00047
67	63.40565	167	158.04094	267	252.67624	367	347.31153	467	441.94682
68	64.35200	168	158.98729	268	253.62259	368	348.25788	468	442.89318
69	65.29835	169	159.93365	269	254.56894	369	349.20424	469	443.83953
70	66.24471	170	160.88000	270	255.51529	370	350.15059	470	444.78588
71	67.19106	171	161.82635	271	256.46165	371	351.09694	471	445.73223
72	68.13741	172	162.77271	272	257.40800	372	352.04329	472	446.67859
73	69.08376	173	163.71906	273	258.35435	373	352.98965	473	447.62494
74	70.03012	174	164.66541	274	259.30070	374	353.93600	474	448.57129
75	70.97647	175	165.61176	275	260.24706	375	354.88235	475	449.51765
76	71.92282	176	166.55812	276	261.19341	376	355.82870	476	450.46400
77	72.86918	177	167.50447	277	262.13976	377	356.77506	477	451.41035
78	73.81553	178	168.45082	278	263.08612	378	357.72141	478	452.35670
79	74.76188	179	169.39718	279	264.03247	379	358.66776	479	453.30306
80	75.70823	180	170.34353	280	264.97882	380	359.61412	480	454.24941
81	76.65459	181	171.28988	281	265.92517	381	360.56047	481	455.19576
82	77.60094	182	172.23623	282	266.87153	382	361.50682	482	456.14212
83	78.54729	183	173.18259	283	267.81788	383	362.45317	483	457.08847
84	79.49365	184	174.12894	284	268.76423	384	363.39953	484	458.03482
85	80.44000	185	175.07529	285	269.71059	385	364.34588	485	458.98117
86	81.38635	186	176.02165	286	270.65694	386	365.29223	486	459.92753
87	82.33271	187	176.96800	287	271.60329	387	366.23859	487	460.87388
88	83.27906	188	177.91435	288	272.54964	388	367.18494	488	461.82024
89	84.22541	189	178.86071	289	273.49600	389	368.13129	489	462.76659
90	85.17176	190	179.80706	290	274.44235	390	369.07764	490	463.71294
91	86.11812	191	180.75341	291	275.38871	391	370.02400	491	464.65929
92	87.06447	192	181.69976	292	276.33506	392	370.97035	492	465.60565
93	88.01082	193	182.64612	293	277.28141	393	371.91671	493	466.55200
94	88.95718	194	183.59247	294	278.22776	394	372.86306	494	467.49835
95	89.90353	195	184.53882	295	279.17412	395	373.80941	495	468.44471
96	90.84988	196	185.48518	296	280.12047	396	374.75576	496	469.39106
97	91.79624	197	186.43153	297	281.06682	397	375.70212	497	470.33741
98	92.74259	198	187.37788	298	282.01318	398	376.64847	498	471.28376
99	93.68894	199	188.32423	299	282.95953	399	377.59482	499	472.23012

Continued - LIQUID QUARTS TO LITERS 1 LIQUID QUART = 0.946 352 946 LITER

LIQUID QUARTS	LITERS	LIQUID QUARTS	LITERS	LIQUID QUARTS	LITERS	LIQUID QUARTS	LITERS	LIQUID QUARTS	LITERS
500	473.17647	600	567.81176	700	662.44706	800	757.08235	900	851.71764
501	474.12282	601	568.75812	701	663.39341	801	758.02870	901	852.66400
502	475.06918	602	569.70447	702	664.33976	802	758.97506	902	853.61035
503	476.01553	603	570.65082	703	665.28612	803	759.92141	903	854.55670
504	476.96188	604	571.59718	704	666.23247	804	760.86776	904	855.50306
505	477.90823	605	572.54353	705	667.17882	805	761.81412	905	856.44941
506	478.85459	606	573.48988	706	668.12518	806	762.76047	906	857.39576
507	479.80094	607	574.43623	707	669.07153	807	763.70683	907	858.34212
508	480.74730	608	575.38258	708	670.01788	808	764.65318	908	859.28847
509	481.69365	609	576.32894	709	670.96423	809	765.59953	909	860.23483
510	482.64000	610	577.27529	710	671.91058	810	766.54588	910	861.18118
511	483.58635	611	578.22164	711	672.85694	811	767.49223	911	862.12753
512	484.53271	612	579.16800	712	673.80329	812	768.43858	912	863.07388
513	485.47906	613	580.11435	713	674.74964	813	769.38494	913	864.02023
514	486.42541	614	581.06071	714	675.69600	814	770.33129	914	864.96658
515	487.37177	615	582.00706	715	676.64235	815	771.27765	915	865.91294
516	488.31812	616	582.95341	716	677.58871	816	772.22400	916	866.85929
517	489.26447	617	583.89977	717	678.53506	817	773.17035	917	867.80565
518	490.21082	618	584.84612	718	679.48141	818	774.11671	918	868.75200
519	491.15718	619	585.79247	719	680.42776	819	775.06306	919	869.69835
520	492.10353	620	586.73882	720	681.37411	820	776.00941	920	870.64471
521	493.04988	621	587.68517	721	682.32047	821	776.95576	921	871.59106
522	493.99623	622	588.63153	722	683.26682	822	777.90211	922	872.53741
523	494.94259	623	589.57788	723	684.21317	823	778.84847	923	873.48376
524	495.88894	624	590.52423	724	685.15953	824	779.79482	924	874.43011
525	496.83529	625	591.47059	725	686.10588	825	780.74117	925	875.37647
526	497.78165	626	592.41694	726	687.05223	826	781.68753	926	876.32282
527	498.72800	627	593.36329	727	687.99859	827	782.63388	927	877.26917
528	499.67435	628	594.30965	728	688.94494	828	783.58023	928	878.21553
529	500.62070	629	595.25600	729	689.89129	829	784.52659	929	879.16188
530	501.56706	630	596.20235	730	690.83765	830	785.47294	930	880.10823
531	502.51341	631	597.14870	731	691.78400	831	786.41930	931	881.05459
532	503.45976	632	598.09505	732	692.73035	832	787.36565	932	882.00094
533	504.40612	633	599.04141	733	693.67670	833	788.31200	933	882.94730
534	505.35247	634	599.98776	734	694.62305	834	789.25835	934	883.89365
535	506.29882	635	600.93411	735	695.56941	835	790.20470	935	884.84000
536	507.24517	636	601.88047	736	696.51576	836	791.15105	936	885.78635
537	508.19153	637	602.82682	737	697.46211	837	792.09741	937	886.73270
538	509.13788	638	603.77318	738	698.40847	838	793.04376	938	887.67905
539	510.08424	639	604.71953	739	699.35482	839	793.99011	939	888.62541
540	511.03059	640	605.66588	740	700.30118	840	794.93647	940	889.57176
541	511.97694	641	606.61224	741	701.24753	841	795.88282	941	890.51812
542	512.92329	642	607.55859	742	702.19388	842	796.82918	942	891.46447
543	513.86964	643	608.50494	743	703.14024	843	797.77553	943	892.41082
544	514.81599	644	609.45129	744	704.08659	844	798.72188	944	893.35718
545	515.76235	645	610.39764	745	705.03294	845	799.66824	945	894.30353
546	516.70870	646	611.34399	746	705.97929	846	800.61459	946	895.24988
547	517.65506	647	612.29035	747	706.92564	847	801.56094	947	896.19624
548	518.60141	648	613.23670	748	707.87200	848	802.50729	948	897.14259
549	519.54776	649	614.18306	749	708.81835	849	803.45364	949	898.08894

Continued - LIQUID QUARTS TO LITERS 1 LIQUID QUART = 0.946 352 946 LITER

LIQUID QUARTS	LITERS	LIQUID QUARTS	LITERS	LIQUID QUARTS	LITERS	LIQUID QUARTS	LITERS	LIQUID QUARTS	LITERS
550	520.49412	650	615.12941	750	709.76470	850	804.40000	950	899.03529
551	521.44047	651	616.07576	751	710.71106	851	805.34635	951	899.98164
552	522.38682	652	617.02212	752	711.65741	852	806.29270	952	900.92800
553	523.33318	653	617.96847	753	712.60376	853	807.23906	953	901.87435
554	524.27953	654	618.91482	754	713.55012	854	808.18541	954	902.82070
555	525.22588	655	619.86118	755	714.49647	855	809.13176	955	903.76706
556	526.17223	656	620.80753	756	715.44283	856	810.07812	956	904.71341
557	527.11858	657	621.75388	757	716.38918	857	811.02447	957	905.65977
558	528.06494	658	622.70023	758	717.33553	858	811.97083	958	906.60612
559	529.01129	659	623.64658	759	718.28188	859	812.91718	959	907.55247
560	529.95764	660	624.59294	760	719.22823	860	813.86353	960	908.49883
561	530.90400	661	625.53929	761	720.17458	861	814.80988	961	909.44518
562	531.85035	662	626.48564	762	721.12094	862	815.75623	962	910.39153
563	532.79671	663	627.43200	763	722.06729	863	816.70258	963	911.33788
564	533.74306	664	628.37835	764	723.01365	864	817.64894	964	912.28423
565	534.68941	665	629.32471	765	723.96000	865	818.59529	965	913.23058
566	535.63577	666	630.27106	766	724.90635	866	819.54165	966	914.17694
567	536.58212	667	631.21741	767	725.85271	867	820.48800	967	915.12329
568	537.52847	668	632.16376	768	726.79906	868	821.43435	968	916.06965
569	538.47482	669	633.11012	769	727.74541	869	822.38071	969	917.01600
570	539.42117	670	634.05647	770	728.69176	870	823.32706	970	917.96235
571	540.36753	671	635.00282	771	729.63811	871	824.27341	971	918.90871
572	541.31388	672	635.94917	772	730.58447	872	825.21976	972	919.85506
573	542.26023	673	636.89553	773	731.53082	873	826.16611	973	920.80141
574	543.20659	674	637.84188	774	732.47717	874	827.11247	974	921.74776
575	544.15294	675	638.78823	775	733.42353	875	828.05882	975	922.69411
576	545.09929	676	639.73459	776	734.36988	876	829.00517	976	923.64047
577	546.04565	677	640.68094	777	735.31623	877	829.95153	977	924.58682
578	546.99200	678	641.62729	778	736.26259	878	830.89788	978	925.53317
579	547.93835	679	642.57365	779	737.20894	879	831.84423	979	926.47953
580	548.88470	680	643.52000	780	738.15529	880	832.79059	980	927.42588
581	549.83105	681	644.46635	781	739.10165	881	833.73694	981	928.37223
582	550.77741	682	645.41270	782	740.04800	882	834.68330	982	929.31859
583	551.72376	683	646.35905	783	740.99435	883	835.62965	983	930.26494
584	552.67011	684	647.30541	784	741.94070	884	836.57600	984	931.21130
585	553.61647	685	648.25176	785	742.88705	885	837.52235	985	932.15765
586	554.56282	686	649.19811	786	743.83341	886	838.46870	986	933.10400
587	555.50917	687	650.14447	787	744.77976	887	839.41505	987	934.05035
588	556.45553	688	651.09082	788	745.72611	888	840.36141	988	934.99670
589	557.40188	689	652.03718	789	746.67247	889	841.30776	989	935.94305
590	558.34824	690	652.98353	790	747.61882	890	842.25412	990	936.88941
591	559.29459	691	653.92988	791	748.56518	891	843.20047	991	937.83576
592	560.24094	692	654.87624	792	749.51153	892	844.14682	992	938.78212
593	561.18729	693	655.82259	793	750.45788	893	845.09318	993	939.72847
594	562.13364	694	656.76894	794	751.40424	894	846.03953	994	940.67482
595	563.07999	695	657.71529	795	752.35059	895	846.98588	995	941.62118
596	564.02635	696	658.66164	796	753.29694	896	847.93224	996	942.56753
597	564.97270	697	659.60800	797	754.24329	897	848.87859	997	943.51388
598	565.91906	698	660.55435	798	755.18964	898	849.82494	998	944.46024
599	566.86541	699	661.50070	799	756.13600	899	850.77129	999	945.40659

260-260 O - 67 - 14

VOLUME-CAPACITY - LITERS TO LIQUID QUARTS 1 LITER = 1.056 6882 LIQUID QUARTS

LITERS	LIQUID QUARTS	LITERS	LIQUID QUARTS	LITERS	LIQUID QUARTS	LITERS	LIQUID QUARTS	LITERS	LIQUID QUARTS
0	0.0000	100	105.6688	200	211.3376	300	317.0065	400	422.6753
1	1.0567	101	106.7255	201	212.3943	301	318.0631	401	423.7320
2	2.1134	102	107.7822	202	213.4510	302	319.1198	402	424.7887
3	3.1701	103	108.8389	203	214.5077	303	320.1765	403	425.8453
4	4.2268	104	109.8956	204	215.5644	304	321.2332	404	426.9020
5	5.2834	105	110.9523	205	216.6211	305	322.2899	405	427.9587
6	6.3401	106	112.0089	206	217.6778	306	323.3466	406	429.0154
7	7.3968	107	113.0656	207	218.7345	307	324.4033	407	430.0721
8	8.4535	108	114.1223	208	219.7911	308	325.4600	408	431.1288
9	9.5102	109	115.1790	209	220.8478	309	326.5167	409	432.1855
10	10.5669	110	116.2357	210	221.9045	310	327.5733	410	433.2422
11	11.6236	111	117.2924	211	222.9612	311	328.6300	411	434.2989
12	12.6803	112	118.3491	212	224.0179	312	329.6867	412	435.3555
13	13.7369	113	119.4058	213	225.0746	313	330.7434	413	436.4122
14	14.7936	114	120.4625	214	226.1313	314	331.8001	414	437.4689
15	15.8503	115	121.5191	215	227.1880	315	332.8568	415	438.5256
16	16.9070	116	122.5758	216	228.2447	316	333.9135	416	439.5823
17	17.9637	117	123.6325	217	229.3013	317	334.9702	417	440.6390
18	19.0204	118	124.6892	218	230.3580	318	336.0268	418	441.6957
19	20.0771	119	125.7459	219	231.4147	319	337.0835	419	442.7524
20	21.1338	120	126.8026	220	232.4714	320	338.1402	420	443.8090
21	22.1905	121	127.8593	221	233.5281	321	339.1969	421	444.8657
22	23.2471	122	128.9160	222	234.5848	322	340.2536	422	445.9224
23	24.3038	123	129.9726	223	235.6415	323	341.3103	423	446.9791
24	25.3605	124	131.0293	224	236.6982	324	342.3670	424	448.0358
25	26.4172	125	132.0860	225	237.7548	325	343.4237	425	449.0925
26	27.4739	126	133.1427	226	238.8115	326	344.4804	426	450.1492
27	28.5306	127	134.1994	227	239.8682	327	345.5370	427	451.2059
28	29.5873	128	135.2561	228	240.9249	328	346.5937	428	452.2626
29	30.6440	129	136.3128	229	241.9816	329	347.6504	429	453.3192
30	31.7006	130	137.3695	230	243.0383	330	348.7071	430	454.3759
31	32.7573	131	138.4262	231	244.0950	331	349.7638	431	455.4326
32	33.8140	132	139.4828	232	245.1517	332	350.8205	432	456.4893
33	34.8707	133	140.5395	233	246.2084	333	351.8772	433	457.5460
34	35.9274	134	141.5962	234	247.2650	334	352.9339	434	458.6027
35	36.9841	135	142.6529	235	248.3217	335	353.9905	435	459.6594
36	38.0408	136	143.7096	236	249.3784	336	355.0472	436	460.7161
37	39.0975	137	144.7663	237	250.4351	337	356.1039	437	461.7727
38	40.1542	138	145.8230	238	251.4918	338	357.1606	438	462.8294
39	41.2108	139	146.8797	239	252.5485	339	358.2173	439	463.8861
40	42.2675	140	147.9363	240	253.6052	340	359.2740	440	464.9428
41	43.3242	141	148.9930	241	254.6619	341	360.3307	441	465.9995
42	44.3809	142	150.0497	242	255.7185	342	361.3874	442	467.0562
43	45.4376	143	151.1064	243	256.7752	343	362.4441	443	468.1129
44	46.4943	144	152.1631	244	257.8319	344	363.5007	444	469.1696
45	47.5510	145	153.2198	245	258.8886	345	364.5574	445	470.2262
46	48.6077	146	154.2765	246	259.9453	346	365.6141	446	471.2829
47	49.6643	147	155.3332	247	261.0020	347	366.6708	447	472.3396
48	50.7210	148	156.3899	248	262.0587	348	367.7275	448	473.3963
49	51.7777	149	157.4465	249	263.1154	349	368.7842	449	474.4530

Continued - LITERS TO LIQUID QUARTS 1 LITER = 1.056 6882 LIQUID QUARTS

LITERS	LIQUID QUARTS	LITERS	LIQUID QUARTS	LITERS	LIQUID QUARTS	LITERS	LIQUID QUARTS	LITERS	LIQUID QUARTS
50	52.8344	150	158.5032	250	264.1721	350	369.8409	450	475.5097
51	53.8911	151	159.5599	251	265.2287	351	370.8976	451	476.5664
52	54.9478	152	160.6166	252	266.2854	352	371.9542	452	477.6231
53	56.0045	153	161.6733	253	267.3421	353	373.0109	453	478.6798
54	57.0612	154	162.7300	254	268.3988	354	374.0676	454	479.7364
55	58.1179	155	163.7867	255	269.4555	355	375.1243	455	480.7931
56	59.1745	156	164.8434	256	270.5122	356	376.1810	456	481.8498
57	60.2312	157	165.9000	257	271.5689	357	377.2377	457	482.9065
58	61.2879	158	166.9567	258	272.6256	358	378.2944	458	483.9632
59	62.3446	159	168.0134	259	273.6822	359	379.3511	459	485.0199
60	63.4013	160	169.0701	260	274.7389	360	380.4078	460	486.0766
61	64.4580	161	170.1268	261	275.7956	361	381.4644	461	487.1333
62	65.5147	162	171.1835	262	276.8523	362	382.5211	462	488.1899
63	66.5714	163	172.2402	263	277.9090	363	383.5778	463	489.2466
64	67.6280	164	173.2969	264	278.9657	364	384.6345	464	490.3033
65	68.6847	165	174.3536	265	280.0224	365	385.6912	465	491.3600
66	69.7414	166	175.4102	266	281.0791	366	386.7479	466	492.4167
67	70.7981	167	176.4669	267	282.1357	367	387.8046	467	493.4734
68	71.8548	168	177.5236	268	283.1924	368	388.8613	468	494.5301
69	72.9115	169	178.5803	269	284.2491	369	389.9179	469	495.5868
70	73.9682	170	179.6370	270	285.3058	370	390.9746	470	496.6435
71	75.0249	171	180.6937	271	286.3625	371	392.0313	471	497.7001
72	76.0816	172	181.7504	272	287.4192	372	393.0880	472	498.7568
73	77.1382	173	182.8071	273	288.4759	373	394.1447	473	499.8135
74	78.1949	174	183.8637	274	289.5326	374	395.2014	474	500.8702
75	79.2516	175	184.9204	275	290.5893	375	396.2581	475	501.9269
76	80.3083	176	185.9771	276	291.6459	376	397.3148	476	502.9836
77	81.3650	177	187.0338	277	292.7026	377	398.3715	477	504.0403
78	82.4217	178	188.0905	278	293.7593	378	399.4281	478	505.0970
79	83.4784	179	189.1472	279	294.8160	379	400.4848	479	506.1536
80	84.5351	180	190.2039	280	295.8727	380	401.5415	480	507.2103
81	85.5917	181	191.2606	281	296.9294	381	402.5982	481	508.2670
82	86.6484	182	192.3173	282	297.9861	382	403.6549	482	509.3237
83	87.7051	183	193.3739	283	299.0428	383	404.7116	483	510.3804
84	88.7618	184	194.4306	284	300.0994	384	405.7683	484	511.4371
85	89.8185	185	195.4873	285	301.1561	385	406.8250	485	512.4938
86	90.8752	186	196.5440	286	302.2128	386	407.8816	486	513.5505
87	91.9319	187	197.6007	287	303.2695	387	408.9383	487	514.6072
88	92.9886	188	198.6574	288	304.3262	388	409.9950	488	515.6638
89	94.0452	189	199.7141	289	305.3829	389	411.0517	489	516.7205
90	95.1019	190	200.7708	290	306.4396	390	412.1084	490	517.7772
91	96.1586	191	201.8274	291	307.4963	391	413.1651	491	518.8339
92	97.2153	192	202.8841	292	308.5530	392	414.2218	492	519.8906
93	98.2720	193	203.9408	293	309.6096	393	415.2785	493	520.9473
94	99.3287	194	204.9975	294	310.6663	394	416.3352	494	522.0040
95	100.3854	195	206.0542	295	311.7230	395	417.3918	495	523.0607
96	101.4421	196	207.1109	296	312.7797	396	418.4485	496	524.1173
97	102.4988	197	208.1676	297	313.8364	397	419.5052	497	525.1740
98	103.5554	198	209.2243	298	314.8931	398	420.5619	498	526.2307
99	104.6121	199	210.2810	299	315.9498	399	421.6186	499	527.2874

Continued - LITERS TO LIQUID QUARTS 1 LITER = 1.056 6882 LIQUID QUARTS

LITERS	LIQUID QUARTS	LITERS	LIQUID QUARTS	LITERS	LIQUID QUARTS	LITERS	LIQUID QUARTS	LITERS	LIQUID QUARTS
500	528.3441	600	634.0129	700	739.6817	800	845.3506	900	951.0194
501	529.4008	601	635.0696	701	740.7384	801	846.4072	901	952.0761
502	530.4575	602	636.1263	702	741.7951	802	847.4639	902	953.1328
503	531.5142	603	637.1830	703	742.8518	803	848.5206	903	954.1894
504	532.5709	604	638.2397	704	743.9085	804	849.5773	904	955.2461
505	533.6275	605	639.2964	705	744.9652	805	850.6340	905	956.3028
506	534.6842	606	640.3531	706	746.0219	806	851.6907	906	957.3595
507	535.7409	607	641.4097	707	747.0786	807	852.7474	907	958.4162
508	536.7976	608	642.4664	708	748.1352	808	853.8041	908	959.4729
509	537.8543	609	643.5231	709	749.1919	809	854.8608	909	960.5296
510	538.9110	610	644.5798	710	750.2486	810	855.9174	910	961.5863
511	539.9677	611	645.6365	711	751.3053	811	856.9741	911	962.6430
512	541.0244	612	646.6932	712	752.3620	812	858.0308	912	963.6996
513	542.0810	613	647.7499	713	753.4187	813	859.0875	913	964.7563
514	543.1377	614	648.8066	714	754.4754	814	860.1442	914	965.8130
515	544.1944	615	649.8632	715	755.5321	815	861.2009	915	966.8697
516	545.2511	616	650.9199	716	756.5888	816	862.2576	916	967.9264
517	546.3078	617	651.9766	717	757.6454	817	863.3143	917	968.9831
518	547.3645	618	653.0333	718	758.7021	818	864.3709	918	970.0398
519	548.4212	619	654.0900	719	759.7588	819	865.4276	919	971.0965
520	549.4779	620	655.1467	720	760.8155	820	866.4843	920	972.1531
521	550.5346	621	656.2034	721	761.8722	821	867.5410	921	973.2098
522	551.5912	622	657.2601	722	762.9289	822	868.5977	922	974.2665
523	552.6479	623	658.3167	723	763.9856	823	869.6544	923	975.3232
524	553.7046	624	659.3734	724	765.0423	824	870.7111	924	976.3799
525	554.7613	625	660.4301	725	766.0989	825	871.7678	925	977.4366
526	555.8180	626	661.4868	726	767.1556	826	872.8245	926	978.4933
527	556.8747	627	662.5435	727	768.2123	827	873.8811	927	979.5500
528	557.9314	628	663.6002	728	769.2690	828	874.9378	928	980.6067
529	558.9881	629	664.6569	729	770.3257	829	875.9945	929	981.6633
530	560.0447	630	665.7136	730	771.3824	830	877.0512	930	982.7200
531	561.1014	631	666.7703	731	772.4391	831	878.1079	931	983.7767
532	562.1581	632	667.8269	732	773.4958	832	879.1646	932	984.8334
533	563.2148	633	668.8836	733	774.5525	833	880.2213	933	985.8901
534	564.2715	634	669.9403	734	775.6091	834	881.2780	934	986.9468
535	565.3282	635	670.9970	735	776.6658	835	882.3346	935	988.0035
536	566.3849	636	672.0537	736	777.7225	836	883.3913	936	989.0602
537	567.4416	637	673.1104	737	778.7792	837	884.4480	937	990.1168
538	568.4983	638	674.1671	738	779.8359	838	885.5047	938	991.1735
539	569.5549	639	675.2238	739	780.8926	839	886.5614	939	992.2302
540	570.6116	640	676.2804	740	781.9493	840	887.6181	940	993.2869
541	571.6683	641	677.3371	741	783.0060	841	888.6748	941	994.3436
542	572.7250	642	678.3938	742	784.0626	842	889.7315	942	995.4003
543	573.7817	643	679.4505	743	785.1193	843	890.7882	943	996.4570
544	574.8384	644	680.5072	744	786.1760	844	891.8448	944	997.5137
545	575.8951	645	681.5639	745	787.2327	845	892.9015	945	998.5704
546	576.9518	646	682.6206	746	788.2894	846	893.9582	946	999.6270
547	578.0084	647	683.6773	747	789.3461	847	895.0149	947	1000.6837
548	579.0651	648	684.7340	748	790.4028	848	896.0716	948	1001.7404
549	580.1218	649	685.7906	749	791.4595	849	897.1283	949	1002.7971

Continued - LITERS TO LIQUID QUARTS 1 LITER = 1.056 6882 LIQUID QUARTS

LITERS	LIQUID QUARTS	LITERS	LIQUID QUARTS	LITERS	LIQUID QUARTS	LITERS	LIQUID QUARTS	LITERS	LIQUID QUARTS
550	581.1785	650	686.8473	750	792.5162	850	898.1850	950	1003.8538
551	582.2352	651	687.9040	751	793.5728	851	899.2417	951	1004.9105
552	583.2919	652	688.9607	752	794.6295	852	900.2983	952	1005.9672
553	584.3486	653	690.0174	753	795.6862	853	901.3550	953	1007.0239
554	585.4053	654	691.0741	754	796.7429	854	902.4117	954	1008.0805
555	586.4620	655	692.1308	755	797.7996	855	903.4684	955	1009.1372
556	587.5186	656	693.1875	756	798.8563	856	904.5251	956	1010.1939
557	588.5753	657	694.2441	757	799.9130	857	905.5818	957	1011.2506
558	589.6320	658	695.3008	758	800.9697	858	906.6385	958	1012.3073
559	590.6887	659	696.3575	759	802.0263	859	907.6952	959	1013.3640
560	591.7454	660	697.4142	760	803.0830	860	908.7519	960	1014.4207
561	592.8021	661	698.4709	761	804.1397	861	909.8085	961	1015.4774
562	593.8588	662	699.5276	762	805.1964	862	910.8652	962	1016.5340
563	594.9155	663	700.5843	763	806.2531	863	911.9219	963	1017.5907
564	595.9721	664	701.6410	764	807.3098	864	912.9786	964	1018.6474
565	597.0288	665	702.6977	765	808.3665	865	914.0353	965	1019.7041
566	598.0855	666	703.7543	766	809.4232	866	915.0920	966	1020.7608
567	599.1422	667	704.8110	767	810.4799	867	916.1487	967	1021.8175
568	600.1989	668	705.8677	768	811.5365	868	917.2054	968	1022.8742
569	601.2556	669	706.9244	769	812.5932	869	918.2620	969	1023.9309
570	602.3123	670	707.9811	770	813.6499	870	919.3187	970	1024.9875
571	603.3690	671	709.0378	771	814.7066	871	920.3754	971	1026.0442
572	604.4257	672	710.0945	772	815.7633	872	921.4321	972	1027.1009
573	605.4823	673	711.1512	773	816.8200	873	922.4888	973	1028.1576
574	606.5390	674	712.2078	774	817.8767	874	923.5455	974	1029.2143
575	607.5957	675	713.2645	775	818.9334	875	924.6022	975	1030.2710
576	608.6524	676	714.3212	776	819.9900	876	925.6589	976	1031.3277
577	609.7091	677	715.3779	777	821.0467	877	926.7156	977	1032.3844
578	610.7658	678	716.4346	778	822.1034	878	927.7722	978	1033.4411
579	611.8225	679	717.4913	779	823.1601	879	928.8289	979	1034.4977
580	612.8792	680	718.5480	780	824.2168	880	929.8856	980	1035.5544
581	613.9358	681	719.6047	781	825.2735	881	930.9423	981	1036.6111
582	614.9925	682	720.6614	782	826.3302	882	931.9990	982	1037.6678
583	616.0492	683	721.7180	783	827.3869	883	933.0557	983	1038.7245
584	617.1059	684	722.7747	784	828.4436	884	934.1124	984	1039.7812
585	618.1626	685	723.8314	785	829.5002	885	935.1691	985	1040.8379
586	619.2193	686	724.8881	786	830.5569	886	936.2257	986	1041.8946
587	620.2760	687	725.9448	787	831.6136	887	937.2824	987	1042.9512
588	621.3327	688	727.0015	788	832.6703	888	938.3391	988	1044.0079
589	622.3894	689	728.0582	789	833.7270	889	939.3958	989	1045.0646
590	623.4460	690	729.1149	790	834.7837	890	940.4525	990	1046.1213
591	624.5027	691	730.1715	791	835.8404	891	941.5092	991	1047.1780
592	625.5594	692	731.2282	792	836.8971	892	942.5659	992	1048.2347
593	626.6161	693	732.2849	793	837.9537	893	943.6226	993	1049.2914
594	627.6728	694	733.3416	794	839.0104	894	944.6793	994	1050.3481
595	628.7295	695	734.3983	795	840.0671	895	945.7359	995	1051.4048
596	629.7862	696	735.4550	796	841.1238	896	946.7926	996	1052.4614
597	630.8429	697	736.5117	797	842.1805	897	947.8493	997	1053.5181
598	631.8995	698	737.5684	798	843.2372	898	948.9060	998	1054.5748
599	632.9562	699	738.6251	799	844.2939	899	949.9627	999	1055.6315

VOLUME-CAPACITY - GALLONS TO LITERS 1 GALLON = 3.785 411 784 LITERS

GALLONS	LITERS	GALLONS	LITERS	GALLONS	LITERS	GALLONS	LITERS	GALLONS	LITERS
0	0.0000	100	378.5412	200	757.0824	300	1135.6235	400	1514.1647
1	3.7854	101	382.3266	201	760.8678	301	1139.4089	401	1517.9501
2	7.5708	102	386.1120	202	764.6532	302	1143.1944	402	1521.7355
3	11.3562	103	389.8974	203	768.4386	303	1146.9798	403	1525.5209
4	15.1416	104	393.6828	204	772.2240	304	1150.7652	404	1529.3064
5	18.9271	105	397.4682	205	776.0094	305	1154.5506	405	1533.0918
6	22.7125	106	401.2536	206	779.7948	306	1158.3360	406	1536.8772
7	26.4979	107	405.0391	207	783.5802	307	1162.1214	407	1540.6626
8	30.2833	108	408.8245	208	787.3656	308	1165.9068	408	1544.4480
9	34.0687	109	412.6099	209	791.1511	309	1169.6922	409	1548.2334
10	37.8541	110	416.3953	210	794.9365	310	1173.4776	410	1552.0188
11	41.6395	111	420.1807	211	798.7219	311	1177.2631	411	1555.8042
12	45.4249	112	423.9661	212	802.5073	312	1181.0485	412	1559.5896
13	49.2104	113	427.7515	213	806.2927	313	1184.8339	413	1563.3751
14	52.9958	114	431.5369	214	810.0781	314	1188.6193	414	1567.1605
15	56.7812	115	435.3224	215	813.8635	315	1192.4047	415	1570.9459
16	60.5666	116	439.1078	216	817.6489	316	1196.1901	416	1574.7313
17	64.3520	117	442.8932	217	821.4343	317	1199.9755	417	1578.5167
18	68.1374	118	446.6786	218	825.2198	318	1203.7609	418	1582.3021
19	71.9228	119	450.4640	219	829.0052	319	1207.5464	419	1586.0875
20	75.7082	120	454.2494	220	832.7906	320	1211.3318	420	1589.8729
21	79.4936	121	458.0348	221	836.5760	321	1215.1172	421	1593.6584
22	83.2791	122	461.8202	222	840.3614	322	1218.9026	422	1597.4438
23	87.0645	123	465.6056	223	844.1468	323	1222.6880	423	1601.2292
24	90.8499	124	469.3911	224	847.9322	324	1226.4734	424	1605.0146
25	94.6353	125	473.1765	225	851.7176	325	1230.2588	425	1608.8000
26	98.4207	126	476.9619	226	855.5031	326	1234.0442	426	1612.5854
27	102.2061	127	480.7473	227	859.2885	327	1237.8296	427	1616.3708
28	105.9915	128	484.5327	228	863.0739	328	1241.6151	428	1620.1562
29	109.7769	129	488.3181	229	866.8593	329	1245.4005	429	1623.9417
30	113.5624	130	492.1035	230	870.6447	330	1249.1859	430	1627.7271
31	117.3478	131	495.8889	231	874.4301	331	1252.9713	431	1631.5125
32	121.1332	132	499.6744	232	878.2155	332	1256.7567	432	1635.2979
33	124.9186	133	503.4598	233	882.0009	333	1260.5421	433	1639.0833
34	128.7040	134	507.2452	234	885.7864	334	1264.3275	434	1642.8687
35	132.4894	135	511.0306	235	889.5718	335	1268.1129	435	1646.6541
36	136.2748	136	514.8160	236	893.3572	336	1271.8983	436	1650.4395
37	140.0602	137	518.6014	237	897.1426	337	1275.6838	437	1654.2249
38	143.8456	138	522.3868	238	900.9280	338	1279.4692	438	1658.0103
39	147.6311	139	526.1722	239	904.7134	339	1283.2546	439	1661.7958
40	151.4165	140	529.9576	240	908.4988	340	1287.0400	440	1665.5812
41	155.2019	141	533.7431	241	912.2842	341	1290.8254	441	1669.3666
42	158.9873	142	537.5285	242	916.0696	342	1294.6108	442	1673.1520
43	162.7727	143	541.3139	243	919.8551	343	1298.3962	443	1676.9374
44	166.5581	144	545.0993	244	923.6405	344	1302.1816	444	1680.7228
45	170.3435	145	548.8847	245	927.4259	345	1305.9671	445	1684.5082
46	174.1289	146	552.6701	246	931.2113	346	1309.7525	446	1688.2936
47	177.9144	147	556.4555	247	934.9967	347	1313.5379	447	1692.0791
48	181.6998	148	560.2409	248	938.7821	348	1317.3233	448	1695.8645
49	185.4852	149	564.0264	249	942.5675	349	1321.1087	449	1699.6499

Continued – GALLONS TO LITERS 1 GALLON = 3.785 411 784 LITERS

GALLONS	LITERS	GALLONS	LITERS	GALLONS	LITERS	GALLONS	LITERS	GALLONS	LITERS
50	189.2706	150	567.8118	250	946.3529	350	1324.8941	450	1703.4353
51	193.0560	151	571.5972	251	950.1384	351	1328.6795	451	1707.2207
52	196.8414	152	575.3826	252	953.9238	352	1332.4649	452	1711.0061
53	200.6268	153	579.1680	253	957.7092	353	1336.2504	453	1714.7915
54	204.4122	154	582.9534	254	961.4946	354	1340.0358	454	1718.5769
55	208.1976	155	586.7388	255	965.2800	355	1343.8212	455	1722.3624
56	211.9831	156	590.5242	256	969.0654	356	1347.6066	456	1726.1478
57	215.7685	157	594.3096	257	972.8508	357	1351.3920	457	1729.9332
58	219.5539	158	598.0951	258	976.6362	358	1355.1774	458	1733.7186
59	223.3393	159	601.8805	259	980.4216	359	1358.9628	459	1737.5040
60	227.1247	160	605.6659	260	984.2071	360	1362.7482	460	1741.2894
61	230.9101	161	609.4513	261	987.9925	361	1366.5336	461	1745.0748
62	234.6955	162	613.2367	262	991.7779	362	1370.3191	462	1748.8602
63	238.4809	163	617.0221	263	995.5633	363	1374.1045	463	1752.6456
64	242.2664	164	620.8075	264	999.3487	364	1377.8899	464	1756.4311
65	246.0518	165	624.5929	265	1003.1341	365	1381.6753	465	1760.2165
66	249.8372	166	628.3783	266	1006.9195	366	1385.4607	466	1764.0019
67	253.6226	167	632.1638	267	1010.7049	367	1389.2461	467	1767.7873
68	257.4080	168	635.9492	268	1014.4903	368	1393.0315	468	1771.5727
69	261.1934	169	639.7346	269	1018.2758	369	1396.8169	469	1775.3581
70	264.9788	170	643.5200	270	1022.0612	370	1400.6024	470	1779.1435
71	268.7642	171	647.3054	271	1025.8466	371	1404.3878	471	1782.9289
72	272.5496	172	651.0908	272	1029.6320	372	1408.1732	472	1786.7144
73	276.3351	173	654.8762	273	1033.4174	373	1411.9586	473	1790.4998
74	280.1205	174	658.6616	274	1037.2028	374	1415.7440	474	1794.2852
75	283.9059	175	662.4471	275	1040.9882	375	1419.5294	475	1798.0706
76	287.6913	176	666.2325	276	1044.7736	376	1423.3148	476	1801.8560
77	291.4767	177	670.0179	277	1048.5591	377	1427.1002	477	1805.6414
78	295.2621	178	673.8033	278	1052.3445	378	1430.8857	478	1809.4268
79	299.0475	179	677.5887	279	1056.1299	379	1434.6711	479	1813.2122
80	302.8329	180	681.3741	280	1059.9153	380	1438.4565	480	1816.9977
81	306.6184	181	685.1595	281	1063.7007	381	1442.2419	481	1820.7831
82	310.4038	182	688.9449	282	1067.4861	382	1446.0273	482	1824.5685
83	314.1892	183	692.7304	283	1071.2715	383	1449.8127	483	1828.3539
84	317.9746	184	696.5158	284	1075.0569	384	1453.5981	484	1832.1393
85	321.7600	185	700.3012	285	1078.8423	385	1457.3835	485	1835.9247
86	325.5454	186	704.0866	286	1082.6278	386	1461.1689	486	1839.7101
87	329.3308	187	707.8720	287	1086.4132	387	1464.9543	487	1843.4955
88	333.1162	188	711.6574	288	1090.1986	388	1468.7398	488	1847.2809
89	336.9016	189	715.4428	289	1093.9840	389	1472.5252	489	1851.0663
90	340.6871	190	719.2282	290	1097.7694	390	1476.3106	490	1854.8518
91	344.4725	191	723.0136	291	1101.5548	391	1480.0960	491	1858.6372
92	348.2579	192	726.7991	292	1105.3402	392	1483.8814	492	1862.4226
93	352.0433	193	730.5845	293	1109.1256	393	1487.6668	493	1866.2080
94	355.8287	194	734.3699	294	1112.9111	394	1491.4522	494	1869.9934
95	359.6141	195	738.1553	295	1116.6965	395	1495.2376	495	1873.7788
96	363.3995	196	741.9407	296	1120.4819	396	1499.0231	496	1877.5642
97	367.1849	197	745.7261	297	1124.2673	397	1502.8085	497	1881.3496
98	370.9704	198	749.5115	298	1128.0527	398	1506.5939	498	1885.1351
99	374.7558	199	753.2969	299	1131.8381	399	1510.3793	499	1888.9205

Continued - GALLONS TO LITERS 1 GALLON = 3.785 411 784 LITERS

GALLONS	LITERS	GALLONS	LITERS	GALLONS	LITERS	GALLONS	LITERS	GALLONS	LITERS
500	1892.7059	600	2271.2470	700	2649.7882	800	3028.3294	900	3406.8706
501	1896.4913	601	2275.0325	701	2653.5736	801	3032.1148	901	3410.6560
502	1900.2767	602	2278.8179	702	2657.3590	802	3035.9002	902	3414.4414
503	1904.0621	603	2282.6033	703	2661.1445	803	3039.6856	903	3418.2268
504	1907.8475	604	2286.3887	704	2664.9299	804	3043.4710	904	3422.0122
505	1911.6329	605	2290.1741	705	2668.7153	805	3047.2565	905	3425.7976
506	1915.4184	606	2293.9595	706	2672.5007	806	3051.0419	906	3429.5830
507	1919.2038	607	2297.7449	707	2676.2861	807	3054.8273	907	3433.3685
508	1922.9892	608	2301.5303	708	2680.0715	808	3058.6127	908	3437.1539
509	1926.7746	609	2305.3158	709	2683.8569	809	3062.3981	909	3440.9393
510	1930.5600	610	2309.1012	710	2687.6423	810	3066.1835	910	3444.7247
511	1934.3454	611	2312.8866	711	2691.4278	811	3069.9689	911	3448.5101
512	1938.1308	612	2316.6720	712	2695.2132	812	3073.7543	912	3452.2955
513	1941.9162	613	2320.4574	713	2698.9986	813	3077.5398	913	3456.0809
514	1945.7016	614	2324.2428	714	2702.7840	814	3081.3252	914	3459.8663
515	1949.4871	615	2328.0282	715	2706.5694	815	3085.1106	915	3463.6518
516	1953.2725	616	2331.8136	716	2710.3548	816	3088.8960	916	3467.4372
517	1957.0579	617	2335.5991	717	2714.1402	817	3092.6814	917	3471.2226
518	1960.8433	618	2339.3845	718	2717.9256	818	3096.4668	918	3475.0080
519	1964.6287	619	2343.1699	719	2721.7111	819	3100.2522	919	3478.7934
520	1968.4141	620	2346.9553	720	2725.4965	820	3104.0376	920	3482.5788
521	1972.1995	621	2350.7407	721	2729.2819	821	3107.8231	921	3486.3642
522	1975.9849	622	2354.5261	722	2733.0673	822	3111.6085	922	3490.1496
523	1979.7704	623	2358.3115	723	2736.8527	823	3115.3939	923	3493.9351
524	1983.5558	624	2362.0969	724	2740.6381	824	3119.1793	924	3497.7205
525	1987.3412	625	2365.8824	725	2744.4235	825	3122.9647	925	3501.5059
526	1991.1266	626	2369.6678	726	2748.2089	826	3126.7501	926	3505.2913
527	1994.9120	627	2373.4532	727	2751.9944	827	3130.5355	927	3509.0767
528	1998.6974	628	2377.2386	728	2755.7798	828	3134.3209	928	3512.8621
529	2002.4828	629	2381.0240	729	2759.5652	829	3138.1064	929	3516.6475
530	2006.2682	630	2384.8094	730	2763.3506	830	3141.8918	930	3520.4329
531	2010.0536	631	2388.5948	731	2767.1360	831	3145.6772	931	3524.2184
532	2013.8391	632	2392.3802	732	2770.9214	832	3149.4626	932	3528.0038
533	2017.6245	633	2396.1656	733	2774.7068	833	3153.2480	933	3531.7892
534	2021.4099	634	2399.9510	734	2778.4922	834	3157.0334	934	3535.5746
535	2025.1953	635	2403.7365	735	2782.2776	835	3160.8188	935	3539.3600
536	2028.9807	636	2407.5219	736	2786.0630	836	3164.6042	936	3543.1454
537	2032.7661	637	2411.3073	737	2789.8484	837	3168.3896	937	3546.9308
538	2036.5515	638	2415.0927	738	2793.6339	838	3172.1750	938	3550.7162
539	2040.3369	639	2418.8781	739	2797.4193	839	3175.9604	939	3554.5016
540	2044.1223	640	2422.6635	740	2801.2047	840	3179.7459	940	3558.2870
541	2047.9078	641	2426.4489	741	2804.9901	841	3183.5313	941	3562.0725
542	2051.6932	642	2430.2343	742	2808.7755	842	3187.3167	942	3565.8579
543	2055.4786	643	2434.0197	743	2812.5609	843	3191.1021	943	3569.6433
544	2059.2640	644	2437.8052	744	2816.3463	844	3194.8875	944	3573.4287
545	2063.0494	645	2441.5906	745	2820.1317	845	3198.6729	945	3577.2141
546	2066.8348	646	2445.3760	746	2823.9172	846	3202.4583	946	3580.9995
547	2070.6202	647	2449.1614	747	2827.7026	847	3206.2437	947	3584.7849
548	2074.4056	648	2452.9468	748	2831.4880	848	3210.0292	948	3588.5703
549	2078.1910	649	2456.7322	749	2835.2734	849	3213.8146	949	3592.3558

Continued - GALLONS TO LITERS 1 GALLON = 3.785 411 784 LITERS

GALLONS	LITERS	GALLONS	LITERS	GALLONS	LITERS	GALLONS	LITERS	GALLONS	LITERS
550	2081.9765	650	2460.5176	750	2839.0588	850	3217.6000	950	3596.1412
551	2085.7619	651	2464.3030	751	2842.8442	851	3221.3854	951	3599.9266
552	2089.5473	652	2468.0885	752	2846.6296	852	3225.1708	952	3603.7120
553	2093.3327	653	2471.8739	753	2850.4150	853	3228.9562	953	3607.4974
554	2097.1181	654	2475.6593	754	2854.2005	854	3232.7416	954	3611.2828
555	2100.9035	655	2479.4447	755	2857.9859	855	3236.5270	955	3615.0682
556	2104.6889	656	2483.2301	756	2861.7713	856	3240.3125	956	3618.8536
557	2108.4743	657	2487.0155	757	2865.5567	857	3244.0979	957	3622.6391
558	2112.2598	658	2490.8009	758	2869.3421	858	3247.8833	958	3626.4245
559	2116.0452	659	2494.5863	759	2873.1275	859	3251.6687	959	3630.2099
560	2119.8306	660	2498.3718	760	2876.9129	860	3255.4541	960	3633.9953
561	2123.6160	661	2502.1572	761	2880.6983	861	3259.2395	961	3637.7807
562	2127.4014	662	2505.9426	762	2884.4838	862	3263.0249	962	3641.5661
563	2131.1868	663	2509.7280	763	2888.2692	863	3266.8103	963	3645.3515
564	2134.9722	664	2513.5134	764	2892.0546	864	3270.5958	964	3649.1369
565	2138.7576	665	2517.2988	765	2895.8400	865	3274.3812	965	3652.9223
566	2142.5431	666	2521.0842	766	2899.6254	866	3278.1666	966	3656.7078
567	2146.3285	667	2524.8696	767	2903.4108	867	3281.9520	967	3660.4932
568	2150.1139	668	2528.6551	768	2907.1962	868	3285.7374	968	3664.2786
569	2153.8993	669	2532.4405	769	2910.9816	869	3289.5228	969	3668.0640
570	2157.6847	670	2536.2259	770	2914.7671	870	3293.3082	970	3671.8494
571	2161.4701	671	2540.0113	771	2918.5525	871	3297.0936	971	3675.6348
572	2165.2555	672	2543.7967	772	2922.3379	872	3300.8791	972	3679.4202
573	2169.0409	673	2547.5821	773	2926.1233	873	3304.6645	973	3683.2056
574	2172.8264	674	2551.3675	774	2929.9087	874	3308.4499	974	3686.9911
575	2176.6118	675	2555.1529	775	2933.6941	875	3312.2353	975	3690.7765
576	2180.3972	676	2558.9384	776	2937.4795	876	3316.0207	976	3694.5619
577	2184.1826	677	2562.7238	777	2941.2649	877	3319.8061	977	3698.3473
578	2187.9680	678	2566.5092	778	2945.0504	878	3323.5915	978	3702.1327
579	2191.7534	679	2570.2946	779	2948.8358	879	3327.3769	979	3705.9181
580	2195.5388	680	2574.0800	780	2952.6212	880	3331.1624	980	3709.7035
581	2199.3242	681	2577.8654	781	2956.4066	881	3334.9478	981	3713.4889
582	2203.1096	682	2581.6508	782	2960.1920	882	3338.7332	982	3717.2744
583	2206.8951	683	2585.4362	783	2963.9774	883	3342.5186	983	3721.0598
584	2210.6805	684	2589.2216	784	2967.7628	884	3346.3040	984	3724.8452
585	2214.4659	685	2593.0070	785	2971.5482	885	3350.0894	985	3728.6306
586	2218.2513	686	2596.7924	786	2975.3336	886	3353.8748	986	3732.4160
587	2222.0367	687	2600.5779	787	2979.1190	887	3357.6602	987	3736.2014
588	2225.8221	688	2604.3633	788	2982.9044	888	3361.4456	988	3739.9868
589	2229.6075	689	2608.1487	789	2986.6899	889	3365.2310	989	3743.7722
590	2233.3929	690	2611.9341	790	2990.4753	890	3369.0165	990	3747.5576
591	2237.1783	691	2615.7195	791	2994.2607	891	3372.8019	991	3751.3430
592	2240.9637	692	2619.5049	792	2998.0461	892	3376.5873	992	3755.1285
593	2244.7492	693	2623.2903	793	3001.8315	893	3380.3727	993	3758.9139
594	2248.5346	694	2627.0757	794	3005.6169	894	3384.1581	994	3762.6993
595	2252.3200	695	2630.8612	795	3009.4023	895	3387.9435	995	3766.4847
596	2256.1054	696	2634.6466	796	3013.1877	896	3391.7289	996	3770.2701
597	2259.8908	697	2638.4320	797	3016.9732	897	3395.5143	997	3774.0555
598	2263.6762	698	2642.2174	798	3020.7586	898	3399.2997	998	3777.8409
599	2267.4616	699	2646.0028	799	3024.5440	899	3403.0852	999	3781.6263

VOLUME-CAPACITY - LITERS TO GALLONS 1 LITER = 0.264 172 052 GALLON

LITERS	GALLONS	LITERS	GALLONS	LITERS	GALLONS	LITERS	GALLONS	LITERS	GALLONS
0	0.00000	100	26.41721	200	52.83441	300	79.25161	400	105.66882
1	0.26417	101	26.68138	201	53.09858	301	79.51579	401	105.93299
2	0.52834	102	26.94555	202	53.36275	302	79.77996	402	106.19716
3	0.79252	103	27.20972	203	53.62693	303	80.04413	403	106.46134
4	1.05669	104	27.47389	204	53.89110	304	80.30830	404	106.72551
5	1.32086	105	27.73807	205	54.15527	305	80.57248	405	106.98968
6	1.58503	106	28.00224	206	54.41944	306	80.83665	406	107.25385
7	1.84920	107	28.26641	207	54.68361	307	81.10082	407	107.51802
8	2.11338	108	28.53058	208	54.94779	308	81.36499	408	107.78220
9	2.37755	109	28.79475	209	55.21196	309	81.62916	409	108.04637
10	2.64172	110	29.05893	210	55.47613	310	81.89334	410	108.31054
11	2.90589	111	29.32310	211	55.74030	311	82.15751	411	108.57471
12	3.17006	112	29.58727	212	56.00447	312	82.42168	412	108.83888
13	3.43424	113	29.85144	213	56.26865	313	82.68585	413	109.10306
14	3.69841	114	30.11561	214	56.53282	314	82.95002	414	109.36723
15	3.96258	115	30.37979	215	56.79699	315	83.21420	415	109.63140
16	4.22675	116	30.64396	216	57.06116	316	83.47837	416	109.89557
17	4.49092	117	30.90813	217	57.32534	317	83.74254	417	110.15975
18	4.75510	118	31.17230	218	57.58951	318	84.00671	418	110.42392
19	5.01927	119	31.43647	219	57.85368	319	84.27088	419	110.68809
20	5.28344	120	31.70065	220	58.11785	320	84.53506	420	110.95226
21	5.54761	121	31.96482	221	58.38202	321	84.79923	421	111.21643
22	5.81179	122	32.22899	222	58.64619	322	85.06340	422	111.48061
23	6.07596	123	32.49316	223	58.91037	323	85.32757	423	111.74478
24	6.34013	124	32.75733	224	59.17454	324	85.59174	424	112.00895
25	6.60430	125	33.02151	225	59.43871	325	85.85592	425	112.27312
26	6.86847	126	33.28568	226	59.70288	326	86.12009	426	112.53729
27	7.13265	127	33.54985	227	59.96706	327	86.38426	427	112.80147
28	7.39682	128	33.81402	228	60.23123	328	86.64843	428	113.06564
29	7.66099	129	34.07819	229	60.49540	329	86.91260	429	113.32981
30	7.92516	130	34.34237	230	60.75957	330	87.17678	430	113.59398
31	8.18933	131	34.60654	231	61.02374	331	87.44095	431	113.85815
32	8.45351	132	34.87071	232	61.28792	332	87.70512	432	114.12233
33	8.71768	133	35.13488	233	61.55209	333	87.96929	433	114.38650
34	8.98185	134	35.39905	234	61.81626	334	88.23346	434	114.65067
35	9.24602	135	35.66323	235	62.08043	335	88.49764	435	114.91484
36	9.51019	136	35.92740	236	62.34460	336	88.76181	436	115.17901
37	9.77437	137	36.19157	237	62.60878	337	89.02598	437	115.44319
38	10.03854	138	36.45574	238	62.87295	338	89.29015	438	115.70736
39	10.30271	139	36.71991	239	63.13712	339	89.55433	439	115.97153
40	10.56688	140	36.98409	240	63.40129	340	89.81850	440	116.23570
41	10.83105	141	37.24826	241	63.66546	341	90.08267	441	116.49987
42	11.09523	142	37.51243	242	63.92964	342	90.34684	442	116.76405
43	11.35940	143	37.77660	243	64.19381	343	90.61101	443	117.02822
44	11.62357	144	38.04078	244	64.45798	344	90.87519	444	117.29239
45	11.88774	145	38.30495	245	64.72215	345	91.13936	445	117.55656
46	12.15191	146	38.56912	246	64.98632	346	91.40353	446	117.82073
47	12.41609	147	38.83329	247	65.25050	347	91.66770	447	118.08491
48	12.68026	148	39.09746	248	65.51467	348	91.93187	448	118.34908
49	12.94443	149	39.36164	249	65.77884	349	92.19604	449	118.61325

Continued - LITERS TO GALLONS 1 LITER = 0.264 172 052 GALLON

LITERS	GALLONS	LITERS	GALLONS	LITERS	GALLONS	LITERS	GALLONS	LITERS	GALLONS
50	13.20860	150	39.62581	250	66.04301	350	92.46022	450	118.87742
51	13.47277	151	39.88998	251	66.30718	351	92.72439	451	119.14159
52	13.73695	152	40.15415	252	66.57136	352	92.98856	452	119.40577
53	14.00112	153	40.41832	253	66.83553	353	93.25273	453	119.66994
54	14.26529	154	40.68250	254	67.09970	354	93.51691	454	119.93411
55	14.52946	155	40.94667	255	67.36387	355	93.78108	455	120.19828
56	14.79363	156	41.21084	256	67.62805	356	94.04525	456	120.46245
57	15.05781	157	41.47501	257	67.89222	357	94.30942	457	120.72663
58	15.32198	158	41.73918	258	68.15639	358	94.57359	458	120.99080
59	15.58615	159	42.00336	259	68.42056	359	94.83777	459	121.25497
60	15.85032	160	42.26753	260	68.68473	360	95.10194	460	121.51914
61	16.11450	161	42.53170	261	68.94890	361	95.36611	461	121.78331
62	16.37867	162	42.79587	262	69.21308	362	95.63028	462	122.04749
63	16.64284	163	43.06004	263	69.47725	363	95.89445	463	122.31166
64	16.90701	164	43.32422	264	69.74142	364	96.15863	464	122.57583
65	17.17118	165	43.58839	265	70.00559	365	96.42280	465	122.84000
66	17.43536	166	43.85256	266	70.26976	366	96.68697	466	123.10418
67	17.69953	167	44.11673	267	70.53394	367	96.95114	467	123.36835
68	17.96370	168	44.38090	268	70.79811	368	97.21531	468	123.63252
69	18.22787	169	44.64508	269	71.06228	369	97.47949	469	123.89669
70	18.49204	170	44.90925	270	71.32645	370	97.74366	470	124.16086
71	18.75622	171	45.17342	271	71.59063	371	98.00783	471	124.42504
72	19.02039	172	45.43759	272	71.85480	372	98.27200	472	124.68921
73	19.28456	173	45.70176	273	72.11897	373	98.53617	473	124.95338
74	19.54873	174	45.96594	274	72.38314	374	98.80035	474	125.21755
75	19.81290	175	46.23011	275	72.64731	375	99.06452	475	125.48172
76	20.07708	176	46.49428	276	72.91149	376	99.32869	476	125.74590
77	20.34125	177	46.75845	277	73.17566	377	99.59286	477	126.01007
78	20.60542	178	47.02262	278	73.43983	378	99.85703	478	126.27424
79	20.86959	179	47.28680	279	73.70400	379	100.12121	479	126.53841
80	21.13376	180	47.55097	280	73.96817	380	100.38538	480	126.80258
81	21.39794	181	47.81514	281	74.23235	381	100.64955	481	127.06676
82	21.66211	182	48.07931	282	74.49652	382	100.91372	482	127.33093
83	21.92628	183	48.34348	283	74.76069	383	101.17790	483	127.59510
84	22.19045	184	48.60766	284	75.02486	384	101.44207	484	127.85927
85	22.45462	185	48.87183	285	75.28903	385	101.70624	485	128.12344
86	22.71880	186	49.13600	286	75.55321	386	101.97041	486	128.38762
87	22.98297	187	49.40017	287	75.81738	387	102.23458	487	128.65179
88	23.24714	188	49.66435	288	76.08155	388	102.49876	488	128.91596
89	23.51131	189	49.92852	289	76.34572	389	102.76293	489	129.18013
90	23.77548	190	50.19269	290	76.60989	390	103.02710	490	129.44430
91	24.03966	191	50.45686	291	76.87407	391	103.29127	491	129.70848
92	24.30383	192	50.72103	292	77.13824	392	103.55544	492	129.97265
93	24.56800	193	50.98521	293	77.40241	393	103.81962	493	130.23682
94	24.83217	194	51.24938	294	77.66658	394	104.08379	494	130.50099
95	25.09634	195	51.51355	295	77.93075	395	104.34796	495	130.76516
96	25.36052	196	51.77772	296	78.19493	396	104.61213	496	131.02934
97	25.62469	197	52.04189	297	78.45910	397	104.87630	497	131.29351
98	25.88886	198	52.30607	298	78.72327	398	105.14048	498	131.55768
99	26.15303	199	52.57024	299	78.98744	399	105.40465	499	131.82185

Continued - LITERS TO GALLONS 1 LITER = 0.264 172 052 GALLON

LITERS	GALLONS	LITERS	GALLONS	LITERS	GALLONS	LITERS	GALLONS	LITERS	GALLONS
500	132.08603	600	158.50323	700	184.92043	800	211.33764	900	237.75484
501	132.35020	601	158.76740	701	185.18461	801	211.60181	901	238.01902
502	132.61437	602	159.03157	702	185.44878	802	211.86598	902	238.28319
503	132.87854	603	159.29575	703	185.71295	803	212.13016	903	238.54736
504	133.14271	604	159.55992	704	185.97712	804	212.39433	904	238.81153
505	133.40689	605	159.82409	705	186.24129	805	212.65850	905	239.07570
506	133.67106	606	160.08826	706	186.50547	806	212.92267	906	239.33988
507	133.93523	607	160.35243	707	186.76964	807	213.18684	907	239.60405
508	134.19940	608	160.61661	708	187.03381	808	213.45102	908	239.86822
509	134.46357	609	160.88078	709	187.29798	809	213.71519	909	240.13239
510	134.72775	610	161.14495	710	187.56215	810	213.97936	910	240.39657
511	134.99192	611	161.40912	711	187.82633	811	214.24353	911	240.66074
512	135.25609	612	161.67329	712	188.09050	812	214.50770	912	240.92491
513	135.52026	613	161.93747	713	188.35467	813	214.77188	913	241.18908
514	135.78443	614	162.20164	714	188.61884	814	215.03605	914	241.45325
515	136.04860	615	162.46581	715	188.88301	815	215.30022	915	241.71743
516	136.31278	616	162.72998	716	189.14719	816	215.56439	916	241.98160
517	136.57695	617	162.99415	717	189.41136	817	215.82857	917	242.24577
518	136.84112	618	163.25833	718	189.67553	818	216.09274	918	242.50994
519	137.10529	619	163.52250	719	189.93970	819	216.35691	919	242.77411
520	137.36946	620	163.78667	720	190.20388	820	216.62108	920	243.03829
521	137.63364	621	164.05084	721	190.46805	821	216.88525	921	243.30246
522	137.89781	622	164.31501	722	190.73222	822	217.14943	922	243.56663
523	138.16198	623	164.57919	723	190.99639	823	217.41360	923	243.83080
524	138.42615	624	164.84336	724	191.26056	824	217.67777	924	244.09497
525	138.69033	625	165.10753	725	191.52474	825	217.94194	925	244.35915
526	138.95450	626	165.37170	726	191.78891	826	218.20611	926	244.62332
527	139.21867	627	165.63588	727	192.05308	827	218.47029	927	244.88749
528	139.48284	628	165.90005	728	192.31725	828	218.73446	928	245.15166
529	139.74701	629	166.16422	729	192.58142	829	218.99863	929	245.41583
530	140.01119	630	166.42839	730	192.84560	830	219.26280	930	245.68001
531	140.27536	631	166.69256	731	193.10977	831	219.52697	931	245.94418
532	140.53953	632	166.95674	732	193.37394	832	219.79115	932	246.20835
533	140.80370	633	167.22091	733	193.63811	833	220.05532	933	246.47252
534	141.06787	634	167.48508	734	193.90228	834	220.31949	934	246.73669
535	141.33205	635	167.74925	735	194.16646	835	220.58366	935	247.00087
536	141.59622	636	168.01342	736	194.43063	836	220.84783	936	247.26504
537	141.86039	637	168.27760	737	194.69480	837	221.11201	937	247.52921
538	142.12456	638	168.54177	738	194.95897	838	221.37618	938	247.79338
539	142.38873	639	168.80594	739	195.22314	839	221.64035	939	248.05755
540	142.65291	640	169.07011	740	195.48732	840	221.90452	940	248.32173
541	142.91708	641	169.33428	741	195.75149	841	222.16869	941	248.58590
542	143.18125	642	169.59846	742	196.01566	842	222.43287	942	248.85007
543	143.44542	643	169.86263	743	196.27983	843	222.69704	943	249.11424
544	143.70959	644	170.12680	744	196.54400	844	222.96121	944	249.37842
545	143.97377	645	170.39097	745	196.80818	845	223.22538	945	249.64259
546	144.23794	646	170.65514	746	197.07235	846	223.48955	946	249.90676
547	144.50211	647	170.91932	747	197.33652	847	223.75373	947	250.17093
548	144.76628	648	171.18349	748	197.60069	848	224.01790	948	250.43510
549	145.03045	649	171.44766	749	197.86487	849	224.28207	949	250.69928

Continued - LITERS TO GALLONS 1 LITER = 0.264 172 052 GALLON

LITERS	GALLONS	LITERS	GALLONS	LITERS	GALLONS	LITERS	GALLONS	LITERS	GALLONS
550	145.29463	650	171.71183	750	198.12904	850	224.54624	950	250.96345
551	145.55880	651	171.97600	751	198.39321	851	224.81042	951	251.22762
552	145.82297	652	172.24018	752	198.65738	852	225.07459	952	251.49179
553	146.08714	653	172.50435	753	198.92155	853	225.33876	953	251.75596
554	146.35131	654	172.76852	754	199.18573	854	225.60293	954	252.02014
555	146.61549	655	173.03269	755	199.44990	855	225.86710	955	252.28431
556	146.87966	656	173.29687	756	199.71407	856	226.13128	956	252.54848
557	147.14383	657	173.56104	757	199.97824	857	226.39545	957	252.81265
558	147.40800	658	173.82521	758	200.24241	858	226.65962	958	253.07682
559	147.67218	659	174.08938	759	200.50659	859	226.92379	959	253.34100
560	147.93635	660	174.35355	760	200.77076	860	227.18796	960	253.60517
561	148.20052	661	174.61773	761	201.03493	861	227.45214	961	253.86934
562	148.46469	662	174.88190	762	201.29910	862	227.71631	962	254.13351
563	148.72886	663	175.14607	763	201.56327	863	227.98048	963	254.39768
564	148.99304	664	175.41024	764	201.82745	864	228.24465	964	254.66186
565	149.25721	665	175.67441	765	202.09162	865	228.50882	965	254.92603
566	149.52138	666	175.93859	766	202.35579	866	228.77299	966	255.19020
567	149.78555	667	176.20276	767	202.61996	867	229.03717	967	255.45437
568	150.04972	668	176.46693	768	202.88413	868	229.30134	968	255.71854
569	150.31390	669	176.73110	769	203.14831	869	229.56551	969	255.98272
570	150.57807	670	176.99527	770	203.41248	870	229.82968	970	256.24689
571	150.84224	671	177.25945	771	203.67665	871	230.09385	971	256.51106
572	151.10641	672	177.52362	772	203.94082	872	230.35803	972	256.77523
573	151.37058	673	177.78779	773	204.20499	873	230.62220	973	257.03940
574	151.63476	674	178.05196	774	204.46917	874	230.88637	974	257.30358
575	151.89893	675	178.31613	775	204.73334	875	231.15054	975	257.56775
576	152.16310	676	178.58031	776	204.99751	876	231.41472	976	257.83192
577	152.42727	677	178.84448	777	205.26168	877	231.67889	977	258.09609
578	152.69144	678	179.10865	778	205.52585	878	231.94306	978	258.36026
579	152.95562	679	179.37282	779	205.79003	879	232.20723	979	258.62444
580	153.21979	680	179.63699	780	206.05420	880	232.47140	980	258.88861
581	153.48396	681	179.90117	781	206.31837	881	232.73558	981	259.15278
582	153.74813	682	180.16534	782	206.58254	882	232.99975	982	259.41695
583	154.01230	683	180.42951	783	206.84672	883	233.26392	983	259.68113
584	154.27648	684	180.69368	784	207.11089	884	233.52809	984	259.94530
585	154.54065	685	180.95785	785	207.37506	885	233.79226	985	260.20947
586	154.80482	686	181.22203	786	207.63923	886	234.05644	986	260.47364
587	155.06899	687	181.48620	787	207.90340	887	234.32061	987	260.73781
588	155.33316	688	181.75037	788	208.16758	888	234.58478	988	261.00198
589	155.59734	689	182.01454	789	208.43175	889	234.84895	989	261.26616
590	155.86151	690	182.27872	790	208.69592	890	235.11312	990	261.53033
591	156.12568	691	182.54289	791	208.96009	891	235.37730	991	261.79450
592	156.38985	692	182.80706	792	209.22426	892	235.64147	992	262.05867
593	156.65403	693	183.07123	793	209.48844	893	235.90564	993	262.32285
594	156.91820	694	183.33540	794	209.75261	894	236.16981	994	262.58702
595	157.18237	695	183.59958	795	210.01678	895	236.43398	995	262.85119
596	157.44654	696	183.86375	796	210.28095	896	236.69816	996	263.11536
597	157.71071	697	184.12792	797	210.54512	897	236.96233	997	263.37953
598	157.97489	698	184.39209	798	210.80930	898	237.22650	998	263.64370
599	158.23906	699	184.65626	799	211.07347	899	237.49067	999	263.90788

Area

AREA - SQUARE FEET TO SQUARE METERS 1 SQUARE FOOT = 0.092 903 04 SQUARE METER

SQUARE FEET	SQUARE METERS	SQUARE FEET	SQUARE METERS	SQUARE FEET	SQUARE METERS	SQUARE FEET	SQUARE METERS	SQUARE FEET	SQUARE METERS
0	0.00000	100	9.29030	200	18.58061	300	27.87091	400	37.16122
1	0.09290	101	9.38321	201	18.67351	301	27.96381	401	37.25412
2	0.18581	102	9.47611	202	18.76641	302	28.05672	402	37.34702
3	0.27871	103	9.56901	203	18.85932	303	28.14962	403	37.43993
4	0.37161	104	9.66192	204	18.95222	304	28.24252	404	37.53283
5	0.46452	105	9.75482	205	19.04512	305	28.33543	405	37.62573
6	0.55742	106	9.84772	206	19.13803	306	28.42833	406	37.71863
7	0.65032	107	9.94063	207	19.23093	307	28.52123	407	37.81154
8	0.74322	108	10.03353	208	19.32383	308	28.61414	408	37.90444
9	0.83613	109	10.12643	209	19.41674	309	28.70704	409	37.99734
10	0.92903	110	10.21933	210	19.50964	310	28.79994	410	38.09025
11	1.02193	111	10.31224	211	19.60254	311	28.89285	411	38.18315
12	1.11484	112	10.40514	212	19.69544	312	28.98575	412	38.27605
13	1.20774	113	10.49804	213	19.78835	313	29.07865	413	38.36896
14	1.30064	114	10.59095	214	19.88125	314	29.17155	414	38.46186
15	1.39355	115	10.68385	215	19.97415	315	29.26446	415	38.55476
16	1.48645	116	10.77675	216	20.06706	316	29.35736	416	38.64766
17	1.57935	117	10.86966	217	20.15996	317	29.45026	417	38.74057
18	1.67225	118	10.96256	218	20.25286	318	29.54317	418	38.83347
19	1.76516	119	11.05546	219	20.34577	319	29.63607	419	38.92637
20	1.85806	120	11.14836	220	20.43867	320	29.72897	420	39.01928
21	1.95096	121	11.24127	221	20.53157	321	29.82188	421	39.11218
22	2.04387	122	11.33417	222	20.62447	322	29.91478	422	39.20508
23	2.13677	123	11.42707	223	20.71738	323	30.00768	423	39.29799
24	2.22967	124	11.51998	224	20.81028	324	30.10058	424	39.39089
25	2.32258	125	11.61288	225	20.90318	325	30.19349	425	39.48379
26	2.41548	126	11.70578	226	20.99609	326	30.28639	426	39.57669
27	2.50838	127	11.79869	227	21.08899	327	30.37929	427	39.66960
28	2.60129	128	11.89159	228	21.18189	328	30.47220	428	39.76250
29	2.69419	129	11.98449	229	21.27480	329	30.56510	429	39.85540
30	2.78709	130	12.07740	230	21.36770	330	30.65800	430	39.94831
31	2.87999	131	12.17030	231	21.46060	331	30.75091	431	40.04121
32	2.97290	132	12.26320	232	21.55351	332	30.84381	432	40.13411
33	3.06580	133	12.35610	233	21.64641	333	30.93671	433	40.22702
34	3.15870	134	12.44901	234	21.73931	334	31.02962	434	40.31992
35	3.25161	135	12.54191	235	21.83221	335	31.12252	435	40.41282
36	3.34451	136	12.63481	236	21.92512	336	31.21542	436	40.50573
37	3.43741	137	12.72772	237	22.01802	337	31.30832	437	40.59863
38	3.53032	138	12.82062	238	22.11092	338	31.40123	438	40.69153
39	3.62322	139	12.91352	239	22.20383	339	31.49413	439	40.78443
40	3.71612	140	13.00643	240	22.29673	340	31.58703	440	40.87734
41	3.80902	141	13.09933	241	22.38963	341	31.67994	441	40.97024
42	3.90193	142	13.19223	242	22.48254	342	31.77284	442	41.06314
43	3.99483	143	13.28513	243	22.57544	343	31.86574	443	41.15605
44	4.08773	144	13.37804	244	22.66834	344	31.95865	444	41.24895
45	4.18064	145	13.47094	245	22.76124	345	32.05155	445	41.34185
46	4.27354	146	13.56384	246	22.85415	346	32.14445	446	41.43476
47	4.36644	147	13.65675	247	22.94705	347	32.23735	447	41.52766
48	4.45935	148	13.74965	248	23.03995	348	32.33026	448	41.62056
49	4.55225	149	13.84255	249	23.13286	349	32.42316	449	41.71346

Continued - SQUARE FEET TO SQUARE METERS 1 SQUARE FOOT = 0.092 903 04 SQUARE METER

SQUARE FEET	SQUARE METERS	SQUARE FEET	SQUARE METERS	SQUARE FEET	SQUARE METERS	SQUARE FEET	SQUARE METERS	SQUARE FEET	SQUARE METERS
50	4.64515	150	13.93546	250	23.22576	350	32.51606	450	41.80637
51	4.73806	151	14.02836	251	23.31866	351	32.60897	451	41.89927
52	4.83096	152	14.12126	252	23.41157	352	32.70187	452	41.99217
53	4.92386	153	14.21417	253	23.50447	353	32.79477	453	42.08508
54	5.01676	154	14.30707	254	23.59737	354	32.88768	454	42.17798
55	5.10967	155	14.39997	255	23.69028	355	32.98058	455	42.27088
56	5.20257	156	14.49287	256	23.78318	356	33.07348	456	42.36379
57	5.29547	157	14.58578	257	23.87608	357	33.16639	457	42.45669
58	5.38838	158	14.67868	258	23.96898	358	33.25929	458	42.54959
59	5.48128	159	14.77158	259	24.06189	359	33.35219	459	42.64250
60	5.57418	160	14.86449	260	24.15479	360	33.44509	460	42.73540
61	5.66709	161	14.95739	261	24.24769	361	33.53800	461	42.82830
62	5.75999	162	15.05029	262	24.34060	362	33.63090	462	42.92120
63	5.85289	163	15.14320	263	24.43350	363	33.72380	463	43.01411
64	5.94579	164	15.23610	264	24.52640	364	33.81671	464	43.10701
65	6.03870	165	15.32900	265	24.61931	365	33.90961	465	43.19991
66	6.13160	166	15.42190	266	24.71221	366	34.00251	466	43.29282
67	6.22450	167	15.51481	267	24.80511	367	34.09542	467	43.38572
68	6.31741	168	15.60771	268	24.89801	368	34.18832	468	43.47862
69	6.41031	169	15.70061	269	24.99092	369	34.28122	469	43.57153
70	6.50321	170	15.79352	270	25.08382	370	34.37412	470	43.66443
71	6.59612	171	15.88642	271	25.17672	371	34.46703	471	43.75733
72	6.68902	172	15.97932	272	25.26963	372	34.55993	472	43.85023
73	6.78192	173	16.07223	273	25.36253	373	34.65283	473	43.94314
74	6.87482	174	16.16513	274	25.45543	374	34.74574	474	44.03604
75	6.96773	175	16.25803	275	25.54834	375	34.83864	475	44.12894
76	7.06063	176	16.35093	276	25.64124	376	34.93154	476	44.22185
77	7.15353	177	16.44384	277	25.73414	377	35.02445	477	44.31475
78	7.24644	178	16.53674	278	25.82705	378	35.11735	478	44.40765
79	7.33934	179	16.62964	279	25.91995	379	35.21025	479	44.50056
80	7.43224	180	16.72255	280	26.01285	380	35.30315	480	44.59346
81	7.52515	181	16.81545	281	26.10575	381	35.39606	481	44.68636
82	7.61805	182	16.90835	282	26.19866	382	35.48896	482	44.77927
83	7.71095	183	17.00126	283	26.29156	383	35.58186	483	44.87217
84	7.80386	184	17.09416	284	26.38446	384	35.67477	484	44.96507
85	7.89676	185	17.18706	285	26.47737	385	35.76767	485	45.05797
86	7.98966	186	17.27997	286	26.57027	386	35.86057	486	45.15088
87	8.08256	187	17.37287	287	26.66317	387	35.95348	487	45.24378
88	8.17547	188	17.46577	288	26.75608	388	36.04638	488	45.33668
89	8.26837	189	17.55867	289	26.84898	389	36.13928	489	45.42959
90	8.36127	190	17.65158	290	26.94188	390	36.23219	490	45.52249
91	8.45418	191	17.74448	291	27.03478	391	36.32509	491	45.61539
92	8.54708	192	17.83738	292	27.12769	392	36.41799	492	45.70830
93	8.63998	193	17.93029	293	27.22059	393	36.51089	493	45.80120
94	8.73289	194	18.02319	294	27.31349	394	36.60380	494	45.89410
95	8.82579	195	18.11609	295	27.40640	395	36.69670	495	45.98700
96	8.91869	196	18.20900	296	27.49930	396	36.78960	496	46.07991
97	9.01159	197	18.30190	297	27.59220	397	36.88251	497	46.17281
98	9.10450	198	18.39480	298	27.68511	398	36.97541	498	46.26571
99	9.19740	199	18.48770	299	27.77801	399	37.06831	499	46.35862

Continued - SQUARE FEET TO SQUARE METERS 1 SQUARE FOCT = 0.092 903 04 SQUARE METER

SQUARE FEET	SQUARE METERS	SQUARE FEET	SQUARE METERS	SQUARE FEET	SQUARE METERS	SQUARE FEET	SQUARE METERS	SQUARE FEET	SQUARE METERS
500	46.45152	600	55.74182	700	65.03213	800	74.32243	900	83.61274
501	46.54442	601	55.83473	701	65.12503	801	74.41533	901	83.70564
502	46.63733	602	55.92763	702	65.21793	802	74.50824	902	83.79854
503	46.73023	603	56.02053	703	65.31084	803	74.60114	903	83.89145
504	46.82313	604	56.11344	704	65.40374	804	74.69404	904	83.98435
505	46.91604	605	56.20634	705	65.49664	805	74.78695	905	84.07725
506	47.00894	606	56.29924	706	65.58955	806	74.87985	906	84.17015
507	47.10184	607	56.39215	707	65.68245	807	74.97275	907	84.26306
508	47.19474	608	56.48505	708	65.77535	808	75.06566	908	84.35596
509	47.28765	609	56.57795	709	65.86825	809	75.15856	909	84.44886
510	47.38055	610	56.67085	710	65.96116	810	75.25146	910	84.54177
511	47.47345	611	56.76376	711	66.05406	811	75.34437	911	84.63467
512	47.56636	612	56.85666	712	66.14696	812	75.43727	912	84.72757
513	47.65926	613	56.94956	713	66.23987	813	75.53017	913	84.82048
514	47.75216	614	57.04247	714	66.33277	814	75.62307	914	84.91338
515	47.84507	615	57.13537	715	66.42567	815	75.71598	915	85.00628
516	47.93797	616	57.22827	716	66.51858	816	75.80888	916	85.09918
517	48.03087	617	57.32118	717	66.61148	817	75.90178	917	85.19209
518	48.12377	618	57.41408	718	66.70438	818	75.99469	918	85.28499
519	48.21668	619	57.50698	719	66.79729	819	76.08759	919	85.37789
520	48.30958	620	57.59988	720	66.89019	820	76.18049	920	85.47080
521	48.40248	621	57.69279	721	66.98309	821	76.27340	921	85.56370
522	48.49539	622	57.78569	722	67.07599	822	76.36630	922	85.65660
523	48.58829	623	57.87859	723	67.16890	823	76.45920	923	85.74951
524	48.68119	624	57.97150	724	67.26180	824	76.55210	924	85.84241
525	48.77410	625	58.06440	725	67.35470	825	76.64501	925	85.93531
526	48.86700	626	58.15730	726	67.44761	826	76.73791	926	86.02821
527	48.95990	627	58.25021	727	67.54051	827	76.83081	927	86.12112
528	49.05280	628	58.34311	728	67.63341	828	76.92372	928	86.21402
529	49.14571	629	58.43601	729	67.72632	829	77.01662	929	86.30692
530	49.23861	630	58.52892	730	67.81922	830	77.10952	930	86.39983
531	49.33151	631	58.62182	731	67.91212	831	77.20243	931	86.49273
532	49.42442	632	58.71472	732	68.00502	832	77.29533	932	86.58563
533	49.51732	633	58.80762	733	68.09793	833	77.38823	933	86.67854
534	49.61022	634	58.90053	734	68.19083	834	77.48114	934	86.77144
535	49.70313	635	58.99343	735	68.28373	835	77.57404	935	86.86434
536	49.79603	636	59.08633	736	68.37664	836	77.66694	936	86.95724
537	49.88893	637	59.17924	737	68.46954	837	77.75984	937	87.05015
538	49.98184	638	59.27214	738	68.56244	838	77.85275	938	87.14305
539	50.07474	639	59.36504	739	68.65535	839	77.94565	939	87.23595
540	50.16764	640	59.45795	740	68.74825	840	78.03855	940	87.32886
541	50.26054	641	59.55085	741	68.84115	841	78.13146	941	87.42176
542	50.35345	642	59.64375	742	68.93406	842	78.22436	942	87.51466
543	50.44635	643	59.73665	743	69.02696	843	78.31726	943	87.60757
544	50.53925	644	59.82956	744	69.11986	844	78.41017	944	87.70047
545	50.63216	645	59.92246	745	69.21276	845	78.50307	945	87.79337
546	50.72506	646	60.01536	746	69.30567	846	78.59597	946	87.88628
547	50.81796	647	60.10827	747	69.39857	847	78.68887	947	87.97918
548	50.91087	648	60.20117	748	69.49147	848	78.78178	948	88.07208
549	51.00377	649	60.29407	749	69.58438	849	78.87468	949	88.16498

Continued - SQUARE FEET TO SQUARE METERS 1 SQUARE FOCT = 0.092 903 04 SQUARE METER

SQUARE FEET	SQUARE METERS	SQUARE FEET	SQUARE METERS	SQUARE FEET	SQUARE METERS	SQUARE FEET	SQUARE METERS	SQUARE FEET	SQUARE METERS
550	51.09667	650	60.38698	750	69.67728	850	78.96758	950	88.25789
551	51.18958	651	60.47988	751	69.77018	851	79.06049	951	88.35079
552	51.28248	652	60.57278	752	69.86309	852	79.15339	952	88.44369
553	51.37538	653	60.66569	753	69.95599	853	79.24629	953	88.53660
554	51.46828	654	60.75859	754	70.04889	854	79.33920	954	88.62950
555	51.56119	655	60.85149	755	70.14180	855	79.43210	955	88.72240
556	51.65409	656	60.94439	756	70.23470	856	79.52500	956	88.81531
557	51.74699	657	61.03730	757	70.32760	857	79.61790	957	88.90821
558	51.83990	658	61.13020	758	70.42050	858	79.71081	958	89.00111
559	51.93280	659	61.22310	759	70.51341	859	79.80371	959	89.09402
560	52.02570	660	61.31601	760	70.60631	860	79.89661	960	89.18692
561	52.11861	661	61.40891	761	70.69921	861	79.98952	961	89.27982
562	52.21151	662	61.50181	762	70.79212	862	80.08242	962	89.37272
563	52.30441	663	61.59472	763	70.88502	863	80.17532	963	89.46563
564	52.39731	664	61.68762	764	70.97792	864	80.26823	964	89.55853
565	52.49022	665	61.78052	765	71.07083	865	80.36113	965	89.65143
566	52.58312	666	61.87342	766	71.16373	866	80.45403	966	89.74434
567	52.67602	667	61.96633	767	71.25663	867	80.54694	967	89.83724
568	52.76893	668	62.05923	768	71.34953	868	80.63984	968	89.93014
569	52.86183	669	62.15213	769	71.44244	869	80.73274	969	90.02305
570	52.95473	670	62.24504	770	71.53534	870	80.82564	970	90.11595
571	53.04764	671	62.33794	771	71.62824	871	80.91855	971	90.20885
572	53.14054	672	62.43084	772	71.72115	872	81.01145	972	90.30175
573	53.23344	673	62.52375	773	71.81405	873	81.10435	973	90.39466
574	53.32634	674	62.61665	774	71.90695	874	81.19726	974	90.48756
575	53.41925	675	62.70955	775	71.99986	875	81.29016	975	90.58046
576	53.51215	676	62.80245	776	72.09276	876	81.38306	976	90.67337
577	53.60505	677	62.89536	777	72.18566	877	81.47597	977	90.76627
578	53.69796	678	62.98826	778	72.27856	878	81.56887	978	90.85917
579	53.79086	679	63.08116	779	72.37147	879	81.66177	979	90.95208
580	53.88376	680	63.17407	780	72.46437	880	81.75467	980	91.04498
581	53.97667	681	63.26697	781	72.55727	881	81.84758	981	91.13788
582	54.06957	682	63.35987	782	72.65018	882	81.94048	982	91.23079
583	54.16247	683	63.45278	783	72.74308	883	82.03338	983	91.32369
584	54.25538	684	63.54568	784	72.83598	884	82.12629	984	91.41659
585	54.34828	685	63.63858	785	72.92889	885	82.21919	985	91.50949
586	54.44118	686	63.73149	786	73.02179	886	82.31209	986	91.60240
587	54.53408	687	63.82439	787	73.11469	887	82.40500	987	91.69530
588	54.62699	688	63.91729	788	73.20759	888	82.49790	988	91.78820
589	54.71989	689	64.01019	789	73.30050	889	82.59080	989	91.88111
590	54.81279	690	64.10310	790	73.39340	890	82.68371	990	91.97401
591	54.90570	691	64.19600	791	73.48630	891	82.77661	991	92.06691
592	54.99860	692	64.28890	792	73.57921	892	82.86951	992	92.15982
593	55.09150	693	64.38181	793	73.67211	893	82.96241	993	92.25272
594	55.18441	694	64.47471	794	73.76501	894	83.05532	994	92.34562
595	55.27731	695	64.56761	795	73.85792	895	83.14822	995	92.43852
596	55.37021	696	64.66052	796	73.95082	896	83.24112	996	92.53143
597	55.46311	697	64.75342	797	74.04372	897	83.33403	997	92.62433
598	55.55602	698	64.84632	798	74.13663	898	83.42693	998	92.71723
599	55.64892	699	64.93923	799	74.22953	899	83.51983	999	92.81014

AREA - SQUARE METERS TO SQUARE FEET 1 SQUARE METER = 10.763 9104 SQUARE FEET

SQUARE METERS	SQUARE FEET	SQUARE METERS	SQUARE FEET	SQUARE METERS	SQUARE FEET	SQUARE METERS	SQUARE FEET	SQUARE METERS	SQUARE FEET
0	0.000	100	1076.391	200	2152.782	300	3229.173	400	4305.564
1	10.764	101	1087.155	201	2163.546	301	3239.937	401	4316.328
2	21.528	102	1097.919	202	2174.310	302	3250.701	402	4327.092
3	32.292	103	1108.683	203	2185.074	303	3261.465	403	4337.856
4	43.056	104	1119.447	204	2195.838	304	3272.229	404	4348.620
5	53.820	105	1130.211	205	2206.602	305	3282.993	405	4359.384
6	64.583	106	1140.975	206	2217.366	306	3293.757	406	4370.148
7	75.347	107	1151.738	207	2228.129	307	3304.520	407	4380.911
8	86.111	108	1162.502	208	2238.893	308	3315.284	408	4391.675
9	96.875	109	1173.266	209	2249.657	309	3326.048	409	4402.439
10	107.639	110	1184.030	210	2260.421	310	3336.812	410	4413.203
11	118.403	111	1194.794	211	2271.185	311	3347.576	411	4423.967
12	129.167	112	1205.558	212	2281.949	312	3358.340	412	4434.731
13	139.931	113	1216.322	213	2292.713	313	3369.104	413	4445.495
14	150.695	114	1227.086	214	2303.477	314	3379.868	414	4456.259
15	161.459	115	1237.850	215	2314.241	315	3390.632	415	4467.023
16	172.223	116	1248.614	216	2325.005	316	3401.396	416	4477.787
17	182.986	117	1259.378	217	2335.769	317	3412.160	417	4488.551
18	193.750	118	1270.141	218	2346.532	318	3422.923	418	4499.315
19	204.514	119	1280.905	219	2357.296	319	3433.687	419	4510.078
20	215.278	120	1291.669	220	2368.060	320	3444.451	420	4520.842
21	226.042	121	1302.433	221	2378.824	321	3455.215	421	4531.606
22	236.806	122	1313.197	222	2389.588	322	3465.979	422	4542.370
23	247.570	123	1323.961	223	2400.352	323	3476.743	423	4553.134
24	258.334	124	1334.725	224	2411.116	324	3487.507	424	4563.898
25	269.098	125	1345.489	225	2421.880	325	3498.271	425	4574.662
26	279.862	126	1356.253	226	2432.644	326	3509.035	426	4585.426
27	290.626	127	1367.017	227	2443.408	327	3519.799	427	4596.190
28	301.389	128	1377.781	228	2454.172	328	3530.563	428	4606.954
29	312.153	129	1388.544	229	2464.935	329	3541.327	429	4617.718
30	322.917	130	1399.308	230	2475.699	330	3552.090	430	4628.481
31	333.681	131	1410.072	231	2486.463	331	3562.854	431	4639.245
32	344.445	132	1420.836	232	2497.227	332	3573.618	432	4650.009
33	355.209	133	1431.600	233	2507.991	333	3584.382	433	4660.773
34	365.973	134	1442.364	234	2518.755	334	3595.146	434	4671.537
35	376.737	135	1453.128	235	2529.519	335	3605.910	435	4682.301
36	387.501	136	1463.892	236	2540.283	336	3616.674	436	4693.065
37	398.265	137	1474.656	237	2551.047	337	3627.438	437	4703.829
38	409.029	138	1485.420	238	2561.811	338	3638.202	438	4714.593
39	419.793	139	1496.184	239	2572.575	339	3648.966	439	4725.357
40	430.556	140	1506.947	240	2583.338	340	3659.730	440	4736.121
41	441.320	141	1517.711	241	2594.102	341	3670.493	441	4746.884
42	452.084	142	1528.475	242	2604.866	342	3681.257	442	4757.648
43	462.848	143	1539.239	243	2615.630	343	3692.021	443	4768.412
44	473.612	144	1550.003	244	2626.394	344	3702.785	444	4779.176
45	484.376	145	1560.767	245	2637.158	345	3713.549	445	4789.940
46	495.140	146	1571.531	246	2647.922	346	3724.313	446	4800.704
47	505.904	147	1582.295	247	2658.686	347	3735.077	447	4811.468
48	516.668	148	1593.059	248	2669.450	348	3745.841	448	4822.232
49	527.432	149	1603.823	249	2680.214	349	3756.605	449	4832.996

Continued - SQUARE METERS TO SQUARE FEET 1 SQUARE METER = 10.763 9104 SQUARE FEET

SQUARE METERS	SQUARE FEET	SQUARE METERS	SQUARE FEET	SQUARE METERS	SQUARE FEET	SQUARE METERS	SQUARE FEET	SQUARE METERS	SQUARE FEET
50	538.196	150	1614.587	250	2690.978	350	3767.369	450	4843.760
51	548.959	151	1625.350	251	2701.741	351	3778.133	451	4854.524
52	559.723	152	1636.114	252	2712.505	352	3788.896	452	4865.287
53	570.487	153	1646.878	253	2723.269	353	3799.660	453	4876.051
54	581.251	154	1657.642	254	2734.033	354	3810.424	454	4886.815
55	592.015	155	1668.406	255	2744.797	355	3821.188	455	4897.579
56	602.779	156	1679.170	256	2755.561	356	3831.952	456	4908.343
57	613.543	157	1689.934	257	2766.325	357	3842.716	457	4919.107
58	624.307	158	1700.698	258	2777.089	358	3853.480	458	4929.871
59	635.071	159	1711.462	259	2787.853	359	3864.244	459	4940.635
60	645.835	160	1722.226	260	2798.617	360	3875.008	460	4951.399
61	656.599	161	1732.990	261	2809.381	361	3885.772	461	4962.163
62	667.362	162	1743.753	262	2820.145	362	3896.536	462	4972.927
63	678.126	163	1754.517	263	2830.908	363	3907.299	463	4983.690
64	688.890	164	1765.281	264	2841.672	364	3918.063	464	4994.454
65	699.654	165	1776.045	265	2852.436	365	3928.827	465	5005.218
66	710.418	166	1786.809	266	2863.200	366	3939.591	466	5015.982
67	721.182	167	1797.573	267	2873.964	367	3950.355	467	5026.746
68	731.946	168	1808.337	268	2884.728	368	3961.119	468	5037.510
69	742.710	169	1819.101	269	2895.492	369	3971.883	469	5048.274
70	753.474	170	1829.865	270	2906.256	370	3982.647	470	5059.038
71	764.238	171	1840.629	271	2917.020	371	3993.411	471	5069.802
72	775.002	172	1851.393	272	2927.784	372	4004.175	472	5080.566
73	785.765	173	1862.156	273	2938.548	373	4014.939	473	5091.330
74	796.529	174	1872.920	274	2949.311	374	4025.702	474	5102.094
75	807.293	175	1883.684	275	2960.075	375	4036.466	475	5112.857
76	818.057	176	1894.448	276	2970.839	376	4047.230	476	5123.621
77	828.821	177	1905.212	277	2981.603	377	4057.994	477	5134.385
78	839.585	178	1915.976	278	2992.367	378	4068.758	478	5145.149
79	850.349	179	1926.740	279	3003.131	379	4079.522	479	5155.913
80	861.113	180	1937.504	280	3013.895	380	4090.286	480	5166.677
81	871.877	181	1948.268	281	3024.659	381	4101.050	481	5177.441
82	882.641	182	1959.032	282	3035.423	382	4111.814	482	5188.205
83	893.405	183	1969.796	283	3046.187	383	4122.578	483	5198.969
84	904.168	184	1980.560	284	3056.951	384	4133.342	484	5209.733
85	914.932	185	1991.323	285	3067.714	385	4144.105	485	5220.497
86	925.696	186	2002.087	286	3078.478	386	4154.869	486	5231.260
87	936.460	187	2012.851	287	3089.242	387	4165.633	487	5242.024
88	947.224	188	2023.615	288	3100.006	388	4176.397	488	5252.788
89	957.988	189	2034.379	289	3110.770	389	4187.161	489	5263.552
90	968.752	190	2045.143	290	3121.534	390	4197.925	490	5274.316
91	979.516	191	2055.907	291	3132.298	391	4208.689	491	5285.080
92	990.280	192	2066.671	292	3143.062	392	4219.453	492	5295.844
93	1001.044	193	2077.435	293	3153.826	393	4230.217	493	5306.608
94	1011.808	194	2088.199	294	3164.590	394	4240.981	494	5317.372
95	1022.571	195	2098.963	295	3175.354	395	4251.745	495	5328.136
96	1033.335	196	2109.726	296	3186.117	396	4262.508	496	5338.900
97	1044.099	197	2120.490	297	3196.881	397	4273.272	497	5349.663
98	1054.863	198	2131.254	298	3207.645	398	4284.036	498	5360.427
99	1065.627	199	2142.018	299	3218.409	399	4294.800	499	5371.191

Continued - SQUARE METERS TO SQUARE FEET 1 SQUARE METER = 10.763 9104 SQUARE FEET

SQUARE METERS	SQUARE FEET	SQUARE METERS	SQUARE FEET	SQUARE METERS	SQUARE FEET	SQUARE METERS	SQUARE FEET	SQUARE METERS	SQUARE FEET
500	5381.955	600	6458.346	700	7534.737	800	8611.128	900	9687.519
501	5392.719	601	6469.110	701	7545.501	801	8621.892	901	9698.283
502	5403.483	602	6479.874	702	7556.265	802	8632.656	902	9709.047
503	5414.247	603	6490.638	703	7567.029	803	8643.420	903	9719.811
504	5425.011	604	6501.402	704	7577.793	804	8654.184	904	9730.575
505	5435.775	605	6512.166	705	7588.557	805	8664.948	905	9741.339
506	5446.539	606	6522.930	706	7599.321	806	8675.712	906	9752.103
507	5457.303	607	6533.694	707	7610.085	807	8686.476	907	9762.867
508	5468.066	608	6544.458	708	7620.849	808	8697.240	908	9773.631
509	5478.830	609	6555.221	709	7631.612	809	8708.003	909	9784.395
510	5489.594	610	6565.985	710	7642.376	810	8718.767	910	9795.158
511	5500.358	611	6576.749	711	7653.140	811	8729.531	911	9805.922
512	5511.122	612	6587.513	712	7663.904	812	8740.295	912	9816.686
513	5521.886	613	6598.277	713	7674.668	813	8751.059	913	9827.450
514	5532.650	614	6609.041	714	7685.432	814	8761.823	914	9838.214
515	5543.414	615	6619.805	715	7696.196	815	8772.587	915	9848.978
516	5554.178	616	6630.569	716	7706.960	816	8783.351	916	9859.742
517	5564.942	617	6641.333	717	7717.724	817	8794.115	917	9870.506
518	5575.706	618	6652.097	718	7728.488	818	8804.879	918	9881.270
519	5586.469	619	6662.861	719	7739.252	819	8815.643	919	9892.034
520	5597.233	620	6673.624	720	7750.015	820	8826.406	920	9902.797
521	5607.997	621	6684.388	721	7760.779	821	8837.170	921	9913.561
522	5618.761	622	6695.152	722	7771.543	822	8847.934	922	9924.325
523	5629.525	623	6705.916	723	7782.307	823	8858.698	923	9935.089
524	5640.289	624	6716.680	724	7793.071	824	8869.462	924	9945.853
525	5651.053	625	6727.444	725	7803.835	825	8880.226	925	9956.617
526	5661.817	626	6738.208	726	7814.599	826	8890.990	926	9967.381
527	5672.581	627	6748.972	727	7825.363	827	8901.754	927	9978.145
528	5683.345	628	6759.736	728	7836.127	828	8912.518	928	9988.909
529	5694.109	629	6770.500	729	7846.891	829	8923.282	929	9999.673
530	5704.872	630	6781.264	730	7857.655	830	8934.046	930	10010.437
531	5715.636	631	6792.027	731	7868.418	831	8944.809	931	10021.201
532	5726.400	632	6802.791	732	7879.182	832	8955.573	932	10031.964
533	5737.164	633	6813.555	733	7889.946	833	8966.337	933	10042.728
534	5747.928	634	6824.319	734	7900.710	834	8977.101	934	10053.492
535	5758.692	635	6835.083	735	7911.474	835	8987.865	935	10064.256
536	5769.456	636	6845.847	736	7922.238	836	8998.629	936	10075.020
537	5780.220	637	6856.611	737	7933.002	837	9009.393	937	10085.784
538	5790.984	638	6867.375	738	7943.766	838	9020.157	938	10096.548
539	5801.748	639	6878.139	739	7954.530	839	9030.921	939	10107.312
540	5812.512	640	6888.903	740	7965.294	840	9041.685	940	10118.076
541	5823.276	641	6899.667	741	7976.058	841	9052.449	941	10128.840
542	5834.039	642	6910.430	742	7986.821	842	9063.213	942	10139.604
543	5844.803	643	6921.194	743	7997.585	843	9073.976	943	10150.367
544	5855.567	644	6931.958	744	8008.349	844	9084.740	944	10161.131
545	5866.331	645	6942.722	745	8019.113	845	9095.504	945	10171.895
546	5877.095	646	6953.486	746	8029.877	846	9106.268	946	10182.659
547	5887.859	647	6964.250	747	8040.641	847	9117.032	947	10193.423
548	5898.623	648	6975.014	748	8051.405	848	9127.796	948	10204.187
549	5909.387	649	6985.778	749	8062.169	849	9138.560	949	10214.951

Continued - SQUARE METERS TO SQUARE FEET 1 SQUARE METER = 10.763 9104 SQUARE FEET

SQUARE METERS	SQUARE FEET	SQUARE METERS	SQUARE FEET	SQUARE METERS	SQUARE FEET	SQUARE METERS	SQUARE FEET	SQUARE METERS	SQUARE FEET
550	5920.151	650	6996.542	750	8072.933	850	9149.324	950	10225.715
551	5930.915	651	7007.306	751	8083.697	851	9160.088	951	10236.479
552	5941.679	652	7018.070	752	8094.461	852	9170.852	952	10247.243
553	5952.442	653	7028.833	753	8105.224	853	9181.615	953	10258.007
554	5963.206	654	7039.597	754	8115.988	854	9192.379	954	10268.771
555	5973.970	655	7050.361	755	8126.752	855	9203.143	955	10279.534
556	5984.734	656	7061.125	756	8137.516	856	9213.907	956	10290.298
557	5995.498	657	7071.889	757	8148.280	857	9224.671	957	10301.062
558	6006.262	658	7082.653	758	8159.044	858	9235.435	958	10311.826
559	6017.026	659	7093.417	759	8169.808	859	9246.199	959	10322.590
560	6027.790	660	7104.181	760	8180.572	860	9256.963	960	10333.354
561	6038.554	661	7114.945	761	8191.336	861	9267.727	961	10344.118
562	6049.318	662	7125.709	762	8202.100	862	9278.491	962	10354.882
563	6060.082	663	7136.473	763	8212.864	863	9289.255	963	10365.646
564	6070.845	664	7147.237	764	8223.627	864	9300.019	964	10376.410
565	6081.609	665	7158.000	765	8234.391	865	9310.782	965	10387.173
566	6092.373	666	7168.764	766	8245.155	866	9321.546	966	10397.937
567	6103.137	667	7179.528	767	8255.919	867	9332.310	967	10408.701
568	6113.901	668	7190.292	768	8266.683	868	9343.074	968	10419.465
569	6124.665	669	7201.056	769	8277.447	869	9353.838	969	10430.229
570	6135.429	670	7211.820	770	8288.211	870	9364.602	970	10440.993
571	6146.193	671	7222.584	771	8298.975	871	9375.366	971	10451.757
572	6156.957	672	7233.348	772	8309.739	872	9386.130	972	10462.521
573	6167.721	673	7244.112	773	8320.503	873	9396.894	973	10473.285
574	6178.485	674	7254.876	774	8331.267	874	9407.658	974	10484.049
575	6189.248	675	7265.640	775	8342.031	875	9418.422	975	10494.813
576	6200.012	676	7276.403	776	8352.794	876	9429.185	976	10505.577
577	6210.776	677	7287.167	777	8363.558	877	9439.949	977	10516.340
578	6221.540	678	7297.931	778	8374.322	878	9450.713	978	10527.104
579	6232.304	679	7308.695	779	8385.086	879	9461.477	979	10537.868
580	6243.068	680	7319.459	780	8395.850	880	9472.241	980	10548.632
581	6253.832	681	7330.223	781	8406.614	881	9483.005	981	10559.396
582	6264.596	682	7340.987	782	8417.378	882	9493.769	982	10570.160
583	6275.360	683	7351.751	783	8428.142	883	9504.533	983	10580.924
584	6286.124	684	7362.515	784	8438.906	884	9515.297	984	10591.688
585	6296.888	685	7373.279	785	8449.670	885	9526.061	985	10602.452
586	6307.651	686	7384.043	786	8460.433	886	9536.825	986	10613.216
587	6318.415	687	7394.806	787	8471.197	887	9547.589	987	10623.979
588	6329.179	688	7405.570	788	8481.961	888	9558.352	988	10634.743
589	6339.943	689	7416.334	789	8492.725	889	9569.116	989	10645.507
590	6350.707	690	7427.098	790	8503.489	890	9579.880	990	10656.271
591	6361.471	691	7437.862	791	8514.253	891	9590.644	991	10667.035
592	6372.235	692	7448.626	792	8525.017	892	9601.408	992	10677.799
593	6382.999	693	7459.390	793	8535.781	893	9612.172	993	10688.563
594	6393.763	694	7470.154	794	8546.545	894	9622.936	994	10699.327
595	6404.527	695	7480.918	795	8557.309	895	9633.700	995	10710.091
596	6415.291	696	7491.682	796	8568.073	896	9644.464	996	10720.855
597	6426.055	697	7502.446	797	8578.837	897	9655.228	997	10731.619
598	6436.818	698	7513.209	798	8589.600	898	9665.991	998	10742.383
599	6447.582	699	7523.973	799	8600.364	899	9676.755	999	10753.146

AREA - SQUARE YARDS TO SQUARE METERS 1 SQUARE YARD = 0.836 127 36 SQUARE METER

SQUARE YARDS	SQUARE METERS	SQUARE YARDS	SQUARE METERS	SQUARE YARDS	SQUARE METERS	SQUARE YARDS	SQUARE METERS	SQUARE YARDS	SQUARE METERS
0	0.00000	100	83.61274	200	167.22547	300	250.83821	400	334.45094
1	0.83613	101	84.44886	201	168.06160	301	251.67434	401	335.28707
2	1.67225	102	85.28499	202	168.89773	302	252.51046	402	336.12320
3	2.50838	103	86.12112	203	169.73385	303	253.34659	403	336.95932
4	3.34451	104	86.95724	204	170.56998	304	254.18272	404	337.79545
5	4.18064	105	87.79337	205	171.40611	305	255.01884	405	338.63158
6	5.01676	106	88.62950	206	172.24224	306	255.85497	406	339.46771
7	5.85289	107	89.46563	207	173.07836	307	256.69110	407	340.30383
8	6.68902	108	90.30175	208	173.91449	308	257.52723	408	341.13996
9	7.52515	109	91.13788	209	174.75062	309	258.36335	409	341.97609
10	8.36127	110	91.97401	210	175.58675	310	259.19948	410	342.81222
11	9.19740	111	92.81014	211	176.42287	311	260.03561	411	343.64835
12	10.03353	112	93.64626	212	177.25900	312	260.87173	412	344.48447
13	10.86966	113	94.48239	213	178.09513	313	261.70786	413	345.32060
14	11.70578	114	95.31852	214	178.93126	314	262.54399	414	346.15673
15	12.54191	115	96.15465	215	179.76738	315	263.38012	415	346.99286
16	13.37804	116	96.99077	216	180.60351	316	264.21624	416	347.82898
17	14.21417	117	97.82690	217	181.43964	317	265.05237	417	348.66511
18	15.05029	118	98.66303	218	182.27576	318	265.88850	418	349.50124
19	15.88642	119	99.49916	219	183.11189	319	266.72463	419	350.33736
20	16.72255	120	100.33528	220	183.94802	320	267.56075	420	351.17349
21	17.55867	121	101.17141	221	184.78415	321	268.39688	421	352.00962
22	18.39480	122	102.00754	222	185.62027	322	269.23301	422	352.84575
23	19.23093	123	102.84367	223	186.45640	323	270.06914	423	353.68187
24	20.06706	124	103.67979	224	187.29253	324	270.90526	424	354.51800
25	20.90318	125	104.51592	225	188.12866	325	271.74139	425	355.35413
26	21.73931	126	105.35205	226	188.96478	326	272.57752	426	356.19025
27	22.57544	127	106.18817	227	189.80091	327	273.41365	427	357.02638
28	23.41157	128	107.02430	228	190.63704	328	274.24977	428	357.86251
29	24.24769	129	107.86043	229	191.47317	329	275.08590	429	358.69864
30	25.08382	130	108.69656	230	192.30929	330	275.92203	430	359.53476
31	25.91995	131	109.53268	231	193.14542	331	276.75816	431	360.37089
32	26.75608	132	110.36881	232	193.98155	332	277.59428	432	361.20702
33	27.59220	133	111.20494	233	194.81767	333	278.43041	433	362.04315
34	28.42833	134	112.04107	234	195.65380	334	279.26654	434	362.87927
35	29.26446	135	112.87719	235	196.48993	335	280.10266	435	363.71540
36	30.10058	136	113.71332	236	197.32606	336	280.93879	436	364.55153
37	30.93671	137	114.54945	237	198.16218	337	281.77492	437	365.38766
38	31.77284	138	115.38558	238	198.99831	338	282.61105	438	366.22378
39	32.60897	139	116.22170	239	199.83444	339	283.44717	439	367.05991
40	33.44509	140	117.05783	240	200.67057	340	284.28330	440	367.89604
41	34.28122	141	117.89396	241	201.50669	341	285.11943	441	368.73217
42	35.11735	142	118.73009	242	202.34282	342	285.95555	442	369.56829
43	35.95348	143	119.56621	243	203.17895	343	286.79168	443	370.40442
44	36.78960	144	120.40234	244	204.01508	344	287.62781	444	371.24055
45	37.62573	145	121.23847	245	204.85120	345	288.46394	445	372.07668
46	38.46186	146	122.07459	246	205.68733	346	289.30006	446	372.91280
47	39.29799	147	122.91072	247	206.52346	347	290.13619	447	373.74893
48	40.13411	148	123.74685	248	207.35958	348	290.97232	448	374.58506
49	40.97024	149	124.58298	249	208.19571	349	291.80845	449	375.42118

Continued - SQUARE YARDS TO SQUARE METERS 1 SQUARE YARD = 0.836 127 36 SQUARE METER

SQUARE YARDS	SQUARE METERS	SQUARE YARDS	SQUARE METERS	SQUARE YARDS	SQUARE METERS	SQUARE YARDS	SQUARE METERS	SQUARE YARDS	SQUARE METERS
50	41.80637	150	125.41910	250	209.03184	350	292.64458	450	376.25731
51	42.64250	151	126.25523	251	209.86797	351	293.48070	451	377.09344
52	43.47862	152	127.09136	252	210.70409	352	294.31683	452	377.92957
53	44.31475	153	127.92749	253	211.54022	353	295.15296	453	378.76569
54	45.15088	154	128.76361	254	212.37635	354	295.98909	454	379.60182
55	45.98700	155	129.59974	255	213.21248	355	296.82521	455	380.43795
56	46.82313	156	130.43587	256	214.04860	356	297.66134	456	381.27407
57	47.65926	157	131.27200	257	214.88473	357	298.49747	457	382.11020
58	48.49539	158	132.10812	258	215.72086	358	299.33360	458	382.94633
59	49.33151	159	132.94425	259	216.55699	359	300.16972	459	383.78246
60	50.16764	160	133.78038	260	217.39311	360	301.00585	460	384.61858
61	51.00377	161	134.61650	261	218.22924	361	301.84198	461	385.45471
62	51.83990	162	135.45263	262	219.06537	362	302.67810	462	386.29084
63	52.67602	163	136.28876	263	219.90149	363	303.51423	463	387.12697
64	53.51215	164	137.12489	264	220.73762	364	304.35036	464	387.96309
65	54.34828	165	137.96101	265	221.57375	365	305.18649	465	388.79922
66	55.18441	166	138.79714	266	222.40988	366	306.02261	466	389.63535
67	56.02053	167	139.63327	267	223.24600	367	306.85874	467	390.47148
68	56.85666	168	140.46940	268	224.08213	368	307.69487	468	391.30761
69	57.69279	169	141.30552	269	224.91826	369	308.53099	469	392.14373
70	58.52892	170	142.14165	270	225.75439	370	309.36712	470	392.97986
71	59.36504	171	142.97778	271	226.59052	371	310.20325	471	393.81599
72	60.20117	172	143.81391	272	227.42664	372	311.03938	472	394.65211
73	61.03730	173	144.65003	273	228.26277	373	311.87550	473	395.48824
74	61.87342	174	145.48616	274	229.09890	374	312.71163	474	396.32437
75	62.70955	175	146.32229	275	229.93502	375	313.54776	475	397.16050
76	63.54568	176	147.15841	276	230.77115	376	314.38389	476	397.99662
77	64.38181	177	147.99454	277	231.60728	377	315.22001	477	398.83275
78	65.21793	178	148.83067	278	232.44341	378	316.05614	478	399.66888
79	66.05406	179	149.66680	279	233.27953	379	316.89227	479	400.50500
80	66.89019	180	150.50292	280	234.11566	380	317.72840	480	401.34113
81	67.72632	181	151.33905	281	234.95179	381	318.56452	481	402.17726
82	68.56244	182	152.17518	282	235.78792	382	319.40065	482	403.01339
83	69.39857	183	153.01131	283	236.62404	383	320.23678	483	403.84951
84	70.23470	184	153.84743	284	237.46017	384	321.07291	484	404.68564
85	71.07083	185	154.68356	285	238.29630	385	321.90903	485	405.52177
86	71.90695	186	155.51969	286	239.13243	386	322.74516	486	406.35789
87	72.74308	187	156.35582	287	239.96855	387	323.58129	487	407.19402
88	73.57921	188	157.19194	288	240.80468	388	324.41742	488	408.03015
89	74.41533	189	158.02807	289	241.64081	389	325.25354	489	408.86628
90	75.25146	190	158.86420	290	242.47693	390	326.08967	490	409.70241
91	76.08759	191	159.70033	291	243.31306	391	326.92580	491	410.53853
92	76.92372	192	160.53645	292	244.14919	392	327.76192	492	411.37466
93	77.75984	193	161.37258	293	244.98532	393	328.59805	493	412.21079
94	78.59597	194	162.20871	294	245.82144	394	329.43418	494	413.04692
95	79.43210	195	163.04483	295	246.65757	395	330.27031	495	413.88304
96	80.26823	196	163.88096	296	247.49370	396	331.10643	496	414.71917
97	81.10435	197	164.71709	297	248.32983	397	331.94256	497	415.55530
98	81.94048	198	165.55322	298	249.16595	398	332.77869	498	416.39143
99	82.77661	199	166.38935	299	250.00208	399	333.61481	499	417.22755

Continued – SQUARE YARDS TO SQUARE METERS 1 SQUARE YARD = 0.836 127 36 SQUARE METER

SQUARE YARDS	SQUARE METERS	SQUARE YARDS	SQUARE METERS	SQUARE YARDS	SQUARE METERS	SQUARE YARDS	SQUARE METERS	SQUARE YARDS	SQUARE METERS
500	418.06368	600	501.67641	700	585.28915	800	668.90189	900	752.51463
501	418.89981	601	502.51254	701	586.12527	801	669.73801	901	753.35075
502	419.73594	602	503.34867	702	586.96140	802	670.57414	902	754.18687
503	420.57206	603	504.18480	703	587.79753	803	671.41027	903	755.02300
504	421.40819	604	505.02092	704	588.63366	804	672.24640	904	755.85913
505	422.24432	605	505.85705	705	589.46979	805	673.08253	905	756.69526
506	423.08044	606	506.69318	706	590.30592	806	673.91865	906	757.53139
507	423.91657	607	507.52931	707	591.14204	807	674.75478	907	758.36752
508	424.75270	608	508.36544	708	591.97817	808	675.59090	908	759.20364
509	425.58883	609	509.20156	709	592.81429	809	676.42703	909	760.03977
510	426.42495	610	510.03769	710	593.65042	810	677.26316	910	760.87590
511	427.26108	611	510.87382	711	594.48655	811	678.09929	911	761.71202
512	428.09721	612	511.70995	712	595.32268	812	678.93542	912	762.54815
513	428.93333	613	512.54607	713	596.15881	813	679.77155	913	763.38428
514	429.76946	614	513.38219	714	596.99493	814	680.60767	914	764.22041
515	430.60559	615	514.21832	715	597.83106	815	681.44379	915	765.05653
516	431.44172	616	515.05445	716	598.66719	816	682.27992	916	765.89266
517	432.27784	617	515.89058	717	599.50332	817	683.11605	917	766.72879
518	433.11397	618	516.72671	718	600.33944	818	683.95218	918	767.56492
519	433.95010	619	517.56284	719	601.17557	819	684.78831	919	768.40104
520	434.78623	620	518.39896	720	602.01170	820	685.62444	920	769.23717
521	435.62235	621	519.23509	721	602.84782	821	686.46056	921	770.07330
522	436.45848	622	520.07121	722	603.68395	822	687.29669	922	770.90942
523	437.29461	623	520.90734	723	604.52008	823	688.13281	923	771.74555
524	438.13074	624	521.74347	724	605.35621	824	688.96894	924	772.58168
525	438.96687	625	522.57960	725	606.19234	825	689.80507	925	773.41781
526	439.80299	626	523.41573	726	607.02847	826	690.64120	926	774.25394
527	440.63912	627	524.25185	727	607.86459	827	691.47733	927	775.09007
528	441.47525	628	525.08798	728	608.70071	828	692.31345	928	775.92619
529	442.31137	629	525.92411	729	609.53684	829	693.14958	929	776.76231
530	443.14750	630	526.76024	730	610.37297	830	693.98571	930	777.59844
531	443.98363	631	527.59636	731	611.20910	831	694.82184	931	778.43457
532	444.81976	632	528.43249	732	612.04523	832	695.65796	932	779.27070
533	445.65588	633	529.26862	733	612.88136	833	696.49409	933	780.10683
534	446.49201	634	530.10474	734	613.71748	834	697.33022	934	780.94296
535	447.32814	635	530.94087	735	614.55361	835	698.16634	935	781.77908
536	448.16426	636	531.77700	736	615.38973	836	699.00247	936	782.61521
537	449.00039	637	532.61313	737	616.22586	837	699.83860	937	783.45133
538	449.83652	638	533.44926	738	617.06199	838	700.67473	938	784.28746
539	450.67265	639	534.28538	739	617.89812	839	701.51086	939	785.12359
540	451.50877	640	535.12151	740	618.73425	840	702.34698	940	785.95972
541	452.34490	641	535.95763	741	619.57037	841	703.18311	941	786.79585
542	453.18103	642	536.79376	742	620.40650	842	704.01923	942	787.63197
543	454.01715	643	537.62989	743	621.24263	843	704.85536	943	788.46810
544	454.85328	644	538.46602	744	622.07875	844	705.69149	944	789.30423
545	455.68941	645	539.30215	745	622.91488	845	706.52762	945	790.14036
546	456.52554	646	540.13828	746	623.75101	846	707.36375	946	790.97648
547	457.36167	647	540.97440	747	624.58714	847	708.19987	947	791.81261
548	458.19779	648	541.81052	748	625.42326	848	709.03600	948	792.64874
549	459.03392	649	542.64665	749	626.25939	849	709.87212	949	793.48486

Continued - SQUARE YARDS TO SQUARE METERS 1 SQUARE YARD = 0.836 127 36 SQUARE METER

SQUARE YARDS	SQUARE METERS	SQUARE YARDS	SQUARE METERS	SQUARE YARDS	SQUARE METERS	SQUARE YARDS	SQUARE METERS	SQUARE YARDS	SQUARE METERS
550	459.87005	650	543.48278	750	627.09552	850	710.70825	950	794.32099
551	460.70618	651	544.31891	751	627.93165	851	711.54438	951	795.15712
552	461.54230	652	545.15504	752	628.76778	852	712.38051	952	795.99325
553	462.37843	653	545.99117	753	629.60390	853	713.21664	953	796.82938
554	463.21456	654	546.82729	754	630.44003	854	714.05276	954	797.66550
555	464.05069	655	547.66342	755	631.27615	855	714.88889	955	798.50163
556	464.88681	656	548.49955	756	632.11228	856	715.72502	956	799.33775
557	465.72294	657	549.33567	757	632.94841	857	716.56115	957	800.17388
558	466.55907	658	550.17180	758	633.78454	858	717.39727	958	801.01001
559	467.39520	659	551.00793	759	634.62067	859	718.23340	959	801.84614
560	468.23132	660	551.84406	760	635.45679	860	719.06953	960	802.68227
561	469.06745	661	552.68018	761	636.29292	861	719.90565	961	803.51839
562	469.90358	662	553.51631	762	637.12904	862	720.74178	962	804.35452
563	470.73970	663	554.35244	763	637.96517	863	721.57791	963	805.19064
564	471.57583	664	555.18857	764	638.80130	864	722.41404	964	806.02677
565	472.41196	665	556.02470	765	639.63743	865	723.25017	965	806.86290
566	473.24809	666	556.86082	766	640.47356	866	724.08630	966	807.69903
567	474.08421	667	557.69695	767	641.30968	867	724.92242	967	808.53516
568	474.92034	668	558.53307	768	642.14581	868	725.75854	968	809.37128
569	475.75647	669	559.36920	769	642.98194	869	726.59467	969	810.20741
570	476.59259	670	560.20533	770	643.81807	870	727.43080	970	811.04354
571	477.42872	671	561.04146	771	644.65419	871	728.26693	971	811.87967
572	478.26485	672	561.87759	772	645.49032	872	729.10306	972	812.71579
573	479.10098	673	562.71371	773	646.32645	873	729.93919	973	813.55192
574	479.93710	674	563.54984	774	647.16257	874	730.77531	974	814.38805
575	480.77323	675	564.38596	775	647.99870	875	731.61144	975	815.22417
576	481.60936	676	565.22209	776	648.83483	876	732.44756	976	816.06030
577	482.44549	677	566.05822	777	649.67096	877	733.28369	977	816.89643
578	483.28161	678	566.89435	778	650.50709	878	734.11982	978	817.73256
579	484.11774	679	567.73048	779	651.34321	879	734.95595	979	818.56869
580	484.95387	680	568.56660	780	652.17934	880	735.79208	980	819.40482
581	485.79000	681	569.40273	781	653.01546	881	736.62820	981	820.24094
582	486.62613	682	570.23886	782	653.85159	882	737.46433	982	821.07706
583	487.46225	683	571.07498	783	654.68772	883	738.30046	983	821.91319
584	488.29838	684	571.91111	784	655.52385	884	739.13658	984	822.74932
585	489.13451	685	572.74724	785	656.35998	885	739.97271	985	823.58545
586	489.97063	686	573.58337	786	657.19611	886	740.80884	986	824.42158
587	490.80676	687	574.41949	787	658.03223	887	741.64497	987	825.25771
588	491.64289	688	575.25562	788	658.86835	888	742.48109	988	826.09383
589	492.47902	689	576.09175	789	659.70448	889	743.31722	989	826.92995
590	493.31514	690	576.92788	790	660.54061	890	744.15335	990	827.76608
591	494.15127	691	577.76401	791	661.37674	891	744.98948	991	828.60221
592	494.98740	692	578.60013	792	662.21287	892	745.82561	992	829.43834
593	495.82352	693	579.43626	793	663.04900	893	746.66173	993	830.27447
594	496.65965	694	580.27238	794	663.88512	894	747.49786	994	831.11060
595	497.49578	695	581.10851	795	664.72125	895	748.33398	995	831.94672
596	498.33191	696	581.94464	796	665.55738	896	749.17011	996	832.78285
597	499.16803	697	582.78077	797	666.39350	897	750.00624	997	833.61898
598	500.00416	698	583.61690	798	667.22963	898	750.84237	998	834.45510
599	500.84029	699	584.45303	799	668.06576	899	751.67850	999	835.29123

AREA - SQUARE METERS TO SQUARE YARDS 1 SQUARE METER = 1.195 990 046 SQUARE YARDS

SQUARE METERS	SQUARE YARDS	SQUARE METERS	SQUARE YARDS	SQUARE METERS	SQUARE YARDS	SQUARE METERS	SQUARE YARDS	SQUARE METERS	SQUARE YARDS
0	0.0000	100	119.5990	200	239.1980	300	358.7970	400	478.3960
1	1.1960	101	120.7950	201	240.3940	301	359.9930	401	479.5920
2	2.3920	102	121.9910	202	241.5900	302	361.1890	402	480.7880
3	3.5880	103	123.1870	203	242.7860	303	362.3850	403	481.9840
4	4.7840	104	124.3830	204	243.9820	304	363.5810	404	483.1800
5	5.9800	105	125.5790	205	245.1780	305	364.7770	405	484.3760
6	7.1759	106	126.7749	206	246.3739	306	365.9729	406	485.5720
7	8.3719	107	127.9709	207	247.5699	307	367.1689	407	486.7679
8	9.5679	108	129.1669	208	248.7659	308	368.3649	408	487.9639
9	10.7639	109	130.3629	209	249.9619	309	369.5609	409	489.1599
10	11.9599	110	131.5589	210	251.1579	310	370.7569	410	490.3559
11	13.1559	111	132.7549	211	252.3539	311	371.9529	411	491.5519
12	14.3519	112	133.9509	212	253.5499	312	373.1489	412	492.7479
13	15.5479	113	135.1469	213	254.7459	313	374.3449	413	493.9439
14	16.7439	114	136.3429	214	255.9419	314	375.5409	414	495.1399
15	17.9399	115	137.5389	215	257.1379	315	376.7369	415	496.3359
16	19.1358	116	138.7348	216	258.3338	316	377.9328	416	497.5319
17	20.3318	117	139.9308	217	259.5298	317	379.1288	417	498.7278
18	21.5278	118	141.1268	218	260.7258	318	380.3248	418	499.9238
19	22.7238	119	142.3228	219	261.9218	319	381.5208	419	501.1198
20	23.9198	120	143.5188	220	263.1178	320	382.7168	420	502.3158
21	25.1158	121	144.7148	221	264.3138	321	383.9128	421	503.5118
22	26.3118	122	145.9108	222	265.5098	322	385.1088	422	504.7078
23	27.5078	123	147.1068	223	266.7058	323	386.3048	423	505.9038
24	28.7038	124	148.3028	224	267.9018	324	387.5008	424	507.0998
25	29.8998	125	149.4988	225	269.0978	325	388.6968	425	508.2958
26	31.0957	126	150.6947	226	270.2937	326	389.8927	426	509.4918
27	32.2917	127	151.8907	227	271.4897	327	391.0887	427	510.6877
28	33.4877	128	153.0867	228	272.6857	328	392.2847	428	511.8837
29	34.6837	129	154.2827	229	273.8817	329	393.4807	429	513.0797
30	35.8797	130	155.4787	230	275.0777	330	394.6767	430	514.2757
31	37.0757	131	156.6747	231	276.2737	331	395.8727	431	515.4717
32	38.2717	132	157.8707	232	277.4697	332	397.0687	432	516.6677
33	39.4677	133	159.0667	233	278.6657	333	398.2647	433	517.8637
34	40.6637	134	160.2627	234	279.8617	334	399.4607	434	519.0597
35	41.8597	135	161.4587	235	281.0577	335	400.6567	435	520.2557
36	43.0556	136	162.6546	236	282.2536	336	401.8527	436	521.4517
37	44.2516	137	163.8506	237	283.4496	337	403.0486	437	522.6476
38	45.4476	138	165.0466	238	284.6456	338	404.2446	438	523.8436
39	46.6436	139	166.2426	239	285.8416	339	405.4406	439	525.0396
40	47.8396	140	167.4386	240	287.0376	340	406.6366	440	526.2356
41	49.0356	141	168.6346	241	288.2336	341	407.8326	441	527.4316
42	50.2316	142	169.8306	242	289.4296	342	409.0286	442	528.6276
43	51.4276	143	171.0266	243	290.6256	343	410.2246	443	529.8236
44	52.6236	144	172.2226	244	291.8216	344	411.4206	444	531.0196
45	53.8196	145	173.4186	245	293.0176	345	412.6166	445	532.2156
46	55.0155	146	174.6145	246	294.2135	346	413.8126	446	533.4116
47	56.2115	147	175.8105	247	295.4095	347	415.0085	447	534.6075
48	57.4075	148	177.0065	248	296.6055	348	416.2045	448	535.8035
49	58.6035	149	178.2025	249	297.8015	349	417.4005	449	536.9995

Continued - SQUARE METERS TO SQUARE YARDS 1 SQUARE METER = 1.195 990 046 SQUARE YARDS

SQUARE METERS	SQUARE YARDS	SQUARE METERS	SQUARE YARDS	SQUARE METERS	SQUARE YARDS	SQUARE METERS	SQUARE YARDS	SQUARE METERS	SQUARE YARDS
50	59.7995	150	179.3985	250	298.9975	350	418.5965	450	538.1955
51	60.9955	151	180.5945	251	300.1935	351	419.7925	451	539.3915
52	62.1915	152	181.7905	252	301.3895	352	420.9885	452	540.5875
53	63.3875	153	182.9865	253	302.5855	353	422.1845	453	541.7835
54	64.5835	154	184.1825	254	303.7815	354	423.3805	454	542.9795
55	65.7795	155	185.3785	255	304.9775	355	424.5765	455	544.1755
56	66.9754	156	186.5744	256	306.1735	356	425.7725	456	545.3715
57	68.1714	157	187.7704	257	307.3694	357	426.9684	457	546.5674
58	69.3674	158	188.9664	258	308.5654	358	428.1644	458	547.7634
59	70.5634	159	190.1624	259	309.7614	359	429.3604	459	548.9594
60	71.7594	160	191.3584	260	310.9574	360	430.5564	460	550.1554
61	72.9554	161	192.5544	261	312.1534	361	431.7524	461	551.3514
62	74.1514	162	193.7504	262	313.3494	362	432.9484	462	552.5474
63	75.3474	163	194.9464	263	314.5454	363	434.1444	463	553.7434
64	76.5434	164	196.1424	264	315.7414	364	435.3404	464	554.9394
65	77.7394	165	197.3384	265	316.9374	365	436.5364	465	556.1354
66	78.9353	166	198.5343	266	318.1334	366	437.7324	466	557.3314
67	80.1313	167	199.7303	267	319.3293	367	438.9283	467	558.5273
68	81.3273	168	200.9263	268	320.5253	368	440.1243	468	559.7233
69	82.5233	169	202.1223	269	321.7213	369	441.3203	469	560.9193
70	83.7193	170	203.3183	270	322.9173	370	442.5163	470	562.1153
71	84.9153	171	204.5143	271	324.1133	371	443.7123	471	563.3113
72	86.1113	172	205.7103	272	325.3093	372	444.9083	472	564.5073
73	87.3073	173	206.9063	273	326.5053	373	446.1043	473	565.7033
74	88.5033	174	208.1023	274	327.7013	374	447.3003	474	566.8993
75	89.6993	175	209.2983	275	328.8973	375	448.4963	475	568.0953
76	90.8952	176	210.4942	276	330.0933	376	449.6923	476	569.2913
77	92.0912	177	211.6902	277	331.2892	377	450.8882	477	570.4872
78	93.2872	178	212.8862	278	332.4852	378	452.0842	478	571.6832
79	94.4832	179	214.0822	279	333.6812	379	453.2802	479	572.8792
80	95.6792	180	215.2782	280	334.8772	380	454.4762	480	574.0752
81	96.8752	181	216.4742	281	336.0732	381	455.6722	481	575.2712
82	98.0712	182	217.6702	282	337.2692	382	456.8682	482	576.4672
83	99.2672	183	218.8662	283	338.4652	383	458.0642	483	577.6632
84	100.4632	184	220.0622	284	339.6612	384	459.2602	484	578.8592
85	101.6592	185	221.2582	285	340.8572	385	460.4562	485	580.0552
86	102.8551	186	222.4541	286	342.0532	386	461.6522	486	581.2512
87	104.0511	187	223.6501	287	343.2491	387	462.8481	487	582.4471
88	105.2471	188	224.8461	288	344.4451	388	464.0441	488	583.6431
89	106.4431	189	226.0421	289	345.6411	389	465.2401	489	584.8391
90	107.6391	190	227.2381	290	346.8371	390	466.4361	490	586.0351
91	108.8351	191	228.4341	291	348.0331	391	467.6321	491	587.2311
92	110.0311	192	229.6301	292	349.2291	392	468.8281	492	588.4271
93	111.2271	193	230.8261	293	350.4251	393	470.0241	493	589.6231
94	112.4231	194	232.0221	294	351.6211	394	471.2201	494	590.8191
95	113.6191	195	233.2181	295	352.8171	395	472.4161	495	592.0151
96	114.8150	196	234.4140	296	354.0131	396	473.6121	496	593.2111
97	116.0110	197	235.6100	297	355.2090	397	474.8080	497	594.4070
98	117.2070	198	236.8060	298	356.4050	398	476.0040	498	595.6030
99	118.4030	199	238.0020	299	357.6010	399	477.2000	499	596.7990

Continued - SQUARE METERS TO SQUARE YARDS 1 SQUARE METER = 1.195 990 046 SQUARE YARDS

SQUARE METERS	SQUARE YARDS	SQUARE METERS	SQUARE YARDS	SQUARE METERS	SQUARE YARDS	SQUARE METERS	SQUARE YARDS	SQUARE METERS	SQUARE YARDS
500	597.9950	600	717.5940	700	837.1930	800	956.7920	900	1076.3910
501	599.1910	601	718.7900	701	838.3890	801	957.9880	901	1077.5870
502	600.3870	602	719.9860	702	839.5850	802	959.1840	902	1078.7830
503	601.5830	603	721.1820	703	840.7810	803	960.3800	903	1079.9790
504	602.7790	604	722.3780	704	841.9770	804	961.5760	904	1081.1750
505	603.9750	605	723.5740	705	843.1730	805	962.7720	905	1082.3710
506	605.1710	606	724.7700	706	844.3690	806	963.9680	906	1083.5670
507	606.3669	607	725.9660	707	845.5650	807	965.1640	907	1084.7630
508	607.5629	608	727.1619	708	846.7609	808	966.3599	908	1085.9590
509	608.7589	609	728.3579	709	847.9569	809	967.5559	909	1087.1549
510	609.9549	610	729.5539	710	849.1529	810	968.7519	910	1088.3509
511	611.1509	611	730.7499	711	850.3489	811	969.9479	911	1089.5469
512	612.3469	612	731.9459	712	851.5449	812	971.1439	912	1090.7429
513	613.5429	613	733.1419	713	852.7409	813	972.3399	913	1091.9389
514	614.7389	614	734.3379	714	853.9369	814	973.5359	914	1093.1349
515	615.9349	615	735.5339	715	855.1329	815	974.7319	915	1094.3309
516	617.1309	616	736.7299	716	856.3289	816	975.9279	916	1095.5269
517	618.3269	617	737.9258	717	857.5249	817	977.1239	917	1096.7229
518	619.5228	618	739.1218	718	858.7208	818	978.3198	918	1097.9189
519	620.7188	619	740.3178	719	859.9168	819	979.5158	919	1099.1148
520	621.9148	620	741.5138	720	861.1128	820	980.7118	920	1100.3108
521	623.1108	621	742.7098	721	862.3088	821	981.9078	921	1101.5068
522	624.3068	622	743.9058	722	863.5048	822	983.1038	922	1102.7028
523	625.5028	623	745.1018	723	864.7008	823	984.2998	923	1103.8988
524	626.6988	624	746.2978	724	865.8968	824	985.4958	924	1105.0948
525	627.8948	625	747.4938	725	867.0928	825	986.6918	925	1106.2908
526	629.0908	626	748.6898	726	868.2888	826	987.8878	926	1107.4868
527	630.2868	627	749.8857	727	869.4848	827	989.0838	927	1108.6828
528	631.4827	628	751.0817	728	870.6807	828	990.2797	928	1109.8788
529	632.6787	629	752.2777	729	871.8767	829	991.4757	929	1111.0747
530	633.8747	630	753.4737	730	873.0727	830	992.6717	930	1112.2707
531	635.0707	631	754.6697	731	874.2687	831	993.8677	931	1113.4667
532	636.2667	632	755.8657	732	875.4647	832	995.0637	932	1114.6627
533	637.4627	633	757.0617	733	876.6607	833	996.2597	933	1115.8587
534	638.6587	634	758.2577	734	877.8567	834	997.4557	934	1117.0547
535	639.8547	635	759.4537	735	879.0527	835	998.6517	935	1118.2507
536	641.0507	636	760.6497	736	880.2487	836	999.8477	936	1119.4467
537	642.2467	637	761.8456	737	881.4447	837	1001.0437	937	1120.6427
538	643.4426	638	763.0416	738	882.6406	838	1002.2396	938	1121.8387
539	644.6386	639	764.2376	739	883.8366	839	1003.4356	939	1123.0346
540	645.8346	640	765.4336	740	885.0326	840	1004.6316	940	1124.2306
541	647.0306	641	766.6296	741	886.2286	841	1005.8276	941	1125.4266
542	648.2266	642	767.8256	742	887.4246	842	1007.0236	942	1126.6226
543	649.4226	643	769.0216	743	888.6206	843	1008.2196	943	1127.8186
544	650.6186	644	770.2176	744	889.8166	844	1009.4156	944	1129.0146
545	651.8146	645	771.4136	745	891.0126	845	1010.6116	945	1130.2106
546	653.0106	646	772.6096	746	892.2086	846	1011.8076	946	1131.4066
547	654.2066	647	773.8055	747	893.4046	847	1013.0036	947	1132.6026
548	655.4025	648	775.0015	748	894.6005	848	1014.1996	948	1133.7986
549	656.5985	649	776.1975	749	895.7965	849	1015.3955	949	1134.9945

Continued - SQUARE METERS TO SQUARE YARDS 1 SQUARE METER = 1.195 990 046 SQUARE YARDS

SQUARE METERS	SQUARE YARDS	SQUARE METERS	SQUARE YARDS	SQUARE METERS	SQUARE YARDS	SQUARE METERS	SQUARE YARDS	SQUARE METERS	SQUARE YARDS
550	657.7945	650	777.3935	750	896.9925	850	1016.5915	950	1136.1905
551	658.9905	651	778.5895	751	898.1885	851	1017.7875	951	1137.3865
552	660.1865	652	779.7855	752	899.3845	852	1018.9835	952	1138.5825
553	661.3825	653	780.9815	753	900.5805	853	1020.1795	953	1139.7785
554	662.5785	654	782.1775	754	901.7765	854	1021.3755	954	1140.9745
555	663.7745	655	783.3735	755	902.9725	855	1022.5715	955	1142.1705
556	664.9705	656	784.5695	756	904.1685	856	1023.7675	956	1143.3665
557	666.1665	657	785.7654	757	905.3645	857	1024.9635	957	1144.5625
558	667.3624	658	786.9614	758	906.5604	858	1026.1595	958	1145.7585
559	668.5584	659	788.1574	759	907.7564	859	1027.3554	959	1146.9544
560	669.7544	660	789.3534	760	908.9524	860	1028.5514	960	1148.1504
561	670.9504	661	790.5494	761	910.1484	861	1029.7474	961	1149.3464
562	672.1464	662	791.7454	762	911.3444	862	1030.9434	962	1150.5424
563	673.3424	663	792.9414	763	912.5404	863	1032.1394	963	1151.7384
564	674.5384	664	794.1374	764	913.7364	864	1033.3354	964	1152.9344
565	675.7344	665	795.3334	765	914.9324	865	1034.5314	965	1154.1304
566	676.9304	666	796.5294	766	916.1284	866	1035.7274	966	1155.3264
567	678.1264	667	797.7254	767	917.3244	867	1036.9234	967	1156.5224
568	679.3223	668	798.9213	768	918.5203	868	1038.1194	968	1157.7184
569	680.5183	669	800.1173	769	919.7163	869	1039.3153	969	1158.9143
570	681.7143	670	801.3133	770	920.9123	870	1040.5113	970	1160.1103
571	682.9103	671	802.5093	771	922.1083	871	1041.7073	971	1161.3063
572	684.1063	672	803.7053	772	923.3043	872	1042.9033	972	1162.5023
573	685.3023	673	804.9013	773	924.5003	873	1044.0993	973	1163.6983
574	686.4983	674	806.0973	774	925.6963	874	1045.2953	974	1164.8943
575	687.6943	675	807.2933	775	926.8923	875	1046.4913	975	1166.0903
576	688.8903	676	808.4893	776	928.0883	876	1047.6873	976	1167.2863
577	690.0863	677	809.6853	777	929.2843	877	1048.8833	977	1168.4823
578	691.2822	678	810.8812	778	930.4802	878	1050.0793	978	1169.6783
579	692.4782	679	812.0772	779	931.6762	879	1051.2752	979	1170.8742
580	693.6742	680	813.2732	780	932.8722	880	1052.4712	980	1172.0702
581	694.8702	681	814.4692	781	934.0682	881	1053.6672	981	1173.2662
582	696.0662	682	815.6652	782	935.2642	882	1054.8632	982	1174.4622
583	697.2622	683	816.8612	783	936.4602	883	1056.0592	983	1175.6582
584	698.4582	684	818.0572	784	937.6562	884	1057.2552	984	1176.8542
585	699.6542	685	819.2532	785	938.8522	885	1058.4512	985	1178.0502
586	700.8502	686	820.4492	786	940.0482	886	1059.6472	986	1179.2462
587	702.0462	687	821.6452	787	941.2442	887	1060.8432	987	1180.4422
588	703.2421	688	822.8411	788	942.4401	888	1062.0392	988	1181.6382
589	704.4381	689	824.0371	789	943.6361	889	1063.2351	989	1182.8341
590	705.6341	690	825.2331	790	944.8321	890	1064.4311	990	1184.0301
591	706.8301	691	826.4291	791	946.0281	891	1065.6271	991	1185.2261
592	708.0261	692	827.6251	792	947.2241	892	1066.8231	992	1186.4221
593	709.2221	693	828.8211	793	948.4201	893	1068.0191	993	1187.6181
594	710.4181	694	830.0171	794	949.6161	894	1069.2151	994	1188.8141
595	711.6141	695	831.2131	795	950.8121	895	1070.4111	995	1190.0101
596	712.8101	696	832.4091	796	952.0081	896	1071.6071	996	1191.2061
597	714.0061	697	833.6051	797	953.2041	897	1072.8031	997	1192.4021
598	715.2020	698	834.8010	798	954.4000	898	1073.9991	998	1193.5981
599	716.3980	699	835.9970	799	955.5960	899	1075.1950	999	1194.7940

AREA - ACRES TO HECTARES 1 ACRE = 0.404 685 642 24 HECTARE

ACRES	HECTARES	ACRES	HECTARES	ACRES	HECTARES	ACRES	HECTARES	ACRES	HECTARES
0	0.00000	100	40.46856	200	80.93713	300	121.40569	400	161.87426
1	0.40469	101	40.87325	201	81.34181	301	121.81038	401	162.27894
2	0.80937	102	41.27794	202	81.74650	302	122.21506	402	162.68363
3	1.21406	103	41.68262	203	82.15119	303	122.61975	403	163.08831
4	1.61874	104	42.08731	204	82.55587	304	123.02444	404	163.49300
5	2.02343	105	42.49199	205	82.96056	305	123.42912	405	163.89768
6	2.42811	106	42.89668	206	83.36524	306	123.83381	406	164.30237
7	2.83280	107	43.30136	207	83.76993	307	124.23849	407	164.70706
8	3.23749	108	43.70605	208	84.17461	308	124.64318	408	165.11174
9	3.64217	109	44.11073	209	84.57930	309	125.04786	409	165.51643
10	4.04686	110	44.51542	210	84.98398	310	125.45255	410	165.92111
11	4.45154	111	44.92011	211	85.38867	311	125.85723	411	166.32580
12	4.85623	112	45.32479	212	85.79336	312	126.26192	412	166.73048
13	5.26091	113	45.72948	213	86.19804	313	126.66661	413	167.13517
14	5.66560	114	46.13416	214	86.60273	314	127.07129	414	167.53986
15	6.07028	115	46.53885	215	87.00741	315	127.47598	415	167.94454
16	6.47497	116	46.94353	216	87.41210	316	127.88066	416	168.34923
17	6.87966	117	47.34822	217	87.81678	317	128.28535	417	168.75391
18	7.28434	118	47.75291	218	88.22147	318	128.69003	418	169.15860
19	7.68903	119	48.15759	219	88.62615	319	129.09472	419	169.56328
20	8.09371	120	48.56228	220	89.03084	320	129.49940	420	169.96797
21	8.49840	121	48.96696	221	89.43553	321	129.90409	421	170.37265
22	8.90308	122	49.37165	222	89.84021	322	130.30878	422	170.77734
23	9.30777	123	49.77633	223	90.24490	323	130.71346	423	171.18203
24	9.71246	124	50.18102	224	90.64958	324	131.11815	424	171.58671
25	10.11714	125	50.58571	225	91.05427	325	131.52283	425	171.99140
26	10.52183	126	50.99039	226	91.45895	326	131.92752	426	172.39608
27	10.92651	127	51.39508	227	91.86364	327	132.33220	427	172.80077
28	11.33120	128	51.79976	228	92.26833	328	132.73689	428	173.20545
29	11.73588	129	52.20445	229	92.67301	329	133.14157	429	173.61014
30	12.14057	130	52.60913	230	93.07770	330	133.54626	430	174.01483
31	12.54525	131	53.01382	231	93.48238	331	133.95095	431	174.41951
32	12.94994	132	53.41850	232	93.88707	332	134.35563	432	174.82420
33	13.35463	133	53.82319	233	94.29175	333	134.76032	433	175.22888
34	13.75931	134	54.22788	234	94.69644	334	135.16500	434	175.63357
35	14.16400	135	54.63256	235	95.10113	335	135.56969	435	176.03825
36	14.56868	136	55.03725	236	95.50581	336	135.97437	436	176.44294
37	14.97337	137	55.44193	237	95.91050	337	136.37906	437	176.84763
38	15.37805	138	55.84662	238	96.31518	338	136.78375	438	177.25231
39	15.78274	139	56.25130	239	96.71987	339	137.18843	439	177.65700
40	16.18743	140	56.65599	240	97.12455	340	137.59312	440	178.06168
41	16.59211	141	57.06068	241	97.52924	341	137.99780	441	178.46637
42	16.99680	142	57.46536	242	97.93392	342	138.40249	442	178.87105
43	17.40148	143	57.87005	243	98.33861	343	138.80717	443	179.27574
44	17.80617	144	58.27473	244	98.74330	344	139.21186	444	179.68042
45	18.21085	145	58.67942	245	99.14798	345	139.61655	445	180.08511
46	18.61554	146	59.08410	246	99.55267	346	140.02123	446	180.48980
47	19.02023	147	59.48879	247	99.95735	347	140.42592	447	180.89448
48	19.42491	148	59.89348	248	100.36204	348	140.83060	448	181.29917
49	19.82960	149	60.29816	249	100.76672	349	141.23529	449	181.70385

Continued - ACRES TO HECTARES 1 ACRE = 0.404 685 642 24 HECTARE

ACRES	HECTARES	ACRES	HECTARES	ACRES	HECTARES	ACRES	HECTARES	ACRES	HECTARES
50	20.23428	150	60.70285	250	101.17141	350	141.63997	450	182.10854
51	20.63897	151	61.10753	251	101.57610	351	142.04466	451	182.51322
52	21.04365	152	61.51222	252	101.98078	352	142.44934	452	182.91791
53	21.44834	153	61.91690	253	102.38547	353	142.85403	453	183.32260
54	21.85302	154	62.32159	254	102.79015	354	143.25872	454	183.72728
55	22.25771	155	62.72627	255	103.19484	355	143.66340	455	184.13197
56	22.66240	156	63.13096	256	103.59952	356	144.06809	456	184.53665
57	23.06708	157	63.53565	257	104.00421	357	144.47277	457	184.94134
58	23.47177	158	63.94033	258	104.40890	358	144.87746	458	185.34602
59	23.87645	159	64.34502	259	104.81358	359	145.28214	459	185.75071
60	24.28114	160	64.74970	260	105.21827	360	145.68683	460	186.15540
61	24.68582	161	65.15439	261	105.62295	361	146.09152	461	186.56008
62	25.09051	162	65.55907	262	106.02764	362	146.49620	462	186.96477
63	25.49520	163	65.96376	263	106.43232	363	146.90089	463	187.36945
64	25.89988	164	66.36844	264	106.83701	364	147.30557	464	187.77414
65	26.30457	165	66.77313	265	107.24169	365	147.71026	465	188.17882
66	26.70925	166	67.17782	266	107.64638	366	148.11494	466	188.58351
67	27.11394	167	67.58250	267	108.05107	367	148.51963	467	188.98819
68	27.51862	168	67.98719	268	108.45575	368	148.92432	468	189.39288
69	27.92331	169	68.39187	269	108.86044	369	149.32900	469	189.79757
70	28.32799	170	68.79656	270	109.26512	370	149.73369	470	190.20225
71	28.73268	171	69.20124	271	109.66981	371	150.13837	471	190.60694
72	29.13737	172	69.60593	272	110.07449	372	150.54306	472	191.01162
73	29.54205	173	70.01062	273	110.47918	373	150.94774	473	191.41631
74	29.94674	174	70.41530	274	110.88387	374	151.35243	474	191.82099
75	30.35142	175	70.81999	275	111.28855	375	151.75711	475	192.22568
76	30.75611	176	71.22467	276	111.69324	376	152.16180	476	192.63037
77	31.16079	177	71.62936	277	112.09792	377	152.56649	477	193.03505
78	31.56548	178	72.03404	278	112.50261	378	152.97117	478	193.43974
79	31.97017	179	72.43873	279	112.90729	379	153.37586	479	193.84442
80	32.37485	180	72.84342	280	113.31198	380	153.78054	480	194.24911
81	32.77954	181	73.24810	281	113.71667	381	154.18523	481	194.65379
82	33.18422	182	73.65279	282	114.12135	382	154.58991	482	195.05848
83	33.58891	183	74.05747	283	114.52604	383	154.99460	483	195.46317
84	33.99359	184	74.46216	284	114.93072	384	155.39929	484	195.86785
85	34.39828	185	74.86684	285	115.33541	385	155.80397	485	196.27254
86	34.80297	186	75.27153	286	115.74009	386	156.20866	486	196.67722
87	35.20765	187	75.67621	287	116.14478	387	156.61334	487	197.08191
88	35.61234	188	76.08090	288	116.54946	388	157.01803	488	197.48659
89	36.01702	189	76.48559	289	116.95415	389	157.42271	489	197.89128
90	36.42171	190	76.89027	290	117.35884	390	157.82740	490	198.29596
91	36.82639	191	77.29496	291	117.76352	391	158.23209	491	198.70065
92	37.23108	192	77.69964	292	118.16821	392	158.63677	492	199.10534
93	37.63576	193	78.10433	293	118.57289	393	159.04146	493	199.51002
94	38.04045	194	78.50901	294	118.97758	394	159.44614	494	199.91471
95	38.44514	195	78.91370	295	119.38226	395	159.85083	495	200.31939
96	38.84982	196	79.31839	296	119.78695	396	160.25551	496	200.72408
97	39.25451	197	79.72307	297	120.19164	397	160.66020	497	201.12876
98	39.65919	198	80.12776	298	120.59632	398	161.06488	498	201.53345
99	40.06388	199	80.53244	299	121.00101	399	161.46957	499	201.93814

Continued - ACRES TO HECTARES 1 ACRE = 0.404 685 642 24 HECTARE

ACRES	HECTARES	ACRES	HECTARES	ACRES	HECTARES	ACRES	HECTARES	ACRES	HECTARES
500	202.34282	600	242.81138	700	283.27995	800	323.74851	900	364.21708
501	202.74751	601	243.21607	701	283.68464	801	324.15320	901	364.62176
502	203.15219	602	243.62076	702	284.08932	802	324.55788	902	365.02645
503	203.55688	603	244.02544	703	284.49400	803	324.96257	903	365.43113
504	203.96156	604	244.43013	704	284.89869	804	325.36726	904	365.83582
505	204.36625	605	244.83481	705	285.30338	805	325.77194	905	366.24051
506	204.77094	606	245.23950	706	285.70806	806	326.17662	906	366.64519
507	205.17562	607	245.64418	707	286.11275	807	326.58131	907	367.04988
508	205.58031	608	246.04887	708	286.51743	808	326.98600	908	367.45456
509	205.98499	609	246.45356	709	286.92212	809	327.39068	909	367.85925
510	206.38968	610	246.85824	710	287.32681	810	327.79537	910	368.26393
511	206.79436	611	247.26293	711	287.73149	811	328.20005	911	368.66862
512	207.19905	612	247.66761	712	288.13618	812	328.60474	912	369.07330
513	207.60373	613	248.07230	713	288.54086	813	329.00943	913	369.47799
514	208.00842	614	248.47698	714	288.94555	814	329.41411	914	369.88268
515	208.41311	615	248.88167	715	289.35023	815	329.81880	915	370.28736
516	208.81779	616	249.28635	716	289.75492	816	330.22348	916	370.69205
517	209.22248	617	249.69104	717	290.15960	817	330.62817	917	371.09673
518	209.62716	618	250.09573	718	290.56429	818	331.03285	918	371.50142
519	210.03185	619	250.50041	719	290.96898	819	331.43754	919	371.90611
520	210.43653	620	250.90510	720	291.37366	820	331.84222	920	372.31079
521	210.84122	621	251.30978	721	291.77835	821	332.24691	921	372.71547
522	211.24590	622	251.71447	722	292.18303	822	332.65160	922	373.12016
523	211.65059	623	252.11915	723	292.58772	823	333.05628	923	373.52485
524	212.05527	624	252.52384	724	292.99240	824	333.46097	924	373.92953
525	212.45996	625	252.92853	725	293.39709	825	333.86565	925	374.33422
526	212.86465	626	253.33321	726	293.80177	826	334.27034	926	374.73890
527	213.26933	627	253.73790	727	294.20646	827	334.67503	927	375.14359
528	213.67402	628	254.14258	728	294.61115	828	335.07971	928	375.54827
529	214.07870	629	254.54727	729	295.01583	829	335.48439	929	375.95296
530	214.48339	630	254.95195	730	295.42052	830	335.88908	930	376.35765
531	214.88807	631	255.35664	731	295.82520	831	336.29377	931	376.76233
532	215.29276	632	255.76133	732	296.22989	832	336.69845	932	377.16702
533	215.69745	633	256.16601	733	296.63457	833	337.10314	933	377.57170
534	216.10213	634	256.57069	734	297.03926	834	337.50782	934	377.97639
535	216.50682	635	256.97538	735	297.44395	835	337.91251	935	378.38107
536	216.91150	636	257.38007	736	297.84863	836	338.31720	936	378.78576
537	217.31619	637	257.78475	737	298.25331	837	338.72188	937	379.19044
538	217.72087	638	258.18944	738	298.65800	838	339.12657	938	379.59513
539	218.12556	639	258.59412	739	299.06269	839	339.53125	939	379.99982
540	218.53025	640	258.99881	740	299.46737	840	339.93594	940	380.40450
541	218.93493	641	259.40350	741	299.87206	841	340.34062	941	380.80919
542	219.33962	642	259.80818	742	300.27674	842	340.74531	942	381.21387
543	219.74430	643	260.21287	743	300.68143	843	341.14999	943	381.61856
544	220.14899	644	260.61755	744	301.08612	844	341.55468	944	382.02324
545	220.55367	645	261.02224	745	301.49080	845	341.95937	945	382.42793
546	220.95836	646	261.42692	746	301.89549	846	342.36405	946	382.83261
547	221.36304	647	261.83161	747	302.30017	847	342.76874	947	383.23730
548	221.76773	648	262.23629	748	302.70486	848	343.17342	948	383.64199
549	222.17242	649	262.64098	749	303.10954	849	343.57811	949	384.04667

Continued - ACRES TO HECTARES 1 ACRE = 0.404 685 642 24 HECTARE

ACRES	HECTARES	ACRES	HECTARES	ACRES	HECTARES	ACRES	HECTARES	ACRES	HECTARES
550	222.57710	650	263.04567	750	303.51423	850	343.98280	950	384.45136
551	222.98179	651	263.45035	751	303.91891	851	344.38748	951	384.85604
552	223.38647	652	263.85504	752	304.32360	852	344.79216	952	385.26073
553	223.79116	653	264.25972	753	304.72829	853	345.19685	953	385.66542
554	224.19584	654	264.66441	754	305.13297	854	345.60154	954	386.07010
555	224.60053	655	265.06910	755	305.53766	855	346.00622	955	386.47478
556	225.00522	656	265.47378	756	305.94234	856	346.41091	956	386.87947
557	225.40990	657	265.87846	757	306.34703	857	346.81559	957	387.28416
558	225.81459	658	266.28315	758	306.75172	858	347.22028	958	387.68884
559	226.21927	659	266.68784	759	307.15640	859	347.62497	959	388.09353
560	226.62396	660	267.09252	760	307.56108	860	348.02965	960	388.49821
561	227.02864	661	267.49721	761	307.96577	861	348.43434	961	388.90290
562	227.43333	662	267.90189	762	308.37046	862	348.83902	962	389.30759
563	227.83802	663	268.30658	763	308.77514	863	349.24371	963	389.71227
564	228.24270	664	268.71127	764	309.17983	864	349.64839	964	390.11696
565	228.64739	665	269.11595	765	309.58451	865	350.05308	965	390.52164
566	229.05207	666	269.52064	766	309.98920	866	350.45776	966	390.92633
567	229.45676	667	269.92532	767	310.39389	867	350.86245	967	391.33101
568	229.86144	668	270.33001	768	310.79857	868	351.26714	968	391.73570
569	230.26613	669	270.73469	769	311.20326	869	351.67182	969	392.14038
570	230.67081	670	271.13938	770	311.60794	870	352.07651	970	392.54507
571	231.07550	671	271.54406	771	312.01263	871	352.48119	971	392.94976
572	231.48019	672	271.94875	772	312.41731	872	352.88588	972	393.35444
573	231.88487	673	272.35344	773	312.82200	873	353.29057	973	393.75913
574	232.28956	674	272.75812	774	313.22668	874	353.69525	974	394.16381
575	232.69424	675	273.16281	775	313.63137	875	354.09993	975	394.56850
576	233.09893	676	273.56749	776	314.03606	876	354.50462	976	394.97319
577	233.50361	677	273.97218	777	314.44074	877	354.90931	977	395.37787
578	233.90830	678	274.37687	778	314.84543	878	355.31399	978	395.78255
579	234.31299	679	274.78155	779	315.25011	879	355.71868	979	396.18724
580	234.71767	680	275.18623	780	315.65480	880	356.12336	980	396.59193
581	235.12236	681	275.59092	781	316.05949	881	356.52805	981	396.99661
582	235.52704	682	275.99561	782	316.46417	882	356.93274	982	397.40130
583	235.93173	683	276.40029	783	316.86885	883	357.33742	983	397.80598
584	236.33641	684	276.80498	784	317.27354	884	357.74211	984	398.21067
585	236.74110	685	277.20966	785	317.67823	885	358.14679	985	398.61536
586	237.14579	686	277.61435	786	318.08291	886	358.55148	986	399.02004
587	237.55047	687	278.01904	787	318.48760	887	358.95616	987	399.42473
588	237.95516	688	278.42372	788	318.89228	888	359.36085	988	399.82941
589	238.35984	689	278.82841	789	319.29697	889	359.76553	989	400.23410
590	238.76453	690	279.23309	790	319.70166	890	360.17022	990	400.63878
591	239.16921	691	279.63778	791	320.10634	891	360.57491	991	401.04347
592	239.57390	692	280.04246	792	320.51103	892	360.97959	992	401.44815
593	239.97858	693	280.44715	793	320.91571	893	361.38428	993	401.85284
594	240.38327	694	280.85183	794	321.32040	894	361.78896	994	402.25753
595	240.78796	695	281.25652	795	321.72508	895	362.19365	995	402.66221
596	241.19264	696	281.66121	796	322.12977	896	362.59834	996	403.06690
597	241.59733	697	282.06589	797	322.53445	897	363.00302	997	403.47158
598	242.00201	698	282.47058	798	322.93914	898	363.40770	998	403.87627
599	242.40670	699	282.87526	799	323.34383	899	363.81239	999	404.28096

AREA - HECTARES TO ACRES 1 HECTARE = 2.471 0538 ACRES

HECTARES	ACRES	HECTARES	ACRES	HECTARES	ACRES	HECTARES	ACRES	HECTARES	ACRES
0	0.0000	100	247.1054	200	494.2108	300	741.3161	400	988.4215
1	2.4711	101	249.5764	201	496.6818	301	743.7872	401	990.8926
2	4.9421	102	252.0475	202	499.1529	302	746.2582	402	993.3636
3	7.4132	103	254.5185	203	501.6239	303	748.7293	403	995.8347
4	9.8842	104	256.9896	204	504.0950	304	751.2004	404	998.3057
5	12.3553	105	259.4606	205	506.5660	305	753.6714	405	1000.7768
6	14.8263	106	261.9317	206	509.0371	306	756.1425	406	1003.2478
7	17.2974	107	264.4028	207	511.5081	307	758.6135	407	1005.7189
8	19.7684	108	266.8738	208	513.9792	308	761.0846	408	1008.1899
9	22.2395	109	269.3449	209	516.4502	309	763.5556	409	1010.6610
10	24.7105	110	271.8159	210	518.9213	310	766.0267	410	1013.1321
11	27.1816	111	274.2870	211	521.3923	311	768.4977	411	1015.6031
12	29.6526	112	276.7580	212	523.8634	312	770.9688	412	1018.0742
13	32.1237	113	279.2291	213	526.3345	313	773.4398	413	1020.5452
14	34.5948	114	281.7001	214	528.8055	314	775.9109	414	1023.0163
15	37.0658	115	284.1712	215	531.2766	315	778.3819	415	1025.4873
16	39.5369	116	286.6422	216	533.7476	316	780.8530	416	1027.9584
17	42.0079	117	289.1133	217	536.2187	317	783.3241	417	1030.4294
18	44.4790	118	291.5843	218	538.6897	318	785.7951	418	1032.9005
19	46.9500	119	294.0554	219	541.1608	319	788.2662	419	1035.3715
20	49.4211	120	296.5265	220	543.6318	320	790.7372	420	1037.8426
21	51.8921	121	298.9975	221	546.1029	321	793.2083	421	1040.3136
22	54.3632	122	301.4686	222	548.5739	322	795.6793	422	1042.7847
23	56.8342	123	303.9396	223	551.0450	323	798.1504	423	1045.2558
24	59.3053	124	306.4107	224	553.5161	324	800.6214	424	1047.7268
25	61.7763	125	308.8817	225	555.9871	325	803.0925	425	1050.1979
26	64.2474	126	311.3528	226	558.4582	326	805.5635	426	1052.6689
27	66.7185	127	313.8238	227	560.9292	327	808.0346	427	1055.1400
28	69.1895	128	316.2949	228	563.4003	328	810.5056	428	1057.6110
29	71.6606	129	318.7659	229	565.8713	329	812.9767	429	1060.0821
30	74.1316	130	321.2370	230	568.3424	330	815.4478	430	1062.5531
31	76.6027	131	323.7080	231	570.8134	331	817.9188	431	1065.0242
32	79.0737	132	326.1791	232	573.2845	332	820.3899	432	1067.4952
33	81.5448	133	328.6502	233	575.7555	333	822.8609	433	1069.9663
34	84.0158	134	331.1212	234	578.2266	334	825.3320	434	1072.4373
35	86.4869	135	333.5923	235	580.6976	335	827.8030	435	1074.9084
36	88.9579	136	336.0633	236	583.1687	336	830.2741	436	1077.3795
37	91.4290	137	338.5344	237	585.6397	337	832.7451	437	1079.8505
38	93.9000	138	341.0054	238	588.1108	338	835.2162	438	1082.3216
39	96.3711	139	343.4765	239	590.5819	339	837.6872	439	1084.7926
40	98.8422	140	345.9475	240	593.0529	340	840.1583	440	1087.2637
41	101.3132	141	348.4186	241	595.5240	341	842.6293	441	1089.7347
42	103.7843	142	350.8896	242	597.9950	342	845.1004	442	1092.2058
43	106.2553	143	353.3607	243	600.4661	343	847.5714	443	1094.6768
44	108.7264	144	355.8317	244	602.9371	344	850.0425	444	1097.1479
45	111.1974	145	358.3028	245	605.4082	345	852.5136	445	1099.6189
46	113.6685	146	360.7739	246	607.8792	346	854.9846	446	1102.0900
47	116.1395	147	363.2449	247	610.3503	347	857.4557	447	1104.5611
48	118.6106	148	365.7160	248	612.8213	348	859.9267	448	1107.0321
49	121.0816	149	368.1870	249	615.2924	349	862.3978	449	1109.5032

Continued - HECTARES TO ACRES 1 HECTARE = 2.471 0538 ACRES .

HECTARES	ACRES	HECTARES	ACRES	HECTARES	ACRES	HECTARES	ACRES	HECTARES	ACRES
50	123.5527	150	370.6581	250	617.7635	350	864.8688	450	1111.9742
51	126.0237	151	373.1291	251	620.2345	351	867.3399	451	1114.4453
52	128.4948	152	375.6002	252	622.7056	352	869.8109	452	1116.9163
53	130.9659	153	378.0712	253	625.1766	353	872.2820	453	1119.3874
54	133.4369	154	380.5423	254	627.6477	354	874.7530	454	1121.8584
55	135.9080	155	383.0133	255	630.1187	355	877.2241	455	1124.3295
56	138.3790	156	385.4844	256	632.5898	356	879.6952	456	1126.8005
57	140.8501	157	387.9554	257	635.0608	357	882.1662	457	1129.2716
58	143.3211	158	390.4265	258	637.5319	358	884.6373	458	1131.7426
59	145.7922	159	392.8976	259	640.0029	359	887.1083	459	1134.2137
60	148.2632	160	395.3686	260	642.4740	360	889.5794	460	1136.6847
61	150.7343	161	397.8397	261	644.9450	361	892.0504	461	1139.1558
62	153.2053	162	400.3107	262	647.4161	362	894.5215	462	1141.6268
63	155.6764	163	402.7818	263	649.8871	363	896.9925	463	1144.0979
64	158.1474	164	405.2528	264	652.3582	364	899.4636	464	1146.5690
65	160.6185	165	407.7239	265	654.8293	365	901.9346	465	1149.0400
66	163.0896	166	410.1949	266	657.3003	366	904.4057	466	1151.5111
67	165.5606	167	412.6660	267	659.7714	367	906.8767	467	1153.9821
68	168.0317	168	415.1370	268	662.2424	368	909.3478	468	1156.4532
69	170.5027	169	417.6081	269	664.7135	369	911.8189	469	1158.9242
70	172.9738	170	420.0791	270	667.1845	370	914.2899	470	1161.3953
71	175.4448	171	422.5502	271	669.6556	371	916.7610	471	1163.8663
72	177.9159	172	425.0213	272	672.1266	372	919.2320	472	1166.3374
73	180.3869	173	427.4923	273	674.5977	373	921.7031	473	1168.8084
74	182.8580	174	429.9634	274	677.0687	374	924.1741	474	1171.2795
75	185.3290	175	432.4344	275	679.5398	375	926.6452	475	1173.7505
76	187.8001	176	434.9055	276	682.0108	376	929.1162	476	1176.2216
77	190.2711	177	437.3765	277	684.4819	377	931.5873	477	1178.6927
78	192.7422	178	439.8476	278	686.9530	378	934.0583	478	1181.1637
79	195.2132	179	442.3186	279	689.4240	379	936.5294	479	1183.6348
80	197.6843	180	444.7897	280	691.8951	380	939.0004	480	1186.1058
81	200.1554	181	447.2607	281	694.3661	381	941.4715	481	1188.5769
82	202.6264	182	449.7318	282	696.8372	382	943.9426	482	1191.0479
83	205.0975	183	452.2028	283	699.3082	383	946.4136	483	1193.5190
84	207.5685	184	454.6739	284	701.7793	384	948.8847	484	1195.9900
85	210.0396	185	457.1450	285	704.2503	385	951.3557	485	1198.4611
86	212.5106	186	459.6160	286	706.7214	386	953.8268	486	1200.9321
87	214.9817	187	462.0871	287	709.1924	387	956.2978	487	1203.4032
88	217.4527	188	464.5581	288	711.6635	388	958.7689	488	1205.8743
89	219.9238	189	467.0292	289	714.1345	389	961.2399	489	1208.3453
90	222.3948	190	469.5002	290	716.6056	390	963.7110	490	1210.8164
91	224.8659	191	471.9713	291	719.0767	391	966.1820	491	1213.2874
92	227.3370	192	474.4423	292	721.5477	392	968.6531	492	1215.7585
93	229.8080	193	476.9134	293	724.0188	393	971.1241	493	1218.2295
94	232.2791	194	479.3844	294	726.4898	394	973.5952	494	1220.7006
95	234.7501	195	481.8555	295	728.9609	395	976.0663	495	1223.1716
96	237.2212	196	484.3265	296	731.4319	396	978.5373	496	1225.6427
97	239.6922	197	486.7976	297	733.9030	397	981.0084	497	1228.1137
98	242.1633	198	489.2687	298	736.3740	398	983.4794	498	1230.5848
99	244.6343	199	491.7397	299	738.8451	399	985.9505	499	1233.0558

Continued - HECTARES TO ACRES 1 HECTARE = 2.471 0538 ACRES

HECTARES	ACRES	HECTARES	ACRES	HECTARES	ACRES	HECTARES	ACRES	HECTARES	ACRES
500	1235.5269	600	1482.6323	700	1729.7377	800	1976.8430	900	2223.9484
501	1237.9980	601	1485.1033	701	1732.2087	801	1979.3141	901	2226.4195
502	1240.4690	602	1487.5744	702	1734.6798	802	1981.7851	902	2228.8905
503	1242.9401	603	1490.0454	703	1737.1508	803	1984.2562	903	2231.3616
504	1245.4111	604	1492.5165	704	1739.6219	804	1986.7272	904	2233.8326
505	1247.8822	605	1494.9875	705	1742.0929	805	1989.1983	905	2236.3037
506	1250.3532	606	1497.4586	706	1744.5640	806	1991.6694	906	2238.7747
507	1252.8243	607	1499.9297	707	1747.0350	807	1994.1404	907	2241.2458
508	1255.2953	608	1502.4007	708	1749.5061	808	1996.6115	908	2243.7169
509	1257.7664	609	1504.8718	709	1751.9771	809	1999.0825	909	2246.1879
510	1260.2374	610	1507.3428	710	1754.4482	810	2001.5536	910	2248.6590
511	1262.7085	611	1509.8139	711	1756.9193	811	2004.0246	911	2251.1300
512	1265.1796	612	1512.2849	712	1759.3903	812	2006.4957	912	2253.6010
513	1267.6506	613	1514.7560	713	1761.8614	813	2008.9667	913	2256.0721
514	1270.1216	614	1517.2270	714	1764.3324	814	2011.4378	914	2258.5432
515	1272.5927	615	1519.6981	715	1766.8035	815	2013.9088	915	2261.0142
516	1275.0638	616	1522.1691	716	1769.2745	816	2016.3799	916	2263.4853
517	1277.5348	617	1524.6402	717	1771.7456	817	2018.8510	917	2265.9563
518	1280.0059	618	1527.1113	718	1774.2166	818	2021.3220	918	2268.4274
519	1282.4769	619	1529.5823	719	1776.6877	819	2023.7931	919	2270.8984
520	1284.9480	620	1532.0534	720	1779.1587	820	2026.2641	920	2273.3695
521	1287.4190	621	1534.5244	721	1781.6298	821	2028.7352	921	2275.8405
522	1289.8901	622	1536.9955	722	1784.1008	822	2031.2062	922	2278.3116
523	1292.3611	623	1539.4665	723	1786.5719	823	2033.6773	923	2280.7827
524	1294.8322	624	1541.9376	724	1789.0430	824	2036.1483	924	2283.2537
525	1297.3032	625	1544.4086	725	1791.5140	825	2038.6194	925	2285.7248
526	1299.7743	626	1546.8797	726	1793.9851	826	2041.0904	926	2288.1958
527	1302.2453	627	1549.3507	727	1796.4561	827	2043.5615	927	2290.6669
528	1304.7164	628	1551.8218	728	1798.9272	828	2046.0325	928	2293.1379
529	1307.1875	629	1554.2928	729	1801.3982	829	2048.5036	929	2295.6090
530	1309.6585	630	1556.7639	730	1803.8693	830	2050.9746	930	2298.0800
531	1312.1296	631	1559.2349	731	1806.3403	831	2053.4457	931	2300.5511
532	1314.6006	632	1561.7060	732	1808.8114	832	2055.9167	932	2303.0221
533	1317.0717	633	1564.1770	733	1811.2824	833	2058.3878	933	2305.4932
534	1319.5427	634	1566.6481	734	1813.7535	834	2060.8589	934	2307.9642
535	1322.0138	635	1569.1192	735	1816.2245	835	2063.3299	935	2310.4353
536	1324.4848	636	1571.5902	736	1818.6956	836	2065.8010	936	2312.9063
537	1326.9559	637	1574.0613	737	1821.1667	837	2068.2720	937	2315.3774
538	1329.4269	638	1576.5323	738	1823.6377	838	2070.7431	938	2317.8484
539	1331.8980	639	1579.0034	739	1826.1088	839	2073.2141	939	2320.3195
540	1334.3690	640	1581.4744	740	1828.5798	840	2075.6852	940	2322.7906
541	1336.8401	641	1583.9455	741	1831.0509	841	2078.1563	941	2325.2616
542	1339.3112	642	1586.4165	742	1833.5219	842	2080.6273	942	2327.7327
543	1341.7822	643	1588.8876	743	1835.9930	843	2083.0984	943	2330.2037
544	1344.2533	644	1591.3586	744	1838.4640	844	2085.5694	944	2332.6748
545	1346.7243	645	1593.8297	745	1840.9351	845	2088.0405	945	2335.1458
546	1349.1954	646	1596.3008	746	1843.4061	846	2090.5115	946	2337.6169
547	1351.6664	647	1598.7718	747	1845.8772	847	2092.9826	947	2340.0880
548	1354.1375	648	1601.2429	748	1848.3482	848	2095.4536	948	2342.5590
549	1356.6085	649	1603.7139	749	1850.8193	849	2097.9247	949	2345.0301

Continued - HECTARES TO ACRES 1 HECTARE = 2.471 0538 ACRES

HECTARES	ACRES	HECTARES	ACRES	HECTARES	ACRES	HECTARES	ACRES	HECTARES	ACRES
550	1359.0796	650	1606.1850	750	1853.2903	850	2100.3957	950	2347.5011
551	1361.5506	651	1608.6560	751	1855.7614	851	2102.8668	951	2349.9722
552	1364.0217	652	1611.1271	752	1858.2325	852	2105.3378	952	2352.4432
553	1366.4928	653	1613.5981	753	1860.7035	853	2107.8089	953	2354.9143
554	1368.9638	654	1616.0692	754	1863.1746	854	2110.2799	954	2357.3853
555	1371.4349	655	1618.5402	755	1865.6456	855	2112.7510	955	2359.8564
556	1373.9059	656	1621.0113	756	1868.1167	856	2115.2220	956	2362.3274
557	1376.3770	657	1623.4823	757	1870.5877	857	2117.6931	957	2364.7985
558	1378.8480	658	1625.9534	758	1873.0588	858	2120.1642	958	2367.2695
559	1381.3191	659	1628.4245	759	1875.5298	859	2122.6352	959	2369.7406
560	1383.7901	660	1630.8955	760	1878.0009	860	2125.1063	960	2372.2116
561	1386.2612	661	1633.3666	761	1880.4719	861	2127.5773	961	2374.6827
562	1388.7322	662	1635.8376	762	1882.9430	862	2130.0484	962	2377.1537
563	1391.2033	663	1638.3087	763	1885.4140	863	2132.5194	963	2379.6248
564	1393.6743	664	1640.7797	764	1887.8851	864	2134.9905	964	2382.0959
565	1396.1454	665	1643.2508	765	1890.3562	865	2137.4615	965	2384.5669
566	1398.6165	666	1645.7218	766	1892.8272	866	2139.9326	966	2387.0380
567	1401.0875	667	1648.1929	767	1895.2983	867	2142.4036	967	2389.5090
568	1403.5586	668	1650.6639	768	1897.7693	868	2144.8747	968	2391.9801
569	1406.0296	669	1653.1350	769	1900.2404	869	2147.3457	969	2394.4511
570	1408.5007	670	1655.6060	770	1902.7114	870	2149.8168	970	2396.9222
571	1410.9717	671	1658.0771	771	1905.1825	871	2152.2878	971	2399.3932
572	1413.4428	672	1660.5482	772	1907.6535	872	2154.7589	972	2401.8643
573	1415.9138	673	1663.0192	773	1910.1246	873	2157.2299	973	2404.3353
574	1418.3849	674	1665.4903	774	1912.5956	874	2159.7010	974	2406.8064
575	1420.8559	675	1667.9613	775	1915.0667	875	2162.1721	975	2409.2774
576	1423.3270	676	1670.4324	776	1917.5378	876	2164.6431	976	2411.7485
577	1425.7980	677	1672.9034	777	1920.0088	877	2167.1142	977	2414.2195
578	1428.2691	678	1675.3745	778	1922.4799	878	2169.5852	978	2416.6906
579	1430.7401	679	1677.8455	779	1924.9509	879	2172.0563	979	2419.1617
580	1433.2112	680	1680.3166	780	1927.4220	880	2174.5273	980	2421.6327
581	1435.6823	681	1682.7876	781	1929.8930	881	2176.9984	981	2424.1038
582	1438.1533	682	1685.2587	782	1932.3641	882	2179.4695	982	2426.5748
583	1440.6244	683	1687.7297	783	1934.8351	883	2181.9405	983	2429.0459
584	1443.0954	684	1690.2008	784	1937.3062	884	2184.4116	984	2431.5169
585	1445.5665	685	1692.6718	785	1939.7772	885	2186.8826	985	2433.9880
586	1448.0375	686	1695.1429	786	1942.2483	886	2189.3537	986	2436.4590
587	1450.5086	687	1697.6140	787	1944.7193	887	2191.8247	987	2438.9301
588	1452.9796	688	1700.0850	788	1947.1904	888	2194.2958	988	2441.4012
589	1455.4507	689	1702.5561	789	1949.6615	889	2196.7668	989	2443.8722
590	1457.9217	690	1705.0271	790	1952.1325	890	2199.2379	990	2446.3433
591	1460.3928	691	1707.4982	791	1954.6036	891	2201.7089	991	2448.8143
592	1462.8638	692	1709.9692	792	1957.0746	892	2204.1800	992	2451.2854
593	1465.3349	693	1712.4403	793	1959.5457	893	2206.6510	993	2453.7564
594	1467.8060	694	1714.9113	794	1962.0167	894	2209.1221	994	2456.2275
595	1470.2770	695	1717.3824	795	1964.4878	895	2211.5931	995	2458.6985
596	1472.7481	696	1719.8534	796	1966.9588	896	2214.0642	996	2461.1696
597	1475.2191	697	1722.3245	797	1969.4299	897	2216.5352	997	2463.6406
598	1477.6902	698	1724.7955	798	1971.9009	898	2219.0063	998	2466.1117
599	1480.1612	699	1727.2666	799	1974.3720	899	2221.4774	999	2468.5827

INDEX